INTRODUCTION TO Engineering Thermodynamics

Howard F. Silver
John E. Nydahl
of University of Wyoming

WEST PUBLISHING CO.
St. Paul • New York • Boston
Los Angeles • San Francisco

Dedicated to Prof. E. M. Sparrow without whose
encouragement this book might not have been published.

COPYRIGHT © 1977 By WEST PUBLISHING CO.

All rights reserved
Printed in the United States of America

Library of Congress Cataloging in Publication Data

Silver, Howard F
    Introduction to engineering thermodynamics.

    Bibliography: p.
    Includes index.
    1. Thermodynamics.   I. Nydahl, John, joint author.
II. Title.
TJ265.S54   621.4'021   76–3601
ISBN 0–8299–0053–5

Preface

One of the major accomplishments of thermodynamics has been the reduction of information obtained from countless observations and experiments on energy transformations to a set of four generalized statements called principles or laws. In order to apply these principles to practical problems, the engineer has found it necessary to reduce these generalized statements to a set of mathematical symbols and equations which can be evaluated numerically.

Now there are many excellent undergraduate engineering science textbooks which discuss the engineering application of the laws of thermodynamics. However, many, if not most of these texts appear to reverse the process of generalization and present the first law in its thermostatic form, $Q - W = \Delta U$. The reader of these texts is often left with the impression that different and seemingly unrelated mathematical relationships must be used to represent the same thermodynamic principle, depending upon the specific application under consideration. In this text, we have attempted to avoid this source of confusion by introducing the principles of thermodynamics in the form of generalized balance or conservation equations. These generalized equations are then reduced in a logical manner to apply to a specific problem.

By eliminating unimportant terms in the generalized statements of the principles of thermodynamics, sophomores and juniors are able to readily solve problems such as 2-34 in this text. With the older thermostatic approach such problems are difficult even for advanced students. The generalized balance equations also provide a link between thermodynamics and fluid flow. For example, the Bernouilli equation, which is widely used in fluid mechanics courses, can be readily deduced from the generalized balance equations.

The topics covered in this text have been presented in a two semester sequence at the University of Wyoming. In the first semester, the student is introduced to the principles of thermodynamics and applies these principles to single component systems. These applications involve the use of tabulated data as well as the use of analytical relationships between thermodynamic properties, and are presented in the first part of this text. Although extensive use of the latest scientific S. I. units is made throughout the text, English units which are common in American industrial practice are also used. In the second semester, the application of thermodynamic principles is extended to multi-component systems, phase equilibrium and chemical reactions. These topics are presented in the second part of this text. If the student has been introduced to thermodynamics in a physics or chemistry course or some other engineering course, it may be possible to cover the material in both parts of this text in a single semester.

The overall objective of this text is to present the basic concepts of thermodynamics to undergraduate engineering students in all fields. At the beginning of each chapter in this text, the main points to be covered in that chapter have been listed as a set of specific objectives which the student can use to judge his own progress. It is hoped that by making use of the lists of specific objectives and by repeated application of the thermodynamic principles developed in this text to different types of problems, the student will be able to develop a working understanding of thermodynamics. Although applications are presented in the text as a means of reinforcing basic concepts, detailed coverage of industrial applications is left to more advanced and specialized courses in the respective engineering fields.

The authors are indebted to Dr. E. M. Sparrow, Professor of Mechanical Engineering at the University of Minnesota, for his in-depth review and invaluable comments on the text. They also wish to acknowledge their colleagues at the Electric Power Research Institute in Palo Alto, California and at the University of Wyoming, particularly Dr. Robert D. Gunn and Dr. Reid C. Miller, Professors of Chemical Engineering, who have made many constructive suggestions. Special thanks are also given to Mr. Stephen L. Chandler and Mr. Jay E. Lyon, undergraduate students who prepared the original figures used in this text, to Miss Georgia Wales, and to Mrs. Jean Edeen who typed both the original and the camera ready copies of the manuscript.

HOWARD F. SILVER

JOHN E. NYDAHL

Contents

Preface   *iii*

Nomenclature   *x*

Chapter 1. Definitions and Concepts   **1**

    Objectives   1
    Scope   1
    Background   2
    Systems   3
    Properties   4
        Units and Dimensions   4
        Intensive Properties   9
            Pressure   9
            Temperature   10
        Extensive Properties   11
    Changes of State   12
    Equilibrium   13
    Phase Diagrams   14
    Problems   20

Chapter 2. The Mass Balance and the Energy Balance   **23**

    Objectives   23
    The Mass Balance   23
    The Energy Balance   27
        The First Law of Thermodynamics   27
        Accountability of Energy   28
        Types of Energy   30

Work 30
Heat 35
Kinetic Energy 37
Potential Energy 38
Flow Energy 40
Internal Energy 41
The Energy Balance Equation 41
Ideal Processes 49
Problems 51

Chapter 3. The Entropy Balance 58

Objectives 58
The Second Law of Thermodynamics 58
Entropy and the Second Law 60
The Entropy Balance 62
  Accountability of Entropy 62
  Reversible Processes 63
  The Carnot Cycle 66
  The Entropy Balance Equation 70
    Irreversible Processes 70
    Lost Work 73
    Mechanical Efficiency 79
    Unavailable Energy 82
    Effectiveness 91
  Availability 94
  Absolute Temperature Scale 98
Problems 99

Chapter 4. Mathematical Relationships Between Thermodynamic Properties 106

Objectives 106
Introduction 106
Derivatives of Thermodynamic Functions 107
The Property Relationship 112
Change of Independent Variables 115
Maxwell Relations 117
Problems 126

Chapter 5. The Ideal Gas 129

Objectives 129
Equation of State 129
Thermodynamic Properties of Ideal Gases 131
Isentropic Processes 137
Entropy and Probability 142
Problems 144

Chapter 6. Thermodynamics of Fluid Flows 146

Objectives 146
The Mechanical Energy Balance Equation 147
Compressible Flows 151

Speed of Sound   151
Stagnation Properties   156
Effects of Area Changes on Compressible Flow   163
Normal Shock Waves   169
The Effect of Heat Transfer and Friction on Compressible Flows   177
Performance of Real Nozzles and Diffusers   178
Problems   183

Chapter 7. Thermodynamics of Heat Engines   187

Objectives   187
Introduction   187
Reciprocating Engines   188
  Gas Compressors   188
  The Otto Cycle   196
  Diesel's Carnot Engine   211
  The Diesel Cycle   214
  The Stirling Cycle   220
Turbine Systems   225
  Turbines   225
  Steam Turbine System   231
  The Rankine Cycle   236
  The Working Medium and Binary Systems   252
  Gas Turbine Systems and Brayton Cycle   254
  The Combined Gas and Steam Cycle   272
Reverse Heat Engines: Refrigerators and Heat Pumps   277
  Reverse Carnot Cycle   277
  Vapor-Compression Refrigerator   279
  Gas Refrigeration Cycles   283
  Absorption Refrigeration Cycle   287
Problems   288

Chapter 8. Equations of State   293

Objectives   293
Introduction   293
Equations of State for Gases   295
  van der Waals' Equation   295
    Theorem of Corresponding States   297
      The Compressibility Factor   297
      Estimation of Critical Constants   301
  Other Equations of State   307
Equations of State of Liquids and Solids   319
Problems   324

Chapter 9. Evaluation of Thermodynamic Functions   327

Objectives   327
Graphical Techniques   327
  Differentiation   327
  Integration   333
Generalized Charts   338
Fugacities   345

        Effect of Pressure and Temperature on Fugacity   353
        Problems   355

Chapter 10. Mixtures   360

Objectives   360
Introduction   360
Ideal Gas Mixtures   361
    Dalton's and Amagat's Laws   361
    Thermodynamic Properties   363
Ideal Solutions   368
    Thermodynamic Properties   368
    Empirical Mixing Rules   371
Non-Ideal Solutions   376
    Partial Molal Properties   376
    Excess Functions   380
Problems   382

Chapter 11. Phase Equilibrium   385

Objectives   385
Thermodynamic Equilibrium   386
Phase Equilibrium in Single Component Systems   388
    Clausius-Clapeyron Equation   388
    Enthalpy of Vaporization   391
Phase Equilibrium in Multi-Component Systems   392
    The Gibbs Phase Rule   392
    Partial Pressure as a Criterion of Equilibrium   396
        Raoult's and Dalton's Law   396
        Humidification   399
        Bubble Points and Dew Points   404
    Fugacities as a Criterion of Equilibrium   408
        Fugacities and the Chemical Potential   408
        Fugacities and the Phase Variables   411
        Phase Equilibrium in Simple Systems   415
        Fugacity Coefficients   416
        Activity Coefficients   424
            Activity coefficients and the phase variables   424
            The Gibbs-Duhem equation   425
            Analytical relationships   430
    Further Applications of Phase Equilibrium Relationships   441
Problems   446

Chapter 12. Chemical Reactions   454

Objectives   454
Introduction   454
Mass Balance   455
Heats of Reaction   457
    Energy Balance for Reacting Mixtures   457
    Enthalpy of Formation   460
    Effect of Temperature and Pressure   464
    Enthalpy of Combustion   466

Chemical Equilibrium 473
  Reaction Free Energy 473
  Equilibrium Conversion 476
    Simple Reactions 476
    Complex Reactions 480
  Calculation of Reaction Free Energy 485
  Third Law of Thermodynamics 494
  Effect of Temperature on Chemical Equilibrium 496
  Estimation Methods for Reaction Free Energy Data 499
Electrochemical Cells 505
Problems 512

Appendix 516

Conversion Factors 517
Steam Tables 519
Specific Heat Capacities
  in Ideal Gas State 531
  of Solids 532
Critical Properties 533
Generalized Thermodynamic Properties 534
Vapor-Liquid Equilibrium Coefficients 542

Answers to Selected Problems 544

Bibliography 547

Index 550

Nomenclature

A	Area, Helmholtz' work function $= U - TS$
a	Acceleration, specific Helmholtz' work function activity, molecular stoichiometric coefficient, speed of sound
B	Availability
b	Intercept, specific availability, atomic stoichiometric coefficient
bmep	Brake mean effective pressure
C	Number of components
c	Specific heat capacity or molal heat capacity $= \left(\dfrac{\partial q}{dT}\right)_{rev}$
COP	Coefficient of performance
D	Deviation
d	Differential change of a property
E	Total energy $= U + \tfrac{1}{2}mv^2 + mg(z - z_0)$, modulus of elasticity $= \dfrac{L}{A}\left(\dfrac{\partial \tau}{\partial L}\right)_T$
F	Force, number of independent variables, Faraday constant
f	Fugacity
G	Gibbs free energy $= U + PV - TS = H - TS$
g	Acceleration due to gravity, specific Gibbs free energy, gas
g_c	Conversion factor from mass to force units
H	Enthalpy $= U + PV$, Henry's law constant
h	Specific enthalpy

Nomenclature

I	Moment of inertia
imep	Indicated mean effective pressure
J_A	Activity quotient
k	Constant of proportionality, polytropic coefficient
K_A	Activity quotient at equilibrium
K_i	Phase distribution coefficient of component i.
L	Length
l	Liquid, characteristic length
LW	Lost work
lw	lost work per unit mass
$LW + LW_c$	Unavailable energy
$lw + lw_c$	Unavailable energy per unit mass
M	Mach number $= v/a$
m	Mass, slope, molality
mep	Mean effective pressure
MW	Molecular weight
N	Number of molecules
n	Number of moles, polytropic coefficient
P	Total pressure
P	Probability
p	Partial pressure
Q	Heat absorbed by the system from the surroundings
q	Electric charge, heat absorbed per unit mass
R	Gas constant
R_e	Reynolds number $= \dfrac{vl}{\nu}$
r	Radius, compression ratio $= V_{max}/V_{min}$
r_c	Cut off ratio = fraction of Diesel cycle compression ratio during which fuel is added
r_P	Pressure ratio $= P_{max}/P_{min}$
S	Entropy
s	Solid, stress, specific entropy
Sh	Superheat
T	Absolute temperature
t	Time, relative temperature
U	Internal energy
u	Specific internal energy
V	Volume

v	Specific volume
W	Work done by the system on the surroundings
w	Weight, work done per unit mass
W	Thermodynamic probability
x	Quality = weight percent vapor in a liquid-vapor mixture, mole fraction or mass fraction in liquid phase
Y	Any thermodynamic function, such as A, G, H, or U
y	Mole fraction or mass fraction in vapor phase
z	Elevation, compressibility factor $= \dfrac{Pv}{nRT}$, mole fraction or mass fraction in a mixture of phases, valence
a	Activity $= f/f^\circ$
α	Extent of reaction
α_T	Coefficient of linear expansion $= \dfrac{1}{L}\left(\dfrac{\partial L}{\partial T}\right)_T$
α_R	Residual volume
β	Coefficient of performance, coefficient of thermal expansion $= \dfrac{1}{v}\left(\dfrac{\partial v}{\partial T}\right)_P$, extent of second reaction
γ	Ratio of molal heat capacities $= \dfrac{c_P}{c_V}$, activity coefficient
Δ	Difference, final minus initial or outlet minus inlet
δ	Infinitesimal quantity flowing, solubility parameter
ϵ	Electric potential, unit strain, effectiveness
η	Efficiency
κ_T	Coefficient of isothermal compressibility $= -\dfrac{1}{v}\left(\dfrac{\partial v}{\partial P}\right)_T$
μ	Chemical potential
μ_J	Joule-Thompson coefficient $= \left(\dfrac{\partial T}{\partial P}\right)_H$
ν	Fugacity coefficient $= f/P$, kinematic viscosity, partial volume
π	Number of phases
ρ	Density $= 1/v$
σ	Surface tension
τ	Tension
υ	Velocity
Φ	Function of, volume fraction
ϕ	Potential
ω	Radial velocity, accentric factor

SUBSCRIPTS

$1, 2, 3, \ldots, i, j$	Refer to different states in a process or to different chemical components: also, 1 and 2 or i and f refer to initial and final states respectively
A	Refers to Avogadro's number or activity
act	Refers to actual or irreversible change
adiab	Adiabatic process
avail	Refers to energy available to do useful work
avg	Refers to average value
b	Refers to atmospheric boiling temperature
C	Refers to combustion reaction
c	Refers to critical state or Carnot cycle
comp	Refers to compressor
D	Refers to dead state, diffuser
eff	Refers to effective value
F	Refers to flow energy, formation reaction
f	Refers to a fluid or a final state, atmospheric freezing point
fg	Refers to a vaporization process
fr	Refers to friction
g	Refers to a vapor
gen	Refers to mass or energy generated within a process
int	Refers to internal energy
k	Refers to kinetic energy
M	Refers to a property of a mixture
max	Maximum value
min	Minimum value
mix	Mixing
N	Refers to nozzle
0	Property of surroundings, stagnation property, lower temperature state
p	Refers to potential energy
pump	Refers to a pump
R	Refers to chemical reaction
r	Reduced condition, relative to critical state
reg	Refers to regenerator
rev	Reversible change
S	Refers to isentropic process

sat	Refers to saturation conditions
sf	Refers to melting or fusion process
sys	Refers to system
surr	Refers to surroundings
therm	Refers to thermal process
turb	Refers to turbine
τ	Refers to tension

If any of the properties, such as H, P, V, S, etc., are used as subscripts, these properties are assumed constant

SUPERSCRIPTS

E	Refers to an excess property of a mixture
I	Refers to property of an ideal mixture
I, II, ...	Refers to a general phase
$\square$	Designates saturation pressure or vapor pressure, as $P^{\square}$
o	Refers to standard or reference state, degrees of temperature
*	Refers to sonic conditions, ideal gas condition
— (overbar)	An intensive property of a component in a mixture on a unit mass or mole basis
· (overdot)	Time rate of change $= \dfrac{d}{dt}$

INTRODUCTION TO

Engineering Thermodynamics

1
Definitions and Concepts

OBJECTIVES

After studying this chapter, the student should be able to

(1) describe a homogeneous or heterogeneous system and an open, closed, or isolated system;

(2) define the intensive and extensive properties of a system including pressure, temperature, mass, and volume;

(3) use both the English System and the International System of Units interchangeably;

(4) show major relationships among pressure, temperature, and specific volume of a single-component system using phase diagrams;

(5) explain the meaning of the terms saturated vapor, saturated liquid, quality, degrees of superheat, and critical point;

(6) use tables to determine numerical values for the pressure, temperature, and specific volume of a single-component system;

(7) define change of state and types of thermodynamic processes; and

(8) describe equilibrium and its role in thermodynamics.

SCOPE

Thermodynamics is an applied science which is concerned with all types of engineering problems. Initially, engineers used thermodynamic concepts in the design of such things as steam engines. The application of thermodynamics to the development of other forms of engines soon followed.

Today, engineers are using thermodynamics to design fuel cells and nuclear power plants as well as jet aircraft engines and rockets. Thermodynamics underlies the engineers' efforts to develop new plastics and metallic alloys, water desalinization processes, and artificial hearts, lungs and kidneys. In the future, thermodynamics will play a major role in the engineer's attempts to optimize the utilization of the world's natural resources and to reduce pollution, as well as his efforts to feed the anticipated population of the earth.

BACKGROUND

As a science, thermodynamics has its historical roots firmly imbedded in the principles of mechanics. Two of the more important ideas from mechanics which have been incorporated into thermodynamics are the concept of energy and the concept of equilibrium. Both can be related to the action of a force.

In mechanics, a force is defined implicitly in Sir Isaac Newton's second law of motion:

Change of momentum is proportional to the force and takes place in the direction of a straight line in which the force acts.

Thus, if there are a number of different forces acting on a body, the body will move in the direction of the net unbalanced force. The product of the unbalanced force and the distance through which that net force acts is called work.

The observation that work done, for example, in lifting a weight could be recovered through a suitable pulley and lever system by allowing the weight to fall back to its original position led to the idea that a body must contain the capacity to do work. This idea proved so useful in the analysis of mechanical systems that this capacity of a body to do work was given the name energy. In this context, work can be considered to be a form of energy.

A method commonly used to evaluate the net unbalanced force on a body is to determine what force is necessary to prevent motion. If all forces acting on a body are in balance, the body will remain in a state of rest. In mechanics, this state of rest was given the name equilibrium.

The birth of thermodynamics as a daughter of mechanics dates back to the time when it was first realized that a body which was hotter than its surroundings had the capacity to do work. This led to the concept that not only mechanical work but also heat was a form of energy. As time went on, the existence of other forms of energy which could be used to lift a weight was recognized. Today, the word energy is commonly used in thermodynamics to refer to a capacity, resident in a body, for doing work and for transferring heat.

In addition to this generalization of the concept of energy, the original concept of equilibrium as a balance of mechanical forces has been generalized in thermodynamics to also include the idea of a balance between chemical forces and thermal forces or temperatures. Since processes tend to move toward equilibrium, thermodynamics can tell us something about the direction of the process and of the extent of the change that can be expected during the process.

As we will see, thermodynamics is a science which is built upon precise definitions of certain terms and a minimum of experimentally based generalizations called principles or laws. An important key to the understanding of thermodynamics lies in a knowledge of the definitions of terms and the concepts used to describe a thermodynamic system and its interactions with its surroundings.

SYSTEMS

A system is that portion of the universe set aside for study. It plays a role in thermodynamics similar to the role a free body diagram plays in statics and dynamics. Systems can be classified as homogeneous or heterogeneous. A homogeneous system is one in which the properties either are the same from point to point or change in a continuous manner. For example, even though the properties of the atmosphere above the earth vary with height, the atmosphere can be considered to be a homogeneous system. There are no regions in a homogeneous system where there are discontinuities or abrupt changes. A homogeneous system is also called a phase. The physical boundary of a phase need not be continuous—drops of water dispersed in a gas phase are all considered to be a single phase. A heterogeneous system is one that consists of more than one phase. There will be sudden or abrupt changes in properties at the phase boundaries of a heterogeneous system.

The rest of the universe, separated from the system by a boundary, or control surface, is called the surroundings. For practical purposes, we limit the surroundings to that portion of the universe adjacent to the system.

Systems can also be classified as open, closed or isolated. A system which can exchange both energy and matter with its surroundings is called an open system. The volume occupied by an open system is sometimes called a control volume. A system which can exchange energy but cannot exchange matter with its surroundings is called a closed system or control mass. A system which can exchange neither energy nor mass with its surroundings is known as an isolated system.

Although systems usually are selected to include matter and energy, they may also be selected to include only energy. For example, an electromagnetic wave may well be selected as a system. Ease of obtaining desired results from thermodynamic analysis depends upon the proper choice of a system. In any given situation, the system should be unambiguously identified or understood.

The characteristic of a system we are most interested in is the state of the system. The thermodynamic state of any system is defined when all of its measurable characteristics are specified. Any identifiable characteristic of a system which can be measured either directly or indirectly is called a property. If the properties of the system we select for study do not vary from point to point, specifying a single value for each property will fix the state of the system. However, if the properties of the system vary, we must mentally divide the system into a number of smaller segments, each with constant properties. Then the thermodynamic state of the system is fixed when the properties of all these segments are fixed. In some cases, it may be necessary to consider the system to be made up of an infinite number of parts.

PROPERTIES

Units and Dimensions

The physical nature of thermodynamic properties can be described in terms of dimensions. These dimensions fall into two different classes. The first class consists of fundamental dimensions and includes length, time, mass, temperature, electric current, and luminous intensity. The other class consists of derived dimensions which are a combination of the fundamental dimensions.

Fundamental dimensions are measured by comparison with standards. The magnitudes of these dimensions are expressed in arbitrary units, which have evolved either through the metric system or the English system. The English system which is in common use today in this country, started about 900 A.D. and grew by whimsy. Initially, the foot was standardized as a unit of measure equal to the distance measured along 36 fat barley-corns from the middle of a mature ear of barley, placed end to end. A yard was defined as the distance measured from the king's nose to the middle finger of his outstretched arm and an inch was measured from the knuckle-bone of a man's thumb to its tip.

The metric system, by contrast, was based on the dimensions of the earth itself. The meter was defined as one ten millionth of the distance between the equator and the true North Pole, along a meridian running near Dunkirk, Paris and Barcelona. The International System of Units, abbreviated S.I., currently in use in the sciences, is a modernized version of the metric system. The magnitude of the International System of Units is fixed by primary international standards while the magnitude of English units are related to these international standards by means of conversion factors.

Until recently, the fundamental standard of length, the meter, was defined as the distance measured at 0 °C between two marks on a platinum irridium bar kept by the International Bureau of Weights and Measures at Paris. Today, a meter is defined in the International System of Units as 1,650,763.73 wavelengths in vacuum of the orange-red line of the spectrum of Krypton-86. In the English system, the foot is defined by the exact relationship

$$1 \text{ ft} = 0.3048 \text{ meters} \tag{1-1}$$

or

$$1 \text{ in.} = 2.54 \text{ cm} \tag{1-2}$$

In both the International and the English systems, the fundamental unit of time is the second. For many years, the magnitude of the second was linked to the rotation of the earth. Time measured with respect to the rotation of the earth is described as Ephemeris time. Exact measurements of Ephemeris time require observations of the stars and are inconvenient to make. The development of the atomic clock provided a time measurement which is easier to make and is thus much more convenient. In October, 1967 the International Committee of Weights and Measures defined the second as the duration of 9,192,631,770 cycles of the radiation associated with a specified transition of the cesium atom.

In the International System of Units, the unit of mass is defined with reference to a platinum irridium cylinder called the kilogram kg. This cylinder is kept by the International Bureau of Weights and Measures at Paris. A duplicate is in the custody of the National Bureau of Standards and serves as the standard of mass in the United States. This is the only base unit still defined by an object. The related unit of mass in the English system is the pound mass lb_m defined by the relationship

$$1.0 \ lb_m = 0.45359237 \ kg$$

$$\cong 453.6 \ grams \tag{1-3}$$

Closely related to the concept of mass is the concept of force. The relationship between mass and force is given by Newton's second law of motion,

$$F = \frac{d(mv)}{dt} \tag{1-4}$$

where the velocity of the body is the rate of change of position of the body with respect to an arbitrary reference point in space. If there is no change of mass with changing velocity, Newton's second law of motion takes the familiar form

$$F = ma \tag{1-5}$$

where the acceleration a of the body due to force F is

$$a = \frac{dv}{dt}$$

In the International System of Units, force is a derived dimension. The unit of force used in this system is the newton N. A constant force of one newton, when applied for one second, will give one kilogram of mass a velocity of one meter per second (an acceleration of one meter per second per second).

Although the terms mass and weight are sometimes incorrectly used interchangeably, the weight of a body w is the force exerted on the body by gravity. Weight and mass are related by Newton's second law of motion,

$$wg_c = mg \tag{1-6}$$

where g is the acceleration of mass m due to gravity and g_c is a conversion factor from mass to force units. In the International System of Units,

$$g_c = 1.0 \frac{\text{kg-m}}{\text{N-sec}^2}$$

A unit of force which is more commonly used in engineering work is the pound force lb_f. This is defined as the force which when applied for one second will give one pound mass a velocity of 32.1740 feet per second (an acceleration of 9.80665 meters per second per second). In these units

$$g_c = 32.174 \frac{\text{lb}_m \text{ ft}}{\text{lb}_f \text{ sec}^2} \tag{1-7}$$

The fundamental unit of temperature in the International System of Units is the kelvin K. It is measured with respect to thermal equilibrium states which can be defined as the condition in which the temperature of the system and surroundings are uniform throughout. The temperature at which the solid, liquid and vapor phases of water coexist in thermal equilibrium is called the triple point temperature and is defined by international agreement to be 273.16 K. The magnitude of the kelvin is also defined to be 1/100 of the differences in temperature between freezing water and boiling water at one standard atmosphere.

Temperatures of a number of thermal equilibrium states other than the triple point of water have been established for convenience. The fixed temperature points approved in October, 1968, by the International Committee on Weights and Measures are summarized in Table 1-1.

TABLE 1-1. Fixed Temperature Points in Kelvins on the 1968 International Practical Temperature Scale

	Compound						
Temperature at	Hydrogen	Neon	Oxygen	Water	Zinc	Silver	Gold
Triple point*	13.81		54.361	273.16			
Vapor-Liquid point**	17.042						
Boiling point***	20.28	27.102	90.188	373.15			
Freezing point****					692.73	1235.08	1337.58

*Equilibrium between solid, liquid, and vapor phases

**Measured at a pressure of 33,330.6 newtons/square meter

***Equilibrium between liquid and vapor under a pressure of 101,325 newtons/square meter (one standard atmosphere)

****Equilibrium between liquid and solid under a pressure of 101,325 newtons/square meter (one standard atmosphere)

Temperatures at points other than these fixed thermal equilibrium temperatures are defined by specifying the method of their measurement and the procedure for interpolation between the fixed temperature points. A platinum resistance thermometer is used to measure temperatures from the triple point of hydrogen at 13.81 K to 903.89 K. Temperatures in the range of 903.89 K to the gold point at 1337.58 K are measured using a thermocouple. One wire in the thermocouple is made of platinum and the other is made of an alloy of platinum and rhodium.

For temperatures higher than the gold point, an optical method is used. The intensity of the radiation of any convenient wave length is compared with the intensity of radiation of the same wave length emitted by a black body at the gold point. The temperature is calculated with the aid of Planck's radiation laws.

Closely related to the kelvin is the degree Celsius °C defined by the relationships

$$1\,°C = 1\,K$$

$$T(K) = T(°C) + 273.15 \tag{1-8}$$

The comparable English set of units are degrees Rankine °R and degrees Fahrenheit °F defined by the exact relationships

$$1.8\,°R = 1.0\,K$$

where

$$1.0\,°R = 1.0\,°F$$

and

$$T(°R) = T(°F) + 459.67 \tag{1-9}$$

In thermodynamics, a very important derived dimension is energy or work. In the International System of Units, the basic unit of energy or work is the joule, defined as a force of one newton acting through one meter.

$$1.0\,\text{joule} = 1.0\,\text{newton-meter} \tag{1-10}$$

The unit for power, defined as the rate of change of energy, in the International System of Units is the watt, defined by the relationship

$$1.0\,\text{watt} = 1.0\,\frac{\text{joule}}{\text{second}} \tag{1-11}$$

A unit of energy which is more commonly used in thermodynamics is the calorie. The calorie was once defined as the amount of energy required to increase the temperature of one gram of water from 14.5 °C to 15.5 °C. Today it is defined in terms of its mechanical energy equivalent by the relationship

$$1.0 \text{ calorie} = 4.1868 \text{ joules} \tag{1-12}$$

The comparable unit of energy in the English system is the British Thermal Unit BTU. The BTU was once defined as the amount of energy required to increase the temperature of one pound mass of water from 63 to 64 °F. Today the BTU is defined in terms of the joule by the relationship

$$1 \text{ BTU} = 1055.056 \text{ joules} \tag{1-13}$$

In units commonly used by engineers,

$$1 \text{ BTU} = 778.1693 \text{ ft-lb}_f \tag{1-14}$$

or

$$1 \text{ BTU} = 251.9958 \text{ cal} \tag{1-15}$$

Of the remaining fundamental dimensions in the International System of Units, the base unit of electric current is the ampere and the base unit of luminous intensity is the candela.

The ampere is defined as the magnitude of the current that when flowing through each of two long parallel wires separated by one meter in free space results in a force between the two wires of 2×10^{-7} newtons for each meter of length. Closely related to the ampere is the volt, a measure of electrical potential, defined by the relationship

$$1.0 \text{ volt} = 1.0 \frac{\text{watt}}{\text{ampere}} \tag{1-16}$$

and the ohm, a measure of current resistance, defined by the relationship

$$1.0 \text{ ohm} = 1.0 \frac{\text{volt}}{\text{ampere}} \tag{1-17}$$

The candela is defined as the luminous intensity of 1/600,000 of a square meter of a black body at a temperature of freezing platinum (2042 K). A light source having an intensity of one candela in all directions radiates a light flux of 4π lumens.

Equations involving measured quantities must be balanced dimensionally as well as numerically. Both dimensions and their units of measurement can be multiplied and divided or raised to powers just like ordinary algebraic quantities. When all of the dimensions (or units) in an equation balance, the equation is said to be dimensionally homogeneous. As an example of the manner in which units can be balanced, consider the following example problem.

EXAMPLE PROBLEM 1-1. The energy associated with the motion of a body is given by the formula

$$E = 1/2 \, mv^2$$

What is the energy in joules of an automobile traveling at 60 miles per hour if the mass of the automobile is 3000 lb_m ?

From the formula $E = 1/2 \, mv^2$ we find

$$E(\text{joules}) = \frac{1}{2} \times 3000 \, lb_m \times 60^2 \left(\frac{\text{mile}^2}{\text{hr}^2}\right) \times 5280^2 \left(\frac{\text{ft}^2}{\text{mile}^2}\right) \times \frac{1}{3600^2} \left(\frac{\text{hr}^2}{\text{sec}^2}\right)$$

$$\times 0.3048^2 \left(\frac{\text{m}^2}{\text{ft}^2}\right) \times 0.4536 \left(\frac{\text{kg}}{lb_m}\right) \times 1.0 \left(\frac{\text{N-sec}^2}{\text{kg-m}}\right) \times 1.0 \left(\frac{\text{joule}}{\text{N-m}}\right)$$

$$= 4.9 \times 10^5 \text{ joules}$$

The equations derived in this text will not contain conversion factors required to make them dimensionally consistent. However, the student will be required to make certain of the dimensional correctness of his calculations. Table A-1 in the Appendix contains many commonly used conversion factors which may be of use.

Intensive Properties.

Properties of a system which do not depend upon the size or extent of the system are called intensive properties. Examples of intensive properties would include pressure and temperature.

Pressure. Pressure may be defined as the force exerted by a static fluid perpendicularly on a unit area of any surface. That is,

$$P = \frac{dF}{dA} \tag{1-18}$$

Over a period of many years, three generalizations have evolved that summarize experimental observations about pressure. One of the generalized observations is that pressure may be strongly influenced by position in a static fluid but at a given position is independent of direction. The expected variation of static fluid pressure with fluid depth is directly proportional to the density of the fluid so that

$$dP = -\rho dz \tag{1-19}$$

The pressure exerted by the standard atmosphere above the earth at sea level and a latitude of 45° is 101,325 newtons/square meter (14.696 $lb_f/in.^2$). Absolute pressures used in thermodynamic calculations include the pressure of the atmosphere. Unfortunately, many pressure gages measure only the difference between atmospheric and system pressure. These values can be corrected to absolute values by adding atmosphere pressure to the observed gage pressure.

Another generalized observation is that although pressure is a function of depth, it is unaffected by the shape of the confining boundaries. Also, when pressure is applied to a confined fluid by means of a slowly moving piston, it is transferred undiminished throughout the fluid to all bounding surfaces. We owe much of what we know about pressure to such eminent scientists as Evangelista Toricelli and Blaise Pascal.

Temperature. Temperature is a quantitative measure of the degree of hotness or coldness of a system. Man has probably always been aware of hot and cold as he encountered burning deserts and cool forests. Yet, the percept of hot and cold remained a relative idea until 1592 when Galileo Galilei invented the first temperature-sensing device.

The zero point and the magnitude of the temperature scale divisions used by Galileo and other early workers were not related to any fixed standards. Recognizing this as a problem, Christian Huygens suggested in 1665 that the zero on a thermometer be taken as either the temperature of freezing water or of boiling water.

In 1694, Carlo Renaldine suggested taking the melting point of ice and the boiling point of water for two fixed points on a thermometer scale. He divided the distance between these points on his scale into 12 equal parts. A few years later, in 1701, Newton independently suggested the use of two fixed points to determine the thermometer scale. Newton labeled the melting point of ice as zero and the armpit temperature of a healthy Englishman as 12. On this scale, water boiled at 34.

In 1706, Daniel Gabriel Fahrenheit began making mercury thermometers in Amsterdam. He selected the zero of his thermometer scale as the degree of cold produced by a mixture of salt water and ice, and 96 as the temperature of the blood of a healthy man. Fahrenheit proved that on his scale, the melting and boiling points of water at one atmosphere pressure were 32° and 212° respectively. These temperatures were soon adopted as the standards for the Fahrenheit temperature scale. Later still, in 1742, Anders Celsius proposed a scale with zero at the boiling point of water and 100 at the melting point of ice. The following year, Christin of Lyons independently suggested the familiar centigrade scale which is now called the Celsius scale.

The selection of the ice-water equilibrium temperature or the ice-salt water equilibrium temperature as the zero points of a temperature scale is strictly arbitrary. Fortunately, thermodynamics furnishes us with a basis for an absolute temperature scale and also permits the definition of a zero point or "absolute zero." In 1848, William Thompson, later called Lord Kelvin, introduced a temperature scale based on thermodynamic considerations that has $-273.15\ ^\circ\text{C}$ as its zero and whose temperature units are of the same magnitude as the magnitude of the degree Celsius. This temperature scale is the Kelvin temperature scale.

The underlying principle used in measuring temperatures of different bodies is known as the zeroth law of thermodynamics, which can be stated as follows:

If two systems are in thermal equilibrium with a third system, then they are in thermal equilibrium with each other.

A system is in thermal equilibrium if its temperature is uniform throughout and the same as the temperature of the surroundings. Thus, if two different bodies come to the same thermal equilibrium point with the same temperature sensing device, then the two bodies are at the same temperature.

We should point out that it is not possible to verify the zeroth law, or any of the other three laws of thermodynamics, for that matter, either theoretically or by exhaustive experimentation. Each of the laws of thermodynamics represents a generalization based on the results of limited experimentation. The presumption of their validity is based on the fact that all the consequences derived from them explain observed experiments satisfactorily.

The utility of temperature sensing devices has been extended greatly by first comparing their thermal equilibrium characteristics with other reproducible standards. These calibrated sensing devices can then be used to measure temperatures in other bodies.

Extensive Properties

System properties which are additive and which depend upon the extent of the system are called extensive properties. A commonly encountered extensive property is mass which is a quantity of matter in a body. The mass of each chemical component or constituent in a system must be specified when defining the state of the system. The term chemical component or constituent is used to designate anything that can be represented by a definite chemical formula. In general, we will find it convenient to specify quantities of chemical components in terms of number of molecules or moles, rather than terms of mass. In the International System of Units, the mole is defined as the amount of substance, of specified chemical formula, containing the same number of formula units (molecules, atoms, ions, electrons or other entities) as there are atoms in 12 grams (exactly) of the pure nuclide, carbon-12. One gram mole of any component contains 6.02252×10^{23} molecules and its mass is equal to the formula mass, from the periodic table, of the component in grams. For example, 32 grams of oxygen contains 6.02252×10^{23} molecules.

Another commonly encountered extensive property is volume, the amount of space occupied in three dimensions. Although volume is an extensive property, it can be converted to an intensive property called the specific volume v by dividing its value by the mass of the system.

$$\frac{V}{m} = v \ (ft^3/lb_m) \tag{1-20}$$

Similarly, by dividing by mass, any other extensive property we may encounter can also be converted to intensive variables called specific properties which are independent of the total mass of the system. Extensive properties may also be converted to intensive variables called molar properties by dividing by the total moles in the system.

The intensive equivalent of the mass of a component in a mixture is called the mass or weight fraction of the component and is defined by the relationship

$$x_i = \frac{m_i}{\Sigma m_i} \tag{1-21}$$

where the sum is taken over all components. Further,

$$\Sigma x_i = \frac{\Sigma m_i}{\Sigma m_i} = 1.0$$

Mole fractions are defined in an analagous manner.

CHANGES OF STATE

If potential differences such as a temperature difference exist between a system and its surroundings, there will be a flow of energy. The effect of this energy flow is a change in the state of the system, as indicated by changes in the values of the system's properties. Any series of events which cause a change in the state of a system is called a process.

If on examining a closed system at two different times we find a change in at least one property, we say that a process has taken place during the time interval between observations. The complete sequence of states which the system assumes in changing from its initial state to its final state is called a path. The point of view is somewhat different when we consider an open system. At steady state, the properties of an open system at any particular point do not change with time. However, if we find a difference in at least one property at two different points in the open system, we say that a process has taken place.

There are many different types of processes. An adiabatic process occurs without the transfer of heat between the system and its surroundings. An isothermal process occurs at constant temperature, an isobaric process occurs at constant pressure and an isochoric process occurs at constant volume. Further, a process in which the initial and final states are identical is called a cycle or cyclic process. Steam from a boiler which flows through a turbine is condensed and then pumped back to the boiler, undergoes a cyclic process.

EQUILIBRIUM

If no potential gradient exists within a system or between a system and its surroundings, then the state of the system will not change. A system whose properties do not change with time is said to be in a state of equilibrium. Now in mechanics equilibrium is defined as a state of balance between forces. Equilibrium may also be defined as a state of absolute rest. For example, a stone lying in a hollow upon a hillside, as shown in Fig. 1-1, is considered to be at rest. However,

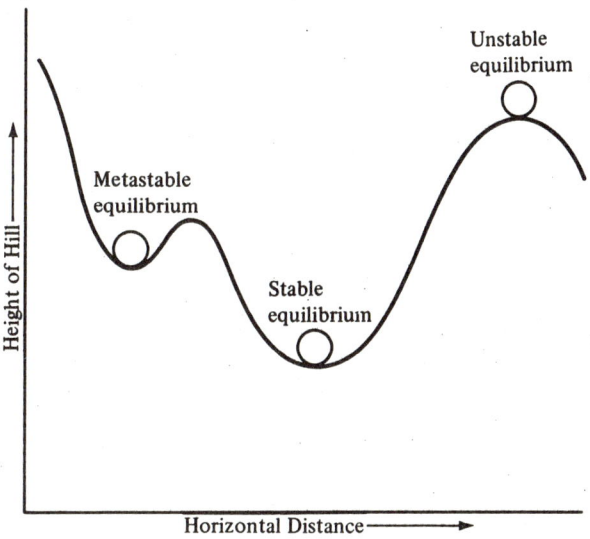

Fig. 1-1. *Types of equilibrium.*

if the stone is given a good shove, it can be made to roll over the lip of the hollow to a position of greater stability at the bottom of the hill. The stone lying in the hollow on the side of the hill is said to be in a state of metastable equilibrium, while the stone at the bottom of the hill is said to be in a state of stable equilibrium. We might also consider a stone lying at rest on the crest of a hill. The slightest push will send this stone rolling to the bottom of the hill. The stone at the crest of the hill is said to be in a state of unstable equilibrium. In addition to these cases, if the stone were lying on a flat, horizontal surface, we would say it was in a state of neutral equilibrium. Thermodynamic analysis provides us with a means of determining the stable equilibrium state of a system--thermodynamic equilibrium.

Stable equilibrium can be distinguished from metastable equilibrium, unstable equilibrium, and neutral equilibrium by displacing a system from its state of rest. Only under conditions of stable equilibrium will the system return to the initial equilibrium state.

If a system is at thermodynamic equilibrium, not only are the mechanical forces such as pressure in balance, but also thermal forces and chemical forces will be in balance. Experimentally determined properties of the system in equilibrium will not change over a long period of time and will return to their stable equilibrium values if the system is perturbed or slightly displaced from its state of rest.

The concept of thermodynamic equilibrium plays a very important role in the definition of the state of a system. Just as there are relationships between the various mechanical forces acting on a mechanical system at rest, so also are there relationships between thermodynamic forces acting on a thermodynamic system at equilibrium. In mechanics, we calculate the forces at one point on a body at equilibrium from a knowledge of forces acting at other points on or within that body. Similarly, in thermodynamics we can estimate the values of thermodynamic properties or potentials at any point within the system at equilibrium in which there are no potential differences from a knowledge of the same or different properties at other points within the same system. As a result, it is not necessary to specify all the properties of a system at equilibrium in order to define its state.

On the other hand, if a thermodynamic system is not at equilibrium, then there are no simple relationships between the various thermodynamic properties. In some cases, analysis may be possible if the system is in a state close to equilibrium or if infinitesimally small segments of the system can be assumed to be at equilibrium. However, for the purposes of this course, we will need only to define the state of a system undergoing a change at those points, either in time or space, along the path of the process on which the system can be assumed to be at equilibrium. A particularly convenient path is a reversible path. During a reversible process, a system is at all times in a state of thermodynamic equilibrium. Such a process is capable of being reversed in direction by an infinitesimal change in the direction of the driving force.

PHASE DIAGRAMS

The intensive thermodynamic properties usually used to define the state of a single component system include pressure, temperature and specific volume. Relationships between these properties can be measured experimentally. Consider, for example, one pound mass of a fluid contained within the cylinder shown in Fig. 1-2. The cylinder is enclosed by a piston at the top and is completely submerged in a large constant temperature reservoir. We can measure the pressure exerted by the fluid on the walls of the cylinder by means of a pressure gage. We can also measure the temperature of the fluid in the cylinder by means of a thermometer. If we know the cross sectional area of the piston face, we can also measure the specific volume occupied by the fluid by measuring the distance between the bottom of the cylinder and the piston face and dividing the calculated volume by the mass of fluid in the cylinder. If we find that at a particular position of the piston, neither the pressure nor the temperature of the fluid changes with time, then these measurements represent an equilibrium state data point.

Now, let us move the piston down a few inches. At first, there may be small pressure and temperature gradients set up in the fluid. But if we wait until the

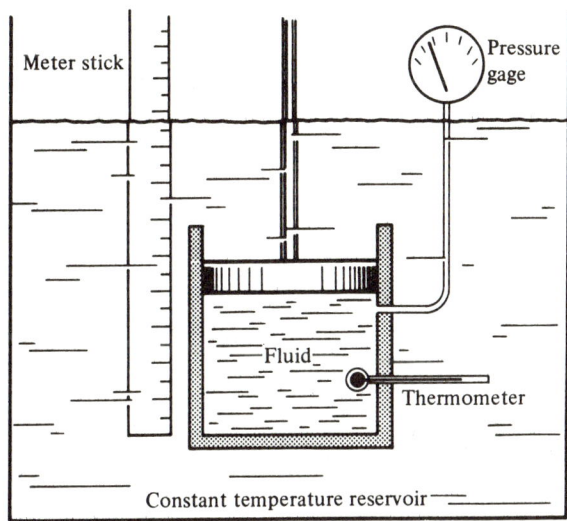

Fig. 1-2. *PVT measurements.*

pressure and temperature stop changing, we can make new equilibrium measurements of pressure, temperature and specific volume. Although the equilibrium temperature of the fluid will not vary because the cylinder is in a constant temperature reservoir, the equilibrium pressure will vary with changes in the equilibrium specific volume. If we now raise the piston back to its original position and wait until the pressure and temperature again stabilize, we will find that not only the temperature and specific volume, but also the pressure has returned to the original equilibrium values. Further, other properties of the system will have the same values as they had in the original state. Thus, we see that it is usually necessary to specify only two independent intensive properties, in this case temperature and specific volume, in order to completely define the state of a single phase, single component system at equilibrium. This greatly simplifies the analysis of thermodynamic processes. The exceptions to this generalization, such as liquid water which at a given pressure exhibits the same specific volume at two different temperatures as a result of the existence of a minimum in its isobars, need not concern us in this text.

An example of equilibrium state data which have been obtained for steam at 500 °F is shown in Table 1-2.

TABLE 1-2. Properties of Steam

$P(lb_f/in.^2)$	$v(ft^3/lb_m)$
5	114.21
10	57.04
15	37.99
20	28.46

A more extensive tabulation of the equilibrium state properties of steam may be found in Table A-2 in the Appendix[1]*. Similar tables of properties for steam and other fluids may be found in the literature. Another widely accepted table of steam properties in both English and metric units has been published by Keenan[2]. Din[3] is the editor of a collection of tables of properties in the metric system of units for ammonia, air, argon, carbon monoxide, carbon dioxide, nitrogen, and the lighter hydrocarbons. These tables are based on a critical examination of experimental data. Canjar[4] and Sage[5] have published similar tables of properties in English units.

In some cases, when we change the position of the piston as shown in Fig. 1-2, the pressure exerted against the piston by the constant temperature system does not change. This can be attributed to a phase change. The liquid-vapor phase change is called vaporization; the liquid-solid phase change is called fusion; and the solid-vapor phase change is called sublimation. Under these conditions, the measured pressure is called the saturation pressure or vapor pressure if a vapor is present.

Somewhat greater insight into the relationships between pressure, temperature and specific volume may be obtained by plotting these variables on a three-dimensional diagram called a phase diagram.

For example, Fig. 1-3 is a portion of a three dimensional diagram for a substance which contracts upon freezing. From our experiments on a single phase, single component fluid at equilibrium, we observed that only one reproducible pressure could be associated with a given specific volume and temperature. Therefore, one could not arbitrarily specify all three coordinates. Graphically this means that points representing the component exist only on the surfaces of the three dimensional diagram shown in Fig. 1-3 and not in the regions above or below the surfaces. These surfaces may be thought of as the elevation given in pressure units as a function of position given in terms of temperature and volume coordinates. The volumes of course are on a unit mass basis since otherwise the diagram would have to show mass as another variable.

The portions of the three dimensional surface through which horizontal lines of constant temperature called isotherms have been drawn, represent regions in which mixtures of two or more phases coexist in equilibrium. The state at which three phases such as solid, liquid and vapor coexist in equilibrium is called the triple point. As can be seen from Fig. 1-3, the triple point temperature and the triple point pressure are not independent of each other. This is the reason that triple point temperatures are convenient points to use as fixed points in defining temperature scales.

The state represented by the highest temperature and the highest pressure at which liquid and vapor can coexist in equilibrium is called the critical point. At the critical point, the properties of the liquid and vapor become identical and indistinguishable. The coordinates of the critical point are called the critical temperature T_c, the critical pressure P_c and the critical specific volume v_c. The critical

---

*Numbers in parentheses refer to entries in the list of references in the Appendix.

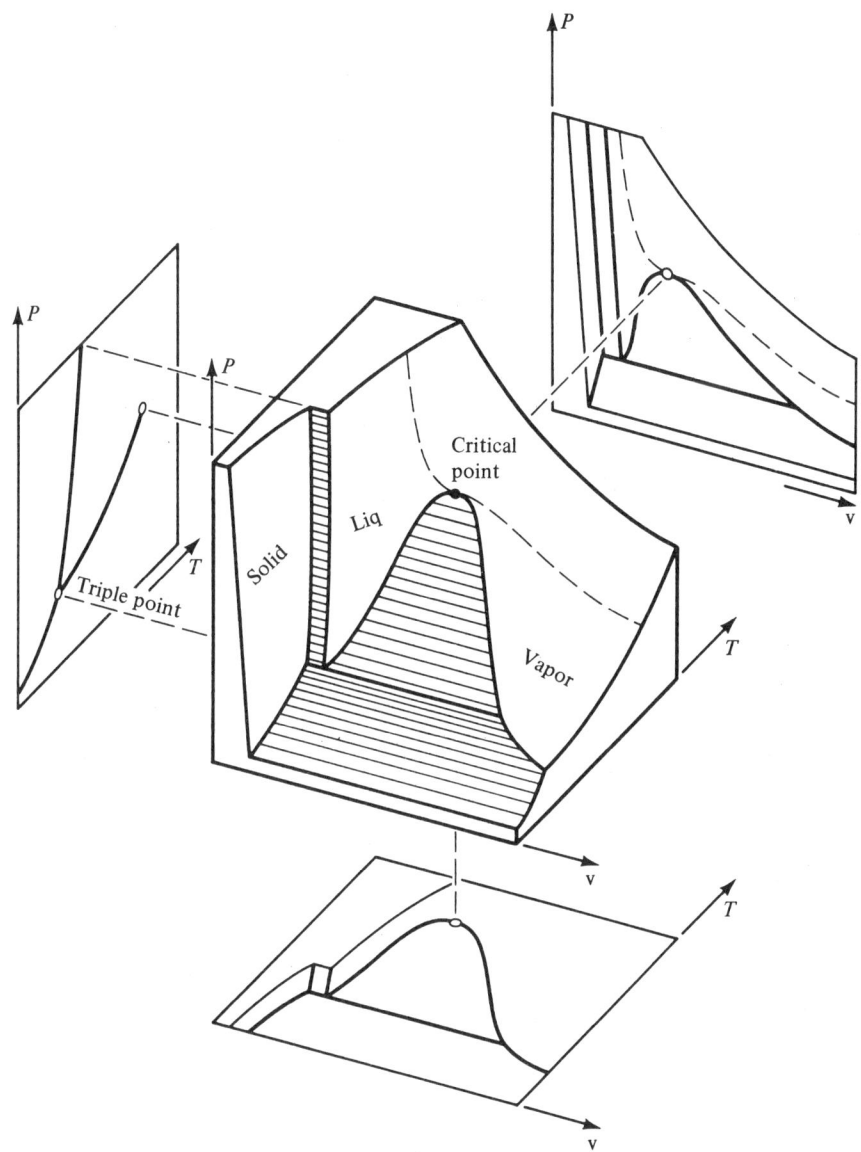

Fig. 1-3. *PVT surface*.

isotherm represented by the dashed line on the three dimensional surface shown in Fig. 1-3 has a horizontal inflection point at the critical point.

Within a coexistence region, for example, the region in which liquid and vapor exist together in thermodynamic equilibrium, the specific volume of a system is independent of the system pressure and temperature. Rather, it depends on the relative quantities of the phases which are in equilibrium. Thus, a triple point, shown in Fig. 1-3, exists as a point only on the two-dimensional *PT* projection of the three dimensional *PvT* surface.

Further insight into the specific volume of a mixture of phases can be obtained by examining the Pv projection of a PvT surface shown in Fig. 1-4. In this figure the coexistence of liquid and vapor is shown as a region under a dome which is made up of the saturated liquid and saturated vapor lines. In this coexistence region, the isotherms are also lines of constant pressure. A mixture of saturated liquid and saturated vapor may be represented by a point M on the constant pressure isotherm FG.

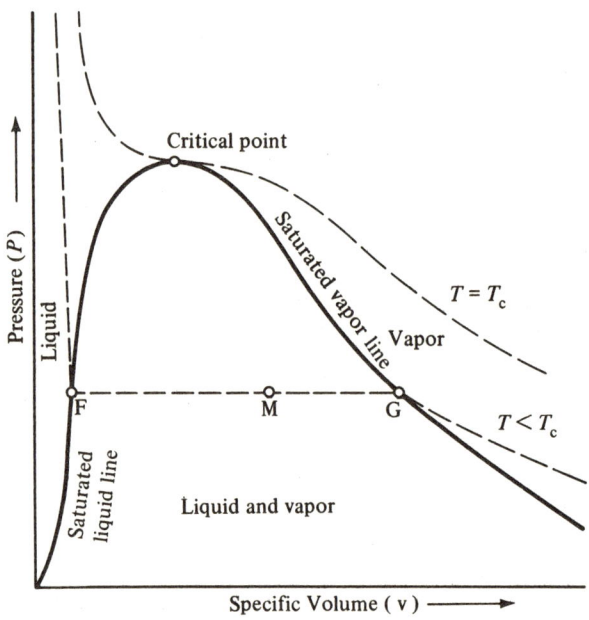

Fig. 1-4. *Pressure volume diagram.*

If the mole fraction vapor in the mixture is x, then the volume of the mixture is given by the relationship

$$v_M = (1-x)v_f + xv_g$$

$$= v_f + x(v_g - v_f)$$

$$= v_g - (1-x)(v_g - v_f) \tag{1-22}$$

These relationships are valid for other thermodynamic properties. When using a slide rule, the second of these relationships gives the most reliable results when x is small, the third gives the most reliable results when x is close to unity. We might note that if Eq. (1-22) is written on a mass basis rather than a mole basis, x, the weight fraction of vapor is called "quality." The term quality arose when early workers recognized that vapor had more energy than liquid and thus had a higher quality.

Graphically, the position of point M representing the mixture is directly related to the percentage of the mixture that is vapor. From Eq. (1-22) we see that

$$x = \frac{v_M - v_f}{v_g - v_f}$$

If we let the distance on Fig. 1-4 representing the difference between v_M and v_f be $\overline{FM}$ and the distance representing the difference between v_g and v_f be $\overline{FG}$, then

$$x = \overline{FM}/\overline{FG}$$

This relationship is called the inverse lever arm principle in that it relates the distance between the points representing the mixture and the liquid to the fraction of vapor in the mixture. As the inverse lever arm principle is nothing more than a graphical material balance, it is applicable to all thermodynamic properties in the coexistence region.

Points representing systems in the vapor phase are sometimes identified by the difference between the system temperature T and the saturation temperature $T^{\square}$ at system pressure P. This temperature difference is called degrees of superheat Sh.

EXAMPLE PROBLEM 1-2. A rigid vessel contains saturated vapor at 300°F. If the temperature of the steam drops to 200°F, calculate the final quality of the steam.

From Table A-2 in the Appendix, we find that in the initial state,

$$v_{\text{initial}} = 6.466 \text{ ft}^3/\text{lb}_m$$

Since neither the volume nor the mass of the steam change,

$$v_{\text{final}} = v_{\text{initial}}$$

Thus, at 200°F, we find using data from Table A-2,

$$x = \frac{v_{\text{final}} - v_f}{v_g - v_f}$$

$$= \frac{6.466 - 0.01664}{33.639 - 0.01664}$$

$$= 0.19$$

Graphical representations of equilibrium phase relationships are extremely useful in presenting an overall picture in a concise and easily understood form. Phase relationships may also be presented in an analytical form called an equation of state. For example, an analytical relationship between the equilibrium pressure and specific volume of a gas at low pressure is

$$P\mathrm{v} = kT \tag{1-23}$$

The constant of proportionality k is selected so that Eq. (1-23) will represent experimental data over a limited range. In Chap. 5 we will undertake a more detailed study of equations of state.

PROBLEMS

1-1. Define the system and the surroundings for the following. Are the systems selected open, closed or isolated?

    (a) A river which is polluted by a nearby manufacturing plant.
    (b) A nuclear power plant that draws cooling water from a nearby bay of the ocean.
    (c) A human heart.

Hint: There may be more than one answer to these questions.

1-2. The distance between Paris and Amsterdam is approximately 500 kilometers. How many hours would it take you to drive this distance if your car averaged 45 miles per hour?

1-3. One lb_f will give a mass of 1.0 slug an acceleration of 1.0 ft/sec². With what force is a mass of 10 slugs attracted to the earth at a point where the gravitational acceleration is 30.6 ft/sec²? What is its weight in lb_f and in newtons? What is its mass in lb_m and in kg?

1-4. A poundal is the force that will give 1.0 lb_m an acceleration of 1.0 ft/sec². How many lb_f are there in a poundal? How many newtons?

1-5. At a certain point on the moon, $g = 5.35$ ft/sec². How much would an astronaut, who weighs 185 lb_f on the earth, weigh at this point on the moon?

1-6. A dyne is the force that will give 1.0 gram an acceleration of 1.0 cm/sec². How many dynes are there in 1.0 lb_f? In a newton? In a poundal?

1-7. (a) Convert $-130\,°F$ to $°R$, $°C$, and K.

  (b) Convert $+300\,°C$ to K, $°R$, and $°F$.

1-8. At what temperature do the numerical values on the Celsius and Fahrenheit scales coincide? The Kelvin and Rankine scales?

1-9. An erg is a unit of energy equivalent to a force of 1.0 dyne, i.e. 1.0 gm cm/sec^2, acting through a distance of 1.0 cm. How many ergs are there in one joule? In 1.0 ft-lb$_f$? In 1.0 BTU? In 1.0 calorie?

1-10. Power is frequently measured in horsepower hp or kilowatts kW. How many kW in 1.0 hp? How many BTU/min in 1.0 hp? In 1.0 kW?

1-11. A pressure gage reads 32.0 lb$_f$/in.^{2} and a barometer reads 28.2 in. of Hg. Calculate the absolute pressure in lb$_f$/in.^{2} and in standard atm.

1-12. What is the height in mm of a column of mercury that will exert a pressure of 14.696 lb$_f$/in.^{2} if the density of mercury is 13.5951 gm/cm^3 and the gravitational acceleration is 980.665 cm/sec^2? Under these conditions, 1.0 mm of Hg is called 1.0 torr.

1-13. What is the maximum height, in meters, that liquid water with a specific volume of 0.01613 ft^3/lb$_m$ can be siphoned if atmospheric pressure is 100,000 newtons/meter^2 and g is 30.0 ft/sec^2?

1-14. What is the density in lb$_m$/ft^3 of a liquid having a density of 1.0 gm/cm^3?

1-15. How many lb-moles are there in 100 lb$_m$ of the following substances: H_2, N_2, O_2, H_2O, NH_3, Fe, and Al? How many gram-moles?

1-16. What is the vapor pressure of water at 300 °F in lb$_f$/in.^{2}? In mm of Hg? In N/m^2?

1-17. Using data from Table A-2 in the Appendix, draw the two-phase dome for steam from 0.0 to 2.0 ft^3/lb$_m$. Superimpose the 600 °F and the 900 °F isotherm on the drawing.

1-18. Find the quality or temperature (if super-heated) of water at the following conditions:

	$P(lb_f/in.^2)$	$v(ft^3/lb_m)$
(a)	1.0	5
(b)	10.0	5
(c)	100.0	5

1-19. The radiator of a heating system has a volume of 2 ft^3 and contains saturated vapor at 20 lb$_f$/in.^{2} The valves are then closed on the radiator, and as result of heat transfer to the room the pressure drops to 15 lb$_f$/in.^{2} Calculate:

    (a) The total mass of steam in the radiator.
    (b) The volume and mass of liquid in the final state.
    (c) The volume and mass of vapor in the final state.

1-20. Steam at the critical state is contained in a rigid vessel. Heat is transferred from the steam until the pressure is 100 lb$_f$/in.^{2} Calculate the final quality.

1-21. A boiler feed pump delivers 500,000 lb$_m$ of water per hr at 2000 lb$_f$/in.^{2}, 560 °F. What is the volume rate of flow in ft^3/min?

1-22. What is the weight percent vapor in an equilibrium vapor-liquid mixture of methane at 150 K if the specific volume of the mixture is 500 cm^3/gm-mole?
Hint: The solution to this problem will require the use of either Ref. 3 or 4.

1-23. An analytical relationship between the vapor pressure of water and temperature is

$$\ln P(\text{atm}) = 70.4346943 - \frac{7362.6981}{T} + 0.006952085T - 9.00000 \ln T$$

where T is in kelvins. This equation is reported to be valid in the range 273 to 398 K.

Using this equation, calculate the vapor pressure of water to three decimal places in units of lb$_f$/in.^{2} from 40 °F to 260 °F in 20 °F steps and prepare a table comparing these results with vapor pressures tabulated in Table A-2 in the Appendix.
Hint: Repetitive calculations of this type are most easily solved using a digital computer.

2

The Mass Balance and the Energy Balance

OBJECTIVES

After studying this chapter, the student should be able to

(1) state the principle of conservation of mass;
(2) apply the principle of conservation of mass, in the form of Eq. (2-1), to new situations using tabulated thermodynamic properties;
(3) state the first law of thermodynamics;
(4) define types of energy and obtain numerical values for examples of each type; and
(5) apply the first law of thermodynamics in the form of Eq. (2-25), combined with the principle of the conservation of mass if necessary, to new situations using tabulated thermodynamic properties.

THE MASS BALANCE

Among the basic principles of thermodynamics are the principles of conservation of mass and energy. As we saw in Chap. 1, mass is a property of matter. Electrons, protons and neutrons have mass; atoms, ions and molecules have mass; and clusters of atoms, ions or molecules in the form of an identifiable object, such as this text, have mass. The quantity of mass in each of these entities can be determined by comparison with reference to an object of known mass. The basic standard is a platinum irridium cylinder called the kilogram and kept in Paris.

The principle of conservation of mass may be stated as follows:

Matter can not be created nor destroyed; it can only be changed to other chemical species or to energy.

In differential form, the principle of conservation of mass may be written as a mass balance equation,

$$\delta m_{in} - \delta m_{out} + \delta m_{gen} = dm_{sys} \qquad (2\text{-}1)$$

This equation applies to a specific system during a fixed interval of time. The terms δm_{in} and δm_{out} represent the mass influx and efflux to an open system or control volume and would reduce to zero for a closed or isolated system. The term δm_{gen} represents the mass converted from one form to another and the term dm_{sys} represents the accumulation of mass within the system, which would be zero for a steady-state, steady-flow process. It should be pointed out that the symbol d means a differential change of a property whereas the symbol δ means either a differential quantity of or flow of a differential quantity between the system and its surroundings.

Now the term representing the generation of mass δm_{gen} requires special attention. Although mass can not be created or destroyed, it can be converted from one form to another. Einstein and others have shown that one form of mass is energy. Thus, for systems involving nuclear reactions, the term δm_{gen} could be used to account for the conversion of mass into energy. In systems involving chemical reactions, components will be converted from one species to another. In this case, the mass balance for an individual chemical species will include the mass conversion term δm_{gen}. However, if we neglect nuclear effects, the total mass of all species produced will be exactly equal to the total mass of all species consumed in the chemical reaction and the total mass balance will not contain the δm_{gen} term.

The effect of time may be taken into account by writing Eq. (2-1) as a rate equation. The rate at which mass accumulates in a system can be found by dividing each term in the mass balance equation by the time of observation dt. If none of the flows vary with time, any interval of time may be used as the basis for analysis. In this case, dt is a constant in all terms and it may be cancelled out. Then, the mass balance equation applied over a fixed interval of time to a system in which N streams of constant flow rate enter and M streams of constant flow rate leave takes the form

$$\sum_{i}^{N} \delta m_{in} - \sum_{i}^{M} \delta m_{out} + \delta m_{gen} = dm_{system} \qquad (2\text{-}2)$$

On the other hand, if the flows change with time, it is more convenient to use a differential interval of time dt as the period of observation. In this case, the instantaneous rate form of Eq. (2-1),

$$\left(\frac{\delta m}{dt}\right)_{in} - \left(\frac{\delta m}{dt}\right)_{out} + \left(\frac{\delta m}{dt}\right)_{gen} = \left(\frac{dm}{dt}\right)_{system} \tag{2-3}$$

is used for the analysis. The total change of mass in the system must be obtained by summing Eq. (2-3) over the total number of streams entering and leaving the system during the total period of observation. Although the mass balance equations can be written in various forms, we will use the form presented in Eq. (2-1) to represent symbolically all possible forms of the mass balance equation, including Eqs. (2-2) and (2-3). In applying this equation to the analysis of a particular system, three points should be kept clearly in mind. First, the mass balance applies to a specific system. The system of interest must be clearly identified before the equation is applied. This is accomplished most readily by first drawing a sketch of the process to be studied and labeling the sketch with all available data. The portion of the process selected as the system can then be shown on the sketch. Second, the mass balance must be integrated over some fixed interval of time. This interval may coincide with the duration of a particular process being investigated or may be selected arbitrarily. In either case, the time interval or basis for the application should be settled upon before the mass balance is applied. Finally, in addition to specifying the system and basis, we must also specify the component or combination of components being considered.

Once a system, a time interval or basis and the component or combination of components have been established, the first step in applying the mass balance equation to any particular problem will normally be the elimination of those terms which do not apply. This is a very critical step and the source of many errors. In a closed system, the terms representing mass in and mass out can be eliminated. In an open system in which there is no change of mass with time (steady-state, steady-flow), the term indicating the change of the mass or accumulation of mass in the system drops out. Further, for systems in which no chemical or nuclear reaction takes place, δm_{gen} is zero. The particular form of the equation which results from the elimination of terms which do not apply is called the reduced mass balance equation.

As an example of the application of the generalized mass balance equation, consider the following problem.

EXAMPLE PROBLEM 2-1. Air in an office building is to be maintained at 70 °F and is to contain 0.01 lb_m of water per pound of water-free air. The air available from a furnace in the basement of the building contains only 0.005 lb_m of water per lb_m of water-free air. It is proposed to continuously pass the air from the furnace through a humidifier before it enters the offices.

If the humid air from the furnace is supplied to the humidifier shown in Fig. 2-1 at a rate of 1000 lb_m/hr, at what rate must water at 70 °F be added to the humidifier?

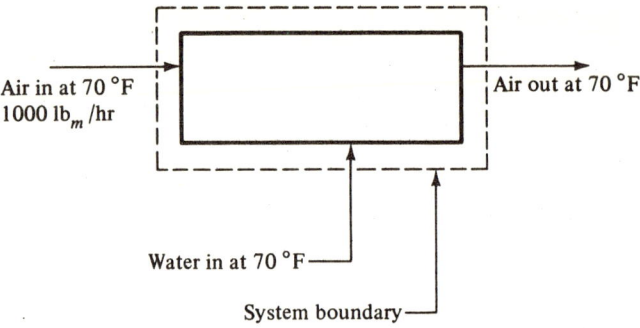

Fig. 2-1. *Humidifier.*

The generalized mass balance equation that will be used to solve this problem is

$$\delta m_{in} - \delta m_{out} + \delta m_{gen} = dm_{sys} \qquad (2\text{-}1)$$

Now, we select the humidifier as our system and base our calculations on one hour of operation. As there are no chemical reactions involved, $\delta m_{gen} = 0$ for any or all of the components in the flowing streams. Further, as the humidifier operates at steady-state, steady-flow, there is no accumulation of mass within the system and $dm_{sys} = 0$.

Thus, in differential form, the generalized mass balance becomes $\delta m_{in} - \delta m_{out} = 0$ or, integrating over one hour of time $m_{in} = m_{out}$. Now a mass balance over all the components flowing into and out of the humidifier gives the result

$$1000 + m_{water} = m_{air\ out}$$

Since this equation contains two unknowns, we will need one more equation. This second equation may either be obtained by making a material balance around the water

$$\frac{(1000)(0.005)}{1.005} + m_{water} = \frac{(m_{air\ out})(0.01)}{1.01}$$

or a material balance around the water-free air

$$\frac{(1000)}{(1.005)} = \frac{m_{air\ out}}{1.01}$$

Solving these equations simultaneously,

$$m_{air\ out} = 1005\ lb_m/hr$$

and our final answer is

$$m_{\text{water in}} = 5 \text{ lb}_m/\text{hr}$$

NOTE: Rigorously, $m_{\text{water in}} = 4.975 \text{ lb}_m/\text{hr}$. However, from the statement of the problem, our answer is probably valid to only one significant figure.

THE ENERGY BALANCE

The First Law of Thermodynamics

The first law of thermodynamics may be summarized in the statement

Energy is conserved.

The concept of energy arose in mechanics and dates back at least to the time of Galileo. He used the concept of energy when he intuitively asserted that the velocity reached by a falling object was capable of raising the object to its original height, but no higher. However, the use of the word "energy" to denote the quantity of work which could be done by a system was not introduced until 1807 by Thomas Young.

The first man generally credited with recognizing that mechanical energy is conserved is Gottfried Wilhelm Leibnitz. In 1695 he observed that the product of the gravitational force times the path equals the increase in mv^2, which can be interpreted to mean that the sum of what we now call the potential and kinetic energies of a mass point remain constant. During the eighteenth century, both Johann Bernouilli and Leonhard Euler expanded upon and contributed to the general acceptance of these ideas. It wasn't until 1798 that Count Rumford of Munich, the former Benjamin Thompson, demonstrated that heat is a flow of energy.

Before this time, heat was thought of as a weightless fluid called caloric which could pass from one body to another. This was called the caloric theory. Joseph Black was one of the principal adherents to this theory. In 1760, he introduced the concept of specific heat, or the capacity for heat fluid, and made a clear distinction between the concepts of temperature and heat.

The first attempt to measure the relationship between heat and other forms of energy was an experiment by Count Rumford. Using horses as the motive power, he rotated a blunt boring tool pressed against a gun metal cylinder and measured the increase in temperature and the amount of metal given off. He showed that the quantity of heat given off was apparently inexhaustible and depended only on the continued operation of the experiment. The amount of heat was so great and the quantity of matter rubbed off so small that there could be no relationship between them. He concluded from his experiments that heat could not possibly be a substance but that it must be a flow of energy. This was the beginning of the end of the caloric theory of heat.

The equivalence of heat and mechanical energy was clearly stated for the first time by Robert Mayer, in 1842. However, it remained for James Prescott Joule to establish beyond question that heat is a form of energy. Working between 1840 and 1850, Joule measured the temperature increase of water at atmospheric pressure in a well insulated cylinder that could be attributed to a given work input. In his original experiments, Joule used an electric generator to convert a known amount of mechanical energy into electric current. The current was passed through a wire coil immersed in the water. Later, Joule supplied energy to the water by means of rotating paddles which were driven by falling weights.

Joule showed that whenever mechanical or electrical work was done, a quantity of heat was generated which always bore a fixed ratio to the work done. His best value of the mechanical equivalent of heat was

$$1 \text{ calorie} = 4.155 \text{ joules}$$

Today, the accepted mechanical equivalent of heat is

$$1 \text{ calorie} = 4.1868 \text{ joules}$$

When it was found that work could be produced from electricity, it was necessary to define electrical energy in terms of electromagnetic fields in order to satisfy the concept of energy conservation. As science progressed, it became necessary to define other forms of energy. For every new type of interaction that has been discovered, including those of nuclear particles, it has been possible to find a characteristic energy function that combined with the principle of conservation of energy has allowed the prediction of changes in one system as a result of changes in another system.

The extension of the principle of energy conservation, first applied to strictly mechanical systems, to heat and other forms of energy constitutes the first law of thermodynamics.

Accountability of Energy

Energy can be divided into forms not associated with mass which flow between a system and its surroundings as a result of a potential difference and other forms which are associated with mass. For convenience, forms of energy not associated with mass are called heat Q and work W. Heat is defined as energy in transition across a system boundary as a result of a temperature gradient. Heat flows from a high temperature source to a low temperature sink and is assumed to be positive if it is absorbed by the system. Work is defined as energy in transition across a system boundary as a result of any potential gradient other than temperature, such as pressure. It is assumed to be positive if it flows from the system to the surroundings. Work flows from a system to its surroundings during a given process when its only effect on the surroundings could conceivably be the raising of a weight by an equivalent weight raising process.

Energy associated with a mass is also a property of the mass and is arbitrarily subdivided into kinetic energy E_k, potential energy E_p, flow energy E_F, and internal energy E_{int}. Each of these types of energy will be discussed in more detail later. But for now, it is sufficient to note that

Kinetic energy is that energy which can be attributed to the motion or velocity of a mass relative to a fixed point in space.

Potential energy is that energy which can be attributed to the position of a mass within a gravitational, an electric, or a magnetic field.

Flow energy is the energy required to push an incremental amount of mass from the surroundings into the system and to push an incremental amount of mass from the system back into the surroundings.

Internal energy is the energy associated with the molecules or other entities which make up a mass.

The total energy which enters a system then is

$$\delta E_{in} = (\delta E_{int} + \delta E_k + \delta E_p + \delta E_F)_{in} + \delta Q$$

the total energy which leaves the system is

$$\delta E_{out} = (\delta E_{int} + \delta E_k + \delta E_p + \delta E_F)_{out} + \delta W$$

and the energy which accumulates in the system is

$$dE_{sys} = d(E_{int} + E_k + E_p)$$

Then, if we include an energy generation term δE_{gen} the first law of thermodynamics may be written as an energy balance equation,

$$(\delta E_{int} + \delta E_k + \delta E_p + \delta E_F)_{in} - (\delta E_{int} + \delta E_k + \delta E_p + \delta E_F)_{out}$$

$$+ \delta Q - \delta W + \delta E_{gen} = d(E_{int} + E_k + E_p)_{sys} \tag{2-4}$$

The term representing the generation of energy δE_{gen}, just as the term δm_{gen} in the mass balance equation, requires special attention. Although energy can not be created nor destroyed, it can be converted from one form to another. One possible form energy can take is the form of mass. Thus, for systems involving nuclear reactions, the term δE_{gen} could be used to account for the conversion of mass into energy. Further, for systems involving chemical reactions, such as burning coal, the term δE_{gen} could be used to account for the conversion of energy from a chemical form to a physical form.

The method of application of the generalized energy balance equation is the same as the method of application of the generalized mass balance equation. However, in order to apply Eq. (2-4) to practical problems, we must first examine each of its terms in more detail. For the present, we shall not consider any problems involving chemical or nuclear reactions which would require the use of the δE_{gen} term.

Types of Energy

Work. Work can be measured in terms of its effect on a system in terms of a force acting through a distance. Numerically, the total work associated with a process is

$$W = \int_{L_1}^{L_2} F \, dL$$

In order to evaluate this integral, we must know how the force F varies as a function of the length L. That is, we must define the path of the process in terms of a curve between the initial and final states. Then work is numerically equal to the area under the curve or path which represents F as a function of L and which is bounded by L_1 and L_2.

Now, we can calculate the work required to compress the gas shown in Fig. 2-2 using the relationship

$$\delta W = F \, dL \quad (2\text{-}5)$$

If we assume that
$$F = PA$$

then
$$\delta W = PA \, dL$$

$$= P \, dV \quad (2\text{-}6)$$

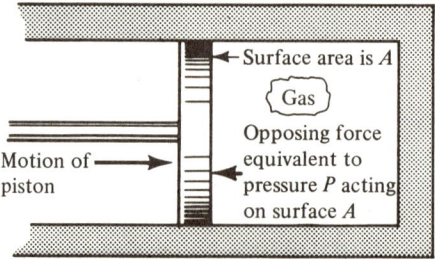

Fig. 2-2. *Gas compressor.*

Note that the pressure term in this relationship is the pressure opposing the motion of the piston. The small δ in Eq. (2-6) refers to a differential flow of energy between the system and its surroundings while the small *d* refers to a differential change in a property.

As we have indicated, the value for work obtained on integrating Eq. (2-6) depends upon the path followed from the initial to the final states. In order to define the path a system takes in undergoing some change, we must define the state of the system at each point along the path. But as we indicated in Chap. 1, the state of a system is defined by the properties of that system, such as pressure, temperature and specific volume, only if the system is in thermodynamic equilibrium. Therefore, it would be particularly convenient if the system were in thermodynamic equilibrium at all times during the process. As we indicated in Chap. 1, such a process is called a reversible process.

Now, recall that in Chap. 1, we also stated that reversible processes are capable of being reversed in direction by an infinitesimal change in the driving forces. This implies that if the process of compressing the gas shown in Fig. 2-2 is carried out reversibly then the pressure of the gas and the pressure opposing the motion would be identical. In such a case, we can use the pressure exerted by the gas to evaluate Eq. (2-6). If the process were not carried out reversibly, for example due to friction of the piston against the cylinder walls, it is extremely difficult, if not impossible, to evaluate force as a function of displacement.

Reversible processes can be carried out in many different ways. For example, they may be isothermal, isobaric or adiabatic. In this text, the specification of the path of a process may be given either explicitly by a statement such as the process is isochoric, or implicitly by a statement such as the process takes place in a rigid vessel. In the latter case the student must realize that a rigid vessel cannot change volume and so the process must be isochoric.

Work or energy may cross system boundaries through other forms of compression processes. For example, consider the reversible stretching of a rubber band. Again, the work required is

$$\delta W = -F\,dL \qquad (2\text{-}7)$$

In this case, however, a minus sign is required because in general tension in the rubber band increases as length increases whereas in compressing a gas, pressure increases as volume decreases.

The reversible stretching of a rubber band may be looked upon as a case of one dimensional expansion work whereas pressure-volume work represents three dimensional expansion work. If a problem involves a change in surface area, such as would occur when a film of plastic is being stretched reversibly, we have a case involving two dimensional expansion work. Then, the work required would be given by the relationship

$$\delta W = -\sigma\,dA \qquad (2\text{-}8)$$

In this case, surface tension σ opposes any increase in surface area and work is done on the system when the change in area dA is positive.

Expansion type work can be recognized when the boundary of a system moves with respect to the surroundings. Energy can also cross a system boundary as work by means of mechanical devices such as a piston, a drive shaft or some other mechanical linkage extending between the system and its surroundings. This type of work is called shaft work.

Expansion type work and shaft work require physical contact with the surroundings. Systems can also interact with their surroundings through forces which act at a distance without any physical contact. For example, it is not essential that two electric charges be in physical contact in order to influence one another. Gravitational and magnetic forces also react through a distance without physical contact. A force that acts through a distance without physical contact is called a body force.

As an example of the work associated with forces that do not require physical contact, consider the problem of determining the work required to separate two unlike electrical charges q_1 and q_2. From physics we learn that any charge q_1 creates around it an electrostatic field of force which can act on a second charge q_2. It has also been found that the magnitude of the electrostatic force acting on charge q_2 is given by Coulomb's inverse square law,

$$F = \frac{kq_1 q_2}{L^2} \tag{2-9}$$

where L is the distance of separation and the force acts along a line joining the two charges. These electrostatic forces can either be repulsive or attractive depending upon whether the charges are of like or unlike kind. Like charges repulse each other and by convention, the sign associated with a repulsive field force is positive.

Associated with an electrostatic field due to charge q_1 is a quantity called the electrostatic potential ϕ_e. This potential is a function of the distance from the charge to the point in the field at which the potential is being measured. It is related to the force experienced by an infinitesimally small test charge a distance L from the charge q_1 by the defining equation

$$\frac{dF}{dq} = -\frac{d\phi_e}{dL} \tag{2-10}$$

where a comparison of Eqs. (2-9) and (2-10) shows that ϕ_e must be defined by the relationship

$$\phi_e = kq_1/L$$

The force per unit charge dF/dq is called the intensity of an electrostatic field.

Now, the path followed by any charge moving through a time invariant electrostatic field is a reversible path. Thus, the work required to move a system consisting of an infinitesimally small charge dq relative to charge q_1 is given by the relationship

$$\delta W = F\, dL \qquad (2\text{-}5)$$

As the force F acting on the charge dq is dF, we can substitute Eq. (2-10) into this relationship to find

$$\delta W = -(\phi_{e_2} - \phi_{e_1})\, dq$$

Further, if we define the difference in electrostatic potentials as

$$\epsilon = \phi_{e_2} - \phi_{e_1}$$

then

$$\delta W = -\epsilon\, dq \qquad (2\text{-}11)$$

If we compare Eqs. (2-6) and (2-11), we see that in both cases work is the product of an intensive property that does not depend upon the extent of the system and the change in an extensive property which does depend upon the extent of the system. In general, we will find it possible to write a relationship for any type of work associated with a reversible process in the form

$$\delta W_{\text{rev}} = \Sigma(\text{intensive property})\, d(\text{extensive property}) \qquad (2\text{-}12)$$

The similarity between this relationship and the equation for mechanical work

$$\delta W = F\, dL \qquad (2\text{-}5)$$

leads us to call all intensive properties "generalized forces" or "intensity factors" and all extensive properties "generalized displacements" or "capacity factors." In this context, the generalized forces and generalized displacements commonly associated with some of the more typical types of work are shown in Table 2-1.

TABLE 2-1. Definitions of Work

Type of Work	"Generalized Force"	"Generalized Displacement"	Element of Work
Volumetric	P	$-V$	$P\, dV$
Length	F	L	$-F\, dL$
Surface	σ	A	$-\sigma\, dA$
Gravitational	gz	m	$-gz\, dm$
Centrifugal	$-\dfrac{r^2\omega^2}{2}$	m	$\dfrac{r^2\omega^2}{2}\, dm$
Electrical	ϵ	q	$-\epsilon\, dq$
Chemical	μ_i	n_i	$-\mu_i\, dn_i$

Then, Eq. (2-12) can be written

$$\delta W_{rev} = P\,dV - F\,dL - \sigma\,dA - \epsilon\,dq - \Sigma \mu_i\,dn_i - \ldots \qquad (2\text{-}13)$$

If the applicable terms in Eq. (2-13) are integrated from the initial to the final state, the total work associated with a given reversible process may be determined. Other relationships must be used to estimate the work associated with irreversible processes.

EXAMPLE PROBLEM 2-2. What is the reversible, adiabatic work required to increase the length of an unstressed steel bar from 10 feet to 10.001 feet? The cross-sectional area of the bar is 0.5 in.^{2}, and the modulus of elasticity E is 3.0×10^7 lb$_f$/in.^{2}

We will take the steel bar as our system and use the time required to stretch the bar 0.001 feet as the basis for our calculations.

The work required is given by the relationship

$$\delta W = -F\,dL \qquad (2\text{-}7)$$

Now,

$$F = sA$$

where

$$s = \text{unit stress on the bar (lb}_f/\text{in.}^2)$$

and

$$A = \text{cross-sectional area of bar (in.}^2)$$

Further,

$$dL = L\,d\epsilon$$

where

$$\epsilon = \text{unit strain on the bar (in./in.)}$$

Thus,

$$\delta W = -ALs\,d\epsilon$$

$$= -Vs\,d\epsilon$$

where

$$V = \text{volume of the bar (in.}^3)$$

Assuming that Hooke's law,

The unit stress in a material is proportional to the accompanying unit strain

is valid for this problem, we have

$$\frac{S}{\epsilon} = E = 3.0 \times 10^7 \text{ lb}_f/\text{in.}^2$$

If we assume that the volume of the bar remains essentially constant, we find

$$W = -VE \int_{\epsilon_1}^{\epsilon_2} \epsilon \, d\epsilon$$

$$= -\frac{1}{2} VE(\epsilon_2^2 - \epsilon_1^2)$$

Now, initially $\epsilon_1 = 0$, and at the end of the process

$$\epsilon_2 = \frac{10.001 - 10.000}{10.000} \times \frac{12}{12} = 0.0001 \text{ in./in.}$$

Substituting these values into our integrated equation for work,

$$W = -\frac{1}{2}(10)(0.5)(3.0 \times 10^7)(0.0001^2 - 0.0^2)$$

or

$$W = -0.75 \text{ ft-lb}_f$$

$$= -1.02 \text{ joules}$$

The negative sign indicates that work must be done on the bar in order to stretch it.

Heat. In some ways, heat is analogous to work. Both are forms of energy, not associated with mass, that are in transition across a system boundary. Just as the flow of energy as work is measured by its effects on a system, so the flow of energy as heat is measured by its effects on a system.

The amount of heat δQ that must be added to a homogeneous unit mass or unit mole in order to change the temperature of that mass or mole an amount dT is called the specific heat capacity or the molal heat capacity of the substance c. That is,

$$c = \frac{\delta q}{dT} \tag{2-14}$$

We will use the term specific heat capacity for both a unit mass and a unit mole of substance.

Specific heat capacities of materials are generally measured in a calorimeter. A calorimeter may be either a rigid closed vessel called a batch calorimeter or an open vessel called a flow calorimeter. The essential components of either type of calorimeter are devices to measure the energy added and devices to measure the resulting temperature change.

Since heat is a path function, the specific heat capacity of a material will depend upon the path of the process by which heat is added. If heat is added reversibly to a homogeneous fluid at constant pressure in a flow calorimeter, the specific heat capacity of the fluid is

$$c_P = \left(\frac{\delta q_{\text{rev}}}{dT}\right)_P \tag{2-15}$$

where c_P is called the specific heat capacity at constant pressure and is a property of the system. Values of c_P for both gases at very low pressures and for solids, on a unit mole basis, are presented as a function of temperature in Table A-3 in the Appendix. Using specific heat capacity data, the heat flow associated with a constant pressure, reversible process is given by integrating Eq. (2-15),

$$q_{\text{rev}} = \int c_P \, dT$$

Note that we must know how c_P varies with T in order to integrate this relationship.

If heat is added reversibly to a homogeneous mass at constant volume in a bomb calorimeter, the specific heat capacity of the mass is given by the relationship

$$c_V = \left(\frac{\delta q_{\text{rev}}}{dT}\right)_V \tag{2-16}$$

where c_V is called the specific heat capacity at constant volume. Heat may also be added reversibly to systems at constant length, constant surface area, constant surface tension, etc. The specific heat capacities associated with each of the different reversible processes will differ.

In some cases, when heat is added to a single component system, the temperature of the system remains unchanged. This can be attributed to a phase change. The amount of heat required to vaporize a unit mass of liquid is called the heat of vaporization or the latent heat of vaporization. The amount of heat required to melt a unit mass of solid is called the heat of fusion. The amount of heat required to vaporize a unit mass of solid is called the heat of sublimation.

Kinetic Energy. A mass in motion relative to a point in space has a definite amount of energy relative to the reference point as a result of this motion. From Eq. (2-4), we see that for a closed system undergoing an adiabatic reversible process at constant potential and internal energy,

$$\Delta E_k = -W$$

Note that since no mass can enter or leave a closed system, there is no change in flow energy. Thus, we may evaluate the kinetic energy of a system by estimating the reversible work required to impart a velocity v to the system.

In this case, the reversible work done on the system is

$$\delta W = -F\,dL \qquad (2\text{-}7)$$

From Newton's second law of motion, the force required is

$$F = \frac{d(mv)}{dt} \qquad (1\text{-}4)$$

and if there is no change of mass with changing velocity

$$F = m\frac{dv}{dt}$$

Applying this relationship and recognizing that

$$\frac{dL}{dt} = v$$

we find

$$\delta W = -mv\,dv$$

As the work has been assumed to be reversible, we can integrate this result between the limits of zero velocity and a velocity v to obtain the result

$$W = -\frac{1}{2}mv^2$$

This quantity of work done on the system is given the name "kinetic energy" and is a thermodynamic property of a system, where

$$E_k = \frac{1}{2}mv^2 \tag{2-17}$$

Here we have assumed that E_k is zero at zero velocity. Note that the velocity term in this equation is a relative velocity, and is the difference in the velocity of a reference point, such as earth, and the velocity of the system. Further, as kinetic energy is a thermodynamic property of the system, it depends only upon the relative velocity of the system and not how that velocity was obtained. We assumed the velocity was attained by means of a reversible process merely for convenience.

Potential Energy. A system also has a definite amount of energy as a result of its position within an external field. From Eq. (2-4), we see that for a closed system undergoing a reversible adiabatic process at constant kinetic, flow and internal energy,

$$\Delta E_P = -W$$

Note that since no mass can enter or leave a closed system, there is no change in flow energy. Thus, we can evaluate the potential energy of a system by estimating the work required to increase the distance of separation between the system and the source of the field. As an example, consider the work required to separate two masses. From physics we learn that any mass m_1 creates around it a gravitational force of attraction for another mass m_2. The magnitude of the gravitational force between two masses is given by Newton's inverse square law,

$$F = -\frac{km_1 m_2}{L^2} \tag{2-18}$$

where L is the distance of separation and the force acts along a line joining the center of masses of the two bodies. While electrostatic forces can be either repulsive or attractive, gravitational forces are always attractive.

Now, associated with a gravitational field due to mass m_1 is a quantity called a gravitational potential which is a function of the distance from the center of mass m_1 to the point in the field at which the potential is being measured. Gravitational potential ϕ_G exerted by mass m_1 is related to the force experienced by an infinitesimally small test mass a distance L from the mass m_1 by the equation

$$\frac{dF}{dm} = -\frac{d\phi_G}{dL} \tag{2-19}$$

The force per unit mass dF/dm is called the intensity of a gravitational field.

Then, since the work required to move the system, mass dm, relative to mass m_1 in a gravitational field is reversible, we find that

$$\delta W = F\,dL \tag{2-5}$$

As the force F acting on the mass dm is dF, we can substitute Eq. (2-19) into this relationship to find

$$\delta W = -(\phi_{G_2} - \phi_{G_1})\,dm$$

We might also observe that the intensity of a gravitational field is defined by Newton's second law of motion to be

$$\frac{dF}{dm} = g$$

or, in integrated form

$$F = mg$$

where g is the acceleration of the mass m due to gravitational attraction.

As the gravitational force is attractive, we must use Eq. (2-7), $\delta W = -F\,dL$, to find that the work done on the system is

$$\delta W = -gm\,dL$$

or

$$W = -g(L_2 - L_1)m$$

Further, if mass m_1 is the earth and L is measured along the line of centers joining the center of mass of two bodies, this result may be written in the more familiar form

$$W = -mg\Delta z$$

Then, the potential energy of the mass, due to its position in a gravitational field is numerically equal to the work done on the system and

$$E_P = \int_{z_0}^{z} gm\, dz \qquad (2\text{-}20)$$

where E_P is zero at zero elevation and g is a function of elevation z. Potential energy may also be considered to be a thermodynamic property of a system. Note that the potential energy of a system is relative quantity which depends upon a difference in positions and not an absolute position.

Flow Energy. Flow energy is the energy required to push an incremental amount of mass from the surroundings into the system or from the system back into the surroundings. From Eq. (2-4), we see that for an open system undergoing a reversible adiabatic process at constant kinetic, potential and internal energy,

$$\delta E_{F_{out}} - \delta E_{F_{in}} = -\delta W$$

Now, we can evaluate the flow energy associated with an incremental mass by estimating the work done on or by a system as a result of fluid flowing into or out of that system.

The work on the system associated with a mass of fluid flowing into a system is expansion type work, and

$$\delta W = -P\, dV \qquad (2\text{-}6)$$

where the minus sign indicates work is done on the system. But the change in volume of the system is equal to the volume of the incremental mass entering the system,

$$dV = v_{in}\, \delta m_{in}$$

Further, as the process is reversible, there are no large pressure gradients between the system and its surroundings. Thus, the pressure at the system boundary and the pressure within the incremental mass are identical, or

$$P = P_{in}$$

Then, the work associated with introducing an incremental mass into the system is given by the relationship

$$\delta W = -P_{in} v_{in}\, \delta m_{in}$$

Similarly, the work required to reject an incremental mass from the system is given by the relationship

$$\delta W = P_{out} v_{out} \delta m_{out}$$

Then the net work associated with the introduction and rejection of an incremental mass from an open system is

$$\delta W = P_{out} v_{out} \delta m_{out} - P_{in} v_{in} \delta m_{in}$$

However, the net work done by the system is the negative of the net work done by the incremental mass. Considering the incremental mass as a system,

$$\delta E_{F_{out}} - \delta E_{F_{in}} = P_{out} v_{out} \delta m_{out} - P_{in} v_{in} \delta m_{in} \qquad (2\text{-}21)$$

The product Pv is a property of the fluid flowing into or out of the system. Thus, it is called the flow energy. Some authors also use the term flow work to describe this quantity.

Internal Energy. All energy associated with a system that is not in the form of kinetic energy, potential energy or flow energy is called internal energy U where

$$E_{int} = U \qquad (2\text{-}22)$$

Internal energy is a property of a system that depends upon the state of that system. It is generally attributed to the motion or interactions of the molecules or other entities within the system. Further, just as the kinetic energy of a system is calculated relative to the velocity of a point in space and just as its potential energy is calculated relative to its position in a field, so its internal energy is calculated relative to the internal energy of the same system in some reference state.

The Energy Balance Equation

The various forms of energy may be combined for a system in a gravitational field using the principle of the conservation of energy to obtain a generalized energy balance over a unit time in differential form,

$$\left[\left(u + Pv + \frac{v^2}{2} + gz\right)\delta m\right]_{in} - \left[\left(u + Pv + \frac{v^2}{2} + gz\right)\delta m\right]_{out} + \delta Q - \delta W$$

$$= d\left[\left(u + \frac{v^2}{2} + gz\right)m\right]_{sys} \qquad (2\text{-}23)$$

Because the sum of the two properties $U + PV$ occur together naturally in Eq. (2-23), it is convenient to define this sum as a new property called enthalpy H where

$$H = U + PV \tag{2-24}$$

Values of the specific enthalpy of water h are presented in Table A-2 of the Appendix, where the specific internal energy of saturated liquid water at the triple point is assigned a value of zero. Any other reference state could have been selected just as easily.

Values of the internal energy of water can be calculated using Eq. (2-24) and enthalpy data from Table A-2. Tabulated values of enthalpy for other substances can be found in the literature[3,4,5]. Enthalpy values are also often presented in graphical form. For example, the effect of pressure on the enthalpy of pure methane is shown in Fig. 2-3[6]. Although internal energy includes all energy associated

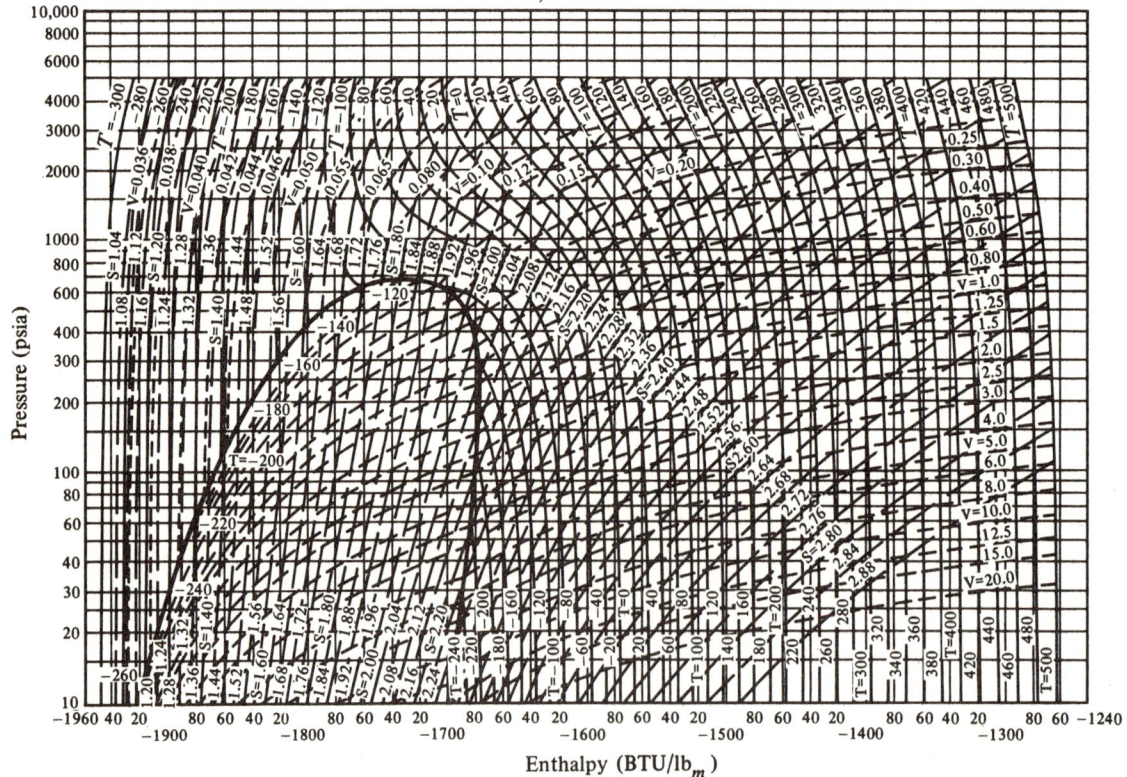

Fig. 2-3. *Methane pressure-enthalpy diagram. Reproduced by permission of* Hydrocarbon Processing, *copyright 1971, Gulf Publishing Co., Houston Texas.*

with mass that is not in the form of kinetic, potential or flow energy, tabulated values of internal energy frequently do not include either the intrinsic energy which binds the protons and neutrons into atoms or the intrinsic energy which binds the atoms into molecules. In such cases, an additional term δE_{gen} may be necessary to represent the net change in intrinsic energy of a system in which nuclear or chemical reactions take place.

Now, if we substitute Eq. (2-24) into Eq. (2-23), we find

$$\left[\left(h + \frac{v^2}{2} + gz\right)\delta m\right]_{in} - \left[\left(h + \frac{v^2}{2} + gz\right)\delta m\right]_{out} + \delta Q - \delta W$$

$$= d\left[\left(u + \frac{v^2}{2} + gz\right)m\right]_{sys} \qquad (2\text{-}25)$$

This is the most generally used form of the generalized energy balance for systems which do not involve nuclear or chemical reactions. As energy is a relative quantity, it is imperative that all similar energy terms in Eq. (2-25), such as U and H, be referred to the same reference state. Note that the integrated form of the right hand side of Eq. (2-25) is

$$\int_i^f d\left[\left(u + \frac{v^2}{2} + gz\right)m\right]_{sys} = (m_f u_f - m_i u_i) + \left(m_f \frac{v_f^2}{2} - m_i \frac{v_i^2}{2}\right)$$

$$+ (m_f gz_f - m_i gz_i) \qquad (2\text{-}26)$$

where subscript f indicates the final state and subscript i indicates the initial state.

Equation (2-25) can also be written in a rate form. The method of application of this energy balance equation is similar to the method of application of the mass balance equation. First, the system must be clearly defined and a time interval or basis selected. Then, the inapplicable terms are eliminated and the resulting equation integrated to provide the desired solution. In a closed system, the terms representing the flow of energy associated with the flow of mass drop out. In an open system in which there is no change of energy with time (steady-state, steady-flow), the term indicating the accumulation of energy within the system drops out. These ideas are considered in more detail in the following example problems.

EXAMPLE PROBLEM 2-3. An uninsulated cylinder having a cross-sectional area of one square foot is fitted with a freely floating piston weighing 763 pounds. The cylinder initially contains one-half pound of steam at 300 °F and as heat is lost slowly through the cylinder wall, the piston drops.

Using data from Table A-2, determine the heat loss corresponding to a one-foot drop of the piston.

The generalized balance equations which apply to this problem are the generalized mass balance equation,

$$\delta m_{in} - \delta m_{out} + \delta m_{gen} = dm_{sys} \qquad (2\text{-}1)$$

and the generalized energy balance equation,

$$\left[\left(h + \frac{v^2}{2} + gz\right)\delta m\right]_{in} - \left[\left(h + \frac{v^2}{2} + gz\right)\delta m\right]_{out} + \delta Q - \delta W$$

$$= d\left[\left(u + \frac{v^2}{2} + gz\right)m\right]_{sys} \qquad (2\text{-}25)$$

In order to solve this problem we select the steam in the cylinder as the system and base our calculations on the period of time required for the piston to drop one foot. As heat is lost slowly, the freely floating piston will drop slowly and reversibly.

Now we can reduce the generalized mass balance to apply to this problem. As the cylinder is a closed system,

$$\delta m_{in} = 0$$

and

$$\delta m_{out} = 0$$

Then, as there is no chemical reaction involved,

$$\delta m_{gen} = 0$$

so that the reduced mass balance equation in differential form is

$$dm_{sys} = 0$$

The generalized energy balance can be reduced in a similar manner. First

$$\left[\left(h + \frac{v^2}{2} + gz\right)\delta m\right]_{in} = 0$$

and

$$\left[\left(h + \frac{v^2}{2} + gz\right)\delta m\right]_{out} = 0$$

since

$$\delta m_{in} = \delta m_{out} = 0$$

and any quantity multiplied by zero is zero. Also, as the system in both its initial and final states is not moving,

$$d\left(m\frac{v^2}{2}\right)_{sys} = 0$$

Finally, although there will be a small change in the center of gravity of the steam in the cylinder, we can neglect this change and set

$$d(mgz)_{sys} = 0$$

Combining these results, we find that the reduced energy balance equation in differential form is

$$\delta Q - \delta W = d(mu)$$

Integrating these reduced mass and energy balance equations over the period of time required for the piston to drop one foot, we find

$$\Delta m_{sys} = 0$$

and

$$Q - W = m\Delta u = \Delta U$$

or

$$Q = \Delta U + W$$

Since the work in this process is reversible, PV work

$$W = \int P\, dV$$

In order to integrate this relationship, we must know $P(V)$. From the problem statement,

$$P_{initial} = P_{final} = \frac{763}{144} + 14.7 = 20 \text{ lb}_f/\text{in}^2$$

Thus, for an isobaric reversible process,

$$W = P\Delta V$$

This equation may be used to calculate the work required to compress the steam in this process. However, as we are not interested in this quantity but rather in the heat transferred, we will find it convenient to combine the internal energy and work terms.

Recall that we defined a property called enthalpy by the relationship

$$H = U + PV \tag{2-24}$$

At constant pressure,

$$\Delta H = \Delta U + P\Delta V$$

Since

$$Q = \Delta U + W = \Delta U + P\Delta V$$

then

$$Q = H_{final} - H_{initial}$$

In order to find the enthalpy of the steam at the initial conditions, we need to determine the values of two independent intensive thermodynamic properties at these conditions. From the problem statement, we find

$$P_{initial} = 20 \text{ lb}_f/\text{in}^2$$

and

$$T_{initial} = 300\,°F$$

At these conditions, we find from Table A-2 that

$$h_{initial} = 1191.4 \text{ BTU/lb}_m$$

To determine the enthalpy of the steam at the final conditions, we need to determine the values of two independent intensive thermodynamic properties at the final conditions. From the statement of the problem, we see that

$$P_{final} = 20 \text{ lb}_f/\text{in}^2$$

Then, to find the second independent intensive thermodynamic property, we note that

$$v_{initial} = 22.36 \text{ ft}^3/\text{lb}_m$$

and since

$$m_{sys} = 0.5 \text{ lb}_m$$

the initial volume of the system is

$$V_{initial} = 11.18 \text{ ft}^3$$

For a cylinder, we note that

$$V = Az$$

where A is the cross-sectional area of the base of the cylinder and z is the height of the piston, so that initially,

$$z = 11.18 \text{ ft}$$

After a one-foot drop in the height of the piston,

$$z = 10.18 \text{ ft}$$

and

$$V_{final} = (10.18)(1) = 10.18 \text{ ft}^3$$

Dividing this volume by the mass gives us the final specific volume of the steam,

$$v_{final} = 20.36 \text{ ft}^3/\text{lb}_m$$

This is the second intensive thermodynamic property we require. Then, at these conditions, from Table A-2,

$$h_{final} = 1160.4 \text{ BTU/lb}_m$$

Substituting these values of enthalpy into our energy balance, we find

$$Q = 0.5 \, (1160.4 - 1191.4)$$

$$= -15.5 \text{ BTU}$$

$$= -16{,}350 \text{ joules}$$

The negative sign indicates that heat flowed from the steam.

48   Chap. 2 $\qquad$ The Mass Balance and the Energy Balance

EXAMPLE PROBLEM 2-4. Reconsider the humidifier discussed in Example Prob. 2-1. If

$$h_{\text{air in}} = 22.2 \text{ BTU/lb}_m$$

$$h_{\text{water in}} = 38.05 \text{ BTU/lb}_m$$

$$h_{\text{air out}} = 27.5 \text{ BTU/lb}_m$$

what is the rate at which heat must be added to the humidifier?

The generalized balance equations which apply to this problem are the generalized mass balance equation

$$\delta m_{\text{in}} - \delta m_{\text{out}} + \delta m_{\text{gen}} = dm_{\text{sys}}$$

and the generalized energy balance equation

$$\left[\left(h + \frac{v^2}{2} + gz\right)\delta m\right]_{\text{in}} - \left[\left(h + \frac{v^2}{2} + gz\right)\delta m\right]_{\text{out}} + \delta Q - \delta W$$

$$= d\left[\left(u + \frac{v^2}{2} + gz\right)m\right]_{\text{sys}} \tag{2-25}$$

Selecting the humidifier as the system and basing our calculations on one hour of operation, we showed in Example Prob. 2-1 that the integrated form of the reduced mass balance equation is

$$m_{\text{in}} = m_{\text{out}}$$

In a similar manner, we can reduce the generalized energy balance equation. As the humidifier operates at steady-state, steady-flow, there is neither an accumulation of mass nor energy within the system over a period of time and

$$d\left[\left(u + \frac{v^2}{2} + gz\right)m\right]_{\text{sys}} = 0$$

Now, as there is no change in the boundaries of the humidifier, we see that there is no expansion type work involved. Further, as there are no mechanical devices extending between the system and its surroundings, there is no shaft work, and

$$\delta W = 0$$

Finally, it is reasonable to assume that the velocities of all streams entering or leaving the humidifier are low and essentially equal and that there is little difference in the elevation of these streams so that

$$\left[\left(\frac{v^2}{2}+gz\right)\delta m\right]_{in} - \left[\left(\frac{v^2}{2}+gz\right)\delta m\right]_{out} = 0$$

Combining these results, we find that for this particular problem the reduced energy balance in differential form is

$$[h\delta m]_{in} - [h\delta m]_{out} + \delta Q = 0$$

Integrating over one hour of operation and noting that

$$m_{in} = m_{out}$$

we find

$$Q = H_{out} - H_{in}$$

Then using the results obtained in Example Prob. 2-1, we find

$$H_{out} = (1005)(27.5) = 27{,}640 \text{ BTU/hr}$$

and

$$H_{in} = (1000)(22.2) + 5(38.05) = 22{,}390 \text{ BTU/hr}$$

Substituting these values into the combined mass and energy balance, we find

$$Q = +5{,}250 \text{ BTU/hr}$$

$$= +5.54 \times 10^6 \text{ joules/hr}$$

The plus sign indicates that heat must be added to the system. This is most conveniently accomplished by supplying either the air or the water or both to the humidifier at a temperature higher than the desired temperature of the leaving humid air.

Ideal Processes

We have previously stated that heat and work are both path functions whose values can be calculated only if the state of the system is known in detail at all points of a process. If we attempt to calculate the work or heat associated with a process as the product of an intensive variable and the change in an extensive variable, we are limited, for all practical purposes, to reversible processes.

At first glance, it may appear to be difficult to conceive of a reversible process for heat transfer and work. For example, in heat transfer, the temperature of the heat source is generally much higher than the temperature of the heat sink. This is necessary in order that the heat be transferred in a reasonable amount of time. Because of this relatively large temperature difference, the system is never infinitesimally close to equilibrium and an infinitesimal increase in the temperature of the heat sink will not reverse the direction of the flow of heat. Similarly, in a compression process, the force driving the piston is frequently much greater than the pressure in the fluid being compressed. This difference in force shows up in overcoming friction and as an acceleration of the piston. In such cases, the system again is not near equilibrium and a small increase in the pressure of the fluid being compressed may not reverse the direction of the piston movement. The same ideas apply to electrical and magnetic systems.

In order to apply the concept of a reversible process to actual or irreversible processes, we must introduce the concept of the ideal or quasi-static process. An ideal process may be thought of as taking place through an infinite number of reversible processes, each having a differential driving force and each contributing an infinitesimal amount to the real process. During an ideal process, the system is at all times infinitesimally near a state of thermodynamic equilibrium.

As an example of the concept of an ideal process, consider a real process, shown in Fig. 2-4(a), involving heat transfer. As shown in Fig. 2-4(b) it is possible to visualize an infinite number of source-sinks between the actual heat source and the actual heat sink, each one differing in temperature from its neighbor by an infinitesimal amount. If the infinite number of source-sinks are ar-

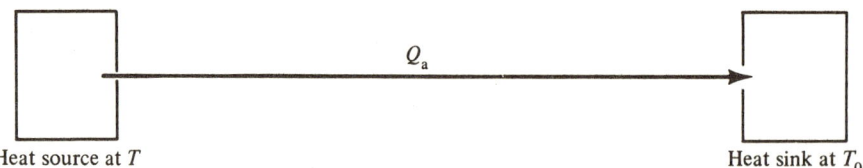

Fig. 2-4(a). *Irreversible heat transfer process.*

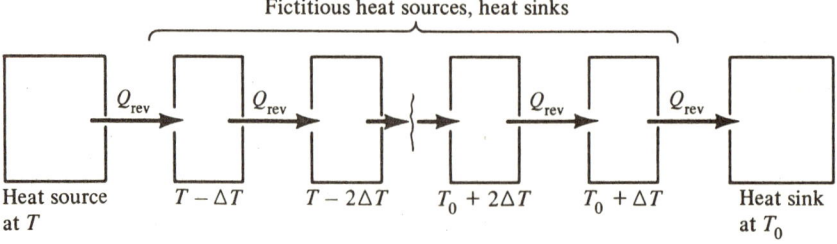

Fig. 2-4(b). *Ideal heat transfer process.*

ranged in order of decreasing temperature, it is possible to transfer heat essentially reversibly between the actual high temperature source and the actual low temperature sink since an incremental increase in the temperature of any intermediate sink will reverse the direction of heat flow at that point.

In a similar manner, we can conceive of reversible compression processes. For example, consider the frictionless compression process shown in Fig. 2-5.

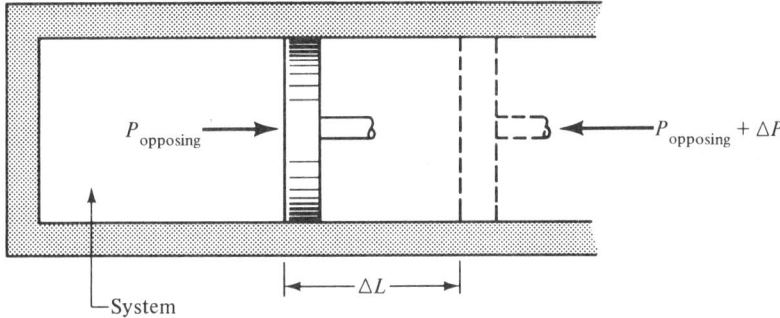

Fig. 2-5. *Ideal compression process.*

Ideal compression process would then consist of compressing the gas within the cylinder in an infinite number of steps. For each step, the pressure producing motion in the piston would be only infinitesimally greater than the pressure opposing the motion for that particular step. Since an incremental increase in the pressure opposing the motion would reverse the motion of the piston, the compression process would be considered to be essentially reversible.

Fortunately, we are not limited to analyzing the flow of heat and work from ideal processes only. Differences in thermodynamic functions, such as internal energy, are independent of the path of the process. One of the most useful results of thermodynamics has been the demonstration that relationships exist between thermodynamic state functions and path functions. Therefore, heat and/or work can be deduced from such relations if the initial and final states of the process are known. For example, in a closed system the energy balance relates Q and W, path functions, to ΔU, a state function which depends only upon the properties of a system in its final and initial states. There are no restrictions as to the path followed in Eq. (2-25). This equation applies equally well to irreversible and reversible processes. It might be said that the primary reason for even introducing thermodynamic functions such as internal energy is that they provide the engineer with a convenient tool to use in calculating the heat or work required to accomplish a desired objective.

PROBLEMS

2-1. A paper mill is fed a slurry containing 2 weight percent pulp, the balance being water. The wet paper produced in the Fourdrinier machine (first stage of plant) contains 50 percent pulp. This wet paper is further dried in the second stage of the plant so that the final product contains 2 weight percent water.

If it is desired to produce 1000 pounds of paper per hour, how much water must be removed in each stage of the process and how much slurry must be fed to the plant?

2-2. Water flowing in a stream contains 1000 ppm salt. A food processing plant is dumping the equivalent of 10 pounds of salt per hour into the stream. If the concentration of salt down stream from this plant is 1100 ppm, what is the flow rate of the stream in GPM?
Hint: ppm = parts salt per million parts of water, by weight.

2-3. A sample of water taken from a reservoir is analyzed and found to contain 250 ppm of salt. One thousand kilograms of salt are then dumped into the reservoir and sufficient time is allowed for the salt to completely mix in the water. A second water analysis shows that the water contains 275 ppm of salt. If the flow rates of the streams flowing into and out of the lake and the rate of evaporation from the lake during this period is negligible, how many liters of water are there in the reservoir?
Hint: ppm = parts of salt per million parts of water, by weight.

2-4. The spent acid from a nitrating process contains 43% H_2SO_4, 36% HNO_3, and 21% H_2O by weight. This acid is to be strengthened by the addition of concentrated sulfuric acid containing 91% H_2SO_4 and concentrated nitric acid containing 88% HNO_3. The strengthened mixed acid is to contain 40% H_2SO_4 and 43% HNO_3. Calculate the quantities of spent and concentrated acids which should be mixed together to yield 1000 lb_m of the desired mixed acid.

2-5. Methane (CH_4) is burned with pure oxygen (O_2) in a calorimeter to form water (H_2O) and carbon dioxide (CO_2). After all the water formed in the reaction is condensed, it is found to weigh 1.3 grams. How many grams of methane were there in the original sample?

2-6. A cylinder fitted with a piston contains 5 lb_m of saturated water vapor at a pressure of 100 lb_f/in^2. The steam is heated until the temperature is 500 °F. During this process the pressure remains constant. Calculate the work done in BTU and in joules by the steam during the process.

2-7. A cylinder in which the piston is restrained by a spring contains 1 ft^3 of superheated steam at a pressure of 15 lb_f/in^2, which just balances the atmospheric pressure of 15 lb_f/in^2. Assume that the weight of the piston is negligible. In this initial state, the spring exerts no force on the piston. The steam is then heated until the volume is doubled. The final pressure of the steam is 50 lb_f/in^2, and during the process the spring exerts a force which is proportional to the displacement of the piston from the initial position. Considering the steam as the system, calculate the total work done by the system in BTU and in joules.

2-8. At 20 °C methanol has a surface tension of 22.6 dynes/cm. Suppose that a film of methanol is maintained on the wire frame as shown below, one side of which can be moved. The original dimensions of the wire frame are as shown. Determine the work done in BTU and in joules (consider the film to be the system), when the wire is moved 1 cm in the direction indicated.

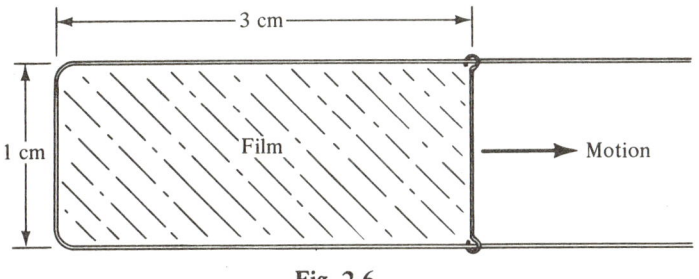

Fig. 2-6.

2-9. Gas in a well insulated cylinder is compressed from a pressure of 1.0 atm and a volume of 10 ft³ to a pressure of 5.0 atm. If the gas follows the relationship

$$PV^{1.3} = \text{constant}$$

what is the work in BTU and joules associated with this process?

2-10. A storage battery is being charged using a charging voltage of 12.3 volts and a current of 24.0 amps. If the storage battery is losing heat at a rate of 20 BTU/hr, at what rate is its internal energy increasing?

2-11. What is the energy, in BTU and in joules, associated with a mass of 10 lb_m

(a) moving horizontally with a velocity of 200 ft/sec?
(b) at an elevation of 3000 meters above a zero datum plane?

2-12. Show that the kinetic energy of a rotating system is given by the relationship

$$E_k = \frac{1}{2} I \omega^2$$

where

I = moment of inertia

$$= \int_0^R r^2 \, dm$$

and

ω = angular velocity (rad/sec)

2-13. What is the internal energy u and the enthalpy h of one lb_m of steam in BTU/lb_m and joules/kg at a pressure of 20 $lb_f/in.^2$ and a specific volume of 15.0 ft^3/lb_m?

2-14. What is the enthalpy of one lb_m of water at 500 °F and 200 $lb_f/in.^2$ referred to

    (a) enthalpy of saturated liquid water at 60 °F is zero
    (b) internal energy of saturated liquid water at 32 °F is zero.

2-15. Using data in Table A-2 of the Appendix, prepare a plot of the vapor-liquid envelope and the 400 $lb_f/in.^2$ isobar from 200 °F to 800 °F for water on a specific enthalpy-temperature diagram. At 400 $lb_f/in.^2$, specific enthalpies for compressed liquids at temperatures below the saturation temperature $T^\square$ are given by the relationship

$$h_f = h_f^\square - 1.04(T^\square - T)$$

2-16. A well insulated cylinder contains 2 lb_m of steam at 35 $lb_f/in.^2$. If the volume of the cylinder is 8.0 ft^3, what is the internal energy U of the steam?

2-17. Steam at 240 psia and having 202.6 degrees of superheat is heated to 800 °F at constant volume. What is its final pressure? Calculate Δh and Δu for the process.

2-18. Steam at 500 °F has an enthalpy of 1180 BTU/lb_m. It is expanded through a well insulated throttling valve to a pressure of 1 atm. What is its final quality or degrees of superheat, whichever is applicable? Calculate the change in internal energy for the process.

2-19. How much heat in joules is required to vaporize 1.0 lb mole of saturated liquid water reversibly at 212 °F?

2-20. A radiator in a steam heating system has a volume of 1.0 ft^3. When the radiator is filled with dry saturated steam at a pressure of 20 $lb_f/in.^2$ all valves to the radiator are closed. How much heat in BTU and in joules will have been transferred to the room when the pressure of the steam is 10 $lb_f/in.^2$?

2-21. A pressure vessel having a volume of 3 ft^3 contains steam at the critical point. Heat is removed until the pressure is 200 $lb_f/in.^2$ Determine the heat transferred from the steam in BTU and joules.

2-22. A vertical cylinder contains 2.0 lb_m of H$_2$O at 100 °F. The initial volume enclosed beneath the piston is 7.0 ft^3. The piston has an area of 60 in.^{2} and a mass of 125 lb_m. Initially the piston rests on the stops. The atmospheric pressure is 14.0 $lb_f/in.^2$ and the gravitational acceleration is 30.9 ft/sec^2. Heat is then transferred to the steam until the cylinder contains saturated vapor.

(a) What is the temperature in °F and K of the H₂O when the piston first rises from the stops?

(b) How much work is done in BTU and in joules by the steam during the entire process?

2-23. A steam boiler has a total volume of 100 ft³. The boiler initially contains 60 ft³ of liquid water and 40 ft³ of vapor in equilibrium at 14.7 lb_f/in^2. The boiler is fired up and heat is transferred to the water and steam in the boiler. Somehow, the valves on the inlet and discharge of the boiler are both left closed. The relief valve lifts when the pressure reaches 600 lb_f/in^2. How much heat in BTU and joules was transferred to the water and steam in the boiler before the relief valve lifted?

2-24. A 100 cubic foot storage tank containing saturated steam at 200 psia is connected to another 100 cubic foot tank containing saturated steam at 50 psia by a short pipeline with a valve. The tanks, valve and line are all well insulated. The volume of the valve and line is negligible.

If the valve is opened, what will the pressure be when complete equilibrium between the two tanks has been reached?

2-25. A closed vessel of 10 ft³ capacity is filled with saturated steam at 250 lb_f/in^2. Heat is lost through the walls of the vessel at a rate of 50 BTU/hr. How long will it take for 25 percent of the steam to condense?

2-26. Methane in a cylinder expands against a piston. Following are the conditions before and after expansion:

	Before Expansion	After Expansion
Pressure	200 lb_f/in^2	100 lb_f/in^2
Temperature	150 °F	
Volume	2.0 ft³	4.0 ft³

The heat transfer during expansion = +60 BTU. Calculate the work done in BTU and joules during this process.
Hint: Use Ref. 3 or 4 or Fig. 2-3.

2-27. Steam is to flow at constant velocity through a vertical line to the bottom of a mine shaft 2334 feet deep. The steam pressure is to be 105 psia at both the surface (mine entrance) and at the bottom of the shaft. Heat losses of 18 BTU per pound of steam flowing are predicted.

If the steam supplied at the mine entrance is saturated vapor, what will be the quality of the steam at the bottom of the mine shaft?

2-28. A stream of saturated liquid water at 150 °F flowing at a rate of 50 GPM is required for a certain process. Saturated liquid water at 60 °F and steam at 30 lb_f/in^2 and 280 °F are available. It is proposed to mix the cold water and the steam to form the hot water. At what rates (lb_m/min) should the cold water and steam be fed to an insulated mixer?

2-29. Dry saturated vapor at 100 °F is to be compressed to 5 lb_f/in^2, 400 °F in a centrifugal compressor. Heat is transferred from the vapor during the compression process at the rate of 1000 BTU/hr. The rate of flow of vapor is 100 lb_m/hr. Calculate the horsepower and the kilowatts required to drive the compressor.

2-30. Steam flowing through an insulated one-inch diameter line has a pressure of 800 psia and a quality of 95 percent. If this steam is throttled adiabatically to 14.7 psia through a valve in a horizontal section of the line, what should be the diameter of the downstream line if there is to be no change in the velocity of the steam?

2-31. A turbine is driven by 10,000 pounds per hour of steam at 650 psia and 850 °F, entering at a velocity of 200 feet per second. The steam leaves the turbine exhaust at a point 10 feet below the inlet with a velocity of 1200 feet per second. The measured shaft work of the turbine is 943 horsepower, and the heat loss has been determined to be 100,000 BTU per hour.

A small portion of the exhaust steam from the turbine is passed through a throttling valve and discharges at atmospheric pressure. Velocity change in passing through the valve may be neglected.

(a) What is the temperature of the steam leaving the valve?
(b) What is the quality or degrees of superheat (whichever is applicable) of the steam leaving the valve?

2-32. In this problem we compare two different systems that involve flow into a vessel from a line in which steam at 100 lb_f/in^2, 600 °F is flowing. In each case the process is adiabatic. Determine the final temperature of the steam in each of the following:

(a) The vessel is initially evacuated, and steam flows in until the pressure is 100 lb_f/in^2.
(b) Steam flows into a cylinder with a piston which is so weighted that a constant pressure of 100 lb_f/in^2 is maintained in the cylinder. Steam flows into the cylinder causing the piston to rise.

2-33. A pressure vessel having a volume of 30 ft^3 contains saturated steam at 500 °F. The vessel initially contains 50% vapor and 50% liquid by volume. Liquid is withdrawn slowly from the bottom of the tank, and heat is transferred to the tank in order to maintain constant temperature. Determine the heat transfer to the tank when half of the contents of the tank has been removed.

2-34. An insulated tank initially contains 1000 lb_m of steam and water at 500 psia. Fifty percent of the tank volume is occupied by liquid and fifty percent by vapor. Fifty pounds of moisture-free vapor are slowly withdrawn from the tank so that the pressure and temperature are always uniform throughout the tank.

Calculate the pressure in the tank after the fifty pounds of steam are withdrawn.

3

The Entropy Balance

OBJECTIVES

After studying this chapter, the student should be able to

(1) define entropy;
(2) state the second law of thermodynamics;
(3) apply the second law of thermodynamics in the form of Eq. (3-16) to new situations involving the concepts of lost work or efficiency, using tabulated thermodynamic properties;
(4) apply the second law of thermodynamics in the form of Eq. (3-23) to new situations involving the concepts of unavailable energy or effectiveness, using tabulated thermodynamic properties; and
(5) define the concept of availability.

THE SECOND LAW OF THERMODYNAMICS

The first law of thermodynamics states that energy is conserved. However, this law places no restrictions whatever on the direction of the flow of energy. The fall or the rise of a mass initially at rest and acted upon only by gravity are both compatible with the first law. Similarly the first law can be satisfied whether heat flows from a hot cup of coffee to the room or from the room to the hot cup of coffee.

However, experience indicates that processes will proceed spontaneously in one direction but not in the opposite direction. For example, a mass acted upon only by gravity will fall, not rise, and the hot cup of coffee will in time cool to room temperature rather than get hotter. Obviously, something more than the first law is needed to indicate the direction a process will take. This need led scientists to the second law.

The second law of thermodynamics may be summarized in the following statement:

> *Every system which is left to itself will, on the average, change toward a condition in which its ability to do work will have decreased.*

Although the beginnings of the first law of thermodynamics were based primarily upon experimental evidence, the beginnings of the second law of thermodynamics were based almost entirely on inductive reasoning.

The first step in the formulation of the second law occurred when Nicolas Leonard Sadi Carnot attempted to determine the maximum amount of work that could be obtained from a given quantity of heat in a heat engine. Carnot reasoned that a heat engine must not only have a source of heat, such as a boiler, but also a receiver of heat, such as a condenser. He felt that work would be produced by the transfer of heat from a hot body through a heat engine to a cold body in the same manner as work would be produced by the fall of water through a turbine to a point of lower elevation. In making this analysis, Carnot used the caloric theory of heat which assumed that heat was a substance or material which was conserved just as mass is conserved. This requires that as much heat be rejected from the heat engine to the low temperature sink as had been supplied to the heat engine from the high temperature heat source.

Now just as work would be produced from a heat engine by allowing heat to flow from a high temperature source to a low temperature sink, work would be required by a refrigerator to make the same quantity of heat flow back from the low temperature sink to the same high temperature source. Carnot reasoned that if differences in the mechanism or working fluid in the heat engine and the refrigerator would permit the work produced from the heat engine as the heat flowed from the high temperature to the low temperature to be greater than the work required by the refrigerator to make the same quantity of heat flow from the same low temperature back to the original high temperature, he would have produced a perpetual motion machine. But Carnot recognized that it is not possible to create work in this manner; that is, he recognized the impossibility of "the unlimited creation of motive power without consumption of either caloric or of any other agent whatsoever." A machine which creates work from nothing is called a perpetual motion machine of the *first kind*. Thus, in 1824 Carnot introduced the second law of thermodynamics to the scientists of his day with the statement that

> *the work which can be obtained from a given quantity of heat in an ideal heat engine depends only upon the temperature of the heat source and heat sink and is entirely independent of the mechanism or working substance employed.*

When Joule demonstrated conclusively that heat was a form of energy, the basis for Carnot's analysis of heat engines seemed to be lost. Caloric was not conserved and Carnot's proof no longer held. However, in 1850, Rudolph Clausius was able to demonstrate that the second law of thermodynamics as stated by Carnot was correct, even though work and heat are merely different forms of energy. Clausius repeated Carnot's indirect reasoning. He showed that if it were possible to extract unlimited work from a heat engine, without the addition of any energy from the surroundings, then heat could flow spontaneously from a colder body to a hotter body. As a result of this work, Clausius presented the first precise statement of the second law.

> *It is impossible to construct a device that operates in a cycle and produces no effect other than the transfer of heat from a cooler body to a hotter body.*

In 1851, Lord Kelvin, the former William Thomson, working independently from Clausius, also recognized that Carnot's conclusions regarding the maximum work that could be obtained from a heat engine were valid. Kelvin showed that if Carnot's premise of the validity of the second law of thermodynamics were incorrect and that it were possible to extract unlimited work from a heat engine, without the addition of any work from the surroundings, then it would be possible to perform work solely by cooling a body. An engine which can perform work solely by cooling a body is known as a perpetual motion machine of the *second kind*. A perpetual motion machine of the second kind would not contradict the first law of thermodynamics which says energy is conserved but says nothing about the direction energy must flow. Further, a perpetual motion machine of the second kind would afford the same advantages as a perpetual motion machine of the first kind in that the surroundings here on earth would be an unlimited source of energy for its operation.

The second law of thermodynamics, as formulated by Kelvin and later amplified by Max Planck is:

> *It is impossible to construct a device that will operate in a cycle and produces no effect other than the raising of a weight and exchange of heat with a single reservoir.*

Both the Clausius statement and the Kelvin-Planck statement of the second law are logically equivalent. In order to demonstrate this conclusion, we must first examine a property called entropy.

ENTROPY AND THE SECOND LAW

As we have indicated, the second law of thermodynamics arose from a study of the conversion of heat to work. Because of the many similarities between heat and work, it would seem logical that if

$$\delta W_{rev} = P\,dV - F\,dL - \sigma\,dA - \epsilon\,dq - \Sigma \mu_i dn_i - \ldots \qquad (2\text{-}13)$$

$$= \Sigma \text{ (intensive property) } d\text{(extensive property)} \qquad (2\text{-}12)$$

then we should be able to write

$$\delta Q_{\text{rev}} = \text{(intensive property) } d\text{(extensive property)}$$

In 1862, Clausius identified the intensive property associated with heat as the temperature of the system and called the extensive property the entropy S of the system. Thus he wrote

$$\delta Q_{\text{rev}} = T \, dS \qquad (3\text{-}1)$$

In Clausius' work, he showed that the utility of the concept of entropy as a thermodynamic property lay in the fact that entropy is a quantitative measure of degraded energy within a system. In other words an increase in entropy is a measure of the decrease in the ability of a system to do work. Then, as the energy which is available to do work decreases in an isolated system, according to the second law the entropy of that isolated system must increase. Unfortunately, most systems are not isolated so we must consider the change in entropy in both the system and its surroundings which taken together form an isolated system. Then the entropy of any system plus its surroundings must always increase or, in the case of a reversible process in which energy is not degraded, remain constant. Thus, we can write the second law of thermodynamics in the following form:

$$\Delta S \text{ (system and surroundings)} \geqslant 0 \qquad (3\text{-}2)$$

That is to say, the change in entropy of a system plus its surroundings cannot decrease.

Now just as the energy of a system is a thermodynamic property whose value is relative to the energy of that system in some arbitrary reference state selected for convenience, the entropy of a system is also a thermodynamic property whose value is relative to the entropy of that system in some reference state. Values of entropy tabulated in Table A-2 in the Appendix are referred to the entropy of saturated liquid water at the triple point[1]. These conditions are called a standard or reference state and the assigned value of entropy of water in this state is zero. Tabulated values for other components may be found in the literature[3,4,5]

Values of entropy, relative to a reference state, cannot be measured directly but must be calculated using equations such as

$$dS = \frac{\delta Q_{\text{rev}}}{T} \qquad (3\text{-}1)$$

and

$$c = \frac{\delta q_{\text{rev}}}{dT} \qquad (2\text{-}14)$$

If values of specific heat capacity are available, we can calculate the variation of entropy of a single phase system with temperature. For example, if heat is added reversibly and isobarically to a system which does not change phase, the resulting change in entropy can be found by combining Eqs. (2-15) and (3-1) to give

$$\Delta s_P = \int_{T_1}^{T_2} \frac{c_P}{T} dT$$

Since changes of phase can take place reversibly, entropy changes associated with phase changes can be calculated directly from Eq. (3-1).

THE ENTROPY BALANCE

Accountability of Entropy

In analyzing thermodynamic processes, it would be convenient if we could write the second law of thermodynamics in the form of an equation for the conservation of entropy. This in fact, is possible. Entropy is associated with mass entering or leaving the system, it can cross system boundaries with heat and it can be generated within the system. For example, the change of entropy in a system due to a differential mass flow is given by the relationship

$$d(sm)_{sys} = (s\delta m)_{in} - (s\delta m)_{out}$$

Entropy can also be exchanged across the system boundaries as heat is transferred. The change of entropy in a system as a result of heat transfer is given by the relationship

$$d(sm)_{sys} = \frac{\delta Q}{T}$$

where T is the temperature at which the heat transfer takes place. Finally, as we will discuss in more detail later, energy is degraded and entropy is generated in all processes which do not occur reversibly.

Then, the difference between the entropy which enters the system and the entropy which leaves the system plus the entropy generated within a system must equal the change in entropy of that system. That is,

$$(s\delta m)_{in} - (s\delta m)_{out} + \frac{\delta Q}{T} + \delta S_{gen} = d(sm)_{sys} \tag{3-3}$$

Equation (3-3) is the generalized entropy balance equation written in differential form. In order to examine its use more closely, let us first consider its application to systems undergoing processes in which there is no entropy generated.

Reversible Processes

Since there is no degradation of energy during a reversible process which could result in the creation of entropy, the generalized entropy balance equation for a reversible process taking place in unit time has the form

$$(s\delta m)_{in} - (s\delta m)_{out} + \frac{\delta Q_{rev}}{T} = d(sm)_{sys} \tag{3-4}$$

Equation (3-4) can be used to demonstrate the equivalence between the Clausius statement of the second law and Eq. (3-2). Consider a device shown in Fig. 3-1. The device is similar to a steam engine in that it operates in a cycle which consists of a series of processes through which a fluid passes, such as is shown in Fig. 3-2, a schematic diagram of an idealized cyclic device, the fluid ultimately returning to its initial state. However the device shown in Fig. 3-1 produces no effect other than the transfer of heat from a cooler body at T_0 to a hotter body at T. That is, no work is added to nor extracted from this device. Now we may ask, what is the total change in entropy of a universe which includes the device as a system and the cooler and hotter bodies as the surroundings? Since the device

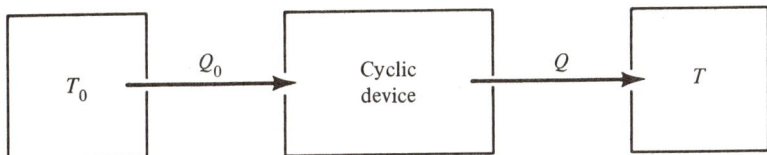

Fig. 3-1. *A Clausius device.*

operates in a cycle, there are no net changes of thermodynamic properties in the device as a system and after integration over a unit period of time, the right hand side of Eq. (3-4) reduces to

$$\Delta S_{sys} = 0$$

Now, if we consider the surroundings which consist only of the hotter and cooler bodies to be a system, an energy balance made around the surroundings reduces to

$$Q + Q_0 = 0$$

since there is no flow of mass from the system into or out of the surroundings, there is no work involved, and there is no accumulation of energy in the surroundings. Further, Eq. (3-4) applied to the surroundings reduces to

$$\frac{Q}{T} + \frac{Q_0}{T_0} = \Delta S_{surr}$$

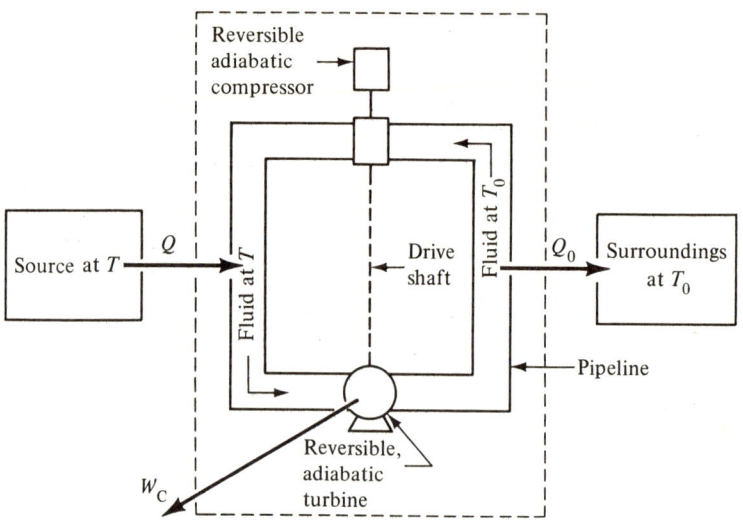

Fig. 3-2. *Ideal thermodynamic machine.*

If we use the reduced energy balance to eliminate Q_0 from the reduced entropy balance, we find

$$\Delta S_{surr} = \frac{Q}{T} - \frac{Q}{T_0}$$

Then, since

$$T > T_0$$

and since Q is a positive quantity as it enters the surroundings,

$$\Delta S_{surr} < 0$$

Thus, if in contradiction to the Clausius statement of the second law, it were possible to construct a device that operates in a cycle and produces no effect other than the transfer of heat from a cooler body to a hotter body, we would have a situation in which

$$\Delta S_{sys} + \Delta S_{surr} = \Delta S_{univ} < 0$$

clearly in disagreement with Eq. (3-2).

Equations (3-2) and (3-4) may also be used to demonstrate the validity of the Kelvin-Planck statement of the second law of thermodynamics. In this case, consider the device shown in Fig. 3-3. This device operates in a cycle and produces no effect other than the raising of a weight and the exchange of heat with a

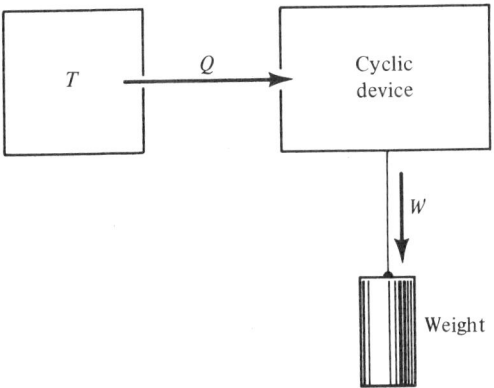

Fig. 3-3. *A Kelvin-Planck device.*

single reservoir at T. Again we ask, what is the total change in entropy of a universe which includes this device as a system and a heat source and weight as the surroundings? Since the device operates in a cycle, there are no net changes of thermodynamic properties in the device and after integration over a unit period of time, the right-hand side of Eq. (3-4) reduces to

$$\Delta S_{sys} = 0$$

Now, if we consider the surroundings to be a system, an energy balance made around the surroundings reduces to

$$Q - W = 0$$

since there is no flow of mass from the system into or out of the surroundings and since there is no accumulation of energy in the surroundings. Further, Eq. (3-4) applied to the surroundings reduces to

$$\frac{Q}{T} = \Delta S_{surr}$$

Now Q is a negative quantity since it is leaving the surroundings and

$$\Delta S_{surr} < 0$$

Thus, if in contradiction to the Kelvin-Planck statement of the second law, it were possible to construct a device that operates in a cycle and produces no effect other than the raising of a weight and the exchange of heat with a single reservoir, we would have a situation in which

$$\Delta S_{sys} + \Delta S_{surr} = \Delta S_{univ} < 0$$

clearly in disagreement with Eq. (3-2).

The Carnot Cycle

The entropy balance for a reversible process, Eq. (3-4) can be used as a basis to determine the maximum amount of work that can be obtained from a given quantity of heat in an ideal thermodynamic machine. This machine, shown schematically in Fig. 3-2 and originally conceived by Carnot, converts a portion of the heat supplied to it reversibly at temperature T to work and reversibly rejects the unused portion of this heat to the surroundings at a low temperature T_0. The Carnot engine operates in a cycle along the paths $ABCD$ shown in the PV diagram, Fig. 3-4(a) or alternatively shown in the TS diagram, Fig. 3-4(b). The cycle consists of the following four steps:

A–B Isothermal reversible expansion carried out at the temperature of the hot source T

B–C Adiabatic reversible expansion bringing the fluid to the temperature of the cold source T_0

C–D Isothermal reversible compression at the temperature of the cold source T_0

D–A Adiabatic reversible compression restoring the fluid to its initial state.

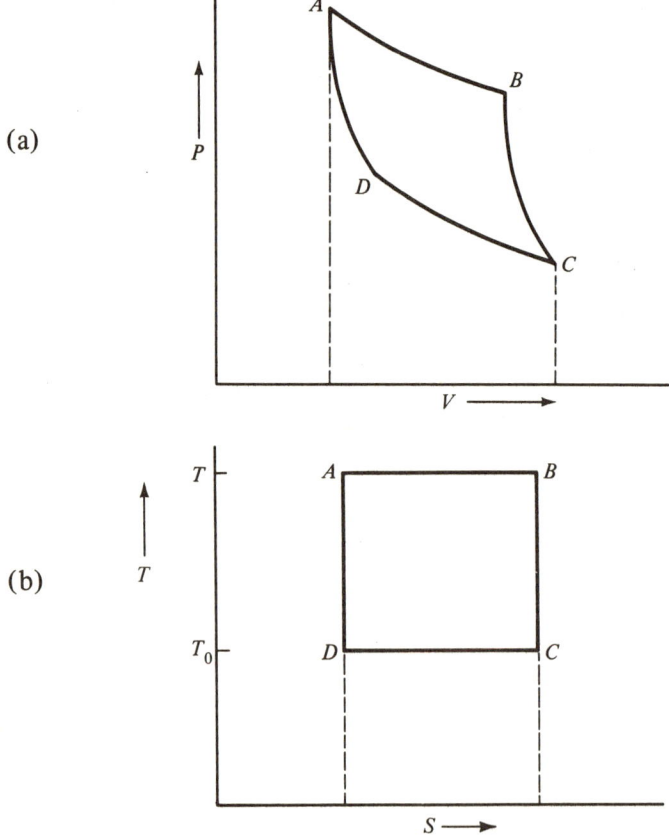

Fig. 3-4. *The Carnot cycle.*

Now, we can use Fig. 3-4(a) to determine the net work that will be produced in the Carnot engine. First, the total amount of work produced during the expansion of the working fluid is

$$W = \int_A^C P\,dV$$

$$= \text{area under curve } ABC$$

Further, the total amount of work requires to compress the working fluid is

$$W = \int_C^A P\,dV$$

$$= \text{area under curve } CDA$$

Then, the net work produced by a Carnot engine is

$$W = \text{area under curve } ABC \text{ less area under curve } CDA$$

$$= \text{area enclosed by curve } ABCD$$

We can also estimate the net work that will be produced in the Carnot engine using the reduced balance equations.

Since there is no flow of mass into or out of the thermodynamic machine and since there is no accumulation of energy within the machine, the combined mass and energy balance equations, Eqs. (2-1) and (2-25), around the machine as a system reduce to

$$\delta Q + \delta Q_0 - \delta W_c = 0$$

Further, since the machine is operating reversibly, the entropy balance equation (3-4) reduces to

$$\frac{\delta Q}{T} + \frac{\delta Q_0}{T_0} = 0$$

Then, we can eliminate δQ_0 to obtain the result

$$\frac{\delta Q}{T} + \frac{\delta W_c - \delta Q}{T_0} = 0$$

Rearranging, we find that

$$\delta W_c = \delta Q \left(1 - \frac{T_0}{T}\right) \tag{3-5}$$

Or, since

$$\delta Q = T\, dS$$

$$\delta W_c = T\, dS - T_0\, dS \tag{3-6}$$

Equations (3-5) and (3-6) represent the maximum work obtainable from a heat engine supplied with δQ and operating between T and T_0. Since T is greater than T_0, we find that when heat is supplied from a high temperature and rejected at a lower temperature, we can obtain work from the engine. If we integrate Eq. (3-6) over the entire Carnot cycle, we find that the net work produced by a Carnot engine,

$$W_c = T\Delta S - T_0 \Delta S$$

is merely the area enclosed by the rectangle $ABCD$ on the TS diagram shown in Fig. 3-4(b).

This analysis demonstrates that only a portion of the heat supplied to a heat engine can be converted to work. That fraction of the heat supplied to a heat engine which is converted to work is called the thermal efficiency η_{therm} where

$$\eta_{\text{therm}} = \frac{W}{Q} \tag{3-7}$$

$$= \frac{\text{work performed by engine}}{\text{heat supplied to engine}}$$

From Eq. (3-5) we see that the thermal efficiency of a Carnot engine is

$$\eta_{\text{therm},c} = 1 - \frac{T_0}{T} \tag{3-8}$$

Note that it is possible to increase the thermal efficiency of the Carnot engine either by decreasing the temperature at which heat is rejected T_0 or increasing the temperature at which heat is supplied T.

Carnot engines can also be operated as refrigerators. In this case, heat is supplied reversibly to a Carnot engine from a low temperature T_0 and is rejected reversibly at a high temperature T. The minimum amount of work that would have to be supplied to this thermodynamic refrigerator may be found by reduction and rearrangement of the balance equations to be

$$\delta W_c = \delta Q_0 \left(1 - \frac{T}{T_0}\right) \tag{3-9}$$

$$= T_0 \, dS - T \, dS \tag{3-10}$$

The thermal efficiency of a refrigerator is expressed in terms of a coefficient of performance β where

$$\beta = \frac{-Q_0}{W} \tag{3-11}$$

$$= \frac{\text{heat removed from low temperature source}}{\text{work required by refrigerator}}$$

From Eq. (3-9), we see that the coefficient of performance of a Carnot refrigerator is

$$\beta_c = -\frac{1}{\left(1 - \frac{T}{T_0}\right)}$$

$$= \frac{T_0}{(T - T_0)} \tag{3-12}$$

EXAMPLE PROBLEM 3-1. What is the maximum amount of work that can be produced from an engine which is supplied 1000 BTU/hour at a temperature of 1000 °F and which rejects the unused portion of heat to the surroundings at 70 °F?

The maximum amount of work can be obtained from a heat engine if the heat engine operates as a Carnot engine. Taking a Carnot engine as our system and basing our calculations on one hour of operation, the combined mass, energy and entropy balances, Eq. (3-5), reduces to

$$W = Q\left(1 - \frac{T_0}{T}\right)$$

$$= 1000\left(1 - \frac{530}{1460}\right)$$

$$= 637 \text{ BTU/hr}$$

The thermal efficiency of this engine is

$$\eta_{\text{therm}} = \frac{W}{Q} \tag{3-7}$$

$$= \frac{637}{1000}$$

$$= 0.637$$

The Entropy Balance Equation

Irreversible Processes. Although entropy is not generated in a reversible process, actual processes are generally not reversible. In order to examine entropy generation, we must recall our definition of entropy as a quantitative measure of the degradation of energy. By degradation of energy, we mean that the energy has been changed from a form from which work can be obtained to a form which cannot be used to produce work.

As an example of energy degradation, consider a case in which a high pressure gas is allowed to expand to atmospheric pressure through a turbine which must operate against friction in its bearings. We won't get as much work from this turbine as we would from another turbine that operated without friction — that is, reversibly. The difference between the work we could have obtained from the hypothetical frictionless turbine operating reversibly and the actual turbine is the work expended against the friction in the actual turbine. As this difference may be considered to be useful work which could have been done, but wasn't, it is called degraded energy or lost work LW and

$$\delta LW = \delta W_{\text{rev}} - \delta W_{\text{act}} \tag{3-13}$$

where the subscript "act" indicates an actual or real process and not a reversible process. Lost work or energy degradation due to friction or any other cause can never be negative, as the second law of thermodynamics tells us that the ability of a system to do work must decrease. Thus there is no such thing as negative lost work or gained work.

As another example of energy degradation, consider a second case in which the high pressure gas is allowed to expand to atmospheric pressure through a throttling valve. In this second case, we will obtain no work from the valve as there are no mechanical devices extending between the valve and its surroundings through which work could be transmitted and the boundaries of the valve do not expand during the process. Once the pressure of the gas has dropped to that of the surrounding atmosphere, it is impossible to obtain further work from it. In this case, we see from Eq. (3-13) that the lost work associated with expanding the gas through a throttling valve is the total work we could have obtained had we passed this same gas through a reversible turbine. That is,

$$\delta LW = \delta W_{\text{rev}}$$

It should be understood that this energy which became unavailable for work is not energy which no longer exists, for the first law statement that energy is conserved is always valid. We might also observe that the lost work in this second case is equal in magnitude but of opposite sign to the work required to reversibly recompress the expanded gas from atmospheric pressure back to its original pressure.

Now we might examine the relationship between lost work and entropy by considering the compression process shown in Fig. 3-5. First, let us compress the

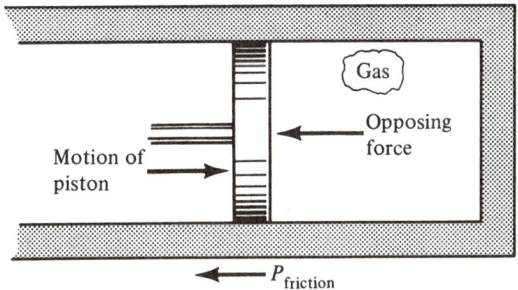

Fig. 3-5. *A compression process.*

gas reversibly from some initial state to some final state. The mass balance tells us there will be no change in the mass of the gas contained within the cylinder and the energy balance for reversible compression reduces to

$$dU_{rev} = \delta Q_{rev} - \delta W_{rev}$$

where we have assumed that heat can be transferred through the walls of the cylinder.

Now let us add friction to the compression process. Friction will appear as a force acting along the edge of the piston in a direction that opposes the motion of the piston. For the compression process involving friction, the mass and energy balance equations reduce to

$$dU_{act} = \delta Q_{act} - \delta W_{act}$$

Using Eq. (3-13), the reduced energy balance equation for the compression process involving friction may be written

$$dU_{act} = \delta Q_{act} - (\delta W_{rev} - \delta LW)$$

Since internal energy is a point function, if the reversible and the actual process take place between the same initial and final states,

$$dU_{act} = dU_{rev}$$

and

$$dQ_{act} - \delta W_{rev} + \delta LW = \delta Q_{rev} - \delta W_{rev}$$

If we cancel the reversible work terms, we find

$$\delta Q_{act} + \delta LW = \delta Q_{rev} \tag{3-14}$$

Dividing by the temperature of the system T we see that

$$\frac{\delta Q_{act}}{T} + \frac{\delta LW}{T} = \frac{\delta Q_{rev}}{T}$$

But

$$\frac{\delta Q_{rev}}{T} = dS \tag{3-1}$$

so

$$\frac{\delta Q_{act}}{T} + \frac{\delta LW}{T} = dS \tag{3-15}$$

This last equation is an entropy balance equation for a closed system. The fact that dS has a unique value regardless of path is not unexpected since entropy is a state function. The heat transferred between the system and its surroundings divided by the absolute temperature of the system,

$$\frac{\delta Q_{act}}{T}$$

represents the contribution of heat transfer to the entropy of the closed system. The lost work resulting from all irreversibilities within the system divided by the absolute temperature,

$$\frac{\delta LW}{T}$$

represents entropy generated within the system δS_{gen}.

In the case of an open system, the entropy associated with mass is also exchanged between the system and its surroundings so that the total entropy exchanged is

$$(s\delta m)_{in} - (s\delta m)_{out} + \frac{\delta Q_{act}}{T}$$

If these terms are included, the generalized entropy balance equation over a unit period of time takes the form

$$(s\delta m)_{in} - (s\delta m)_{out} + \frac{\delta Q_{act}}{T} + \frac{\delta LW}{T} = d(sm)_{sys} \tag{3-16}$$

This equation reduces to Eq. (3-4) for reversible processes. We might note that in integrated form, the right hand side of Eq. (3-16) is

$$\int d(sm)_{sys} = s_f m_f - s_i m_i$$

Equation (3-16) can also be written in a rate form.

The method of application of Eq. (3-16) is similar to the method of application of the generalized mass and energy balance equations. First, the system must be clearly identified and the time period or basis for the analysis established. Then, the terms which do not apply are eliminated. The solution of most thermodynamic problems will require the simultaneous application of the mass, energy and entropy balance equations to the system of interest.

Lost Work. If we wish to calculate the amount of entropy generated *within* a system, we must calculate the ratio of lost work to the temperature at which the lost work is produced. However, if the temperature of the system varies during the process, it is inconvenient to use Eq. (3-16) for this purpose. Even so, we may be able to calculate the lost work associated with any process if we observe from Eq. (3-13) that lost work is the difference between the reversible and actual work associated with a process; that is

$$\delta LW = \delta W_{rev} - \delta W_{act} \tag{3-13}$$

From this relationship, we can see that if the actual process is reversible, lost work and thus entropy generation is zero. In order to calculate lost work, we must compare the work we could have obtained from a reversible process with the work we did obtain from the actual process; both processes taking place between the same initial and final states.

EXAMPLE PROBLEM 3-2. Steam at 300 lb_f/in^2 and 700 °F is flowing through a valve at steady state in an insulated line. If the pressure on the downstream side of the valve is 250 lb_f/in^2, what is the lost work LW?

The generalized balance equations which apply to this problem are the generalized mass balance equation,

$$\delta m_{in} - \delta m_{out} + \delta m_{gen} = dm_{sys} \tag{2-1}$$

the generalized energy balance equation,

$$\left[\left(h + \frac{v^2}{2} + gz\right)\delta m\right]_{in} - \left[\left(h + \frac{v^2}{2} + gz\right)\delta m\right]_{out} + \delta Q - \delta W$$

$$= d\left[\left(u + \frac{v^2}{2} + gz\right)m\right]_{sys} \qquad (2\text{-}25)$$

and the generalized entropy balance equation

$$(s\delta m)_{in} - (s\delta m)_{out} + \frac{\delta Q}{T} + \frac{\delta LW}{T} = d(sm)_{sys} \qquad (3\text{-}16)$$

We can take the valve as the boundary of our system and base our calculations on the time required for one pound mass of fluid to flow through the valve. Then, as the valve operates under steady flow, steady state conditions, there is no accumulation of mass, energy or entropy within the valve and

$$dm_{sys} = 0$$

$$d\left[\left(u + \frac{v^2}{2} + gz\right)m\right]_{sys} = 0$$

and

$$d(sm)_{sys} = 0$$

Now, as there is no chemical reaction involved,

$$\delta m_{gen} = 0$$

and the reduced mass balance equation in integrated form becomes

$$m_{in} = m_{out} = 1.0 \text{ lb}_m$$

Next, turning our attention to the energy balance equation we see that as the boundaries of the valve do not change during the process and as there are no mechanical devices such as shafts or wires extending between the valve and its surroundings,

$$\delta W = 0$$

Further, as the valve and pipeline are insulated,

$$\delta Q = 0$$

Neglecting the effect of changes in kinetic energy and potential energy, which would be small in this problem, we find that the reduced energy balance becomes

$$(h\delta m)_{in} - (h\delta m)_{out} = 0$$

or, for one pound mass flowing,

$$h_{in} - h_{out} = 0$$

Finally, since $\delta Q = 0$, the reduced entropy balance becomes

$$(s\delta m)_{in} - (s\delta m)_{out} + \frac{\delta LW}{T} = 0$$

or, for one pound mass flowing,

$$lw = T(s_{out} - s_{in})$$

Now we need values of thermodynamic properties to substitute into the reduced balance equation. From Table A-2, we find at

$$P_{in} = 300 \text{ lb}_f/\text{in}^2$$

and

$$T_{in} = 700 \,°\text{F}$$

that

$$h_{in} = 1368.9 \text{ BTU/lb}_m$$

and

$$s_{in} = 1.6758 \text{ BTU/lb}_m \,°\text{R}$$

At the outlet of the valve,

$$P_{out} = 250 \text{ lb}_f/\text{in}^2$$

and from the reduced energy balance, we find

$$h_{out} = h_{in} = 1367.4 \text{ BTU/lb}_m$$

At these conditions

$$T_{out} = 695\,°F$$

and

$$s_{out} = 1.6952 \text{ BTU/lb}_m\,°R$$

Now, if the temperature in the system had not varied and T_{out} had been the same as T_{in}, we could have substituted these values into the reduced entropy balance to find lost work. However, since the temperature across the valve varies, we will have to estimate lost work from the relationship

$$lw = w_{rev} - w_{act}$$

In order to calculate w_{rev}, let us first replace the valve in our system with an imaginary reversible turbine followed by an isobaric reversible heat exchanger. Since the valve is well insulated, we will also insulate this turbine. Now, we ask, what is the work we could obtain from this adiabatic, reversible turbine? Taking the turbine as our system and basing our calculations on one pound mass flowing and again neglecting changes in kinetic and potential energy, the generalized mass and energy balance reduce to

$$h_{in} - h_{out} - w_{rev} = 0$$

Then, the entropy balance reduces to

$$s_{in} - s_{out} = 0$$

At the turbine outlet,

$$s_{out} = s_{in} = 1.6758 \text{ BTU/lb}_m\,°R$$

and

$$P_{out} = 250 \text{ lb}_f/\text{in}^2$$

We find from the steam tables that

$$T_{out} = 654.0\,°F$$

and

$$h_{out} = 1347.4 \text{ BTU/lb}_m$$

Then, for one pound mass flowing,

$$-w_{rev} = h_{out} - h_{in}$$

$$= 1347.4 - 1368.9$$

$$w_{rev} = 21.5 \text{ BTU}$$

This is the work we could have obtained from a turbine operating reversibly and adiabatically. Now, in order to raise the adiabatic, reversible turbine effluent temperature of 654.0 °F back up to the valve effluent temperature of 695 °F, we must add heat to the steam in an isobaric reversible heat exchanger. Neglecting kinetic and potential energy changes, the generalized energy balance around this heat exchanger reduces to

$$h_{in} - h_{out} + q_{rev} = 0$$

or

$$q_{rev} = 1368.9 - 1347.4$$

$$= 21.5 \text{ BTU/lb}_m$$

As there is no work associated with the reversible, isobaric heat exchanger, the total work associated with the overall process then is

$$w_{rev} = 21.5 \text{ BTU/lb}_m$$

Then, since

$$w_{act} = 0$$

we find that

$$lw = w_{rev}$$

$$= 21.5 \text{ BTU/lb}_m$$

$$= 22.7 \times 10^3 \text{ joules/lb}_m$$

It is also possible to conceive of another reversible process to accomplish the same objective. Let us replace the valve with an isothermal reversible turbine. Basing our calculations on one pound mass flowing and neglecting kinetic and potential energy changes, the generalized energy balance reduces to

$$h_{in} - h_{out} + q_{rev} - w_{rev} = 0$$

and the generalized entropy balance reduces to

$$s_{in} - s_{out} + \frac{q_{rev}}{T} = 0$$

or

$$q_{rev} = T(s_{out} - s_{in})$$

At the isothermal, reversible turbine effluent conditions of 250 $lb_f/in.^2$ and 700 °F,

$$h_{out} = 1371.6 \text{ BTU/lb}_m$$

$$s_{out} = 1.6976 \text{ BTU/lb}_m \text{ °R}$$

Substituting these values into the reduced entropy balance equation,

$$q_{rev} = 1160(1.6976 - 1.6758)$$

$$= 25.3 \text{ BTU/lb}_m$$

Thus, we find on substituting this value into the reduced energy balance equation that for the isothermal, reversible turbine

$$-w_{rev} = 1371.6 - 1368.9 - 25.3$$

or

$$w_{rev} = 22.6 \text{ BTU/lb}_m$$

Now, in order to reduce the isothermal reversible turbine effluent temperature from 700 °F to the valve effluent temperature of 695 °F, we must cool the steam in an isobaric reversible heat exchanger. The generalized energy balance around this heat exchanger reduces to

$$h_{in} - h_{out} + q_{rev} = 0$$

or

$$q_{rev} = 1368.9 - 1371.6$$

$$= -2.7 \text{ BTU/lb}_m$$

As there is no work associated with the reversible, isobaric heat exchanger, the total reversible work associated with the overall process is

$$w_{rev} = 22.6 \text{ BTU/lb}_m$$

Then, as

$$w_{act} = 0$$

we find that in this case,

$$lw = w_{rev}$$

$$= 22.6 \text{ BTU/lb}_m$$

$$= 23.8 \times 10^3 \text{ joules/lb}_m$$

This result is not the same as the result we obtained from an adiabatic reversible turbine.

From the preceding example, we see that lost work is not a uniquely defined parameter since W_{rev} depends upon which particular reversible path is followed. In this respect, lost work is similar to work or heat which also depend upon the path followed.

Mechanical Efficiency. In practice, the work that does not accomplish any useful objective is generally defined in terms of efficiency rather than in terms of lost work. The efficiency of a work producing device is defined by the ratio

$$\eta = \frac{W_{act}}{W_{rev}} \qquad (3\text{-}17)$$

while the efficiency of a work consuming device is defined by the ratio

$$\eta = \frac{W_{rev}}{W_{act}} \qquad (3\text{-}18)$$

Note that in both cases, the numerical value of efficiency ranges from 0.0 to 1.0.

Again we are faced with the dilemma of which reversible path to select in evaluating W_{rev}. Therefore, the definition of efficiency for any particular device must include a description of the reversible path to be followed. In general, the reversible path selected is that path which most nearly represents the actual process. For example, the quantity of heat transferred from either a turbine or a compressor is generally quite small in relation to the work transferred. Therefore the efficiency of a turbine is generally defined as

$$\eta_{turb} = \frac{W_{act}}{W_{adiab., rev}} \tag{3-19}$$

while the efficiency of a compressor is generally defined as

$$\eta_{comp} = \frac{W_{adiab., rev}}{W_{act}} \tag{3-20}$$

EXAMPLE PROBLEM 3-3. Saturated water vapor at 200 lb_f/in^2 flows at a rate of 1000 lb_m/hr through a well insulated turbine, where it expands to a pressure of 20 lb_f/in^2. If the efficiency of the turbine is 80 percent, determine the actual work produced by the turbine, and the quality of the steam leaving the turbine.

Taking the turbine as the system and basing our calculations on one hour of operation, the integrated balance equations reduce to

$$m_{in} - m_{out} = 0$$

$$h_{in}m_{in} - h_{out}m_{out} - W = 0$$

where we have neglected changes in kinetic and potential energy, and

$$s_{in}m_{in} - s_{out}m_{out} + \frac{LW}{T} = 0$$

Then

$$W = m\Delta h$$

From Table A-2,

$$h_{in} = 1198.3 \text{ BTU/lb}_m$$

Now, in order to determine h_{out} at 20 lb_f/in^2, we need an additional independent intensive property at the turbine outlet. However, since the efficiency defined in Eq. (3-19) relates W_{act} to $W_{adiab., rev}$, let us first assume that the turbine operates adiabatically and reversibly. In this case, the reduced energy balance for the actual turbine still applies and

$$LW = W_{rev} - W_{act} \tag{3-13}$$

$$= 0$$

so, from the reduced entropy balance we see that

$$s_{in} = s_{out} = 1.5454 \text{ BTU/lb}_m \text{ °R}$$

Under these conditions, we find that at the turbine outlet

$$x = \frac{s_{out} - s_f}{s_g - s_f}$$

$$= \frac{1.5454 - 0.3358}{1.7320 - 0.3358} = 0.866$$

and

$$h_{out} = h_g - (1-x)(h_g - h_f)$$

or

$$h_{out} = 1156.3 - (0.134)(1156.3 - 196.27)$$

$$= 1027.7 \text{ BTU/lb}_m$$

Thus, for the adiabatic, reversible turbine

$$W_{rev} = 1000(1198.3 - 1027.7)$$

$$= 170,600 \text{ BTU/hr}$$

But the efficiency of the actual turbine is 80 percent of an adiabatic, reversible turbine. So, from Eq. (3-19)

$$W_{act} = 170,600 \times 0.8 = 136,480 \text{ BTU/hr}$$

$$= 1440 \times 10^5 \text{ joules/hr}$$

To find the specific enthalpy of the steam leaving the actual turbine, we reapply the reduced energy balance equation to find

$$h_{out} = h_{in} - \frac{W_{act}}{m}$$

$$= 1198.3 - \frac{136,480}{1000}$$

$$= 1061.8 \text{ BTU/lb}_m$$

Then, at $P = 20 \text{ lb}_f/\text{in}^2$

$$x = \frac{h_{out} - h_f}{h_g - h_f}$$

$$x = \frac{1061.8 - 196.27}{1156.3 - 196.27} = 0.90$$

Unavailable Energy. As we have defined the term, lost work accounts for energy degradation resulting from irreversibilities within a system. Another similar, but less frequently used concept is that of unavailable energy. In order to determine unavailable energy, we must not only determine the lost work associated with the process but also we must take into account that portion of energy transferred from the system in the form of heat which could be but was not converted to work. In order to determine the portion of heat which is unavailable to do work, we conceive of a second, additional process by which heat leaving the original system is transferred reversibly to a Carnot engine. Then the total amount of work theoretically available from the combined system which is operating at temperature T and rejecting heat to temperature T_0 is the algebraic sum of the work we can obtain by carrying out the original process reversibly and the work we can obtain in the second additional process from the heat rejected from the original process and transferred reversibly to the surroundings at T_0 through a Carnot engine. That is,

$$\delta W_{avail} = \delta W_{rev} + \delta W_c \qquad (3\text{-}21)$$

If the original process is isothermal so that T is constant,

$$\delta W_{avail} = \delta W_{rev} - \delta Q_{rev}\left(1 - \frac{T_0}{T}\right)$$

$$= -dU_{rev} + T_0 \frac{\delta Q_{rev}}{T}$$

$$= -dU_{rev} + T_0 \, dS_{sys}$$

where we have added the minus sign in front of δQ_{rev} to denote that heat has flowed from the system to the reversible thermodynamic machine. Now, the lost work plus the work that could have been obtained from a Carnot engine but wasn't $\delta(LW + LW_c)$, is the total energy that is unavailable for doing work. The quantity $\delta(LW + LW_c)$ is called unavailable energy or irreversibility. Unavailable energy may also be considered to be the difference between the available work and the actual work that is realized.

$$\delta(LW + LW_c) = \delta W_{avail} - \delta W_{act} \tag{3-22}$$

$$= -dU_{rev} + T_0\, dS_{sys} + dU_{act} - \delta Q_{act}$$

$$= T_0\, dS - \delta Q_{act}$$

since

$$dU_{rev} = dU_{act}$$

This result may be rearranged to give

$$dS_{sys} = \frac{\delta(LW + LW_c)}{T_0} + \frac{\delta Q_{act}}{T_0}$$

If we combine this result with the change in entropy of our system that would result from mass flow, we obtain another form of the general entropy balance equation, over a unit period of time,

$$(s\delta m)_{in} - (s\delta m)_{out} + \frac{\delta Q_{act}}{T_0} + \frac{\delta(LW + LW_c)}{T_0} = d(sm)_{sys} \tag{3-23}$$

Equation (3-23) can also be written in a rate form. The method of application of this equation is similar to the method of application of other generalized balance equations.

The unavailable energy term in this equation $\delta(LW + LW_c)$ plays a role analogous to the lost work term δLW in Eq. (3-16). However, as will be shown in Example Prob. 3-4, at a fixed T_0, $\delta(LW + LW_c)$ has a unique value independent of the path used in its evaluation. In this respect, the change in unavailable energy is similar to changes in thermodynamic properties such as internal energy.

On the other hand, the magnitude of the lost work associated with a process depends upon an arbitrary selection of some reversible path between the same initial and final conditions. In this respect, lost work is similar to work and heat, both path functions.

Unavailable energy can be calculated by applying Eq. (3-23) to any system of interest. The method of application of this equation is similar to the method of application of the other forms of the entropy balance equation. Then, unavailable energy is readily determined since the temperature of the surroundings can generally be assumed to be constant. For example, on applying Eq. (3-23) to the surroundings of the system, we find

$$-(s\delta m)_{in} + (s\delta m)_{out} - \frac{\delta Q_{act}}{T_0} = d(sm)_{surroundings}$$

where the signs on the left hand side of this equation have been reversed since quantities flowing into the surroundings are the same as quantities flowing from the system. Further, if we assume that all irreversibilities take place within the system, then in the surroundings, $\delta(LW + LW_c)$ is zero. Adding this result to Eq. (3-23), we find

$$\frac{\delta(LW + LW_c)}{T_0} = d(sm)_{sys} + d(sm)_{surroundings}$$

$$= dS \text{ (system + surroundings)} \qquad (3\text{-}24)$$

Thus we see that unavailable energy is a measure of the total change in entropy of the universe.

Just as lost work can be determined from the relationship

$$\delta LW = \delta W_{rev} - \delta W_{act} \qquad (3\text{-}13)$$

unavailable energy can be estimated from the analogous relationship

$$\delta(LW + LW_c) = \delta W_{avail} - \delta W_{act} \qquad (3\text{-}22)$$

For processes in which no actual work is performed, Eq. (3-22) reduces to

$$\delta(LW + LW_c) = \delta W_{avail}$$

where δW_{avail} is the sum of the reversible work and the Carnot work which could have been obtained from the process. Unavailable work is also equal in magnitude but of opposite sign to the available work required to restore the system to its original state.

EXAMPLE PROBLEM 3-4. Oxygen is flowing irreversibly through a horizontal, uninsulated pipe, shown in Fig. 3-6(a). If the oxygen enters the pipe at 200 °F and 300 lb_f/in^2 and leaves the pipe at 160 °F and 250 lb_f/in^2, what is the unavailable

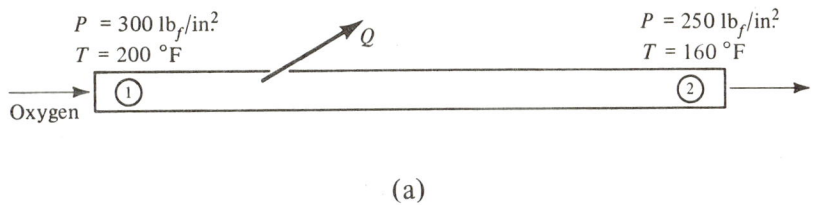

(a)

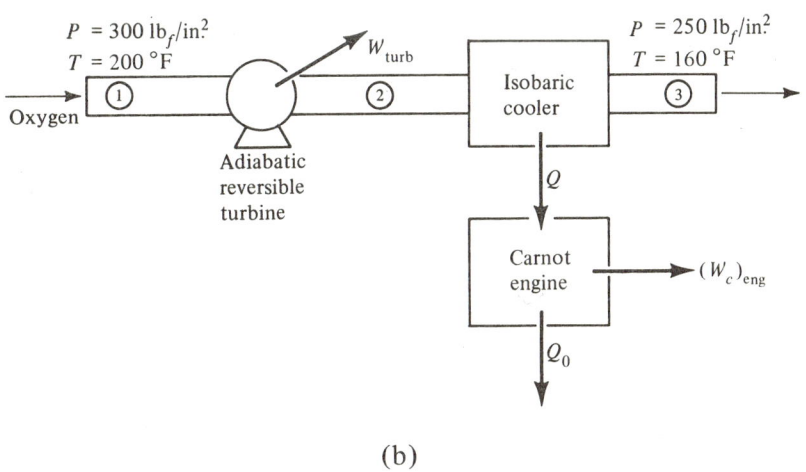

(b)

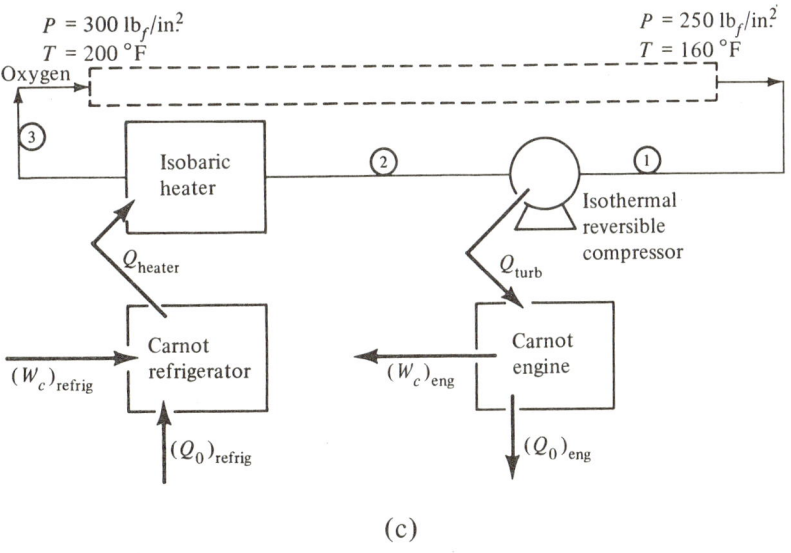

(c)

Fig. 3-6. *Unavailable energy in a flow system.*

energy $(LW + LW_c)$ in the process per pound of oxygen flowing? The temperature of the surroundings is 60 °F and kinetic energy changes may be neglected. The following data on oxygen is available in tables in the literature[4].

$P(\text{lb}_f/\text{in}^2)$	300	250	250	300
$T(°F)$	200	169	160	160
$h(\text{BTU}/\text{lb}_m)$	142.3	135.4	133.4	133.2
$s(\text{BTU}/\text{lb}_m \, °R)$	1.396	1.396	1.393	1.382

Taking the pipe as the system and basing all our calculations on one pound of oxygen flowing, the mass balance reduces to

$$m_{in} - m_{out} = 0$$

where

$$m = 1.0 \text{ lb}_m \text{ of oxygen}$$

As the pipe is horizontal, there is no change in potential energy and the combined mass and energy balance reduces to

$$h_{in} - h_{out} + q = 0$$

while the combined mass and entropy balance reduces to

$$s_{in} - s_{out} + \frac{q}{T_0} + \frac{(lw + lw_c)}{T_0} = 0$$

Substituting values into the reduced energy balance equation, we find

$$q = 133.4 - 142.3 = -8.9 \text{ BTU}/\text{lb}_m$$

This result may be used in the reduced entropy balance equation to show

$$(lw + lw_c) = 520(1.393 - 1.396) + 8.9$$

$$= 7.34 \text{ BTU}/\text{lb}_m$$

$$= 7.74 \times 10^3 \text{ joules}/\text{lb}_m$$

It may prove instructive to calculate the unavailable energy by some other methods.

Let us estimate the work we could have obtained had the flow process been reversible and the heat from the system been used to produce work in a Carnot engine. In order to do this, we first allow the oxygen at 200 °F to expand from 300 lb_f/in^2 to 250 lb_f/in^2 through a reversible, adiabatic turbine, as shown in Fig. 3-6(b). Then, we will remove sufficient heat from the expanded products to cool the oxygen reversibly and isobarically to 160 °F and use this energy to run a Carnot engine.

Taking the turbine as our system, the combined mass and energy balances reduce to

$$h_1 - h_2 - w_{turb} = 0$$

for one pound of oxygen flowing, and the combined mass and entropy balances reduce to

$$s_1 - s_2 = 0$$

so that

$$s_2 = 1.396 \text{ BTU/lb}_m \text{ °R}$$

For a turbine outlet pressure of 250 lb_f/in^2 and an outlet specific entropy of 1.396 BTU/lb_m °R, we find from the given data

$$T_2 = 169 \text{ °F}$$

and

$$h_2 = 135.4 \text{ BTU/lb}_m$$

Substituting this result into the combined mass and energy balance,

$$-w_{turb} = 135.4 - 142.3$$

$$w_{turb} = +6.9 \text{ BTU/lb}_m$$

Next, we must consider the isobaric cooler. For the exchanger as a system, the combined mass and energy balances reduce to

$$h_2 - h_3 + q = 0$$

for one pound mass of oxygen flowing. Substituting into this equation, we find

$$q_{cooler} = 133.4 - 135.4$$

$$= -2.0 \text{ BTU/lb}_m$$

This heat can be used to run a Carnot engine. In order to do this, however, we must devise an ideal process. Let us first remove δq from the oxygen at 169 °F and use this energy to operate a Carnot engine running between 169 °F and 60 °F. After we have removed δq from the oxygen, its temperature will be 169 °F less dT. Let us now remove a second δq from the oxygen, and use it to operate a second Carnot engine operating between 169 °F less dT and 60 °F. If we continue this procedure until the temperature of the oxygen is 160 °F, the total work we have obtained from the Carnot engine is

$$w_{engine} = \int \left(1 - \frac{T_0}{T}\right)(-\delta q)$$

$$= -q + T_0(s_3 - s_2)$$

since the integral of $\delta q / T$ is the total change in entropy of the oxygen in cooling from 169 °F to 160 °F. The minus sign in front of q is necessary as the heat leaving the system is added to the Carnot engine and q refers to the system. Then,

$$w_{engine} = 2.0 + 520(1.393 - 1.396)$$

$$= 0.44 \text{ BTU/lb}_m$$

Now, the total work we could have obtained from the oxygen is

$$w_{total} = w_{turbine} + w_{engine}$$

$$= 6.9 + 0.44$$

$$= 7.34 \text{ BTU/lb}_m$$

But

$$(lw + lw_c) = w_{avail} - w_{act}$$

and since

$$w_{act} = 0$$

we find, once again, that

$$(lw + lw_c) = 7.34 \text{ BTU/lb}_m$$

$$= 7.74 \times 10^3 \text{ joules/lb}_m$$

Finally, let us derive a reversible process to restore the oxygen to its initial state. This process, shown in Fig. 3-6(c), consists of an isothermal, reversible compressor followed by an isobaric, reversible heater.

Taking the compressor as our system, for one pound mass of oxygen flowing, the combined mass and energy balances reduce to

$$h_1 - h_2 + q_{comp} - w_{comp} = 0$$

while the combined mass and entropy balances reduce to

$$s_1 - s_2 + \frac{q_{comp}}{T_1} = 0$$

At 300 $lb_f/in.^2$ and 160 °F, the conditions found at the compressor outlet,

$$h_2 = 133.2 \text{ BTU/lb}_m$$

and

$$s_2 = 1.382 \text{ BTU/lb}_m \text{ °R}$$

Substituting this value into our reduced entropy balance equation, we find that for one pound of oxygen flowing,

$$q_{comp} = 620(1.382 - 1.393)$$

$$= 6.82 \text{ BTU/lb}_m$$

Using this value of q, the reduced energy balance equation yields

$$w_{comp} = -(133.2 - 133.4) - 6.82$$

$$= -6.62 \text{ BTU/lb}_m$$

Now, the heat rejected from the compressor can be used to operate a Carnot engine. This work is given by the relationship

$$w_{engine} = -q_{comp}\left(1 - \frac{T_0}{T}\right)$$

since the heat added to the Carnot engine is the negative of the heat from the reversible, isothermal compressor. Then,

$$w_{engine} = 6.82\left(1 - \frac{520}{620}\right)$$

$$= +1.10 \text{ BTU/lb}_m$$

Next, we must heat the oxygen at 300 $lb_f/in.^2$ reversibly and isobarically from 160 °F to 200 °F. For one pound of oxygen flowing, the combined mass and energy balances around the heater reduce to

$$h_2 - h_3 + q_{heater} = 0$$

or

$$q_{heater} = 142.3 - 133.2$$

$$= +9.1 \text{ BTU/lb}_m$$

Now, this heat must be extracted from the atmosphere by means of a Carnot refrigerator. Further, as the temperature of the oxygen is increasing, we will have to conceive another ideal process whereby the heat is added to the oxygen incrementally.

The combined mass, energy, and entropy balances show that the work required by a Carnot refrigerator operating between T_0 and T to reject δq at T is

$$\delta w_{refrigerator} = \delta q_{refrigerator}\left(1 - \frac{T_0}{T}\right)$$

Since T_0 is less than T and since $\delta q_{refrigerator}$ leaving the Carnot refrigerator is negative, the work calculated using this equation will be negative. This indicates that work has to be added to a Carnot refrigerator.

Then, as the heat rejected from the Carnot refrigerator is the negative of the heat added to the oxygen, the work that must be added to a Carnot refrigerator is

$$\delta w_{refrigerator} = -\delta q_{heater}\left(1 - \frac{T_0}{T}\right)$$

where δq is the incremental heat added to the oxygen. The total work required by the Carnot refrigerator then is

$$w_{\text{refrigerator}} = \int -\left(1 - \frac{T_0}{T}\right) \delta q_{\text{heater}}$$

$$= -q_{\text{heater}} + T_0(s_3 - s_2)$$

since the integral of $\delta q/T$ is the change in entropy of the oxygen at 300 $\text{lb}_f/\text{in.}^2$ in going from 160 °F to 200 °F. Evaluating this equation,

$$w_{\text{refrigerator}} = -9.1 + 520(1.396 - 1.382)$$

$$= 1.82 \text{ BTU/lb}_m$$

Then, the total work of restoration is

$$w_{\text{total}} = w_{\text{turbine}} + w_{\text{engine}} + w_{\text{refrigerator}}$$

$$= -6.62 + 1.10 - 1.82$$

$$= -7.34 \text{ BTU/lb}_m$$

But unavailable energy is the negative of the work of restoration. Thus, once again we find

$$(lw + lw_c) = +7.34 \text{ BTU/lb}_m$$

$$= 7.74 \times 10^3 \text{ joules/lb}_m$$

Effectiveness. Whereas lost work is a measure of the efficiency of a process in converting energy other than heat into work, unavailable energy is a measure of how effective a process is in converting all forms of energy to work. The ratio of the actual work to the work available from a work producing device ranges from 0 to 1.0 and is called the effectiveness ϵ where

$$\epsilon = \frac{W_{\text{act}}}{W_{\text{avail}}} \qquad (3\text{-}25)$$

On the other hand, the ratio of the available work to the actual work of a work consuming device ranges from 0 to 1.0 and

$$\epsilon = \frac{W_{\text{avail}}}{W_{\text{act}}} \qquad (3\text{-}26)$$

Note that these definitions are similar to the definitions of efficiency.

EXAMPLE PROBLEM 3-5. In Example Prob. 3-3, saturated water vapor at 200 lb_f/in^2 entered a well insulated turbine at a rate of 1000 lb_m/hr. The efficiency of this turbine was 80 percent and the steam leaving the turbine was at 20 lb_f/in^2. If the temperature of the atmosphere around the turbine is 70 °F, what is the effectiveness of the turbine?

Taking the turbine as our system and basing our calculations on one hour of operation, the balance equations reduce to

$$m_{in} - m_{out} = 0$$

$$h_{in}m_{in} - h_{out}m_{out} - W = 0$$

$$s_{in}m_{in} - s_{out}m_{out} + \frac{LW + LW_c}{T_0} = 0$$

Now, the effectiveness of the turbine is given by the relationship

$$\epsilon_{turb} = \frac{W_{act}}{W_{avail}} \tag{3-25}$$

From Example Prob. 3-3, we find

$$W_{act} = 136{,}480 \text{ BTU/hr}$$

$$= 1440 \times 10^5 \text{ joules/hr}$$

From Eq. (3-22), we see that

$$W_{avail} = (LW + LW_c) + W_{act}$$

In order to evaluate W_{avail}, we must consider a new process which operates between the same initial and final states as the actual process, but for which the unavailable energy $(LW + LW_c)$ is zero. In this new case, the generalized balance equations reduce to

$$m_{in} - m_{out} = 0$$

$$h_{in}m_{in} - h_{out}m_{out} + Q - W_{avail} = 0$$

$$s_{in}m_{in} - s_{out}m_{out} + \frac{Q}{T_0} = 0$$

From Example Prob. 3-3

$$h_{in} = 1198.3 \text{ BTU/lb}_m$$

$$s_{in} = 1.5454 \text{ BTU/lb}_m$$

and

$$h_{out} = 1061.8 \text{ BTU/lb}_m$$

Since

$$P_{out} = 20 \text{ lb}_f/\text{in}^2$$

we find from the steam tables that

$$s_{out} = 1.5946 \text{ BTU/lb}_m \text{ °R}$$

so that

$$Q = m\, T_0(s_{out} - s_{in})$$

$$= (1000)(530)(1.5946 - 1.5454)$$

$$= 26{,}080 \text{ BTU/hr}$$

$$= 275 \times 10^5 \text{ joules/hr}$$

Then

$$W_{avail} = m_{in}h_{in} - m_{out}h_{out} + Q$$

$$= (1000)(1198.3) - (1000)(1061.8) + 26{,}080$$

$$= 162{,}580 \text{ BTU/hr}$$

$$= 1715 \times 10^5 \text{ joules/hr}$$

and from Eq. (3-25), the turbine effectiveness ϵ_{turb} is

$$\epsilon_{turb} = \frac{136{,}480}{162{,}580}$$

$$= 0.84$$

Availability

We have stated that the total work available from any system at temperature T that is rejecting heat to the surroundings at temperature T_0 is given by the relationship

$$\delta W_{avail} = \delta W_{rev} + \delta W_c \qquad (3\text{-}21)$$

The term δW_{avail} can also be related to a change in thermodynamic properties of the system. If we combine the general energy balance, Eq. (2-25), and the entropy balance as given by Eq. (3-23), we find that the actual work associated with a given process is

$$\delta W_{act} = \left[\left(h - T_0 s + \frac{v^2}{2} + gz\right)\delta m\right]_{in} - \left[\left(h - T_0 s + \frac{v^2}{2} + gz\right)\delta m\right]_{out}$$

$$- \delta(LW + LW_c) - d\left[\left(u - T_0 s + \frac{v^2}{2} + gz\right)m\right]_{sys} \qquad (3\text{-}27)$$

If this particular process is carried out reversibly,

$$\delta W_{act} = \delta W_{rev}$$

and

$$\delta LW = 0$$

Further, if the heat rejected from the system at temperature T is used to operate a Carnot engine, then by eliminating terms, Eq. (3-27) reduces to

$$\delta W_{avail} = \left[\left(h - T_0 s + \frac{v^2}{2} + gz\right)\delta m\right]_{in} - \left[\left(h - T_0 s + \frac{v^2}{2} + gz\right)\delta m\right]_{out}$$

$$- d\left[\left(u - T_0 s + \frac{v^2}{2} + gz\right)m\right]_{sys} \qquad (3\text{-}28)$$

In certain types of thermodynamic analyses, it is desirable to be able to calculate the maximum work available from a unit mass in a given state. The available work between any two states is given by Eq. (3-28). However, the question arises, what must be the final state in order to obtain maximum work from a unit mass? The answer must be that state from which no further work can be extracted from the mass. This state is called the dead state. A mass in the dead state would have no kinetic or potential energy and would be in thermodynamic equilibrium with its surroundings.

If we consider a unit mass at a given state flowing through an open system to the dead state, Eq. (3-28) reduces to

$$(w_{avail})_{max} = \left(h - T_0 s + \frac{v^2}{2} + gz\right)_{in} - (h_D - T_0 s_D + gz_D)_{out}$$

where the subscript D denotes the dead state. For any specified dead state, the quantity $(\delta w_{avail})_{max}$ has the characteristics of a thermodynamic property since its value depends only on the initial state of the system. This property is denoted by the symbol b and is given the name availability. Then

$$b_{open\ system} = (w_{avail})_{max}$$

or

$$b_{open\ system} = (h - h_D) - T_0(s - s_D) + \frac{v^2}{2} + g(z - z_D) \qquad (3\text{-}29)$$

Equation (3-28) can also be used to develop an expression for the availability associated with a closed system. For a closed system of unit mass, we find on integrating Eq. (3-28) from the actual state to the dead state that

$$(w_{avail})_{max} = (u - u_D) - T_0(s - s_D) + \frac{v^2}{2} + g(z - z_D)$$

Now, at the dead state

$$P_{sys} = P_D = P_{surr}$$

In most cases, it will be necessary to conceive of a process in which the closed system is allowed to expand reversibly in order that the final pressure in the system is the dead state pressure. The work associated with this expansion would be

$$w_{expansion} = P_D(v_D - v)$$

since the pressure opposing the expansion is P_D.

Although this expansion work is in a form which is just as available for actual operation of a pump or turbine as any Carnot work, it is common practice to assume that expansion work is not a part of the availability of a closed system. Thus, for a closed system, availability is defined as

$$b_{\text{closed system}} = (w_{\text{avail}})_{\max} - P_D(v_D - v)$$

or

$$b_{\text{closed system}} = (u - u_D) - T_0(s - s_D) + P_D(v - v_D)$$

$$+ \frac{v^2}{2} + g(z - z_D) \qquad (3\text{-}30)$$

As availability is a measure of the maximum work available from a system, it more truly reflects the amount of work that can be obtained from any process stream than either internal energy or enthalpy. For this reason, availability is frequently used as the basis for assigning a monetary value to the energy in a process stream for cost accounting purposes. It should be emphasized that the dead state has to be carefully stated to insure that cost comparisons are meaningful.

EXAMPLE PROBLEM 3-6. In Example Prob. 3-4, oxygen flowed irreversibly through a horizontal, uninsulated pipe. The oxygen entered the pipe at 200 °F and 300 lb$_f$/in.^{2} and left the pipe at 160 °F and 250 lb$_f$/in.^{2} Using the concept of availability, calculate the unavailable energy associated with this process. The data on oxygen presented in Example Prob. 3-4 may be used and the temperature of the surroundings is 60 °F.

In order to solve this problem, let us first determine the relationship between availability and unavailable work. From Eq. (3-29), we see that the difference in availability of a unit mass at the inlet and outlet of an open system is

$$\Delta b = \left(h - T_0 s + \frac{v^2}{2} + gz\right)_{\text{out}} - \left(h - T_0 s + \frac{v^2}{2} + gz\right)_{\text{in}}$$

But, for a unit mass flowing in an open system, the generalized energy balance can be written in the form

$$\left(h + \frac{v^2}{2} + gz\right)_{\text{out}} - \left(h + \frac{v^2}{2} + gz\right)_{\text{in}} = q_{\text{act}} - w_{\text{act}}$$

Further, the entropy balance can be written

$$(T_0 s)_{\text{out}} - (T_0 s)_{\text{in}} = q_{\text{act}} + (lw + lw_c)$$

Combining these results, we see that

$$-\Delta b = -q_{act} + w_{act} + q_{act} + (lw + lw_c)$$

$$= w_{act} + (lw + lw_c)$$

From Eq. (3-22), we see that

$$w_{act} + (lw + lw_c) = w_{avail}$$

so that

$$-\Delta b = w_{avail}$$

In this case,

$$w_{act} = 0$$

so

$$(lw + lw_c) = -\Delta b$$

With this background, let us solve for the unavailable energy associated with this process by taking the pipe as our system and basing our calculations on one pound mass of oxygen flowing. Neglecting kinetic and potential energy terms, we find that

$$\Delta b = (h - T_0 s)_{out} - (h - T_0 s)_{in}$$

$$= [133.4 - 520(1.393)] - [142.3 - 520(1.396)]$$

$$= -7.34 \text{ BTU/lb}_m$$

where the required data is taken from Example Prob. 3-4. Then

$$(lw + lw_c) = -\Delta b$$

$$= 7.34 \text{ BTU/lb}_m$$

$$= 7.74 \times 10^3 \text{ joules/lb}_m$$

This is the same answer as we obtained previously.

Absolute Temperature Scale

As we mentioned in an earlier chapter, thermodynamics can be used as a basis for constructing an absolute temperature scale. Kelvin was the first to recognize that a Carnot engine provides a basis for such a scale.

Noting that the entropy balance around a Carnot engine reduces to

$$\frac{Q}{T} + \frac{Q_0}{T_0} = 0$$

Kelvin proposed in 1854 that the ratio of absolute temperatures was equal to the ratio of the quantities of heat received by a Carnot engine from a source and rejected to a sink,

$$\frac{T_0}{T} = \frac{-Q_0}{Q} \qquad (3\text{-}31)$$

This ratio is a positive quantity since the heat rejected to the sink Q_0 is a negative quantity. Equation (3-31) gives only the ratio of absolute temperatures and it is necessary to complete the definition by assigning an arbitrary value of temperature to some reproducible state.

Kelvin proposed the temperature interval from the freezing point to the boiling point of water be defined as 100 degrees, consistent with the earlier Celsius scale. Further, an arbitrary value of temperature selected was chosen to represent experimentally determined phase behavior.

Experimentally, it has been determined that the pressure-volume product of a gas at low pressure is proportional to temperature.

$$Pv = kT \qquad (1\text{-}23)$$

Joseph Gay-Lussac, working from 1802-1808, observed that at a low constant pressure, the volume of air increased by 1/267 of its volume at 0 °C for each degree rise in temperature. In 1845, Henri Victor Regnault refined Gay-Lussac's measurements and showed that the volume of a gas at low pressure increased by 1/273 of its volume at 0 °C for each degree rise in temperature. He further found that the pressure of the gas at constant volume also increased by 1/273 of its pressure at 0 °C for each degree rise in temperature. As a deduction from the results of Regnault's experiments, it followed that the pressure-volume product of a gas at low pressure should be directly proportional to an absolute temperature that had a value of zero at −273 °C. Then, a simple equation such as Eq. (1-23) could be used to represent the phase behavior of a gas at low temperature.

In order to relate his absolute temperature scale to something that could be measured, Kelvin arbitrarily selected the zero of his scale to be −273 °C. More recently, experiments suggest that the pressure volume product of a gas at low pressure is zero at −273.15 °C. Because the ice-water-steam triple point of 0.01

°C is independent of pressure and is thus more readily reproduced in a laboratory than the ice-water equilibrium temperature of 0.0 °C at one standard atmosphere, today we define absolute zero on the Kelvin scale as −273.16 °C below the ice-water-steam triple point.

It is important to note that the thermodynamic temperature scale is derived from the laws of thermodynamics and, therefore, thermodynamic relationships are strictly valid only for temperatures defined by this scale.

PROBLEMS

3-1. What is the quality of an equilibrium liquid-vapor mixture of water at 300 °F if the entropy of the mixture is 1.0 BTU/lb_m °R?

3-2. What is the change in entropy in joules/gm-K of carbon dioxide vapor heated from 100 to 300 °F if c_p of the vapor is 0.1989 kcal/kg-K?

3-3.
(a) Using data in Table A-2 of the Appendix, prepare a plot of the vapor-liquid envelope and the 400 lb_f/in² isobar from 200 to 800 °F for water on a specific entropy-temperature diagram. At 400 lb_f/in², the specific heat capacity of compressed liquid water at constant pressure and temperatures below the saturation temperature is 1.04 BTU/lb_m °R.

(b) On this graph, sketch the path of a Carnot cycle operating between 400 and 300 °F and discuss the significance of the areas enclosed by this path.

3-4. An inventor claims to have developed a device which operates in a cycle and which produces no effects other than the lowering of a weight and the exchange of heat with a single reservoir. Does this device violate the second law of thermodynamics?

3-5. A well insulated vertical cylinder, open at the top and having a cross-sectional area of 1 ft², is fitted with a frictionless piston weighing 2200 pounds. Initially, the piston is held in place by a latch, and the cylinder contains 5 ft³ of steam at 130 psia and 600 °F. The latch is released, and the gas expands isentropically until the piston comes to rest. Calculate the change in enthalpy of the steam for this process.

3-6. One thousand BTU of heat are transferred from a reservoir at 600 °F to an engine that operates on the Carnot cycle. The engine rejects heat to a reservoir at 80 °F. Determine the thermal efficiency of the cycle and the work in BTU and joules done by the engine.

3-7. A refrigerator that operates on a Carnot cycle is required to transfer 10,000 joules/min from a reservoir at 245 K to the atmosphere at 300 K. What is the power required in BTU/hr and in watts?

3-8. It is proposed to heat a house using a heat pump. The heat transfer from the house is 50,000 BTU/hr. The house is to be maintained at 75 °F while the outside air is at a temperature of 20 °F. What is the minimum power required to drive the heat pump?

3-9. On hot summer days, it is estimated that 50,000 BTU/hr of heat will leak into a building. A two horsepower Carnot engine is available for use as an air conditioning unit. What is the maximum permissible temperature outside the building for which a temperature of 70 °F can be maintained inside the building?

3-10. A Carnot refrigerator is to be designed to remove 10 joules/hr from a heat reservoir at 0 K. If the heat from the refrigerator can be discarded at 27 °C, how much work is required?

3-11. Preliminary cost estimates for a 1 million-kW nuclear-powered thermal power station require an estimate of the cooling water demands of the facility. For purposes of this estimate, the heat engine can be assumed to be a Carnot engine operating between 650 and 150 °F. If cooling water enters the process as saturated liquid at 70 °F and leaves as saturated liquid at 120 °F, at what rate (GPM) must cooling water be supplied?

3-12. Saturated water vapor at 100 lb_f/in^2 enters a reversible, isothermal turbine and the steam leaves at 20 lb_f/in^2. Determine the work per lb_m of steam flowing through the turbine.

3-13. Two *identical* bodies of constant heat capacity and at temperatures T_1 and T_2, respectively, are used as reservoirs for a heat engine. If the bodies remain at constant pressure and undergo no phase change, show that the maximum work obtainable is equal to $c_p(T_1 + T_2 - 2\sqrt{T_1 T_2})$.
Hint: Use an ideal process to exchange increments of heat.

3-14. Two *identical* bodies of constant heat capacity are at the same initial temperature T_1. A refrigerator operates between these two bodies until one body is cooled to T_2. If the bodies remain at constant pressure and undergo no phase change, show that the minimum work required is equal to

$$c_p\left(\frac{T_1^2}{T_2} + T_2 - 2T_1\right)$$

Hint: Use an ideal process to exchange increments of heat.

3-15. A well insulated cylinder is divided into two parts by a freely floating piston as shown below. The piston, which is also well insulated, is restrained by a pin.

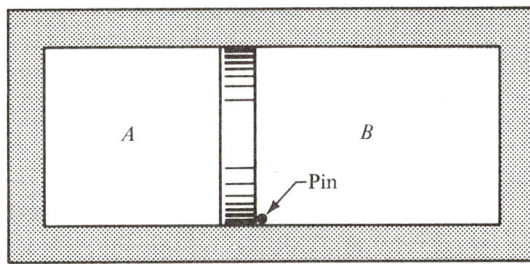

Fig. 3-7.

Initially, chamber A contains 1 lb_m of steam as saturated vapor at 400 °F. Chamber B, which has a volume of 3 ft^3, contains a perfect vacuum.

The pin is removed and the piston moves to a new position of equilibrium.

(a) Considering A as the system, what is the work and the heat in BTU?
(b) Considering A and B as the system, what is the work and the heat in BTU?
(c) Is the process reversible?

3-16. A spring at 27 °C is stretched isothermally and reversibly from its equilibrium length L_0 to $2L_0$. During this reversible stretching, 10 cal of heat is absorbed by the spring. The stretched spring is then released without any restraining back tension and allowed to jump back to its initial length L_0. During this spontaneous process, the spring evolved 25 cal of heat. If the temperature of the surroundings is 27 °C,

(a) What is the entropy change for the stretching of the spring?
(b) What is the entropy change for the collapse of the spring?
(c) What is the entropy change for the universe (spring and surroundings) for the total process, stretching plus return to initial L_0?
(d) How much work was done on the spring in the stretching process?

3-17. Steam at 400 $lb_f/in.^2$, 600 °F expands through a well insulated nozzle to 300 $lb_f/in.^2$ at the rate of 20,000 lb_m/hr. If the pressure at the nozzle throat is 300 $lb_f/in.^2$ and if the expansion occurs isentropically and the initial velocity is low, calculate

(a) the velocity leaving the nozzle, and
(b) the area of the nozzle throat.

3-18. A tank shown in Fig. 3-8 initially contains 10 lb_m of steam at 100 $lb_f/in.^2$, 600 °F, and is connected through a valve to a cylinder fitted with a frictionless piston. A pressure of 20 $lb_f/in.^2$ is required to balance the weight of the piston.

The connecting valve is opened until the pressure in the tank equals 20 $lb_f/in.^2$ Assume the entire process to be adiabatic, and that the steam that finally remains in the tank has undergone a reversible adiabatic process.

Determine the work done against the piston and the final temperature of the steam in the cylinder.

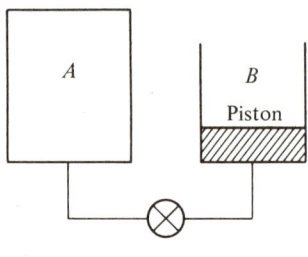

Fig. 3-8.

3-19. A well insulated tank contains 10 ft^3 of steam at 100 $lb_f/in.^2$ at 600 °F. A small leak develops in the tank and steam escapes to the atmosphere which is at 77 °F. When one half the mass of steam has leaked from the tank, what is the pressure inside the tank?

3-20. A five ft^3 tank containing steam at 1 atm pressure, 1 percent quality, is fitted with a relief valve. Heat is transferred to the tank from a large source at 500 °F. When the pressure in the tank reaches 300 $lb_f/in.^2$, the relief valve opens. Saturated vapor at 300 $lb_f/in.^2$ is throttled across the valve and discharged at atmospheric pressure. The process continues until the quality in the tank is 90 percent.

(a) Calculate the mass discharged from the tank.

(b) Determine the heat transfer to the tank during the process.

(c) Considering a system that contains the tank and valve, calculate the entropy change within the system and that of the surroundings. Show that the sum of the change of entropy of the system and the surroundings is equal to or greater than zero.

3-21. Two insulated horizontal cylinders are separated by removable insulation as shown in Fig. 3-9. Each cylinder is divided into two parts by a freely-floating, insulated piston having negligible thickness.

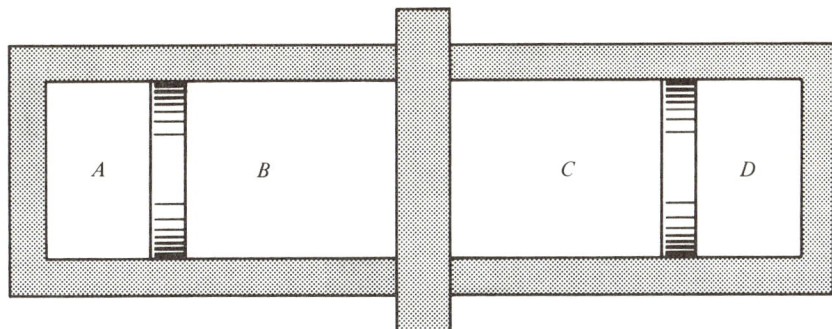

Fig. 3-9.

Initially, chamber A has a volume of five ft³ and contains steam at 490 psia and 520 °F; chamber B, also five ft³, contains steam at 740 °F; chamber C contains 10 pounds of saturated vapor at 320 °F; and chamber D contains air.

The removable insulation is raised, and heat flows slowly from B to C, until the steam in A is saturated vapor. At the same time, air is bled slowly out of D to maintain a constant temperature in C.

(a) What is the final pressure in A?
(b) How much heat flows from B to C?
(c) What is W for the steam in B?
(d) What is the final pressure in C?

3-22. Steam flows in the cycle shown in Fig. 3-10. This cycle is operated under conditions such that the work output of the turbine is just sufficient to drive the compressor.

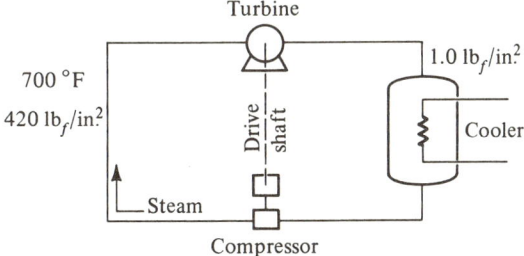

Fig. 3-10.

The compressor may be assumed to operate adiabatically but irreversibly and the drive shaft connecting the compressor and the turbine is frictionless. The turbine operates isothermally and reversibly. The pressure drop of the steam in all lines and across the cooler may be considered to be negligible.

Calculate the heat removed from the steam in the cooler in BTU per pound of steam flowing.

3-23. Steam at 300 lb$_f$/in.^{2} and 1000 °F enters a well insulated turbine at a rate of 1000 lb$_m$/hr. Steam from the turbine is then cooled reversibly and isobarically in a heat exchanger. The exchanger effluent is at 80 °F and 1 atm pressure. If the efficiency of the turbine is 75 percent and if the temperature of the surroundings is 80 °F, calculate

(a) the availability of the steam at turbine inlet,
(b) the lost work associated with this process, and
(c) the unavailable energy associated with this process.

3-24. In a certain chemical plant, shown schematically in Fig. 3-11, a 40 percent solution of organic solids is being evaporated to an 80 percent solution (all percentages on a mass basis). The organic solids cause only a very small boiling point rise, which can be considered negligible. The condenser operates in such a manner that the temperature in the vapor line to the condenser is 130 °F.

At the present time, the only steam available for this operation is that which comes directly from the boiler. Since the tubes in the evaporator cannot withstand high pressure, the boiler steam is throttled to 10 psig before entering the evaporator.

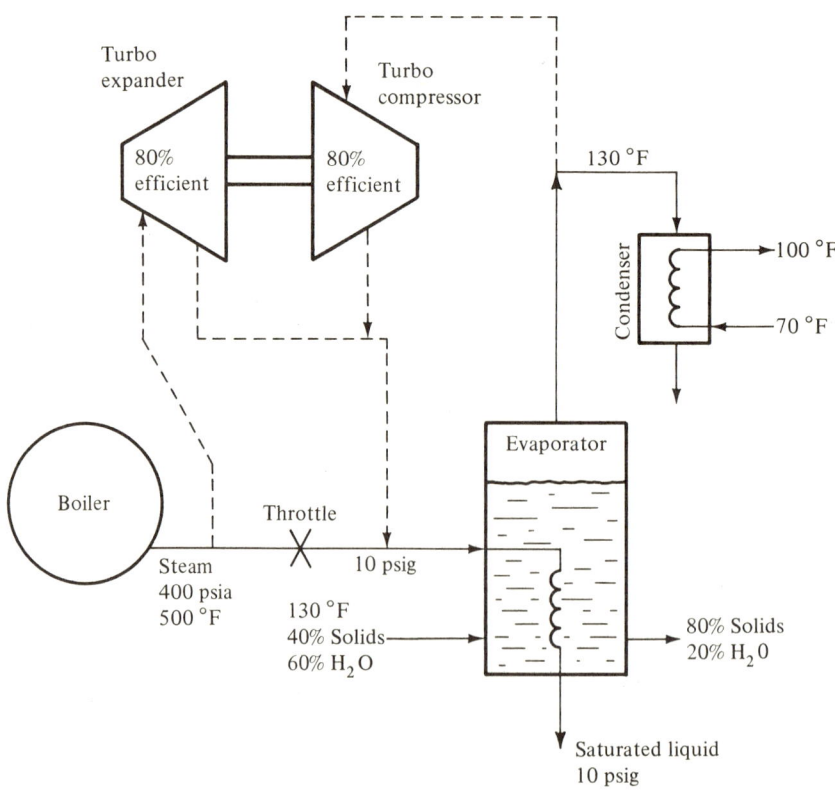

Fig. 3-11. *Chemical plant.*

A manufacturer of turbo expander-compressors suggests improving the steam economy by installing one of his machines as shown by the dashed lines in Fig. 3-11. Per pound of feed solution handled, calculate the saving in the boiler steam which might be made if the turbo expander-compressor operates with an efficiency of 80 percent (based on adiabatic reversible operation) at both ends.

3-25. I. M. Nutty, an industrial engineer, has decided to design a new temperature scale based on the mercury thermometer. Therefore, on the Nutty scale, 0 °N corresponds to the freezing point of mercury (−38.87 °C) and 1000 °N corresponds to the normal boiling point of mercury (356.58 °C). Answer the following questions.

(a) At the freezing point of mercury by what fraction does the pressure of a gas whose properties may be represented by Eq. (1-23) change at constant volume on being heated by 1 °N?

(b) What temperature is absolute zero on the Nutty scale?

(c) What is the normal blood temperature (98.6 °F) of a human being on the Nutty scale?

4

Mathematical Relationships Between Thermodynamic Properties

OBJECTIVES

After studying this chapter, the student should be able to formulate differential relationships for the change of any thermodynamic function in terms of the measurable properties pressure, temperature, specific volume and specific heat capacities.

INTRODUCTION

In the preceding chapters, we have developed the basic balance equations of thermodynamics. One of the most useful features of these equations is that they relate the energy transferred during a process as heat or work to a difference in thermodynamic functions. Both heat and work are called path functions since their magnitudes depend on the path of the process. In many cases, this path is not known in sufficient detail to permit us to calculate directly the heat or work associated with a process. On the other hand, thermodynamic functions are called either analytical or point functions since their difference depends only upon the initial and final states of the system.

Values of thermodynamic functions for water or steam as well as for many other single component systems can be found in tables. It should be noted that tabulated values of thermodynamic functions such as enthalpy and entropy are not measured directly as there is no such thing as an enthalpy meter or an entropy meter. Rather, these functions are calculated from measurable system properties such as pressure, temperature, specific volume and specific heat capacities. An objective of this chapter is to develop the relationships used in such calculations.

In using the steam tables, we discovered that if we specified any two independent intensive thermodynamic properties of a system, all other intensive thermodynamic properties of the system were defined. Thus, we have been led to the conclusion that two *independent* intensive thermodynamic properties completely define the state of a single component system. This result suggests that the thermodynamic properties of a system are not all independent variables. Rather, a set of mathematical relationships must exist between them. In this chapter, we will examine some of these relationships.

DERIVATIVES OF THERMODYNAMIC FUNCTIONS

Experimentally, we find that when two phases of a given substance exist in equilibrium, the vapor pressure $P^\square$ and the system temperature T of a single component system are not independent variables and an additional independent property is required to define the state of the system. This suggests that some sort of mathematical relationship exists between $P^\square$ and T. Thus, over a limited temperature range, we might write

$$P^\square = kT$$

This equation says that the value of the dependent variable $P^\square$ is a linear function of the independent variable T in this temperature range. Now the relationship between $P^\square$ and T does not have to have an analytic form nor is it necessarily linear over a wide temperature range. Figure 4-1 is a graphical representation of the re-

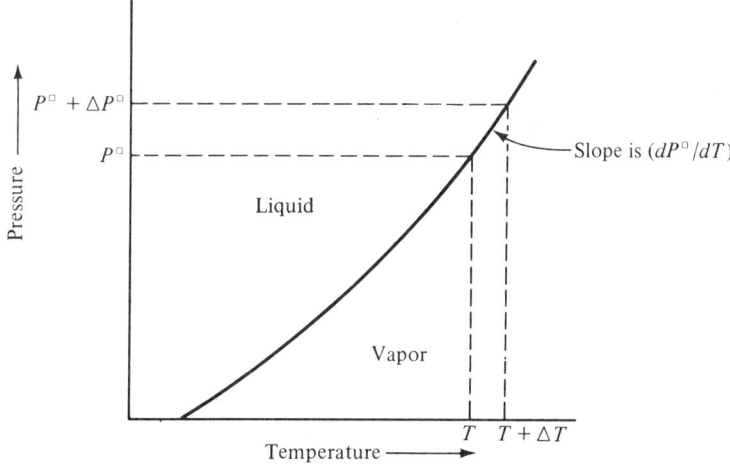

Fig. 4-1. *Vapor pressure of a liquid.*

lationship between vapor pressure and temperature. It is also possible to present the relationship between $P^\square$ and T in tabular form. In all cases, we may indicate that a relationship exists between $P^\square$ and T by writing

$$P^\square = P^\square(T) \tag{4-1}$$

This expression is to be read $P^\square$, the dependent variable, is a function of T, the independent variable.

Sometimes it is necessary to determine the rate of change of $P^\square$ with T. If we increase T by an incremental amount ΔT, as shown in Fig. 4-1, the vapor pressure at $T + \Delta T$ is $P^\square + \Delta P^\square$. The rate of change of $P^\square$ with T over this interval is defined by the relationship

$$\frac{P^\square(T+\Delta T) - P^\square(T)}{(T+\Delta T) - T} = \frac{\Delta P^\square}{\Delta T}$$

The instantaneous rate of change of P with respect to T at a specific T is called a first order ordinary derivative and is found by letting the value of ΔT approach zero,

$$\lim_{\Delta T \to 0} \frac{\Delta P^\square}{\Delta T} = \frac{dP^\square}{dT} \tag{4-2}$$

The rate of change of $dP^\square/dT$ with T is second order ordinary derivative and is written

$$\lim_{\Delta T \to 0} \frac{\frac{dP^\square}{dT}(T+\Delta T) - \frac{dP^\square}{dT}(T)}{(T+\Delta T) - T} = \frac{d^2 P^\square}{dT^2} \tag{4-3}$$

Although the vapor pressure of a single component system depends only on the single independent variable T, we find experimentally that other thermodynamic properties of the system are functions of two independent variables. For example, the pressure exerted by a single phase one-component system depends not only on the system temperature but also the specific volume of the component. That is, P is a function of T and v, or

$$P = P(T, v) \tag{4-4}$$

This relationship is presented schematically in Fig. 4-2. In many instances, we may not know the exact analytical form of the relationship between variables. However, much information can be gained by merely knowing that such relationships exist. The analysis of the relationships between three variables, such as P, T, and v, can be simplified by holding one of the variables constant. For example,

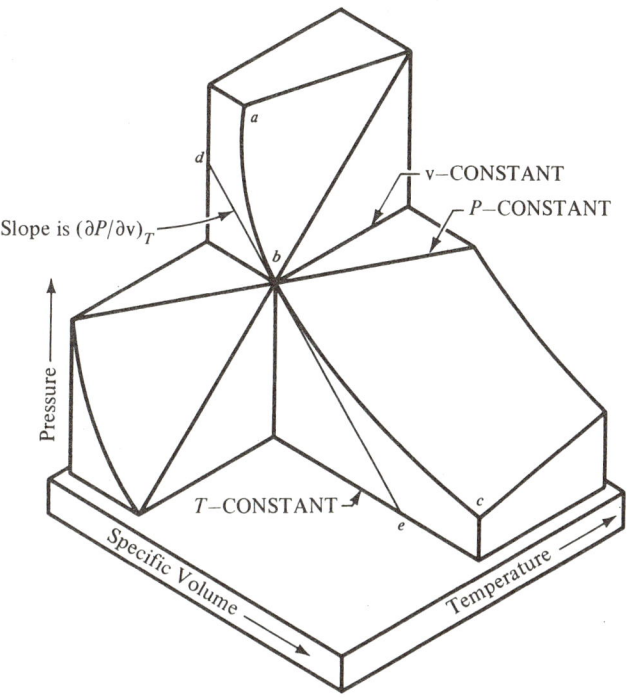

Fig. 4-2. *Schematic representation of partial derivatives.*

the intersection of the plane $T = \text{CONSTANT}$ with the surface $P = P(T,v)$ in Fig. 4-2 results in a curve ac representing P as a function of v on the constant temperature plane. The rate of change of P with v at point b on this plane is the slope of the line de and is given by the relationship

$$\lim_{\Delta v \to 0} \frac{P(T, v + \Delta v) - P(T, v)}{(v + \Delta v) - v} = \left(\frac{\partial P}{\partial v}\right)_T \tag{4-5}$$

where $(\partial P/\partial v)_T$ is a first order partial derivative. A partial derivative is defined as the derivative of the function with respect to one of the independent variables when all other independent variables are held constant. The independent variables which are held constant are denoted by subscripts. Thus, in a constant specific volume plane, the rate of change of P with respect to T is written $(\partial P/\partial T)_V$. In a constant pressure plane, the rate of change of v with respect to T is written $(\partial v/\partial T)_P$.

A first order partial derivative can be differentiated a second time to obtain a second order partial derivative. For example,

$$\lim_{\Delta v \to 0} \frac{\left[\frac{\partial P}{\partial v}(T, v + \Delta v)\right]_T - \left[\frac{\partial P}{\partial v}(T, v)\right]_T}{(v + \Delta v) - v} = \frac{\partial}{\partial v_T}\left(\frac{\partial P}{\partial v}\right)_T = \left(\frac{\partial^2 P}{\partial v^2}\right)_T \tag{4-6}$$

or

$$\lim_{\Delta T \to 0} \frac{\left[\frac{\partial P}{\partial v}(T+\Delta T, v)\right]_{T+\Delta T} - \left[\frac{\partial P}{\partial v}(T, v)\right]_T}{(T+\Delta T) - T} = \frac{\partial}{\partial T}_V \left(\frac{\partial P}{\partial v}\right)_T = \left(\frac{\partial^2 P}{\partial T \partial v}\right)_{V,T} \quad (4\text{-}7)$$

The total differential change of P as a function of T and v is defined by the relationship

$$dP = \left(\frac{\partial P}{\partial T}\right)_V dT + \left(\frac{\partial P}{\partial v}\right)_T dv \quad (4\text{-}8)$$

where dP is the total change in P as a result of differential changes in T and v, $(\partial P/\partial T)_V$ and $(\partial P/\partial v)_T$ are the two first order partial derivatives of the function P, and dT and dv are the differentials of the independent variables T and v respectively.

Any relationship which gives us P as a function of T and v can be rearranged to give us v as a function of T and P. Then, the total differential of $v(T, P)$ is

$$dv = \left(\frac{\partial v}{\partial T}\right)_P dT + \left(\frac{\partial v}{\partial P}\right)_T dP \quad (4\text{-}9)$$

But

$$dP = \left(\frac{\partial P}{\partial T}\right)_V dT + \left(\frac{\partial P}{\partial v}\right)_T dv \quad (4\text{-}8)$$

So

$$dv = \left(\frac{\partial v}{\partial T}\right)_P dT + \left(\frac{\partial v}{\partial P}\right)_T \left(\frac{\partial P}{\partial T}\right)_V dT + \left(\frac{\partial v}{\partial P}\right)_T \left(\frac{\partial P}{\partial v}\right)_T dv$$

or

$$\left[1 - \left(\frac{\partial v}{\partial P}\right)_T \left(\frac{\partial P}{\partial v}\right)_T\right] dv = \left[\left(\frac{\partial v}{\partial P}\right)_T \left(\frac{\partial P}{\partial T}\right)_V + \left(\frac{\partial v}{\partial T}\right)_P\right] dT$$

Since the variables v and T are independent, T can be given any value, and the value of v will not be affected. Let T be a constant. Then,

$$dT = 0$$

and

$$\left[1 - \left(\frac{\partial v}{\partial P}\right)_T \left(\frac{\partial P}{\partial v}\right)_T\right] dv = 0$$

Since dv may have any value, not necessarily zero,

$$1 - \left(\frac{\partial v}{\partial P}\right)_T \left(\frac{\partial P}{\partial v}\right)_T = 0$$

and

$$\left(\frac{\partial v}{\partial P}\right)_T = \frac{1}{\left(\frac{\partial P}{\partial v}\right)_T} \tag{4-10}$$

That is, a partial derivative is equal to the reciprocal of the partial derivative between the same two variables taken in the opposite order provided that the same independent variables are held constant.

By similar arguments, we may assign v a value as a constant so that $dv = 0$ and as dT does not depend on v,

$$\left[\left(\frac{\partial v}{\partial P}\right)_T \left(\frac{\partial P}{\partial T}\right)_V + \left(\frac{\partial v}{\partial T}\right)_P\right] dT = 0$$

Since dT may have any value, not necessarily zero,

$$\left(\frac{\partial v}{\partial P}\right)_T \left(\frac{\partial P}{\partial T}\right)_V + \left(\frac{\partial v}{\partial T}\right)_P = 0$$

This equation may be written in several forms; such as

$$\left(\frac{\partial v}{\partial P}\right)_T \left(\frac{\partial P}{\partial T}\right)_V \left(\frac{\partial T}{\partial v}\right)_P = -1 \tag{4-11}$$

or

$$\left(\frac{\partial v}{\partial T}\right)_P = -\frac{\left(\frac{\partial P}{\partial T}\right)_V}{\left(\frac{\partial P}{\partial v}\right)_T} \tag{4-12}$$

112   Chap. 4   *Mathematical Relationships*

Another very useful tool is the chain rule for differentiation. The rule is written as:

$$\left(\frac{\partial P}{\partial T}\right)_V = \left(\frac{\partial P}{\partial y}\right)_V \left(\frac{\partial y}{\partial T}\right)_V \tag{4-13}$$

where all partial derivatives are taken under the same restrictions, in this case at constant V, so that

$$\left(\frac{\partial y}{\partial y}\right)_V = 1.0$$

Equations (4-10), (4-11), (4-12), and (4-13) will be useful in manipulating thermodynamic functions besides P, v, and T. First, however, we will need to develop relationships between these measurable thermodynamic properties and other, derivable thermodynamic properties.

THE PROPERTY RELATIONSHIP

The energy balance may be used to develop an equation which relates the measurable and derived properties of a system. For a closed system, the energy balance can be reduced to

$$dU = \delta Q - \delta W$$

For a reversible process

$$dU_{rev} = \delta Q_{rev} - \delta W_{rev}$$

where the subscript rev indicates that the process is reversible. If we substitute Eq. (2-13),

$$\delta W_{rev} = P\,dV - F\,dL - \sigma\,dA - \epsilon\,dq - \Sigma\,\mu_i\,dn_i - \ldots \tag{2-13}$$

and Eq. (3-1),

$$\delta Q_{rev} = T\,dS \tag{3-1}$$

into the reduced energy balance, recognizing that

$$dU_{rev} = dU$$

since the change in internal energy depends only upon the initial and final states of the system and not on the path, we find that

$$dU = T\,dS - P\,dV + F\,dL + \sigma\,dA + \epsilon\,dq + \Sigma\,\mu_i\,dn_i + \ldots \quad (4\text{-}14)$$

This last equation is called the property relationship and is a basic equation of thermodynamics. It is valid for infinitesimal displacements of potentials within a system at equilibrium.

Initially, we will only be interested in the first two terms of the property relationship,

$$dU = T\,dS - P\,dV \quad (4\text{-}15)$$

Equation (4-15) relates the derived properties U and S to measurable properties, P, V, and T. In this equation, S and V play the role of mathematically independent variables while T and P play the role of mathematically dependent variables. Other relationships between these properties can be obtained if we assume that U is a function of the independent variables S and V,

$$U = U(S, V)$$

Then the total differential of U is

$$dU = \left(\frac{\partial U}{\partial S}\right)_V dS + \left(\frac{\partial U}{\partial V}\right)_S dV$$

Comparing this result with Eq. (4-15), and equating like coefficients, we see that

$$\left(\frac{\partial U}{\partial S}\right)_V = T \quad (4\text{-}16)$$

and

$$\left(\frac{\partial U}{\partial V}\right)_S = -P \quad (4\text{-}17)$$

It is also possible to develop relationships for other derived thermodynamic properties in terms of measurable thermodynamic properties. For example, by definition,

$$H = U + PV \quad (2\text{-}24)$$

If we first take the total differential of this equation,

$$dH = dU + P\,dV + V\,dP$$

and then using the property relationship in the form of Eq. (4-15), replace dU with $T\,dS - P\,dV$, we find

$$dH = T\,dS + V\,dP \qquad (4\text{-}18)$$

There are two other thermodynamic functions we will find useful. The first is called the Gibbs free energy, G, defined by the relationship

$$G = U + PV - TS$$

$$= H - TS \qquad (4\text{-}19)$$

Proceeding in a manner similar to that used in obtaining Eq. (4-18), we find that

$$dG = V\,dP - S\,dT \qquad (4\text{-}20)$$

A second new thermodynamic function is called the Helmholtz free energy A. This function is defined by the relationships

$$A = U - TS \qquad (4\text{-}21)$$

$$= G - PV$$

Proceeding in a manner similar to that used in obtaining Eq. (4-18), we find

$$dA = -S\,dT - P\,dV \qquad (4\text{-}22)$$

Then, just as we showed that

$$\left(\frac{\partial U}{\partial S}\right)_V = T \qquad (4\text{-}16)$$

and

$$\left(\frac{\partial U}{\partial V}\right)_S = -P \qquad (4\text{-}17)$$

we can also show

$$\left(\frac{\partial H}{\partial S}\right)_P = T \qquad (4\text{-}23)$$

$$\left(\frac{\partial H}{\partial P}\right)_S = V \qquad (4\text{-}24)$$

$$\left(\frac{\partial G}{\partial P}\right)_T = V \qquad (4\text{-}25)$$

$$\left(\frac{\partial G}{\partial T}\right)_P = -S \qquad (4\text{-}26)$$

$$\left(\frac{\partial A}{\partial V}\right)_T = -P \qquad (4\text{-}27)$$

and

$$\left(\frac{\partial A}{\partial T}\right)_V = -S \qquad (4\text{-}28)$$

CHANGE OF INDEPENDENT VARIABLES

We have written the property relationship in the form

$$dU = T\,dS - P\,dV \qquad (4\text{-}15)$$

As we noted, in this form, the extensive variables S and V play the role of the mathematically independent variables, while the intensive variables T and P are the mathematically dependent variables. This form is inconvenient in that no device has yet been manufactured that will measure entropy, an independent variable. However, thermometers to measure temperature are readily available. Thus, internal energy as a function of temperature and volume,

$$u(T, v)$$

would be a much more convenient relationship than internal energy as a function of entropy and volume. In order to find the relationship between u, T, and v, we may write

$$du = \left(\frac{\partial u}{\partial T}\right)_V dT + \left(\frac{\partial u}{\partial v}\right)_T dv$$

The problem at this point is to evaluate the coefficients $(\partial u/\partial T)_V$ and $(\partial u/\partial v)_T$ in terms of measurable properties.

From the property relationship

$$du = T\,ds - P\,dv \qquad (4\text{-}15)$$

we find on holding volume constant and then dividing by the derivative of T,

$$\left(\frac{\partial u}{\partial T}\right)_V = T\left(\frac{\partial s}{\partial T}\right)_V$$

Since

$$\delta q_{rev} = T\,ds \qquad (3\text{-}1)$$

and

$$c_V = \left(\frac{\delta q_{rev}}{dT}\right)_V \qquad (2\text{-}16)$$

we find both that

$$T\left(\frac{\partial s}{\partial T}\right)_V = c_V \qquad (4\text{-}29)$$

and

$$\left(\frac{\partial u}{\partial T}\right)_V = c_V \qquad (4\text{-}30)$$

where c_V is called the specific heat capacity at constant volume. Many authors use Eq. (4-30) as the definition of c_V. Similarly, since

$$c_P = \left(\frac{\delta q_{rev}}{dT}\right)_P \qquad (2\text{-}15)$$

$$T\left(\frac{ds}{dT}\right)_P = c_P \qquad (4\text{-}31)$$

where c_P is called the specific heat capacity at constant pressure. From Eqs. (4-18) and (4-31) it can also be shown that

$$\left(\frac{\partial h}{\partial T}\right)_P = c_P \qquad (4\text{-}32)$$

Many authors use Eq. (4-32) as the definition of c_P.

From kinetic theory, the specific heat capacity of gases at pressures approaching zero are as follows:

	Monatomic Gas	Diatomic Gas
c_V	$3/2\,R$	$5/2\,R$
c_P	$5/2\,R$	$7/2\,R$

where R is a universal gas constant whose value is given in Appendix A-1. A monatomic gas is a gas that consists of single atoms; such as, argon and neon. A diatomic gas is a gas that consists of molecules made up of pairs of atoms; such as, hydrogen, oxygen, and nitrogen. Specific heat capacity values at constant pressure for more complex molecules in the gas at zero pressure (the ideal gas state) and solid state are given in Table A-3 of the Appendix.

Next, we must evaluate $(\partial u/\partial v)_T$. In order to do this, we once again make use of the property relationship in the form

$$du = T\,ds - P\,dv \qquad (4\text{-}15)$$

This time, we hold T constant and then divide by the derivative of v,

$$\left(\frac{\partial u}{\partial v}\right)_T = T\left(\frac{\partial s}{\partial v}\right)_T - P$$

If we could now evaluate the derivative $(\partial s/\partial v)_T$, we would have the required differential relationship between u, T, and v. Before we can do this, we must consider a special mathematical property of thermodynamic functions called the Maxwell reciprocity relations.

MAXWELL RELATIONS

Let us consider the total differential of $P(T, v)$

$$dP = \left(\frac{\partial P}{\partial T}\right)_V dT + \left(\frac{\partial P}{\partial v}\right)_T dv \qquad (4\text{-}8)$$

This can be written

$$dP = m_V\,dT + m_T\,dv$$

where we understand that m_V and m_T are different slopes. Now,

$$\left(\frac{\partial m_V}{\partial v}\right)_T = \frac{\partial^2 P}{\partial v_T \partial T_V}$$

and

$$\left(\frac{\partial m_T}{\partial T}\right)_V = \frac{\partial^2 P}{\partial T_V \, \partial v_T}$$

Now, since the difference in P depends only upon the initial and final states of the system, P is a point function and the order of differentiation makes no difference. Thus,

$$\left(\frac{\partial m_V}{\partial v}\right)_T = \left(\frac{\partial m_T}{\partial T}\right)_V \quad (4\text{-}33)$$

This is frequently called the Maxwell reciprocity relation. Since this relationship is valid for all point functions, it is used as a test to determine whether or not any particular function is a point function.

Since internal energy is known to be a point function, we find on applying the Maxwell relation to the property relationship equation,

$$dU = T \, dS - P \, dV \quad (4\text{-}15)$$

that

$$\left(\frac{\partial T}{\partial V}\right)_S = -\left(\frac{\partial P}{\partial S}\right)_V \quad (4\text{-}34)$$

From the relationship

$$dH = T \, dS + V \, dP \quad (4\text{-}18)$$

we find that since H is a point function,

$$\left(\frac{\partial T}{\partial P}\right)_S = \left(\frac{\partial V}{\partial S}\right)_P \quad (4\text{-}35)$$

From the relationship

$$dG = V \, dP - S \, dT \quad (4\text{-}20)$$

we find that since G is a point function,

$$\left(\frac{\partial V}{\partial T}\right)_P = -\left(\frac{\partial S}{\partial P}\right)_T \quad (4\text{-}36)$$

Further, from the relationship

$$dA = -S \, dT - P \, dV \quad (4\text{-}22)$$

we find that since A is a point function,

$$\left(\frac{\partial P}{\partial T}\right)_V = \left(\frac{\partial S}{\partial V}\right)_T \tag{4-37}$$

The Maxwell relations can also be written on a unit mass basis in terms of the variables s and v.

The Maxwell relations of thermodynamic functions are not limited to those involving P, T, V, and S. If we consider the general property relationship,

$$dU = T\,dS - P\,dV + F\,dL + \sigma\,dA + \epsilon\,dq + \Sigma \mu_i\,dn_i \tag{4-14}$$

we find that at constant S, V, q, and n_i,

$$dU_{S,V,q,n_i} = F\,dL_{S,V,q,n_i} + \sigma\,dA_{S,V,q,n_i}$$

Upon applying the Maxwell relationship to this function, we find

$$\left(\frac{\partial F}{\partial A}\right)_{L,S,V,q,n_i} = \left(\frac{\partial \sigma}{\partial L}\right)_{A,S,V,q,n_i} \tag{4-38}$$

In general, we do not use all the independent variables in Eq. (4-14) as subscripts, as it is understood that all variables except those involved in the problem are held constant.

With these results, we are now in a position to return to the problem of evaluating the function

$$\left(\frac{\partial u}{\partial v}\right)_T = T\left(\frac{\partial s}{\partial v}\right)_T - P$$

Since

$$\left(\frac{\partial s}{\partial v}\right)_T = \left(\frac{\partial P}{\partial T}\right)_V \tag{4-37}$$

we find that

$$\left(\frac{\partial u}{\partial v}\right)_T = T\left(\frac{\partial P}{\partial T}\right)_V - P$$

Combining this result with the previously determined result, that

$$\left(\frac{\partial u}{\partial T}\right)_V = c_V \tag{4-30}$$

we find the desired differential relationship between u, v, and T is

$$du = \left[T\left(\frac{\partial P}{\partial T}\right)_V - P\right]dv + c_V\,dT \tag{4-39}$$

As we will see in the following example problems, similar mathematical manipulations can be applied to other thermodynamic functions.

EXAMPLE PROBLEM 4-1. Find the differential relationship between h and v at constant T in terms of measurable properties, which include P, T, v, c_P and c_V.

Because temperature will not vary, we find it most convenient to write

$$dh = \left(\frac{\partial h}{\partial v}\right)_T dv + \left(\frac{\partial h}{\partial T}\right)_V dT$$

since this equation reduces to

$$dh_T = \left(\frac{\partial h}{\partial v}\right)_T dv_T$$

at constant temperature. In order to evaluate $(\partial h/\partial v)_T$ in terms of measurable properties, we recall that

$$dh = T\,ds + v\,dP \tag{4-18}$$

Holding T constant and dividing by the differential of v,

$$\left(\frac{\partial h}{\partial v}\right)_T = T\left(\frac{\partial s}{\partial v}\right)_T + v\left(\frac{\partial P}{\partial v}\right)_T$$

Then, from the Maxwell relationship,

$$\left(\frac{\partial s}{\partial v}\right)_T = \left(\frac{\partial P}{\partial T}\right)_V \tag{4-37}$$

we find

$$\left(\frac{\partial h}{\partial v}\right)_T = T\left(\frac{\partial P}{\partial T}\right)_V + v\left(\frac{\partial P}{\partial v}\right)_T$$

or

$$dh_T = \left[T\left(\frac{\partial P}{\partial T}\right)_V + v\left(\frac{\partial P}{\partial v}\right)_T\right]dv_T \tag{4-40}$$

EXAMPLE PROBLEM 4-2. Values of c_P presented in Table A-3 in the Appendix are valid for gases at zero pressure. It is frequently necessary to know the value of c_P for gases at higher pressures. What is the change in c_P with pressure?

Differentiating the relationship

$$c_P = T\left(\frac{\partial s}{\partial T}\right)_P \tag{4-31}$$

with respect to P at constant T, we find

$$\left(\frac{\partial c_P}{\partial P}\right)_T = \frac{\partial}{\partial P_T}\left[T\left(\frac{\partial s}{\partial T}\right)_P\right] = T\frac{\partial^2 s}{\partial P_T \partial T_P}$$

But, from the Maxwell relation,

$$\left(\frac{\partial s}{\partial P}\right)_T = -\left(\frac{\partial v}{\partial T}\right)_P \tag{4-36}$$

So

$$\frac{\partial^2 s}{\partial P_T \partial T_P} = \frac{\partial}{\partial T_P}\left[-\left(\frac{\partial v}{\partial T}\right)_P\right] = -\left(\frac{\partial^2 v}{\partial T^2}\right)_P$$

or

$$\left(\frac{\partial c_P}{\partial P}\right)_T = -T\left(\frac{\partial^2 v}{\partial T^2}\right)_P \tag{4-41}$$

By similar procedures, we can show

$$\left(\frac{\partial c_V}{\partial v}\right)_T = T\left(\frac{\partial^2 P}{\partial T^2}\right)_V \tag{4-42}$$

This last result can also be obtained by "taking the Maxwell" of the equation

$$du = \left[T\left(\frac{\partial P}{\partial T}\right)_V - P\right]dv + c_V\,dT \tag{4-39}$$

Taking the "Maxwell," we find

$$\left(\frac{\partial c_V}{\partial v}\right)_T = \frac{\partial}{\partial T_V}\left[T\left(\frac{\partial P}{\partial T}\right)_V - P\right]$$

or

$$\left(\frac{\partial c_V}{\partial v}\right)_T = T\left(\frac{\partial^2 P}{\partial T^2}\right)_P + \left(\frac{\partial P}{\partial T}\right)_V - \left(\frac{\partial P}{\partial T}\right)_V$$

so that

$$\left(\frac{\partial c_V}{\partial v}\right)_T = T\left(\frac{\partial^2 P}{\partial T^2}\right)_V \tag{4-42}$$

EXAMPLE PROBLEM 4-3. Tables of specific heat capacities are generally presented in the form of c_P's. These values can be used to predict values of c_V if we first develop a relationship for $c_P - c_V$. Develop this relationship in terms of P, T, and v.

Subtracting Eq. (4-29) from Eq. (4-31), we see that

$$c_P - c_V = T\left(\frac{\partial s}{\partial T}\right)_P - T\left(\frac{\partial s}{\partial T}\right)_V$$

In order to evaluate the terms on the right hand side of this equation, we assume

$$s = s(T, P)$$

from which we find

$$ds = \left(\frac{\partial s}{\partial T}\right)_P dT + \left(\frac{\partial s}{\partial P}\right)_T dP$$

Further, if we assume

$$s = s(T, v)$$

we find

$$ds = \left(\frac{\partial s}{\partial T}\right)_V dT + \left(\frac{\partial s}{\partial v}\right)_T dv$$

Combining these equations, we find

$$ds = \left(\frac{\partial s}{\partial T}\right)_P dT + \left(\frac{\partial s}{\partial P}\right)_T dP = \left(\frac{\partial s}{\partial T}\right)_V dT + \left(\frac{\partial s}{\partial v}\right)_T dv$$

If we now hold P constant and divide through by dT, we find that

$$\left(\frac{\partial s}{\partial T}\right)_P = \left(\frac{\partial s}{\partial T}\right)_V + \left(\frac{\partial s}{\partial v}\right)_T \left(\frac{\partial v}{\partial T}\right)_P$$

or

$$\left(\frac{\partial s}{\partial T}\right)_P - \left(\frac{\partial s}{\partial T}\right)_V = \left(\frac{\partial s}{\partial v}\right)_T \left(\frac{\partial v}{\partial T}\right)_P$$

Next, we multiply by T to obtain

$$T\left(\frac{\partial s}{\partial T}\right)_P - T\left(\frac{\partial s}{\partial T}\right)_V = T\left(\frac{\partial s}{\partial v}\right)_T \left(\frac{\partial v}{\partial T}\right)_P$$

However,

$$\left(\frac{\partial s}{\partial v}\right)_T = \left(\frac{\partial P}{\partial T}\right)_V \tag{4-37}$$

Thus,

$$c_P - c_V = T\left(\frac{\partial P}{\partial T}\right)_V \left(\frac{\partial v}{\partial T}\right)_P \tag{4-43}$$

Further, since

$$\left(\frac{\partial v}{\partial T}\right)_P = -\frac{\left(\frac{\partial P}{\partial T}\right)_V}{\left(\frac{\partial P}{\partial v}\right)_T} \tag{4-12}$$

we may write

$$c_P - c_V = -\frac{T\left(\frac{\partial P}{\partial T}\right)_V^2}{\left(\frac{\partial P}{\partial v}\right)_T} \tag{4-44}$$

Similarly, we may also show that

$$c_P - c_V = -\frac{T\left(\frac{\partial v}{\partial T}\right)_P^2}{\left(\frac{\partial v}{\partial P}\right)_T} \tag{4-45}$$

EXAMPLE PROBLEM 4-4. Find the differential relationship between s and P at constant v, in terms of measurable properties.

Because specific volume will not vary, we find it most convenient to write

$$ds = \left(\frac{\partial s}{\partial P}\right)_V dP + \left(\frac{\partial s}{\partial v}\right)_P dv$$

since this equation reduces to

$$ds_V = \left(\frac{\partial s}{\partial P}\right)_V dP_V$$

at constant volume. Now, we must evaluate $(\partial s/\partial P)_V$. This can be done by first applying the reciprocity relationship, Eq. (4-10), to obtain

$$\left(\frac{\partial s}{\partial P}\right)_V = \frac{1}{\left(\frac{\partial P}{\partial s}\right)_V}$$

Then, from the Maxwell relationship

$$\left(\frac{\partial P}{\partial s}\right)_V = -\left(\frac{\partial T}{\partial v}\right)_S \tag{4-34}$$

we find

$$\left(\frac{\partial s}{\partial P}\right)_V = \frac{1}{-\left(\frac{\partial T}{\partial v}\right)_S} = -\left(\frac{\partial v}{\partial T}\right)_S$$

Next, we can eliminate the restriction of constant s on this last partial derivative by writing Eq. (4-11) in the form

$$\left(\frac{\partial v}{\partial T}\right)_S \left(\frac{\partial T}{\partial s}\right)_V \left(\frac{\partial s}{\partial v}\right)_T = -1$$

or

$$-\left(\frac{\partial v}{\partial T}\right)_S = \frac{\left(\frac{\partial s}{\partial T}\right)_V}{\left(\frac{\partial s}{\partial v}\right)_T}$$

Now, from Eq. (4-29),

$$\left(\frac{\partial s}{\partial T}\right)_V = \frac{c_V}{T}$$

and from the Maxwell relationship

$$\left(\frac{\partial s}{\partial v}\right)_T = \left(\frac{\partial P}{\partial T}\right)_V \qquad (4\text{-}37)$$

we find

$$-\left(\frac{\partial v}{\partial T}\right)_S = \frac{\left(\frac{c_V}{T}\right)}{\left(\frac{\partial P}{\partial T}\right)_V}$$

Combining these results, we see that

$$\left(\frac{\partial s}{\partial P}\right)_V = \frac{c_V}{T}\left(\frac{\partial T}{\partial P}\right)_V$$

This last result could have been obtained more readily by making use of the chain rule, Eq. (4-13), as follows:

$$\left(\frac{\partial s}{\partial P}\right)_V = \left(\frac{\partial s}{\partial T}\right)_V \left(\frac{\partial T}{\partial P}\right)_V$$

$$= \frac{c_V}{T}\left(\frac{\partial T}{\partial P}\right)_V$$

Finally, we find

$$ds_V = \left(\frac{\partial s}{\partial P}\right)_V dP_V$$

$$= \frac{c_V}{T}\left(\frac{\partial T}{\partial P}\right)_V dP_V$$

$$ds_V = \frac{c_V}{T} dT \qquad (4\text{-}46)$$

The number of manipulations possible using thermodynamic functions is almost limitless. The problems shown here are merely examples. The purpose of these manipulations is to provide us with a means of estimating changes in the derived thermodynamic properties from a knowledge of the changes in measurable thermodynamic properties.

If we know the relationship between P, v and T as well as the relationships between specific heat capacities and T, we can use this information to integrate differential equations, such as those developed in this chapter, between any two states to determine numerical values of differences in thermodynamic functions. This is the procedure which has been used to prepare tables of thermodynamic functions. In the next chapter, we will begin to examine relationships between P, v and T.

PROBLEMS

4-1. An equation which has been developed to represent the phase behavior of a gas over a limited range is

$$\frac{P\text{v}}{RT} = 1 + \frac{B}{\text{v}}$$

where R and B are constants. Using this relationship, evaluate

$$\left(\frac{\partial P}{\partial \text{v}}\right)_T, \left(\frac{\partial P}{\partial T}\right)_V, \left(\frac{\partial \text{v}}{\partial T}\right)_P \text{ and } \frac{\partial^2 P}{\partial \text{v}_T \partial T_V}$$

4-2. Show that

(a) $\left(\dfrac{\partial h}{\partial s}\right)_P = T$

(b) $\left(\dfrac{\partial g}{\partial T}\right)_P = \left(\dfrac{\partial a}{\partial T}\right)_V = -s$

(c) $\left(\dfrac{\partial h}{\partial P}\right)_S = \left(\dfrac{\partial g}{\partial P}\right)_T = \text{v}$

4-3. Considering u as a function of any two of the variables P, v, and T, prove that

$$\left(\frac{\partial u}{\partial T}\right)_P \left(\frac{\partial T}{\partial P}\right)_V = -\left(\frac{\partial u}{\partial \text{v}}\right)_P \left(\frac{\partial \text{v}}{\partial P}\right)_T$$

4-4. The length L of a wire is a function of the temperature T and the tension τ on the wire. The coefficient of linear expansion α_τ is defined by

$$\alpha_\tau = \frac{1}{L}\left(\frac{\partial L}{\partial T}\right)_\tau$$

and is essentially constant over a small range of temperatures. Likewise, the isothermal modulus of elasticity E is defined by the relationship

$$E = \frac{L}{A}\left(\frac{\partial \tau}{\partial L}\right)_T$$

where A, the cross-sectional area of the wire, is essentially constant over a small temperature range. Prove that

$$\left(\frac{\partial \tau}{\partial T}\right)_L = -\alpha_\tau A E$$

4-5. Derive expressions for $(\partial T/\partial v)_U$ and $(\partial h/\partial s)_V$ that do not contain the properties h, u, or s.

4-6. Derive an expression for $(\partial v/\partial T)_S$ that does not contain the properties h, u, or s.

4-7. Find the differential relationship for $u(P, T)$ such that the final expression does not contain the properties s or h.

4-8. Find the differential relationship for $h(P, v)$ such that the final equation does not contain the properties u or s.

4-9. The Joule-Thompson coefficient μ_J is defined by the relationship

$$\mu_J = \left(\frac{\partial T}{\partial P}\right)_H$$

and is used to indicate whether the temperature of a gas being throttled across an expansion valve will rise or fall. Show that

$$\mu_J = \frac{1}{c_P}\left[T\left(\frac{\partial v}{\partial T}\right)_P - v\right]$$

4-10. A new thermodynamic quantity y is defined by the relationship

$$dy = \left(\frac{\partial u}{\partial T}\right)_V dT + \left[\left(\frac{\partial u}{\partial v}\right)_T + P\right]dv$$

(a) Is y a point function? Justify your answer mathematically.

(b) Is dy/T a point function? Justify your answer mathematically.

4-11. The phase behavior of a fluid can be represented by the relationship

$$dP = \frac{R}{v-b} dT - \left[\frac{RT}{(v-b)^2} - \frac{2a}{v^3}\right]dv$$

where R, a, and b are constants. Is P a point function? Justify your answer mathematically.

5
The Ideal Gas

OBJECTIVES

After studying this chapter, the student should be able to apply the ideal gas law, Eq. (5-1), to a single component system in solving new problems involving the application of the principle of conservation of mass, the principle of conservation of energy as well as the second law of thermodynamics.

EQUATION OF STATE

Up to this point, we have used thermodynamic data primarily in the form of tables to solve thermodynamic problems. Another convenient method of presenting experimental information is by means of an equation of state. An equation of state is an analytical representation of observed PVT behavior. For example: in 1662 Robert Boyle reported an observation that at constant temperature, the volume of a given sample of gas varies inversely as the pressure

$$v \propto \frac{1}{P}$$

or

$$Pv = k_{\text{Boyle}}$$

This is Boyle's law, and it represents experimentally determined isothermal PVT behavior of gases at very low density.

Further examination of experimental *PVT* behavior leads to other similar relationships. Joseph Gay-Lussac, working from 1802-1808, reported that if the pressure does not change, the volume of a sample of gas varies directly with the absolute temperature

$$v = k_{\text{Gay-Lussac}} T$$

This is Gay-Lussac's law and is sometimes called Charles' law.

Later, in 1811, Amadeo Avogadro stated that equal volumes of all gases, at the same temperature and pressure, contain the same number of molecules. Since one gram mole contains 6.02252×10^{23} molecules, Avogadro's principle also says that equal volumes of all gases, at the same temperature and pressure, contain the same number of moles.

Since the molar specific volume of a gas is a function of system pressure and temperature, we may write

$$dv = \left(\frac{\partial v}{\partial T}\right)_P dT + \left(\frac{\partial v}{\partial P}\right)_T dP \tag{4-9}$$

From Boyle's law, we find

$$\left(\frac{\partial v}{\partial P}\right)_T = -\frac{k_{\text{Boyle}}}{P^2}$$

Also, from Gay-Lussac's law, we find

$$\left(\frac{\partial v}{\partial T}\right)_P = k_{\text{Gay-Lussac}}$$

Upon substituting the values of the partial derivatives into the total derivative of v, we obtain the equation

$$dv = k_{\text{Gay-Lussac}} dT - \frac{k_{\text{Boyle}}}{P^2} dP$$

But

$$k_{\text{Boyle}} = Pv$$

and

$$k_{\text{Gay-Lussac}} = \frac{v}{T}$$

So,

$$dv = \frac{v}{T} dT - \frac{v}{P} dP$$

or,

$$\frac{dv}{v} = \frac{dT}{T} - \frac{dP}{P}$$

Upon integrating this equation, we find

$$Pv = RT \qquad (5\text{-}1)$$

or

$$PV = nRT \qquad (5\text{-}2)$$

The constant of integration R is called the universal gas constant. If specific volumes are based on a mole basis, Avogadro's principle indicates that the universal gas constant R is the same for all gases. Another similar constant used by many engineers who prefer to work on a mass basis rather than a mole basis is called the specific gas constant, denoted in this text by the symbol R/MW, where MW is the molecular weight of the gas. Values of the universal gas constant R in different sets of units are presented in Table A-1 of the Appendix while values of the molecular weights of compounds can be obtained from readily available periodic tables found in most introductory chemistry textbooks.

Equation (5-1) not only represents experimental data for real gases at low density, but also represents the behavior predicted by molecular theory of a gas consisting of point particles which exhibit neither attractive nor repulsive forces. For this reason, Eq. (5-1) is called the equation of state for an ideal gas.

THERMODYNAMIC PROPERTIES OF IDEAL GASES

We are now in a position to evaluate the changes in thermodynamic functions of an ideal gas with changes in pressure, temperature, and volume.

Reconsider the change in u when the system volume is changed from v_1 to v_2 and the system temperature is changed from T_1 to T_2. We have shown that the differential relationship between u, v, and T is

$$du = \left[T\left(\frac{\partial P}{\partial T}\right)_V - P\right] dv + c_V\, dT \tag{4-39}$$

Now, we can evaluate the value of $u_2 - u_1$ for possible inclusion in the energy balance equation by integration this equation.

In ordinary Riemann integration, we may define the value of the integral

$$\int_{v_1}^{v_2} P\, dv$$

as the area under the curve $P = P(v)$ that is bounded by v_1 and v_2. However, in thermodynamics, we frequently must integrate differential expressions in terms of two independent variables instead of a single independent variable. For example, in order to evaluate Δu using Eq. (4-39), we must integrate

$$du = \Phi(v, T)\, dv + c_V(v, T)\, dT$$

where

$$\Phi(v, T) = \left(\frac{\partial u}{\partial v}\right)_T = T\left(\frac{\partial P}{\partial T}\right)_V - P$$

and

$$c_V(v, T) = \left(\frac{\partial u}{\partial T}\right)_V \tag{4-30}$$

in terms of both v and T.

In this case, suppose the path between the initial state $u(v_1, T_1)$ and the final state $u(v_2, T_2)$ may be represented by a line on the surface $u(v, T)$ as shown in Fig. 5-1. In order to evaluate Δu, a procedure that may be used is first to divide the path between the initial and final states into a number of incremental segments ΔL_i. The change in the independent variables v and T along one of these incremental segments shown in Fig. 5-1 is $(\Delta v_i)_T$ and $(\Delta T_i)_V$.

Now we can evaluate the change in u associated with an incremental step in the v direction at constant T,

$$(\Delta u_i)_T = [\Phi(v, T)\Delta v_i]_T$$

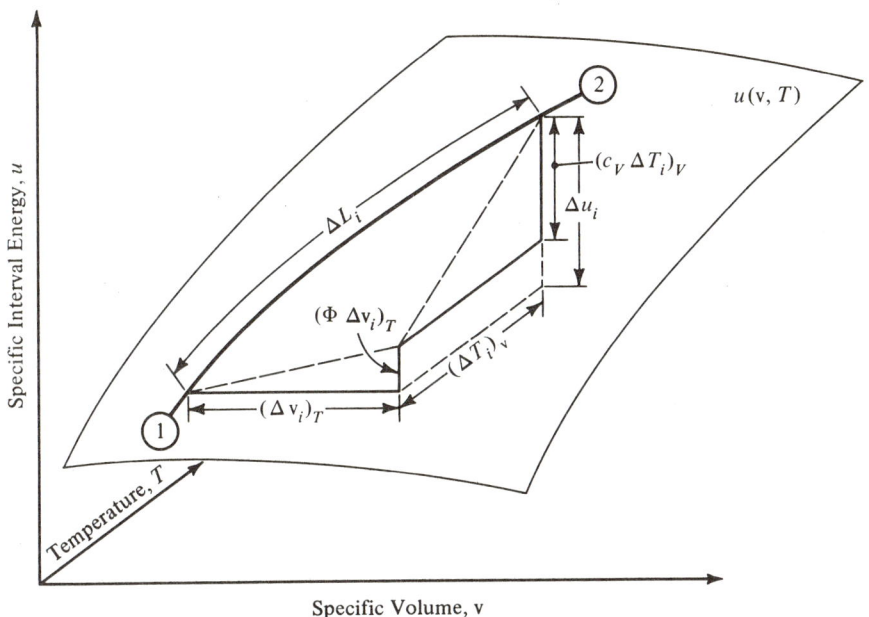

Fig. 5-1. *Integration along a path.*

by substituting a value of v and a value of T along the step $(\Delta v_i)_T$ into the relationship $\Phi(v, T)$. We can also evaluate the change in u associated with an incremental step in the T direction at constant v,

$$(\Delta u_i)_V = [c_V(v, T)\Delta T_i]_V$$

by substituting a value of v and a value of T along the step $(\Delta T_i)_V$ into the relationship $c_V(v, T)$.

Then, the incremental change in u along the segment ΔL_i is

$$\Delta u_i = [\Phi(v, T)\Delta v_i]_T + [c_V(v, T)\Delta T_i]_V$$

and the total change in u, between the initial and final states is the sum of the incremental steps,

$$\Delta u = \Sigma[\Phi(v, T)\Delta v_i]_T + \Sigma[c_V(v, T)\Delta T_i]_V$$

If we let the size of the incremental steps approach zero, we find

$$\lim_{\Delta V \to 0} \Sigma[\Phi(v, T)\Delta v_i]_T = \int_C \Phi(v, T)\, dv_T$$

and

$$\lim_{\Delta T_i \to 0} [\Sigma c_V(v, T)\Delta T_i]_V = \int_C c_V(v, T)\, dT_V$$

The subscript C on an integral sign indicates the integral is a line integral along a curve. In this particular case, the curves associated with the line integrals are in the v, T plane. Further, we find that the total change in u is given by the sum of the two line integrals

$$\Delta u = \int_C \Phi(v, T)\, dV + \int_C c_V(v, T)\, dT$$

Fortunately, we find that changes in the thermodynamic functions are independent of the path. This is why they are called point functions. Thus, in order to integrate du, we might first evaluate $\int \Phi(v, T)\, dv$ in one step at constant T_1, from v_1 to v_2. Since T_1 is constant along this path, $\Phi(v, T)$ depends only on v and we may evaluate the line integral using ordinary Riemann integration. Once we have evaluated $\int \Phi(v, T)\, dv$, we may then evaluate $\int c_V(v, T)\, dT$ in another single step at constant v_2 from T_1 to T_2. The sum of these two integrals in Δu. If it is more convenient, we could select some other, similar path.

Before we integrate dv, however, we need to know both Φ and c_V as functions of v and T. Since $PV = nRT$ for an ideal gas,

$$T\left(\frac{\partial P}{\partial T}\right)_V = \frac{nRT}{V} = P$$

so that

$$\Phi(v, T) = T\left(\frac{\partial P}{\partial T}\right)_V - P = 0$$

Substituting this result into Eq. (4-39) and integrating the first integral from v_1, T_1 to v_2, T_1 and the second integral from v_2, T_1 to v_2, T_2, we find

$$\Delta u_1 = c_V(T_2 - T_1) \tag{5-3}$$

where we have assumed that c_V is independent of temperature. This is not generally true, but it does simplify the math.

Now the c_V used in Eq. (5-3) was evaluated at v_2. In Example Prob. 4-2, we showed that

$$\left(\frac{\partial c_V}{\partial v}\right)_T = T\left(\frac{\partial^2 P}{\partial T^2}\right)_V \tag{4-42}$$

For an ideal gas

$$\left(\frac{\partial^2 P}{\partial T^2}\right)_V = 0$$

so

$$\left(\frac{\partial c_V}{\partial v}\right)_T = 0 \tag{5-4}$$

This relationship says that c_V of an ideal gas is independent of volume. Thus, the value of c_V obtained at v_1 would be the same as the value of c_V obtained at v_2. Examination of Eqs. (5-3) and (5-4) leads us to the important conclusion that the internal energy of an ideal gas is independent of the volume occupied by the gas. In a similar manner, it can be shown that the internal energy of an ideal gas is independent of pressure.

Next, let us re-examine the effect of pressure on the enthalpy of an ideal gas. The differential relationship between h, P, and T can be shown to be

$$dh = \left[v - T\left(\frac{\partial v}{\partial T}\right)_P\right]dP + c_P\, dT$$

But, for an ideal gas,

$$T\left(\frac{\partial v}{\partial T}\right)_P = \frac{RT}{P} = v$$

Thus

$$v - T\left(\frac{\partial v}{\partial T}\right)_P = 0$$

combining these results, and integrating, we find

$$h_2 - h_1 = c_P(T_2 - T_1) \tag{5-5}$$

where we have assumed that c_P is independent of temperature.

Now in Example Prob. 4-2, we also showed that

$$\left(\frac{\partial c_P}{\partial P}\right)_T = -T\left(\frac{\partial^2 v}{\partial T^2}\right)_P \tag{4-41}$$

For an ideal gas

$$\left(\frac{\partial^2 v}{\partial T^2}\right)_P = 0$$

so

$$\left(\frac{\partial c_P}{\partial P}\right)_T = 0 \tag{5-6}$$

This relationship says that c_P of an ideal gas is independent of pressure. Further, examination of Eqs. (5-5) and (5-6) leads us to the important conclusion that the enthalpy of an ideal gas is independent of the pressure of the system. In a similar manner, it can be shown that the enthalpy of an ideal gas is independent of volume.

Next, we might consider the effect of changes in volume and temperature on entropy. In differential form, the relationship between s, v, and T can be shown to be

$$ds = \left(\frac{\partial P}{\partial T}\right)_V dv + \frac{c_V}{T} dT$$

For an ideal gas, note that

$$\left(\frac{\partial P}{\partial T}\right)_V = \frac{R}{v}$$

Substituting this relationship into the equation for ds and integrating,

$$\Delta s = R \ln \frac{v_2}{v_1} + c_V \ln \frac{T_2}{T_1} \tag{5-7}$$

where we have again assumed that c_V is independent of temperature. Here we see that although the internal energy and enthalpy of an ideal gas are independent of volume, the entropy of an ideal gas does depend upon the volume of the gas. By a similar procedure, it can be shown that the effect of pressure on the entropy of an ideal gas is given by the relationship

$$\Delta s = -R \ln \frac{P_2}{P_1} + c_P \ln \frac{T_2}{T_1} \tag{5-8}$$

Finally, we might examine the effects of pressure and volume on c_P and c_V for an ideal gas. In Example Prob. 4-3, we showed that

$$c_P - c_V = T\left(\frac{\partial P}{\partial T}\right)_V \left(\frac{\partial v}{\partial T}\right)_P \tag{4-43}$$

For an ideal gas,

$$\left(\frac{\partial P}{\partial T}\right)_V = \frac{R}{v}$$

and

$$\left(\frac{\partial v}{\partial T}\right)_P = \frac{R}{P}$$

so that

$$c_P - c_V = (T)\left(\frac{R}{v}\right)\left(\frac{R}{P}\right) = R \tag{5-9}$$

ISENTROPIC PROCESSES

A case of particular interest is the isentropic expansion of a gas. We have shown that for a gas which follows the ideal gas equation of state,

$$\Delta s = R \ln \frac{v_2}{v_1} + c_V \ln \frac{T_2}{T_1} \tag{5-7}$$

In an isentropic process, $\Delta s = 0$, thus

$$\left(\frac{T_2}{T_1}\right)^{c_V/R} = \frac{v_1}{v_2}$$

Applying Eq. (5-9) to the exponential,

$$\frac{R}{c_V} = \frac{c_P - c_V}{c_V} = \gamma - 1$$

where γ is the ratio of the specific heat capacities,

$$\gamma = c_P/c_V \tag{5-10}$$

we find

$$\left(\frac{T_2}{T_1}\right)^{1/(\gamma-1)} = \frac{v_1}{v_2}$$

In a similar manner, it can be shown that

$$P_1 v_1^\gamma = P_2 v_2^\gamma = \text{constant} \tag{5-11}$$

so that for an isentropic expansion or compression of an ideal gas,

$$\left(\frac{T_2}{T_1}\right)^{\gamma/(\gamma-1)} = \left(\frac{v_1}{v_2}\right)^\gamma = \frac{P_2}{P_1} \tag{5-12}$$

While real gases obey the ideal gas law only at very low density and while few, if any real processes are isentropic, the concept of an isentropic expansion or compression of an ideal gas does find application in solving practical problems.

EXAMPLE PROBLEM 5-1. Air at 900 mm of Hg pressure and 75 °F is contained within a closed, insulated vessel. A valve at the top of the vessel is cracked open and the pressure of the air within the vessel drops slowly. What is the temperature of the air when its pressure has dropped to 760 mm of Hg?

In this case, we will find it most convenient to select as our system the air remaining in the vessel after the pressure in the vessel has dropped to 760 mm. Then, basing our calculation on the period of time required for the process to take place, the reduced mass balance equation is

$$0 = m_f - m_i$$

since there is no flow of air into or out of the system we have selected. If we now assume that there is no instantaneous temperature difference between the gas remaining in the vessel and the gas being displaced from the insulated vessel,

$$Q = 0$$

and the reduced energy balance equation becomes

$$-W = m_f u_f - m_i u_i$$

where we have neglected the effect of a change in elevation of the system's center of mass. The work term represents the energy expended by our system as it expands.

Finally, the process can be assumed to be reversible, since we could change the direction of flow at any instant by forcing air at a pressure slightly greater than the system pressure back into the vessel. Thus, the reduced entropy balance equation becomes

$$0 = m_f s_f - m_i s_i$$

Now, under the pressure and temperature conditions of this problem, air may be assumed to be a diatomic ideal gas. Then, as the process is isentropic, we can use the relationship

$$\frac{T_f}{T_i} = \left(\frac{P_f}{P_i}\right)^{(\gamma-1)/\gamma} \tag{5-12}$$

where

$$\gamma = \frac{c_P}{c_V} \tag{5-10}$$

$$= \frac{7/2\,R}{5/2\,R} = 1.4$$

Substituting values into Eq. (5-12), we find

$$T_f = 535 \left(\frac{760}{900}\right)^{0.4/1.4}$$

$$= 509.7\,°R = 49.7\,°F$$

EXAMPLE PROBLEM 5-2. Derive an expression for the atmospheric pressure as a function of altitude assuming air to be an ideal gas and (1) the atmosphere to be isothermal or (2) the atmosphere to be adiabatic (i.e., if a parcel of air were to slowly rise, its expansion would be adiabatic).

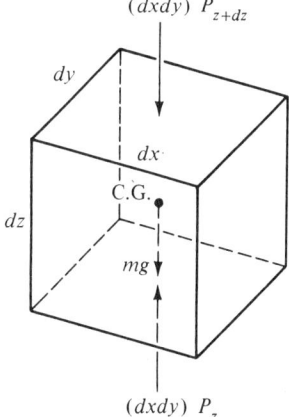

Fig. 5-2. *Small elementary volume of air.*

The pressure change in the atmosphere is due to the force of gravity acting on the air mass. Therefore, to solve this problem, we will utilize both thermodynamics and statics.

Consider a small elemental volume of air in a stationary atmosphere, as shown in Fig. 5-2. The summation of forces acting on that volume must be zero. The forces in the vertical direction are the gravitational force mg acting at the center of gravity and the pressure forces acting on the lower and upper faces. Summing these forces and equating them to zero gives

$$dxdy\,(P_z - P_{z+dz}) - mg = 0$$

Noting that

$$mg = \rho g\, dx\, dy\, dz$$

and

$$P_z - P_{z+dz} = -\frac{dP}{dz} dz$$

for small dz, then

$$\frac{dP}{dz} + \rho g = 0$$

Since we are assuming air to be an ideal gas, $Pv = RT$, or $P = \rho RT/MW$, since $\rho = MW/v$. Thus

$$\frac{dP}{P} = -\frac{MW\, g\, dz}{RT} \tag{5-13}$$

In the case of an isothermal atmosphere, the above equation can be directly integrated

$$\int_{P_0}^{P} \frac{dP}{P} = -\frac{MWg}{RT_0} \int_0^z dz$$

to give

$$P_z = P_0 \exp\left(-\frac{MWgz}{RT_0}\right) \tag{5-14}$$

In the case of an adiabatic atmosphere the temperature will also vary with altitude and we can not integrate Eq. (5-13) until we know how temperature varies with pressure. This relationship is supplied by the isentropic expression

$$T = T_0 \left(\frac{P}{P_0}\right)^{(\gamma-1)/\gamma} \tag{5-12}$$

Substituting for T in Eq. (5-13) and integrating

$$T_0 P_0^{(1-\gamma)/\gamma} \int_{P_0}^{P} P^{-1/\gamma}\, dP = -\frac{MWg}{R} \int_0^z dz$$

gives

$$\left(\frac{\gamma}{\gamma-1}\right) T_0 \left[\left(\frac{P}{P_0}\right)^{(\gamma-1)/\gamma} - 1\right] = -\frac{MWgz}{R}$$

or

$$P = P_0 \left[1 - \frac{(\gamma-1)MWgz}{\gamma R T_0}\right]^{\gamma/(\gamma-1)} \tag{5-15}$$

Figure 5-3 is a plot of these results where the standard sea-level values, $P_0 = 14.7$ lb$_f$/in.^{2} and $T_0 = 519\,°R$ (59 °F), are used. A comparison of these curves with the U. S. Standard Atmosphere shows that the actual atmospheric value lies between the values predicted by these two models.

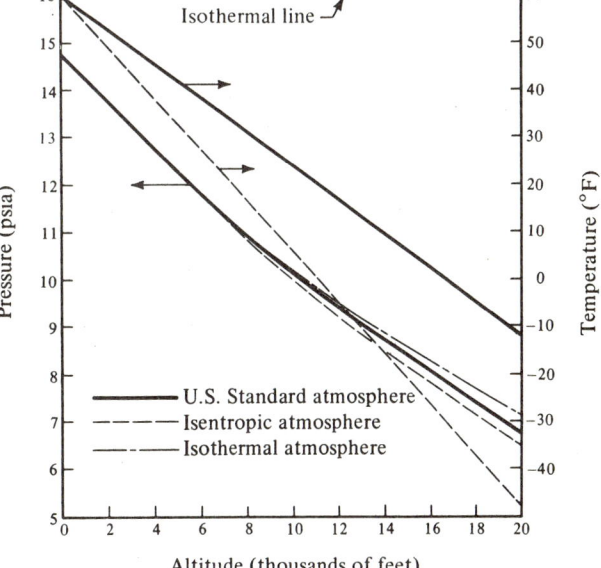

Fig. 5-3. *Atmospheric temperature and pressure vs altitude.*

ENTROPY AND PROBABILITY

It is sometimes stated that the entropy of a system is related to the probability that the system can exist in the given state. It may be of interest to examine the relationship between entropy and probability.

Consider two vessels of volumes V_1 and V_2, shown in Fig. 5-4, connected by

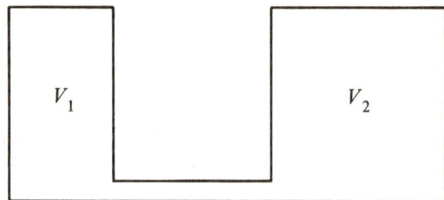

Fig. 5-4. *Ideal gas container.*

a tube of negligible volume. Let us put into this system N_A molecules of an ideal gas—N_A being Avogadro's number, 6.02252×10^{23}. Now, the most probable distribution of molecules between the two containers will correspond to an equal concentration of molecules in both containers. (Minor fluctuations from a state of uniform concentration may occur, but they will be insignificant and of no consequence in the following argument.)

Let us now ask what is the probability of finding all N_A molecules in V_1, keeping in mind the fact that the two containers are connected and are both accessible.

Since the time a molecule spends in one particular container is proportional to the relative volume of that container, the probability of finding any one specific molecule in volume V_1 is

$$P_1 = \frac{V_1}{V_1 + V_2}$$

which is simply the fraction of the total volume represented by the container of interest. Then, the probability of finding two specific molecules in volume V_1 is

$$P_2 = \left(\frac{V_1}{V_1 + V_2}\right)^2$$

and the probability of finding all N_A molecules in V_1 is

$$\frac{1}{P_{N_A}} = \left(\frac{V_1}{V_1 + V_2}\right)^{N_A}$$

Now, all N_A molecules are located somewhere in $V_1 + V_2$. If we consider the ratio of the probability of finding all the molecules somewhere in the system,

$$P = 1 \text{ (certainty)}$$

to that of finding all the molecules in V_1, we find

$$\frac{1}{P_{N_A}} = \left(\frac{V_1 + V_2}{V_1}\right)^{N_A}$$

Further, if we take the logarithm of this ratio and multiply by Boltzman's constant k_B where

$$k_B = \frac{R}{N_A}$$

$$= 3.2971 \times 10^{-24} \text{ cal/molecule-K}$$

we find

$$k_B \ln \frac{1}{P_{N_A}} = k_B \ln \left(\frac{V_1 + V_2}{V_1}\right)^{N_A} = N_A k_B \ln \frac{V_1 + V_2}{V_1}$$

Since $R = N_A k_B$,

$$k_B \ln \frac{1}{P_{N_A}} = R \ln\left(\frac{V_1 + V_2}{V_1}\right)$$

But, from Eq. (5-7), we see that for the isothermal expansion of an ideal gas from V_1 to $V_1 + V_2$,

$$\Delta S = R \ln\left(\frac{V_1 + V_2}{V_1}\right)$$

Thus, for an ideal gas

$$\Delta S = -k_B \ln P_{N_A}$$

Ludwig Boltzmann proposed the relationship

$$\Delta S = k_B \ln W \tag{5-16}$$

where W is the thermodynamic probability, defined as the number of different ways in which a thermodynamic state can be realized. Since a system having a high degree of order is an improbable state, it has a low entropy. Conversely, a state of high entropy or high probability will appear disorganized or random.

PROBLEMS

5-1. Assuming that $u(P, T)$ show that for an ideal gas Δu is independent of pressure.

5-2. Assuming that $h(v, T)$, show that for an ideal gas Δh is independent of volume.

5-3. The cylinder shown below is initially evacuated. It is closed at the top by a heavy frictionless piston that initially rests on the vessel shoulders as shown. Air at 300 K is fed to the vessel until the volume doubles. If no heat is transferred, what is the final temperature? Consider air to be a diatomic ideal gas.

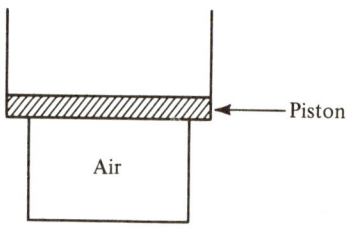

Fig. 5-5

5-4. Propane (MW = 44) flows through a well insulated pipe of constant cross-sectional area. At the inlet section of this pipe the velocity of the propane is 60 ft/sec, the pressure is 400 $lb_f/in.^2$, and the temperature is 200 °F. At the exit section the pressure is 100 $lb_f/in.^2$ If propane acts as an ideal gas, and c_p = 21.0 BTU/lb mole °R, determine the temperature and velocity of the propane leaving the pipe.

5-5. Show that the work associated with the isentropic compression of an ideal gas in a closed system is given by the relationship

$$W = \frac{P_2 V_2 - P_1 V_1}{1 - \gamma}$$

5-6. Propane (MW = 44) is compressed in a reversible adiabatic steady-flow process from 1 atm pressure, 700 °R to 40 atm pressure. Determine the work of compression per lb_m of propane, if the propane acts as an ideal gas.

5-7. Air ($MW = 29$) enters a well insulated centrifugal compressor at 14.7 lb_f/in^2, 60 °F, and a velocity of 200 ft/sec and leaves at 60 lb_f/in^2, 400 °F and a velocity of 200 ft/sec. Calculate the lost work and unavailable energy per lb_m of air for this process if air acts as a diatomic ideal gas. The temperature of the surroundings is 80 °F

5-8. Ten thousand cubic feet per hour of air ($MW = 29$) at 70 °F and a 1.0 atm is to be compressed in a well insulated centrifugal compressor to 5.0 atm. If the efficiency of the compressor is 75 percent, what is the horsepower requirement of the compressor? What is the temperature of the air leaving the compressor? What is the lost work in BTU/hr? Air may be assumed to be an ideal diatomic gas.

5-9. A pressure vessel has a volume of 1.0 m^3 and contains methane ($MW = 16$) at 1.4×10^6 N/m^2, 600 K. The methane is cooled to 300 K by heat transfer to the surroundings at 300 K. If methane acts as an ideal gas, calculate the availability in the initial and final states and the unavailable energy of this process. The specific heat of methane at constant pressure may be assumed to be constant and equal to 58.62 (joules)/(gm mole)(K).

5-10. An insulated tank having a volume of 2 ft^3 contains carbon dioxide ($MW = 44$) at 14.7 lb_f/in^2 and 80 °F. It is pressurized from a line in which carbon dioxide is flowing at 1000 lb_f/in^2, 80 °F. The valve is closed when the pressure in the tank reaches 1000 lb_f/in^2. Assuming carbon dioxide acts as an ideal gas, determine the amount of carbon dioxide which flows into the tank.

5-11. A cold room in a packing plant is to be maintained at 15 °F even on the hottest days of the year when the ambient temperature is 105 °F. A refrigeration machine using air ($MW = 29$) as a working fluid is to be employed. This machine consists of an adiabatic compressor, an isobaric cooler, an adiabatic expander, and an isobaric cold room heat exchanger. The operation is as follows: air at atmospheric pressure and 10 °F enters the compressor from which it is discharged at 100 psia. The air passes from the compressor to the cooler where it is cooled by the ambient air to 115 °F. The air then enters the expander which discharges at atmospheric pressure. The air from the expander is sent to the cold room heat exchanger to perform the desired cooling job before being returned to the compressor. The work from the expander is used to help drive the compressor. The compressor and expander are 90 percent efficient based on the work of reversible machines operating over the same pressure ranges. Air may be assumed to obey the ideal gas law, and its constant pressure heat capacity may be taken as 6.97 BTU/mole °R.

Evaluate the efficiency of the above machine in terms of a Carnot refrigerator performing the same job.

6
Thermodynamics of Fluid Flows

OBJECTIVES

After studying this chapter, the student should be able to

(1) apply the mechanical energy balance equation, Eq. (6-3), to the analysis of flows in simple systems;
(2) calculate the speed of sound in a compressible medium from the general equation, (6-11), or, in the case of an ideal gas, from Eq. (6-12);
(3) define the concepts of Reynolds number, Mach number, and static and stagnation state variables;
(4) calculate the stagnation properties in ideal gas flows;
(5) calculate the property changes across a normal shock wave in an ideal gas flow;
(6) discuss the effect that area changes, heat transfer and friction have on compressible flows; and
(7) employ the concepts of isentropic nozzle and isentropic diffuser efficiencies.

THE MECHANICAL ENERGY BALANCE EQUATION

The generalized balance equations can be combined in a variety of useful ways; one example of which is the combination used to obtain the Carnot relationships, Eqs. (3-5) and (3-9). As another example, we will again use these balance equations to derive some of the relationships commonly employed in the analysis of fluid flows through open systems having a single entrance and exit such as pipe flows.

If there is a compressor or a pump in the system, the general energy balance, Eq. (2-25), at steady-state reduces to

$$\Delta h + \Delta\left(\frac{v^2}{2}\right) + g\Delta z - \delta q + \delta w = 0 \tag{6-1}$$

where Δ indicates the outlet value minus the inlet value. In the case of an infinitesimally small system, these differences would be differentials.

$$dh + d\left(\frac{v^2}{2}\right) + g\,dz - \delta q + \delta w = 0$$

Similarly, the generalized entropy balance, Eq. (3-16), reduces to

$$ds - \frac{\delta q}{T} - \frac{\delta lw}{T} = 0$$

for this steady-flow case. If we combine the above two equations with Eq. (4-18),

$$dh = T\,ds + v\,dP \tag{4-18}$$

the relationship

$$v\,dP + d\left(\frac{v^2}{2}\right) + g\,dz + \delta lw + \delta w = 0 \tag{6-2}$$

is obtained. Integrating Eq. (6-2) across the system yields

$$\int_{P_1}^{P_2} v\,dP + \Delta\left(\frac{v^2}{2}\right) + g\Delta z + lw + w = 0 \tag{6-3}$$

This relationship is known as the mechanical energy balance equation since it involves only traditional hydrodynamic variables. The lost work term accounts for the mechanical energy dissipated by fluid friction and in general must be determined experimentally. Fanning[7], Moody[8] and others have developed empirical correlations for predicting this "friction factor" for various types of flow conditions. The primary flow parameter in any of these correlations is the dimensionless Reynolds number,

$$\text{Re} = \frac{vl}{\nu}$$

where

v = mean velocity of the fluid

l = some characteristic length of the system

ν = kinematic viscosity of the fluid

For instance, the empirical relation for turbulent flows in smooth pipes of length L and diameter D

$$lw = 0.158 \frac{L}{D} \frac{v^2}{\text{Re}^{1/4}} \tag{6-4}$$

holds well for Reynolds numbers between 2000 and 10^5. The Reynolds number for a pipe is based upon the pipe diameter D. For Re below 2000, the flow is laminar and the lost work can be predicted theoretically to be

$$lw = 32 \frac{L}{D} \frac{v^2}{\text{Re}} \tag{6-5}$$

Another term in Eq. (6-3) which requires special consideration is the integral which cannot be evaluated until the equation of state and the flow process are specified. In the special case that the fluid behaves as though it were incompressible, the integral can be evaluated directly since the specific volume v is constant. The mechanical energy balance for a steady, incompressible flow is, therefore,

$$v\Delta P + \Delta\left(\frac{v^2}{2}\right) + g\Delta z + lw + w = 0 \tag{6-6}$$

For the case of no lost work and no shaft work, Eq. (6-6) reduces to the well known incompressible Bernoulli's equation

$$v\Delta P + \Delta\left(\frac{v^2}{2}\right) + g\Delta z = 0 \tag{6-7}$$

This relationship is used extensively in the analysis of low speed flows if viscous effects can be neglected. Equation (6-7) indicates that an inviscid flow can be accelerated at the expense of pressure and/or elevation. These two incompressible flow equations, (6-6) and (6-7), are frequently used by engineers to predict pressure drops for compressible flows if the change in the fluid density in the system is less than around 10 percent. The following example is an application of the mechanical energy balance equation.

EXAMPLE PROBLEM 6-1. Liquid water at 15 °C (59 °F) is flowing through a smooth horizontal pipe 5 cm (1.97 in.) in diameter and 300 meters (984.3 ft) long at a flow rate of 4 liters/sec (1.06 gal/sec). Calculate the total power loss and the pressure drop in the pipe.

We start by defining the pipe as our system. Equation (6-6) is applicable since the flow is steady state and incompressible. We see that no shaft work is performed and that there is no potential energy change since the pipe is horizontal. If we assume that the velocity profile does not vary down the pipe, there is also no change in kinetic energy in the system. The mechanical energy balance equation reduces in this case to

$$v\Delta P + lw = 0 \qquad (6\text{-}8)$$

The lost work can be calculated from either Eq. (6-4) or (6-5) depending upon the Reynolds number Re. To calculate Re, we must find the average flow velocity v in the pipe. This can be obtained from the rate form of the generalized mass balance equation which, for steady-state, reduces to

$$\left(\frac{\delta m}{dt}\right)_{in} - \left(\frac{\delta m}{dt}\right)_{out} = 0$$

or

$$\dot{m} = \left(\frac{\delta m}{dt}\right)_{in} = \left(\frac{\delta m}{dt}\right)_{out} = Av/\mathrm{v}$$

where A is the cross-sectional area of the pipe and the dot above the m indicates the derivative with respect to time. The volume flow rate Av is 4 liters/sec and the specific volume of water is 1 cm³/gram. Thus

$$v = \frac{Av}{A} = \frac{4 \frac{\text{liter}}{\text{sec}} \times 10^3 \frac{\text{cm}^3}{\text{liter}}}{\pi \frac{(5 \text{ cm})^2}{4}}$$

$$= 203.7 \text{ cm/sec } (6.68 \text{ ft/sec})$$

150 Chap. 6   *Thermodynamics of Fluid Flows*

and

$$\dot{m} = \frac{4 \frac{\text{liter}}{\text{sec}} \times 10^3 \frac{\text{cm}^3}{\text{liter}}}{1 \frac{\text{cm}^3}{\text{gram}}}$$

$$= 4 \times 10^3 \text{ gram/sec}$$

The kinematic viscosity of water ν at 15 °C is found to be 0.0114 cm²/sec from a standard handbook with viscosity tables. The Reynolds number for this pipe flow is

$$\text{Re} = \frac{vD}{\nu} = \frac{203.7 \text{ cm/sec} \times 5 \text{ cm}}{0.0114 \text{ cm}^2/\text{sec}}$$

$$= 89{,}342$$

which implies that the flow is turbulent (Re > 2000). Equation (6-4) is, therefore, used to calculate the lost work per unit mass.

$$lw = 0.158 \frac{L}{D} \frac{v^2}{\text{Re}^{1/4}} \tag{6-4}$$

$$= 0.158 \times \frac{3 \times 10^4 \text{ cm}}{5 \text{ cm}} \times \frac{(203.7 \text{ cm/sec})^2}{(89{,}342)^{1/4}}$$

$$= 2.275 \times 10^6 \frac{\text{cm}^2}{\text{sec}^2}$$

The pressure drop can now be calculated from the simplified mechanical energy balance Eq. (6-8).

$$\Delta P = -lw/v \tag{6-8}$$

$$= \frac{-2.275 \times 10^6 \text{ cm}^2/\text{sec}^2}{1 \text{ cm}^3/\text{gram}}$$

$$= -2.275 \times 10^6 \frac{\text{gram}}{\text{cm sec}^2}$$

$$= -2.275 \times 10^6 \frac{\text{dyne}}{\text{cm}^2} \quad (-33 \text{ psia})$$

The minus sign indicates that the pressure in the pipe decreases from the inlet to the outlet due to friction.

The total mechanical power dissipated by viscosity is found by multiplying the work lost per unit mass lw by the mass flow rate $\dot{m}$.

$$\dot{W} = (\dot{m})(lw)$$

$$= 4 \times 10^3 \, \frac{\text{gram}}{\text{sec}} \times 2.275 \times 10^6 \, \frac{\text{cm}^2}{\text{sec}^2}$$

$$= 9.10 \times 10^9 \, \frac{\text{gram cm}^2}{\text{sec}^3}$$

$$= 910 \text{ watts } (1.22 \text{ horsepower})$$

COMPRESSIBLE FLOWS

Speed of Sound

Neglecting compressible effects in the analysis of low velocity flow systems is often a valid approximation, but compressibility is of major importance in the analysis of high velocity flow systems. The dynamic behavior of compressible flows can be quite different from that of incompressible flows. The analysis of compressible flows in general entails the simultaneous solution of all the balance equations presented thus far, plus the momentum balance. Fortunately, many compressible flow problems are amenable to assumptions which drastically simplify their solution. For instance, many flows are essentially adiabatic, occur in the temperature and pressure range where the medium can be assumed to behave as an ideal gas, and can be regarded as frictionless.

The primary flow parameter in the analysis of compressible flows is the speed with which small pressure disturbances, such as weak acoustic waves, propagate through a compressible medium. This finite velocity is called the speed of sound or sonic velocity and will be designated by the symbol a. To determine the sonic velocity, we will consider the propagation of a pressure wave through the long, gas filled tube depicted in Fig. 6-1.

The external force needed to hold the frictionless piston stationary against the internal pressure P is PA where A is the cross-sectional area of the piston. Increasing the external force on the left by an infinitesimal amount to $(P + dP)A$ will cause the piston to accelerate to the right until the pressure acting on the piston has increased to $P + dP$ to re-establish static equilibrium. After this initial transient acceleration, the piston will continue to move at a constant infinitesimal velocity v toward the stationary piston on the right.

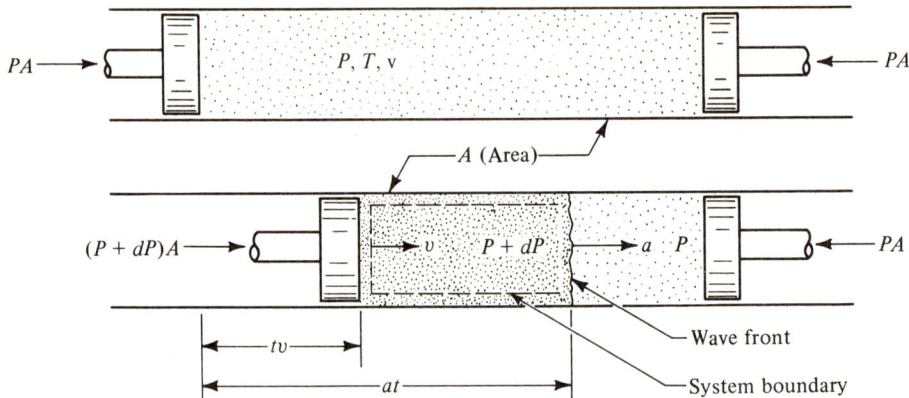

Fig. 6-1. *Propagation of an infinitesimal pressure wave in a tube.*

At the same time, a pressure front which divides the gas at pressure $P + dP$ from the gas at pressure P will move to the right at the speed of sound a. As this pressure front moves through the stationary gas, it imparts the piston velocity v to the compressed gas behind it.

To analyze this process, we will define our system to be the volume that encompasses the moving fluid. This is a fairly complicated system in that it is open and moving. The left boundary is moving with the piston velocity v and the right boundary moves with the wave front at velocity a. Even so, the application of the generalized mass balance equation to this system is straight forward and results in the equation

$$\dot{m}_{sys} = \dot{m}_{in}$$

The mass flux into the system is aA/v where A is the cross-sectional area of the tube and v is the specific volume of the fluid at rest. Thus

$$\dot{m}_{sys} = \left(\frac{dm}{dt}\right)_{sys} = \frac{aA}{\mathrm{v}}$$

Integrating this expression with respect to time gives the total mass in the system,

$$m_{sys} = \frac{aAt}{\mathrm{v}}$$

where t denotes the length of time the piston has been moving.

The change in specific volume dv experienced by the fluid in the system is obtained by dividing the system's mass m_{sys} into the total volume change

$$dV = -At\upsilon$$

that this mass has undergone. Thus,

$$dv = \frac{dV}{m_{sys}} = -\frac{At\upsilon}{\frac{aAt}{\mathrm{v}}} = -\frac{\mathrm{v}\upsilon}{a} \tag{6-9}$$

A second relationship involving the speed of sound a can be obtained from the generalized energy balance equation, but this approach will not be used here. Instead, the generalized momentum balance will be introduced and employed since this equation will be needed for the shock wave analysis to follow. This equation follows from Newton's second law of motion which, when generalized[9], asserts that

$$\text{Momentum Flux}_{in} - \text{Momentum Flux}_{out} + F = \frac{d}{dt}(\text{Momentum})_{sys} \tag{6-10}$$

where

> Momentum Flux$_{in}$ is the rate at which momentum is transferred to the system by the entering mass
> Momentum Flux$_{out}$ is the rate at which the momentum is transferred from the system by the existing mass
> F is the resultant force acting on the system
> d,(Momentum)$_{sys}/dt$ is the rate at which the momentum of the fluid within the system changes.

When this momentum balance equation is applied to the open system depicted in Fig. 6-1, it simplifies to

$$F = \frac{d}{dt}(\text{Momentum})_{sys}$$

since no mass leaves the system and the mass entering the system is at rest and, therefore, carries no momentum. If friction and body forces are neglected, the only force acting on this system is due to the pressure difference across the system.

$$F = (P + dP)A - PA = A\,dP$$

The momentum of the fluid within the system is mv. Thus,

$$A\, dP = \frac{d}{dt}(mv)$$

Integrating this expression from time equal zero when the momentum of the system is zero to time equal t gives

$$(A\, dP)t = mv = (Aat/\mathrm{v})v$$

This relationship can also be obtained by applying the impulse-momentum theorem, an important concept in mechanics, directly to this system. After simplifying the above expression, we find that $dP = av/\mathrm{v}$. When this equation is divided by Eq. (6-9), we obtain the result

$$\frac{dP}{d\mathrm{v}} = \frac{av/\mathrm{v}}{-\mathrm{v}v/a} = -\frac{a^2}{\mathrm{v}^2}$$

This last relationship simplifies to the form

$$\frac{dP}{d\rho} = a^2$$

since

$$d\mathrm{v} = d(1/\rho) = -d\rho/\rho^2 = -\mathrm{v}^2\, d\rho$$

We must now determine how the pressure varies with density before this equation can be used to calculate the speed of sound. Newton, who originally deduced the above expression from dimensional considerations, was familiar with Boyle's constant temperature experiments with gases at low pressures. He theorized that heat would be conducted from the compressed region to the uncompressed region so rapidly that essentially no temperature difference could be maintained across the wave front. Thus, the pressure front would pass through the gas medium isothermally. We will follow this line of reasoning by first rewriting the ideal gas equation in the form

$$P = \frac{RT}{\mathrm{v}} = \frac{\rho RT}{MW}$$

where MW, the molecular weight of the gas, is introduced to account for the fact that most tabulated values of density are in mass units rather than in mole units. If we now differentiate the ideal gas equation at constant temperature, we find

$$dP_T = \frac{RT}{MW} d\rho_T$$

or

$$a^2 = \left(\frac{\partial P}{\partial \rho}\right)_T = \frac{RT}{MW}$$

Newton calculated the velocity of sound in air using this relationship and compared his result with experimental measurements obtained by measuring the time difference between the flash and the sound from a cannon across a known distance. His predicted value underestimated the measured value by some 15 percent, but he ascribed the difference to dust particles and moisture in the air.

The Marquis de Laplace modified Newton's computation by proposing that the process is adiabatic and reversible (isentropic). In this case,

$$a^2 = \left.\frac{\partial P}{\partial \rho}\right)_s \qquad (6\text{-}11)$$

This is a much better assumption since the heat transfer at any one location in the gas is insignificant during the brief time that it takes the wave front to pass. If we now assume that the gas is ideal with constant heat capacities, we can obtain an explicit expression for the speed of sound by differentiating Eq. (5-11)

$$Pv^\gamma = P\rho^{-\gamma} = \text{constant} \qquad (5\text{-}11)$$

to obtain

$$\rho^{-\gamma} dP - \gamma P \rho^{-(\gamma+1)} d\rho = 0$$

Thus,

$$\left(\frac{dP}{d\rho}\right)_s = \gamma \frac{P}{\rho} = \frac{\gamma RT}{MW}$$

or

$$a^2 = \frac{\gamma RT}{MW} \qquad (6\text{-}12)$$

The square of the speed of sound is, therefore, proportional to the local absolute temperature and the ratio of specific heats and inversely proportional to the molecular weight for an ideal gas. Equation (6-12) also provides a convenient formula for determining γ since a, T and MW can all be measured accurately.

EXAMPLE PROBLEM 6-2. Calculate the speed of sound in air ($MW = 28.96$ and $\gamma = 1.4$) at 15.6 °C (60 °F) and one atmosphere.

If air is assumed to be an ideal gas,

$$a = \sqrt{\gamma RT/MW} \qquad (6\text{-}12)$$

$$= \left(\frac{1.4 \times 8.314 \frac{\text{joule}}{\text{(gm mole) (K)}} \times 10^3 \frac{\text{gm meter}^2/\text{sec}^2}{\text{joule}} \times 288.8 \text{ K}}{28.96 \frac{\text{gm}}{\text{gm mole}}} \right)^{1/2}$$

$$= 340.7 \text{ m/sec } (1117.8 \text{ ft/sec})$$

which agrees well with the measured value of 1,120 ft/sec. We see that the old rule of thumb that there is a mile between you and the point where the lightning struck for every five seconds between the sighting and the thunder clap is not a bad approximation. At the same temperature, the speed of sound for He ($\gamma = 5/3$, $MW = 4$) is estimated to be 1,000 m/sec (3,283 ft/sec) which is almost three times faster. Since the speed of sound of air is calculated repeatedly, it is convenient to rewrite the above expression for air in the form

$$a(\text{ft/sec}) = 49.0 \sqrt{T(^\circ R)}$$

or

$$a(\text{m/sec}) = 20.0 \sqrt{T(K)} \tag{6-13}$$

It should be noted that the sonic velocity is a function of the local thermodynamic state and will, therefore, vary from point to point with the local temperature variations in the medium. Equation (6-12) is applicable to the transmission of pressure waves generated by a piston vibrating at a moderate frequency such as a speaker diaphragm. However, for very high frequenceis on the order of 10^9 Hertz, the temperature gradients become too large to assume that the process is adiabatic even though the amplitude of pressure pulses may be quite small. These high frequency pulses are found to propagate more quickly through the medium than predicted by Eq. (6-11).

Stagnation Properties

There are numerous steady-state flow processes which require a fluid stream to be accelerated to a higher velocity or deaccelerated to a lower velocity. Devices used to increase the fluid velocity at the expense of a pressure drop are called nozzles, and the devices used for decelerating the flow through a pressure increase are called diffusers.

As an example of a compressible-gas flow, we will concentrate on the flow through the one-dimensional, convergent-divergent nozzle pictured in Fig. 6-2 both because of its practical importance, and because it illustrates the principal characteristics of compressible flows. An arbitrary section of this nozzle will be taken as our system. We note that this system does no work and it will be assumed that the flow is a steady-state flow experiencing no heat transfer or po-

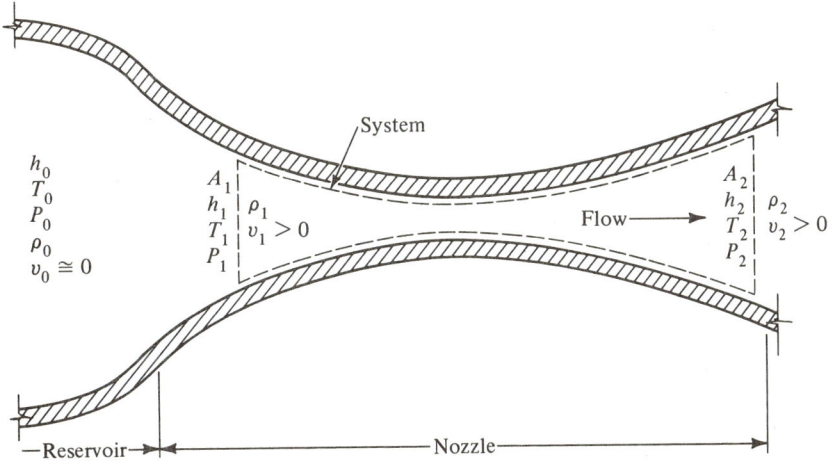

Fig. 6-2. *Converging-diverging nozzle.*

tential energy changes. Under these conditions, the generalized energy balance equation reduces to

$$\left(h_1 + \frac{v_1^2}{2}\right) - \left(h_2 + \frac{v_2^2}{2}\right) = 0$$

or

$$h_1 + \frac{v_1^2}{2} = h_2 + \frac{v_2^2}{2} = \text{constant} = h_0 \tag{6-14}$$

where the subscripts 1 and 2 denote two arbitrary cross-sections in the nozzle.

The quantity h_0 is called the stagnation or total enthalpy because it is the enthalpy that would be obtained if the flow were adiabatically brought to rest. The quantities h_1 and h_2 are called static enthalpies since they are the enthalpies that would be measured at sections 1 and 2 respectively by an observer moving with the flow. Equation (6-14) indicates that the static enthalpy varies from cross-section to cross-section depending upon the local velocity while the stagnation enthalpy remains constant.

In the particular case of an ideal gas, the enthalpy is only a function of temperature as shown by Eqs. (5-5) and (5-6). This implies that there is also a constant stagnation temperature T_0 associated with this adiabatic flow since h_0 is constant. If the heat capacities of the ideal gas are also constant, Eq. (6-14) can be written in the simplified form

$$(h_0 - h) = c_P(T_0 - T) = \frac{v^2}{2}$$

where the subscripts 1 and 2 have been dropped because this relationship is true at any arbitrary cross-section in this nozzle system. Thus,

$$c_p T + \frac{v^2}{2} = c_p T_0 \qquad (6\text{-}15)$$

Note that c_p must be given in mass units since $v^2/2$ is the kinetic energy per unit mass.

Since the absolute static temperature T cannot be negative, the maximum velocity attainable by an adiabatic expansion of an ideal gas is found by setting T equal to zero in Eq. (6-15),

$$v_{max} = \left(2 c_p T_0\right)^{1/2} \qquad (6\text{-}16)$$

Our analysis of the flow in an adiabatic nozzle has indicated that the stagnation enthalpy h_0 is constant along with the stagnation temperature T_0 for a perfect gas. These values, therefore, correspond to the enthalpy and temperature in the reservoir where the velocity is essentially zero. We will next examine what happens to pressure when the velocity is reduced adiabatically to zero.

The stagnation process for a perfect gas is depicted on the hs diagram in Fig. 6-3. Lines of constant pressure have also been superimposed on this diagram.

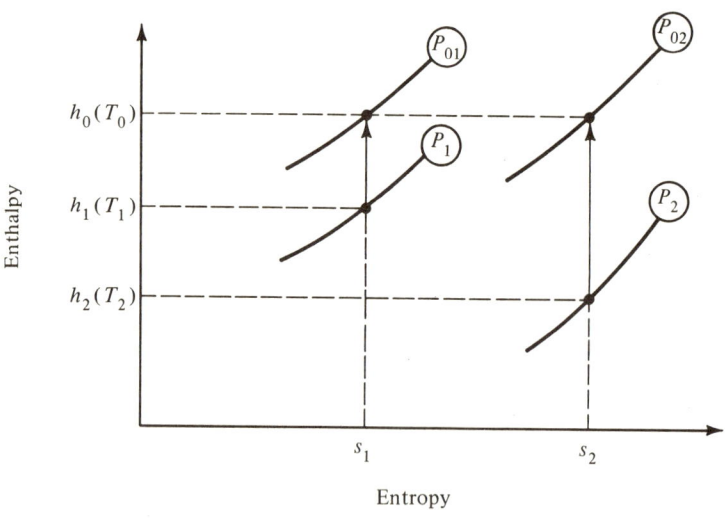

Fig. 6-3. *hs diagram for an adiabatic expansion of an ideal gas.*

If the flow is stopped adiabatically, we know that the stagnation state must lie somewhere along the constant enthalpy line $h = h_0$, but the actual state remains to be specified.

If the generalized entropy balance equation is applied to the adiabatic nozzle system pictured in Fig. 6-2, this balance reduces to the inequality

$$s_2 - s_1 = \frac{lw}{T} \geqslant 0$$

since T is positive and lw is either zero for a reversible process or greater than zero for an irreversible process. This result may be written in the form

$$s_2(P_2, T_2) \geqslant s_1(P_1, T_1)$$

Figure 6-3 depicts the nonisentropic flow case $s_2 > s_1$. If the flows at section 1 and section 2 were both stopped isentropically so that the static entropy and the stagnation entropy were the same at each section, the above inequality would still hold and

$$s_2(P_1, T_1) = s_{02}(P_{02}, T_{02}) \geqslant s_{01}(P_{01}, T_{01}) = s_1(P_1, T_1)$$

The entropy change for an ideal gas between states 1 and 2 can be calculated from Eq. (5-8),

$$s_2 - s_1 = R \ln \frac{P_1}{P_2} + c_P \ln \frac{T_2}{T_1} \tag{5-8}$$

$$= R \ln \frac{P_{01}}{P_{02}} + c_P \ln \frac{T_{02}}{T_{01}}$$

As shown in the derivation of Eq. (6-15), the stagnation temperature is constant in this case,

$$T_{01} = T_{02} = T_0$$

and the above entropy equation reduces to

$$s_2 - s_1 = R \ln \frac{P_{01}}{P_{02}} \tag{6-17}$$

The stagnation pressure must, therefore, decrease along the nozzle during an irreversible adiabatic expansion $P_{01} > P_{02}$ since the entropy must increase and remains constant during an isentropic expansion $P_{01} = P_{02}$. Friction causes the flow in an actual nozzle to be irreversible but the entropy change is small for a well designed nozzle. A nozzle flow is, therefore, usually assumed to be isentropic. The stagnation pressure in a well designed nozzle would remain essentially constant along the nozzle as would all other stagnation thermodynamic properties.

An important relationship between the static and stagnation temperatures at a point in the flow can be obtained by combining Eqs. (6-12) and (6-15),

$$\frac{T_0}{T} = 1 + \frac{v^2}{2c_p T} = 1 + \frac{\gamma R}{2c_p MW} \frac{v^2}{\frac{\gamma RT}{MW}} = 1 + \frac{\gamma - 1}{2}\left(\frac{v}{a}\right)^2$$

The ratio of the local fluid velocity to the local speed of sound that appears in the above equation is the principal parameter for compressible flows and is called the Mach number,

$$M = \frac{v}{a} \qquad (6\text{-}18)$$

after Ernest Mach, a pioneer in supersonic flow. If this definition is substituted into the above expression, we obtain

$$\frac{T_0}{T} = 1 + \frac{\gamma - 1}{2} M^2 \qquad (6\text{-}19)$$

This equation can be combined with the isentropic relations given in Eq. (5-12) to obtain the additional relationships

$$\frac{P_0}{P} = \left(1 + \frac{\gamma - 1}{2} M^2\right)^{\gamma/(\gamma-1)} \qquad (6\text{-}20)$$

$$\frac{\rho_0}{\rho} = \frac{v}{v_0} = \left(1 + \frac{\gamma - 1}{2} M^2\right)^{1/(\gamma-1)} \qquad (6\text{-}21)$$

The ratios T_0/T, P_0/P and ρ_0/ρ are, therefore, only a function of the local Mach number for an ideal gas. Figure 6-4 graphically represents these ratios as a function of Mach number for $\gamma = 1.4$, the specific heat ratio for diatomic gases including air. The corresponding tabulated values for these ratios can be found in most gas dynamic textbooks. Gas tables by Keenan and Kaye[10] or NACA[11] may be consulted for other specific heat ratios.

These simple stagnation relationships are very useful in the analysis of many compressible flow problems, and we will initially utilize them to help answer one of the important questions in fluid mechanics—when is it acceptable to assume a gas flow to be incompressible? Figure 6-4 indicates that the density variation in an adiabatic flow field will be less than 5 percent if the maximum Mach number is below 0.3 but the variations will become significant for Mach numbers much above this value. This value is usually accepted as the upper limit for simplifying a problem with the incompressible flow assumption $\rho \sim \rho_0$.

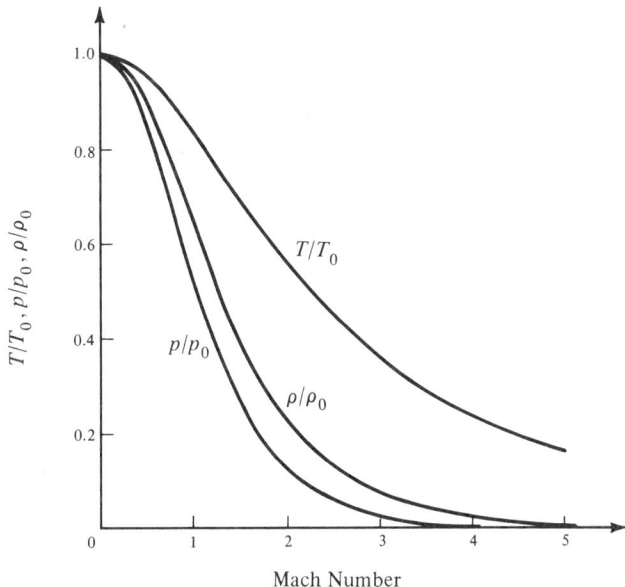

Fig. 6-4. *Isentropic flow chart for* $\gamma = 1.4$.

The following problem will give another example of the usefulness of these compressible flow relations.

EXAMPLE PROBLEM 6-3. A simple wind tunnel can be made from the nozzle pictured in Fig. 6-2 by attaching a constant area duct to the downstream end of the nozzle to act as a constant velocity test section. If this test section has a 0.2 m² (2.15 ft²) cross-sectional area and the reservoir is capable of supplying air at 10 atmospheres and 38 °C (100.4 °F), calculate the velocity, mass flow rate and the static pressure and temperature in the test section for a Mach 2 flow. Assume the expansion through the nozzle to be isentropic.

Since the flow is isentropic, the stagnation temperature and pressure will be equal to the reservoir values where the velocity is assumed to be zero. The static temperature can be calculated from the ideal gas relationships (6-19).

$$T = T_0 \left(1 + \frac{\gamma - 1}{2} M^2\right)^{-1} = (311.2 \text{ K})\left[1 + \frac{1.4 - 1}{2} (2)^2\right]^{-1}$$

$$= \frac{311.2 \text{ K}}{1.8} = 172.9 \text{ K} = -100.3 \,°\text{C} \,(-148.6 \,°\text{F})$$

Note how cold the air gets during this expansion to a Mach 2 flow. This is a serious problem for supersonic wind tunnels because even the small amount of water vapor contained in fairly dry air can condense out at this low temperature.

Since the static temperature in the test section is now known, we can now compute the flow velocity from the Mach number and the local speed of sound which is obtained from Eq. (6-13),

$$v = Ma = (2)(20.0) \ 172.9 = 526.0 \text{ m/sec} (1725 \text{ ft/sec})$$

The static pressure is calculated from Eq. (6-20)

$$P = P_0 \left(1 + \frac{\gamma - 1}{2} M^2\right)^{-\gamma/(\gamma-1)} = 10 \text{ atm} \times 1.8^{-1.4/0.4}$$

$$= 1.28 \text{ atm}$$

and the mass flow rate by

$$\dot{m} = \rho A v = \frac{PMW}{RT} A v$$

$$= \frac{(1.28 \text{ atm})(28.96 \text{ gm/gm mole})(0.2 \text{ m}^2)(526.0 \text{ m/sec})(10^6 \text{ cm}^3/\text{m}^3)}{(82.06 \text{ atm cm}^3/\text{gm mole K})(172.9 \text{ K})(10^3 \text{ gm/kg})}$$

$$= 274.9 \text{ kg/sec} (606 \text{ lb/sec})$$

An enormous amount of air must be handled even though the test cross-sectional flow area and Mach number are not large by modern NASA standards. This gives an indication of why supersonic wind tunnels are so expensive to operate.

Equation (6-19) is often used to roughly estimate the maximum temperatures at the stagnation points on high speed vehicles although these points are not necessarily adiabatic points. For instance, if a model of a vehicle were placed in the test section just considered, the leading edges of the model could see temperatures of the order of 38 °C but, more importantly, if it were flown at Mach 2 in a 38 °C atmosphere, the stagnation temperatures at the leading edges could approach

$$T_0 = T\left(1 + \frac{\gamma - 1}{2} M^2\right) \tag{6-19}$$

$$= (311.2 \text{ K})(1.8) = 560.2 \text{ K} = 287.0 \text{ °C} (549 \text{ °F})$$

Thus, the leading edges of high speed vehicles must be constructed of special high temperature materials, and cooling provisions are made in some cases. Calculating the actual surface temperature is a problem attacked in heat transfer courses which is beyond the object of this book.

Effects of Area Changes on Compressible Flow

The reason for using a converging-diverging nozzle shape to obtain supersonic velocities lies in the fact that the response of a supersonic flow to an area change is opposite that of a subsonic flow. We can demonstrate this phenomena by first applying the generalized mass balance equation to the nozzle system depicted in Fig. 6-2.

$$\dot{m}_{in} - \dot{m}_{out} = 0$$

or

$$(\rho A v)_1 = (\rho A v)_2 = \rho A v$$

A relationship between the variation of area, density and velocity is then obtained by differentiating this last equation and dividing the results by the mass flux $\rho A v$.

$$\frac{dA}{A} + \frac{d\rho}{\rho} + \frac{dv}{v} = 0 \tag{6-22}$$

We can obtain a second equation from the combined energy-entropy balance, Eq. (6-2). For an isentropic expansion ($\delta LW = 0$) with no potential energy change and no work performed, this expression simplifies to

$$\frac{dP}{\rho} + v\,dv = 0 \tag{6-23}$$

The above two equations are combined to obtain

$$\frac{dA}{A} = \frac{dP}{\rho}\left(\frac{1}{v^2} - \frac{d\rho}{dP}\right) \tag{6-24}$$

Since the flow is assumed to be isentropic, we can use Eq. (6-11) in the form

$$\left(\frac{\partial \rho}{\partial P}\right)_s = \frac{1}{a^2}$$

in Eq. (6-24) to obtain

$$dA = \frac{A}{\rho v^2}(1 - M^2)\,dP \tag{6-25}$$

This equation can be rewritten in the form

$$dA = \frac{A}{v}(M^2 - 1)\,dv \tag{6-26}$$

by again using Eq. (6-23).

Equation (6-25) indicates that dP/dA is positive for subsonic flows, $M < 1$, and negative for supersonic flows, $M > 1$. Equation (6-26) indicates that dv/dA is positive for supersonic flows and negative for subsonic flows. Thus, the area must increase to accelerate a supersonic flow and decrease to accelerate a subsonic flow. A similar analysis can be made to determine how the static temperature and Mach number vary with changes in the cross-sectional area.

These conclusions are summarized in Figs. 6-5 and 6-6.

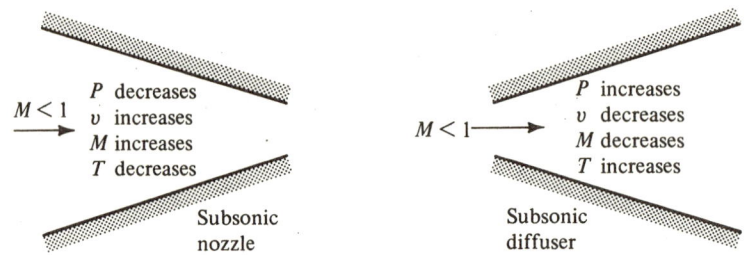

Fig. 6-5. *Effect of area changes on subsonic flows.*

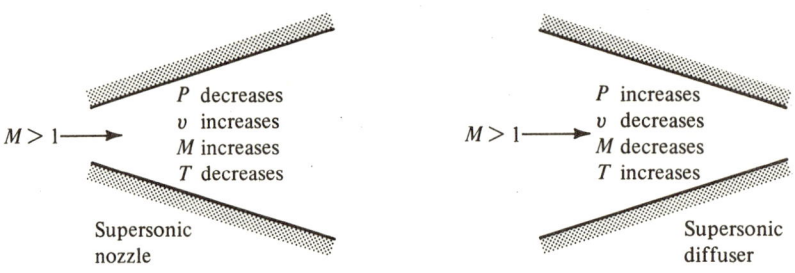

Fig. 6-6. *Effect of area changes on supersonic flows.*

These surprising results explain the reason why nozzles that develop supersonic flows have a convergent-divergent shape. The convergent section is used to accelerate the subsonic flow, but it cannot accelerate the flow to a velocity greater than the speed of sound regardless of the pressure difference imposed across the convergent section. This conclusion follows from Eq. (6-26) which implies that v must decrease in a converging nozzle for Mach numbers greater than one. Thus, a divergent section must be used to expand the flow to supersonic speeds after the convergent section has expanded it to Mach one. In a convergent-divergent nozzle which produces supersonic flow, the sonic velocity must occur at the minimum area which is called the throat and will be denoted here by A^*. It should be noted here that these conclusions for isentropic flows are quite general since they were deduced from first principles and we did not have to specify a particular equation of state.

Let us elaborate on these conclusions a little more by a physical example. The upper diagram of Fig. 6-7 shows a convergent-divergent nozzle between two

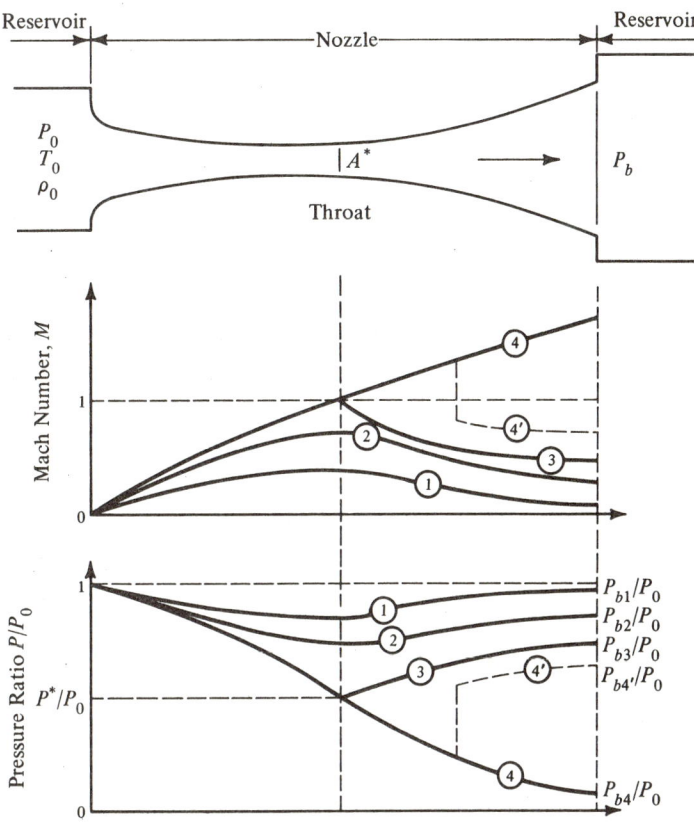

Fig. 6-7. *Flow characteristic of a nozzle.*

reservoirs. Suppose that the pressure and temperature, P_0 and T_0, in the left-hand reservoir are held constant while the back pressure P_b in the right-hand reservoir is decreased slightly below P_0. The small pressure difference will result in a subsonic flow in the nozzle. As shown in Fig. 6-5, the velocity and Mach number will increase and the static pressure and temperature decrease in the convergent section and vice-versa according to Fig. 6-6 in the divergent section. Thus, the Mach number reaches a maximum and the static pressure reaches a minimum at the throat as depicted by curve 1 in Fig. 6-7. Further decreases in P_b will continue to increase the Mach number at the throat as indicated by curve 2 until sonic flow is finally obtained at the throat—curve 3.

At this point, the maximum possible Mach number has been obtained at the throat and any further decrease in the back pressure P_b will have no effect on the flow in the convergent part of the nozzle. For this reason, the nozzle is often referred to as being "choked" since the maximum mass flow rate has been obtained. This maximum mass flow rate and the corresponding sonic properties at the throat can be easily derived for an ideal gas. The mass flow is given by

$$\dot{m} = \rho A v = \frac{PMW}{RT} A v$$

This relationship can be put in the following form:

$$\dot{m} = \left(\frac{P}{P_0}\right) \sqrt{\frac{T_0}{T}} \frac{v}{\sqrt{\gamma RT/MW}} \sqrt{\frac{\gamma MW}{R}} \frac{AP_0}{\sqrt{T_0}}$$

Equations (6-12) and (6-18) may be combined to give the following Mach number expression for an ideal gas:

$$M = \frac{v}{\sqrt{\gamma RT/MW}}$$

We also note from Eqs. (6-19) and (6-20) that the temperature ratio T/T_0 and the pressure ratio P/P_0 are only functions of the local Mach number. When these relationships are substituted into the mass flux equation, we find that

$$\dot{m} = AP_0 \sqrt{\frac{\gamma MW}{RT_0}} \frac{M}{\left(1 + \frac{\gamma-1}{2} M^2\right)^{(\gamma+1)/2(\gamma-1)}} \qquad (6\text{-}27)$$

which is the mass flow rate across any nozzle cross-section. The maximum mass flow rate occurs when the throat A^* is choked, $M = 1$. Thus,

$$\dot{m}^* = \dot{m}_{max} = P_0 A^* \sqrt{\frac{\gamma MW}{RT_0} \left(\frac{2}{\gamma+1}\right)^{(\gamma+1)/(\gamma-1)}} \qquad (6\text{-}28)$$

Similarly, the static to stagnation property ratios at a sonic throat can be evaluated from Eqs. (6-19), (6-20) and (6-21), yielding the results

$$\frac{T^*}{T_0} = \frac{2}{\gamma+1} \qquad (6\text{-}29)$$

$$= 0.833 \text{ for } \gamma = 1.4$$

$$\frac{P^*}{P_0} = \left(\frac{2}{\gamma+1}\right)^{\gamma/(\gamma-1)} \qquad (6\text{-}30)$$

$$= 0.528 \text{ for } \gamma = 1.4$$

and

$$\frac{\rho^*}{\rho_0} = \left(\frac{2}{\gamma+1}\right)^{1/(\gamma-1)} \tag{6-31}$$

$$= 0.634 \text{ for } \gamma = 1.4$$

where the sonic properties at the throat are designated with the superscript *. Thus, the maximum flow rate is proportional to the stagnation pressure and inversely proportional to the square root of the stagnation temperature. From Eq. (6-29) we see that the sonic temperature at the throat is only about 17 percent less than the stagnation temperature for a diatomic gas but from Eq. (6-30) we see that the sonic pressure is about 47 percent less than the stagnation pressure.

An important relationship between the sonic throat area A^* and the area A where the Mach number is M can be derived by equating Eqs. (6-27) and (6-28).

$$\frac{A}{A^*} = \frac{1}{M}\left[\left(\frac{2}{\gamma+1}\right)\left(1 + \frac{\gamma-1}{2}M^2\right)\right]^{(\gamma+1)/2(\gamma-1)} \tag{6-32}$$

We see that the area ratio A/A^* is only a function of the local Mach number and γ as was found to be the case for the static to stagnation property ratios. Equation (6-32) is plotted in Fig. 6-8 and numerical values for A/A^* versus M can be

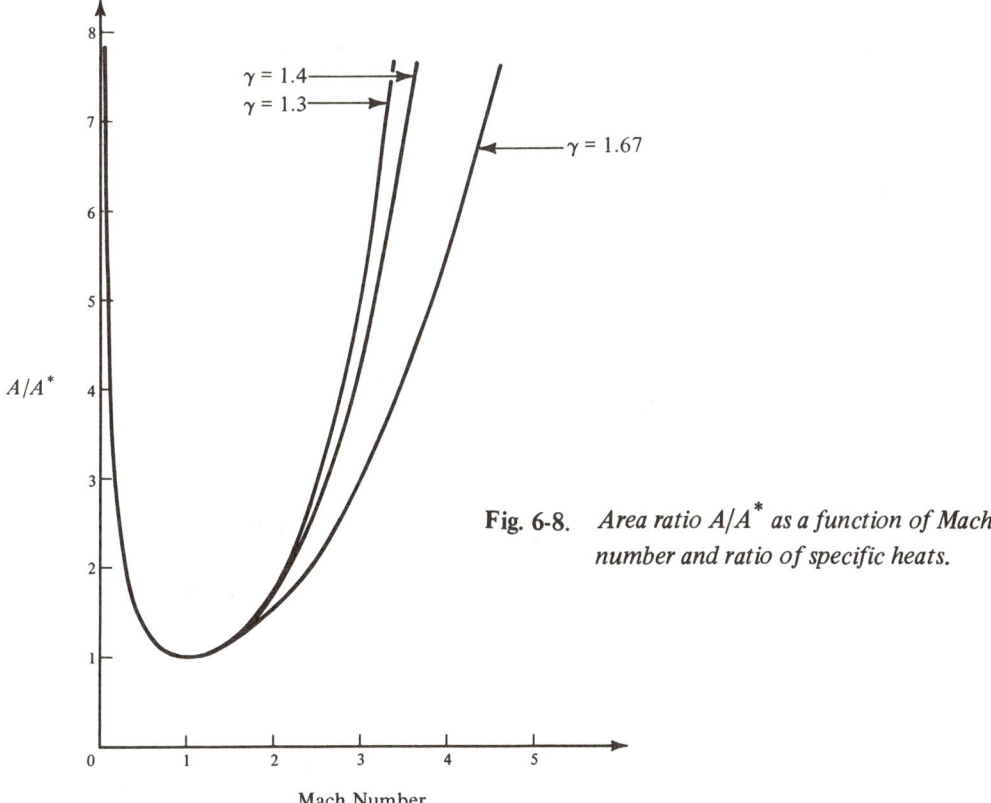

Fig. 6-8. Area ratio A/A^* as a function of Mach number and ratio of specific heats.

found in most textbooks on gasdynamics or in gas tables by Keenan and Kaye[10] and NACA[11]. We note once again in Fig. 6-8 that subsonic nozzles have convergent geometries while supersonic nozzles are divergent. This figure also shows that there are, of course, two possible Mach numbers for a given area ratio. These two cases are depicted in Fig. 6-7 by curve 3 where the divergent section after the sonic throat acts as a subsonic diffuser, and by curve 4 where the divergent section acts as a supersonic nozzle.

The following example is an application of the relationships developed in this section.

EXAMPLE PROBLEM 6-4. Calculate the area, the static temperature and pressure, and the velocity at the throat for the supersonic wind tunnel described in Example Prob. 6-3.

We know from Example Prob. 6-3 that the Mach number is 2 in the 0.2 m² test section. We can now calculate the throat area from Eq. (6-32),

$$\frac{A}{A^*} = \frac{1}{2}\left[\left(\frac{2}{2.4}\right)\left(1 + \frac{0.4}{2} 2^2\right)\right]^{2.4/0.8}$$

$$= 1.69$$

Therefore,

$$A^* = A/1.69 = 0.2 \text{ m}^2/1.69$$

$$= 0.118 \text{ m}^2 \ (1.27 \text{ ft}^2)$$

The stagnation temperature is equal to the reservoir temperature of 38 °C and the corresponding static temperature at the throat can be found from Eq. (6-29).

$$T^* = 0.833 \, T_0 = (0.833)(311.2 \text{ K})$$

$$= 259.3 \text{ K} = -13.9 \, ^\circ\text{C} \ (7.0 \, ^\circ\text{F})$$

Similarly, the stagnation pressure remains constant for an isentropic expansion and the static pressure at the throat is calculated from Eq. (6-30).

$$P^* = 0.528 \, P_0 = (0.528)(10 \text{ atm})$$

$$= 5.28 \text{ atm}$$

Finally, the sonic velocity at the throat can be calculated from Eq. (6-13),

$$v^* = a^* = 20.0\sqrt{T^*} = 20.0\sqrt{259.3}$$

$$= 322.0 \text{ m/sec} \ (1057 \text{ ft/sec})$$

Notice that the velocity only increases from 322.0 to 526.0 m/sec when expanding from the throat ($M = 1$) to the test section ($M = 2$) since the static temperature decreases during the isentropic expansion.

Normal Shock Waves. It was demonstrated that only two possible isentropic flows can exist in the divergent section of a choked nozzle. One of these is the subsonic case depicted by curve 3 in Fig. 6-7 and the other is the supersonic case depicted by curve 4. The respective back pressures in the right-hand reservoir for these two isentropic flows are denoted by P_{b3} and P_{b4}. What happens to the flow when the back pressure lies between these two pressures and the flow can no longer be isentropic?

In this case, the flow is experimentally observed to follow curve 4, the isentropic expansion of a supersonic flow, part way, but there is then a sudden increase in the static pressure as indicated by curve 4′. This pressure increase can be large but occurs in a remarkably thin region—of the order of several molecular mean-free paths. This sudden large pressure increase is called a "shock wave" which you are probably familiar with in the form of "sonic booms" from supersonic aircraft. The flow is decelerated from a supersonic velocity to a subsonic velocity by the large pressure rise across the normal shock as indicated by curve 4′. This means that the rest of the divergent section after the shock acts as a subsonic diffuser (Fig. 6-5) instead of a supersonic nozzle (Fig. 6-6). As the exit pressure P_b is decreased from P_{b3} toward P_{b4}, the shock location in the divergent section will move from the throat towards the exit plane and finally out of the nozzle where it is no longer normal to the flow.

The flow across a shock wave is nonisentropic due to the large temperature and pressure gradients that exist in the shock structure. We cannot, therefore, use the isentropic assumption that we have so freely used in past developments, even though the flow is still adiabatic. It is interesting to note that the first man who tried to analyze this problem, the famous German mathematician Bernhard Rieman (1826-1866), mistakenly assumed the shock to be isentropic.

To determine how the flow properties vary across a normal shock, we will enclose the shock wave in the open system shown in Fig. 6-9. The upstream and

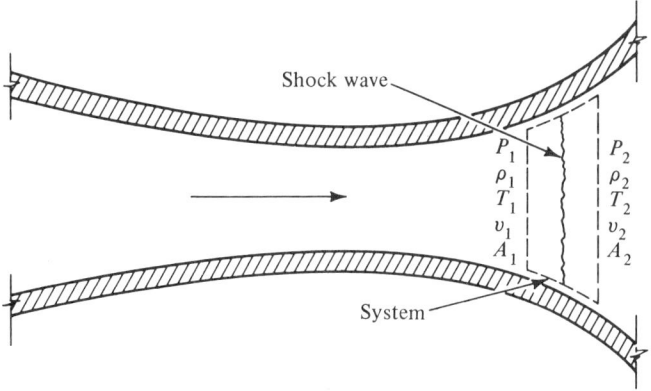

Fig. 6-9. *Flow across a normal shock wave.*

downstream properties are again denoted by the subscripts 1 and 2, respectively. Since the shock has a very thin structure, it can be treated as a discontinuity in flow properties. Thus, the upstream and downstream system boundaries can be placed very close to each other to essentially make the system's cross-sectional area constant.

$$A_1 = A_2 = A$$

We will also simplify the analysis of this system by again assuming the gas to be an ideal gas with constant heat capacities.

$$P = \frac{\rho RT}{MW} \tag{5-1}$$

To complete the analysis of this system, we will have to employ the mass, energy, entropy and momentum balance equations. For a steady state flow in a constant area system, the generalized mass balance equation reduces to the form

$$\rho_1 v_1 = \rho_2 v_2 = \dot{m}/A \tag{6-33}$$

Since this steady state flow is also adiabatic and performs no work, the generalized energy balance equation reduces to the same result that was obtained for the adiabatic nozzle,

$$h_1 + \frac{v_1^2}{2} = h_2 + \frac{v_2^2}{2} = h_0 \tag{6-14}$$

For an ideal gas with constant heat capacities, this equation simplifies to

$$c_P T_1 + \frac{v_1^2}{2} = c_P T_2 + \frac{v_2^2}{2} = c_P T_0 \tag{6-15}$$

If friction is neglected, the generalized momentum balance, Eq. (6-10), reduces to

$$\dot{m} v_1 - \dot{m} v_2 + (P_1 - P_2)A = 0 \tag{6-34}$$

When combined with the simplified mass balance, Eq. (6-33), this equation becomes

$$P_1 + \rho_1 v_1^2 = P_2 + \rho_2 v_2^2 \tag{6-35}$$

The entropy balance equation reduces to the form

$$s_2 - s_1 = \frac{lw}{T}$$

or

$$s_2 \geqslant s_1 \tag{6-36}$$

If we assume that the upstream flow properties P_1, ρ_1, T_1 and v_1 are known, then there are four equations [equation of state (5-1) and the three simplified balance equations—(6-15), (6-33) and (6-35)] and four unknowns (P_2, ρ_2, T_2 and v_2). The solution of this set of equations will call for a little math-a-magics. The first equation to be developed will use the mathematical procedure that is described below.

(1) The temperature ratio T_2/T_1 is formed from the ideal gas equation of state, Eq. (5-1).

(2) The result is combined with the simplified mass balance equation, Eq. (6-33).

(3) The Mach number definition, Eq. (6-18), is introduced.

(4) The speed of sound for an ideal gas, Eq. (6-12), is introduced.

Following these steps,

(1) $\quad \dfrac{T_2}{T_1} = \dfrac{P_2}{P_1} \dfrac{\rho_1}{\rho_2}$

(2) $\quad\quad\, = \dfrac{P_2}{P_1} \dfrac{v_2}{v_1}$

(3) $\quad\quad\, = \dfrac{P_2}{P_1} \dfrac{M_2 a_2}{M_1 a_1}$

(4) $\quad\quad\, = \dfrac{P_2}{P_1} \dfrac{M_2}{M_1} \dfrac{\sqrt{\gamma RT_2/MW}}{\sqrt{\gamma RT_1/MW}} = \dfrac{P_2}{P_1} \dfrac{M_2}{M_1} \sqrt{\dfrac{T_2}{T_1}}$

or

$$\frac{T_2}{T_1} = \left(\frac{P_2}{P_1} \frac{M_2}{M_1}\right)^2 \tag{6-37}$$

A second equation is developed from the simplified energy balance, Eq. (6-15), by first rewriting it in the form of Eq. (6-19),

$$T_1 \left(1 + \frac{\gamma-1}{2} M_1^2\right) = T_2 \left(1 + \frac{\gamma-1}{2} M_2^2\right) = T_0 \qquad (6\text{-}19)$$

When the value of T_2 from Eq. (6-37) is substituted into the above equation and the resulting expression is solved for P_2/P_1, we obtain the result

$$\frac{P_2}{P_1} = \frac{M_1}{M_2} \left[\frac{2 + (\gamma-1)M_1^2}{2 + (\gamma-1)M_2^2}\right]^{1/2} \qquad (6\text{-}38)$$

Since the upstream variables P_1 and M_1 are assumed to be known, this equation is a property relation between the downstream variables P_2 and M_2. Similar relationships can, of course, be developed for the other state variables. For instance, state equation (5.1) indicates that the denisty ratio $\rho_2/\rho_1 = (P_2/P_1)/(T_2/T_1)$ can be obtained from Eqs. (6-19) and (6-38). The curve which Eq. (6-38) traces out on the hs diagram is called the Fanno line and is depicted on Fig. 6-10. It is found that the point of maximum entropy on this curve corresponds to the sonic

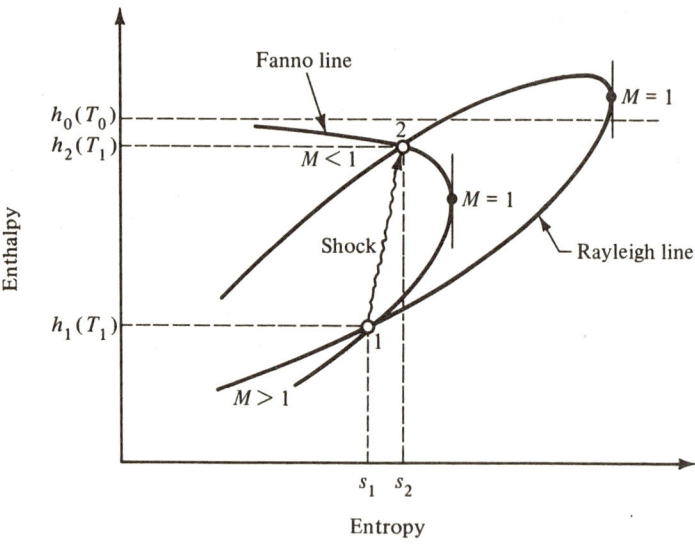

Fig. 6-10. $h-s$ diagram of Fanno line, Rayleigh line and shock process.

condition $M_2 = 1$ while the lower part of this curve corresponds to supersonic flow and the upper part to subsonic flow.

A second relationship for the pressure ratio across the shock can be obtained by combining

(1) the simplified momentum balance, Eq. (6-35) with
(2) the Mach number definition, and
(3) the speed of sound and equation of state for an ideal gas.

Proceeding in this stepwise manner, we find

(1) $P_1 + \rho_1 v_1^2 = P_2 + \rho_2 v_2^2$

(2) $P_1 + \rho_1 M_1^2 a_1^2 = P_2 + \rho_2 M_2^2 a_2^2$

(3) $P_1 + \dfrac{P_1 MW}{RT_1} M_1^2 \dfrac{\gamma RT_1}{MW} = P_2 + \dfrac{P_2 MW}{RT_2} M_2^2 \dfrac{\gamma RT_2}{MW}$

or

$$\frac{P_2}{P_1} = \frac{1 + \gamma M_1^2}{1 + \gamma M_2^2} \tag{6-39}$$

The curve that this equation traces out on the hs diagram is called the Rayleigh line. The Rayleigh line is also plotted on Fig. 6-10.

We note that steady-state flows in constant area systems will satisfy the mass balance and adiabatic energy balance if they lie along a Fanno line, and that they satisfy the mass balance and nonviscous momentum balance if they lie along a Rayleigh line. The shock relations are graphically determined by the intersections of the Fanno and Rayleigh lines as depicted in Fig. 6-10 since the flow across a shock must satisfy all these balance equations simultaneously. Figure 6-10 indicates that there are two solutions, one of which corresponds to $M_2 = M_1$ where no shock occurs and is, therefore, trivial. The other solution indicates that the downstream Mach number M_2 is subsonic when the upstream Mach number M_1 is supersonic. We note that the opposite case, $M_1 < 1$ and $M_2 > 1$, is not possible since this would not satisfy the simplified entropy balance equation (6-36) that requires $s_2 > s_1$.

The corresponding analytic solution is obtained by equating the two values of the pressure ratio P_2/P_1 obtained from the Fanno and Rayleigh relations, Eqs. (6-38) and (6-39). The nontrivial solution of the resulting equation is found after some algebraic manipulation to be

$$M_2^2 = \frac{(\gamma - 1)M_1^2 + 2}{2\gamma M_1^2 + 1 - \gamma} \tag{6-40}$$

The pressure ratio can now be evaluated by substituting the value of M_2 into Eq. (6-39) to obtain

$$\frac{P_2}{P_1} = 1 + \frac{2\gamma}{\gamma + 1}(M_1^2 - 1) \tag{6-41}$$

and Eqs. (6-37), (6-40), and (6-41) can be used to obtain the temperature ratio

$$\frac{T_2}{T_1} = 1 + \frac{2(\gamma - 1)(\gamma M_1^2 + 1)}{(\gamma + 1)^2 M_1^2}(M_1^2 - 1) \tag{6-42}$$

The density ratio and, therefore, the velocity ratio follow from the equation of state and the above temperature and pressure ratio equations

$$\frac{v_1}{v_2} = \frac{\rho_2}{\rho_1} = \frac{P_2}{P_1}\frac{T_1}{T_2} = \frac{(\gamma + 1)M_1^2}{(\gamma - 1)M_1^2 + 2} \tag{6-43}$$

A plot of these ratios as a function of the upstream Mach number M_1 are given in Fig. 6-11. This figure indicates that the pressure ratio P_2/P_1 will increase

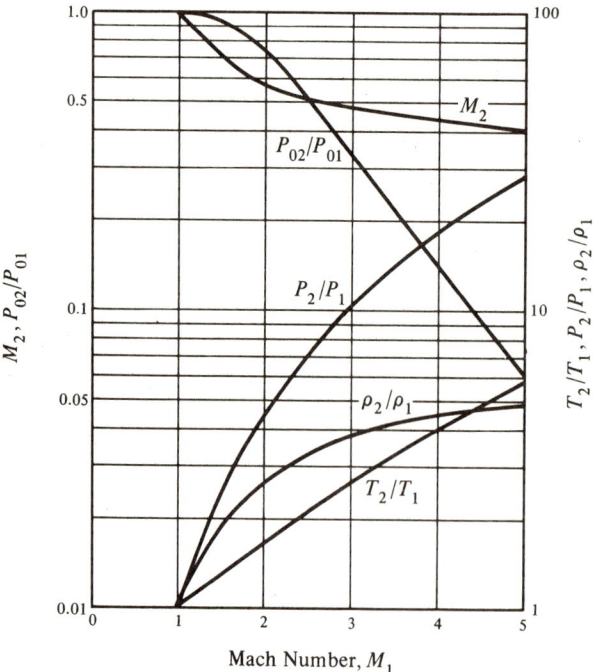

Fig. 6-11. *The normal shock relationships for $\gamma = 1.4$.*

from around 1 at a Mach near 1 to around 29 for a Mach number of 5 and the corresponding temperature ratio T_2/T_1 increases from 1 to 5.8. As mentioned before, the pressure and temperature increase across a normal shock can be enormous.

The stagnation pressure ratio across the shock may be found in terms of M_1 by writing this ratio in the form

$$\frac{P_{02}}{P_{01}} = \frac{P_{02}}{P_2} \frac{P_2}{P_1} \frac{P_1}{P_{01}}$$

The static to stagnation pressure ratios in this expression are given in terms of the local Mach number by Eq. (6-20). The Mach number after the shock M_2 is obtained from Eq. (6-40) and the relationship between the static pressure ratio across the shock and M_1 is given by Eq. (6-41). After substituting these two equations into the above expression for P_{02}/P_{01} and manipulating the results, one may show that

$$\frac{P_{02}}{P_{01}} = \left[1 + \frac{2\gamma}{\gamma+1}(M_1^2 - 1)\right]^{1/(1-\gamma)} \left[\frac{(\gamma+1)M_1^2}{(\gamma-1)M_1^2 + 2}\right]^{\gamma/(\gamma-1)} \quad (6\text{-}44)$$

This result is also plotted on Fig. 6-11 for $\gamma = 1.4$. The tabulated values for all of these ratios can be found in most textbooks on gasdynamics for $\gamma = 1.4$.

The following example illustrates the use of the above shock wave relations.

EXAMPLE PROBLEM 6-5. A normal shock wave occurs in the test section of the Mach 2 supersonic wind tunnel described in Example Probs. 6-3 and 6-4. Calculate the static temperature, static and stagnation pressures, Mach number and velocity after this shock.

The static pressure and temperature in the test section upstream of the normal shock were calculated in Example Prob. 6-3. The static pressure after the shock is obtained from Eq. (6-41)

$$P_2 = P_1 \left[1 + \frac{2\gamma}{\gamma+1}(M_1^2 - 1)\right]$$

$$= 1.28 \left[1 + \frac{(2)(1.4)}{1+1.4}(2^2 - 1)\right]$$

$$= 5.76 \text{ atm}$$

and the static temperature after the shock from Eq. (6-42)

$$T_2 = T_1 \left[1 + \frac{2(\gamma - 1)(\gamma M_1^2 + 1)(M_1^2 - 1)}{(\gamma + 1)^2 M_1^2}\right]$$

$$= 172.9 \left[1 + \frac{0.8(1.4 \times 2^2 + 1)(2^2 - 1)}{2.4^2 \times 2^2}\right]$$

$$= 291.8 \text{ K } (65\,°F)$$

The downstream Mach number is obtained from Eq. (6-40)

$$M_2 = \left[\frac{(\gamma - 1) M_1^2 + 2}{2\gamma M_1^2 + 1 - \gamma}\right]^{1/2}$$

$$= \left(\frac{0.4 \times 2^2 + 2}{2 \times 1.4 \times 2^2 - 0.4}\right)^{1/2}$$

$$= 0.577$$

Equation (6-44) can be used to calculate the stagnation pressure after the shock, but since we know the Mach number and static pressure after the shock, the use of Eq. (6-20) is much more convenient.

$$P_{02} = P_2 \left(1 + \frac{\gamma - 1}{2} M_2^2\right)^{\gamma/(\gamma - 1)}$$

$$= 5.76 \text{ atm} \left(1 + \frac{0.4}{2} \times 0.577^2\right)^{1.4/0.4}$$

$$= 7.21 \text{ atm}$$

The velocity downstream of the shock can be conveniently calculated from Eq. (6-13)

$$v_2 = M_2 a_2 = M_2\, 20.0\, \sqrt{T_2}$$

$$= 0.577 \times 20.0 \sqrt{291.8}$$

$$= 197.1 \text{ m/sec } (647 \text{ ft/sec})$$

We see that in crossing the shock, the flow goes from a Mach 2 supersonic flow to a Mach 0.58 subsonic flow and that the static pressure and temperature increased by 350 and 69 percent respectively while the total pressure and velocity decreased by 28 percent and 63 percent respectively.

These dramatic changes in the flow properties across the shock increase in magnitude with Mach number (Fig. 6-11) until finally real gas effects can no longer be ignored around a Mach number 5. The flows above Mach number 5 are usually classified as hypersonic flows.

It should be noted that a shock wave also precedes an object that moves through an atmosphere at supersonic speeds in order to quickly accelerate the motionless gas out of the object's way. This shock wave, which is called a bow shock, curves around the nose of the object and weakens as the distance from the body increases. The normal shock wave relations developed here would not apply for the curved portion of the shock wave but does apply at the nose where the flow is normal to the shock.

It is also interesting to note that the shock wave spectrum ranges from the weak bow shock which forms in front of the supersonic earth as it moves through very low density space, to the current development of laser implosion techniques to produce shocks so strong that they produce pressure and temperatures comparable to those existing in the core of stars. This would make hydrogen fusion power possible.

The Effect of Heat Transfer and Friction on Compressible Flows

In the previous section, the Rayleigh line was formed by combining the ideal gas equation of state, Eq. (5-1), with the mass balance, Eq. (6-33), and the frictionless momentum balance, Eq. (6-35), for a steady flow in a constant area system. States along this line do not satisfy the adiabatic energy balance, Eq. (6-14), which would require the stagnation enthalpy h_0 to be constant. The Rayleigh line, therefore, corresponds to a constant area, frictionless flow with heat transfer. If we examine the Rayleigh line depicted in Fig. 6-10, we can get some qualitative information on the effect heat transfer has on compressible flows.

It was noted that the top section of this curve corresponds to subsonic flows; the bottom section is supersonic and the point of maximum entropy is the sonic condition. This implies that the Mach number increases as we move along the curve from top to bottom. Since the entropy must increase as heat is added, the flow would be driven towards the sonic velocity from either section of the curve by heating. Once the flow becomes sonic, it will choke. There is, therefore, a maximum amount of heat that can be added to a constant area system with a given mass flow rate and fixed inlet properties. This would be, of course, an important consideration in the design of gas heat exchangers. It should be noted that frictional effects would be very important in a high speed flow. Therefore, any quantitative analysis using the Rayleigh line as its model would be limited to very low Mach numbers in accordance with the frictionless assumption.

The Fanno line employs the adiabatic energy balance, Eq. (6-14), in place of the frictionless momentum balance used in the development of the Rayleigh line.

The Fanno line, therefore, corresponds to a constant area, adiabatic flow with friction. Since friction in an adiabatic flow increases the entropy, the effect of friction is also to drive the flow towards the choked condition. This implies that there is a maximum pipe length for a given flow rate and inlet conditions. In fact, tubes carrying supersonic flows are limited by friction to very short lengths.

PERFORMANCE OF REAL NOZZLES AND DIFFUSERS

The efficiencies of actual nozzles and diffusers are measured in several different ways depending upon the specific application. The purpose of turbine nozzles, for instance, is to deliver linear kinetic energy to a turbine so that it can be converted to the more useful power form of torque applied to a rotating shaft. The performance of these nozzles are based, therefore, upon the kinetic energy they delivered as compared to a reversible nozzle having the same inlet conditions and discharge pressure. Since most nozzles are reasonably adiabatic, the reversible nozzle used for comparison is assumed to be isentropic. Thus,

$$\eta_N = \frac{\text{Actual kinetic energy at exit}}{\substack{\text{Kinetic energy obtained by isentropic} \\ \text{expansion to discharge pressure}}} = \frac{v_2^2/2}{v_{2s}^2/2}$$

The velocity at the inlet is usually small enough that the inlet enthalpy can be assumed to be equal to the stagnation enthalpy h_0. Referring back to Eq. (6-14), the nozzle efficiencies can be expressed by the relationship

$$\eta_N = \frac{h_0 - h_2}{h_0 - h_{2s}} \tag{6-45}$$

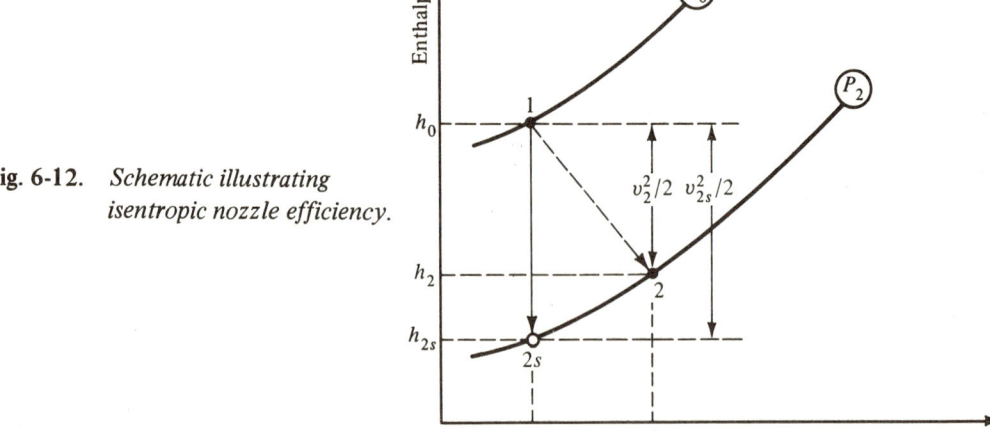

Fig. 6-12. *Schematic illustrating isentropic nozzle efficiency.*

h_2 is the exit enthalpy and h_{2s} is the enthalpy that would be obtained from an isentropic expansion to the same exit pressure as shown in Fig. 6-12. For an ideal gas, Eq. (6-45) can be simplified to

$$\eta_N = \frac{v_2^2/2}{c_P(T_0 - T_{2s})} \tag{6-46}$$

Frictional effects and shocks when they occur cause irreversibilities in actual nozzles and, therefore, loss in efficiency, but proper design can eliminate shocks and minimize the influence of friction. Well designed nozzles operating at their designed pressure ratios can have efficiencies of the order of 90 to 95 percent.

EXAMPLE PROBLEM 6-6. Steam enters a turbine nozzle at 2,500 psia and 1,000 °F and is expanded adiabatically to an exit pressure of 2,400 psia. Calculate the exit velocity assuming a 95 percent efficiency and a negligible inlet velocity.

From Eq. (6-14), we see that

$$h_0 - h_2 = \frac{v_2^2}{2}$$

If this result is introduced into Eq. (6-45), we obtain the expression

$$v_2 = \sqrt{2\eta_N(h_0 - h_{2s})}$$

From the steam tables, Table A-2, we find that at 1,000 °F and 2,500 psia

$$h_1 = h_0 = 1457.5 \text{ BTU/lb}_m$$

$$s_1 = s_0 = 1.527 \text{BTU/lb}_m \text{ °R}$$

Further, at 2,400 psia and an entropy of 1.527 BTU/lb$_m$ °R, the enthalpy is

$$h_{2s} = 1451.9 \text{ BTU/lb}_m$$

Then,

$$v_2 = \sqrt{(2)(0.95)(1457.5 - 1451.9) \text{ BTU/lb}_m \ 778.2 \text{ ft·lb}_f/\text{BTU} \ 32.2 \text{ ft/sec}^2}$$

$$= 516 \text{ ft/sec}$$

If, instead of the above procedure, we treat steam as an ideal gas with constant heat capacity, then Eq. (6-46) gives

$$v_2 = \sqrt{2\eta_N c_P T_0 (1 - T_{2s}/T_0)}$$

Using Eq. (5-12) to change the temperature ratio into the known pressure ratio gives

$$v_2 = \sqrt{2\eta_N c_P T_0 [1 - (P_{2s}/P_0)^{(\gamma-1)/\gamma}]}$$

If we use an average c_P of 0.72 BTU/lb$_m$ °R and an average γ of 1.33, we then have

$$v_2 = \sqrt{2 \times 0.95 \times 0.72 \text{ BTU/lb}_m \text{ R} \times 1460 \text{ R} \left[1 - \left(\frac{2400 \text{ psia}}{2500 \text{ psia}}\right)^{0.33/1.33}\right]}$$

$$\times \sqrt{778.2 \text{ ft·lb}_f/\text{BTU} \times 32.2 \text{ ft/sec}^2}$$

$$= 713 \text{ ft/sec}$$

which overestimates the steam table values by 38 percent. This discrepancy should not be surprising when the ideal gas assumption is used at this high pressure.

In the case of most diffusers, the purpose of the diffuser is to decelerate the flow as reversibly as possible. Most diffusers are essentially adiabatic so that the corresponding reversible process used for comparison is also the isentropic process. The efficiency of a diffuser is calculated by taking the kinetic energy change that occurs in the actual diffuser and dividing it into the kinetic energy change that is required by an isentropic diffuser to produce the same discharge pressure. That is

$$\eta_D = \frac{v_1^2/2 - v_{2s}^2/2}{v_1^2/2 - v_2^2/2}$$

where v_1 is the diffuser inlet velocity, v_2 is the discharge velocity of the actual diffuser, and v_{2s} is the discharge velocity of an adiabatic and reversible diffuser discharging at the same final pressure as the actual diffuser. The energy balance equation applied to the diffuser system will again simplify to Eq. (6-14). Therefore,

$$\eta_D = \frac{h_{2s} - h_1}{h_2 - h_1}$$

The discharge velocity is usually small and the exit enthalpy can be assumed to be equal to the stagnation enthalpy. The above definition then takes the form

$$\eta_D = \frac{h_{2s} - h_1}{v_1^2/2} = \frac{h_{2s} - h_1}{h_0 - h_1} \tag{6-47}$$

The relationship between these variables is shown schematically in Fig. 6-13. The

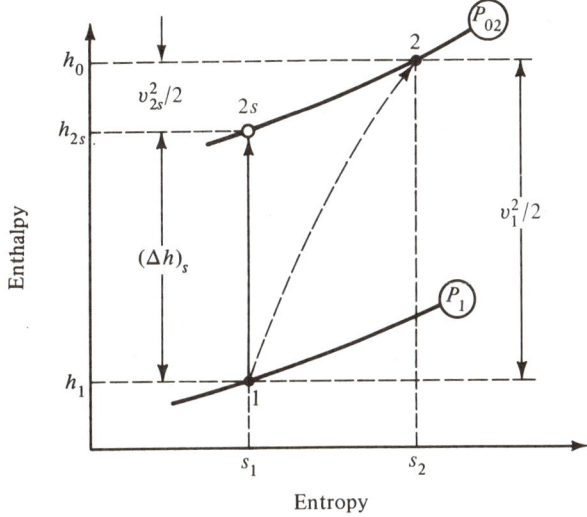

Fig. 6-13. *Schematic illustrating the isentropic diffuser efficiency.*

efficiencies obtained by actual diffusers are much lower than those obtained by nozzles. This is a result of viscous flows being unstable in the increasing pressure field of a diffuser and generally stable in the decreasing pressure field of a nozzle. High diffuser efficiencies can be maintained only for very small diffuser angles which would make them too large and expensive for most practical applications.

EXAMPLE PROBLEM 6-7. An aircraft is flying at a Mach number of 0.9 at 35,000 feet where the temperature is $-70\,°F$ and the pressure is 3 psia. If the inlet diffuser to the jet engine has an isentropic efficiency of 85 percent, what will the pressure be at the compressor face assuming that the specific kinetic energy is neglectable there?

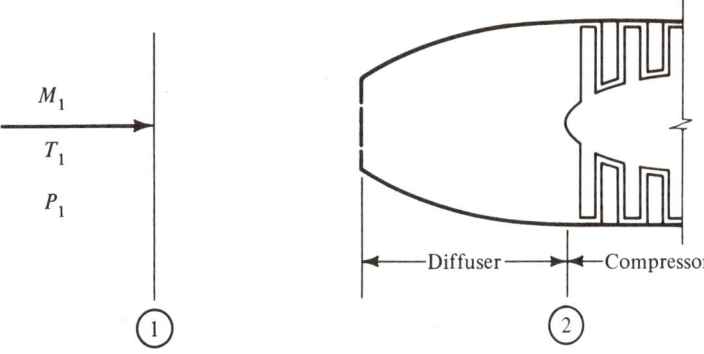

Fig. 6-14.

If we make the ideal gas assumption in the isentropic diffuser efficiency equation, Eq. (6-47), then

$$\eta_D = \frac{h_{2s} - h_1}{h_0 - h_1} = \frac{c_p(T_{2s} - T_1)}{c_p(T_0 - T_1)} = \frac{T_{2s} - T_1}{T_0 - T_1}$$

and

$$T_{2s} = T_1 + \eta_D(T_0 - T_1)$$

The stagnation temperature can be calculated from Eq. (6-19),

$$T_0 = \left(1 + \frac{\gamma - 1}{2} M_1^2\right) T_1$$

$$= \left[1 + \frac{1.4 - 1}{2}(0.9)^2\right](460 - 70)$$

$$= 453 \,°R$$

Thus,

$$T_{2s} = 390 + (0.85)(453 - 390) = 444 \,°R$$

By definition, the discharge pressure of the actual diffuser and the isentropic diffuser are the same. Thus,

$$P_{2s} = P_2$$

so that

$$\frac{P_{2s}}{P_1} = \frac{P_2}{P_1}$$

From the isentropic relationship given by Eq. (5-12),

$$\frac{P_{2s}}{P_1} = \left(\frac{T_{2s}}{T_1}\right)^{\gamma/(\gamma-1)}$$

we find that

$$P_2 = (3 \text{ psia})\left(\frac{444 \,°R}{390 \,°R}\right)^{1.4/0.4}$$

$$= 4.72 \text{ psia}$$

This pressure rise reduces the amount of work that the compressor must do to compress the air to the required pressure. At supersonic speeds, much of the compression is accomplished by a normal shock in front of the diffuser or shocks inside the diffuser. At very high speeds, the pressure rise in the diffuser can be so large that no compressor is needed in the engine. This is the principle of operation of a ram jet.

PROBLEMS

6-1. Recalculate the pressure drop in Example Prob. 6-1 if the elevation of the pipe outlet is 30 meters above the elevation of the pipe entrance.

6-2. One hundred million standard cu ft (60 °F, 1 atm) per day of radioactive waste gas at 1,000 °F must be released at a height of 400 ft above the ground to avoid contamination of the surrounding area. A circular stack of uniform diameter is to be used. A draft at the base of the stack of 1 in. of water will be required (pressure inside stack base is 1 in. H_2O less than barometric pressure). The barometric pressure at the base of the stack is 740 mm Hg and the ambient temperature 60 °F. The gas has a molecular weight of 32 and may be considered an ideal gas. Assume the gas passes through the stack isothermally. What stack diameter will be required?

The lost work of gas flowing through the stack may be approximated by the equation

$$lw = \frac{0.032\, zv^2}{g_c D} \quad \text{where } lw = \text{lost work in ft } lb_f/lb_m$$

$$z = \text{height in ft}$$

$$D = \text{diameter in ft}$$

$$v = \text{velocity in ft/sec}$$

6-3. Despite the high molecular weight of uranium hexafluoride UF_6, its vapor closely approximates the properties of an ideal gas. Calculate the speed of sound in UF_6 and hydrogen at 50 °C. The molar heat capacity at constant pressure for UF_6 at 50 °C is 31.75 cal/mole/°C.

6-4. In deriving the speed of sound equation, Eq. (6-11), use a system that encloses the pressure wave front and moves with the wave instead of the system depicted in Fig. 6-1.

Hint: Write the balance equations in terms of an observer moving with the wave front.

6-5. Do Prob. 6-4 but substitute the energy balance equation, Eq. (2-25), for the momentum balance equation, Eq. (6-10), in the analysis.

Hint: Remember that the velocity change and the thermodynamic property changes across the pressure wave front are assumed to be very small. In the analysis, neglect all second-degree differentials. For example, $(dP)^2$ is neglectable since dP is very small.

6-6. The coefficient of isentropic compressibility

$$\kappa_s = -\frac{1}{v}\left(\frac{\partial v}{\partial P}\right)_s$$

is a property that is tabulated for many liquids. Show that the speed of sound can be calculated from the expression

$$a = \sqrt{1/\rho \kappa_s}$$

Calculate the speed of sound in water at 15 °C if $\rho = 1$ gm/cc and $\kappa_s = 47 \times 10^{-6}$ bar^{-1}.

6-7. Nitrogen flows through a 1 ft^2 cross-section of a nozzle with a Mach number of 0.5. What is the cross-sectional area where the Mach number is 2.0? Assume the flow is adiabatic and reversible.

6-8. A small converging-diverging nozzle has an exit area of 5 cm^2 and produces an exit air flow of Mach 1.5 at atmospheric pressure, and a stagnation temperature of 100 °C. What is the (a) mass flow rate, (b) throat area, (c) exit temperature, and (d) stagnation pressure?

6-9. A converging nozzle with an exit area of 50 cm^2 is attached to a large air reservoir which has a pressure of 10 atm and a temperature of 300 °C. Estimate the mass flow rate through the nozzle if the back pressure outside the reservoir is (a) 2 atm and (b) 7 atm.

6-10. The test section specifications for supersonic wind tunnel are

Mach number	4
Static temperature	−56 °C
Static pressure	0.072 bars
Area	0.1 m^2

(a) Calculate the reservoir temperature and pressure.
(b) What is the mass flow rate?

6-11. Complete the derivation of Eqs. (6-40) and (6-43).

6-12. A converging-diverging nozzle has a designed exit Mach number of 2.0. The nozzle is supplied by an air reservoir at 100 atm. What is the maximum back pressure that will

    (a) choke the flow?
    (b) result in an exit Mach number of 2?

6-13. A blast wave from an atomic bomb travels at 5,500 m/sec in air at 1 atm and 0 °C. What would the pressure behind the wave be if air were to act as an ideal gas?

6-14. Show that the point of maximum entropy on a Fanno line must correspond to sonic flow.
Hint: Write the mass and energy balance equation for a process involving infinitesimal changes. Then show that the velocity is equal to the local speed of sound in the neighborhood of the point of maximum entropy ($ds = 0$).

6-15. Using an ideal gas medium, demonstrate that the friction in a pipe may cause the pressure to increase or decrease along the flow depending upon whether the velocity is greater than or less than speed of sound.
Hint: Use the Fanno line depicted in Fig. 6-10.

6-16. A Mach 2 stream of nitrogen with a pressure of 1 bar and a temperature of 0 °C passes through a normal shock.

    (a) Calculate the final velocity and stagnation pressure.
    (b) Draw the corresponding Fanno and Rayleigh lines.

6-17. The exit area for a converging-diverging nozzle is 2.5 times as large as the throat area. The reservoir pressure and temperature are 10 atm and 100 °C, respectively. What must the back pressure be to produce a normal shock at the exit plane of the nozzle?

6-18. Air is flowing through a duct in which the following properties were measured at a cross-section: $T_1 = 260$ °C, $p_1 = 1$ atm and $v_1 = 46$ m/sec. At a second cross section further downstream, the following properties were also measured: $T_2 = 200$ °C, $P_2 = 1/2$ atm and $v_2 = 386$ m/sec. How much heat, if any, is added to the flow between these two sections?

6-19. Air is expanded through a nozzle from a reservoir pressure and temperature of 100 psia and 1000 °F to a discharge pressure of 1 atm. At what velocity and Mach number does the air leave the nozzle if the nozzle efficiency is 0.92?

6-20. Air enters a diffuser at 50 °C and 1 atm and is discharged at 2 atm. What is the inlet velocity and Mach number if the diffuser is 80 percent efficient?

6-21. A diffuser is to reduce the velocity of an air stream initially at 150 m/sec, 1 atm, and 20 °C down to 50 m/sec. Estimate the exit to inlet area ratio required to accomplish this. Assume the diffuser has an efficiency of 0.85.

6-22. In Example Prob. 6-7, the aircraft is now flying at a Mach number of 2 at the same altitude. What is the pressure at the compressor face if a normal shock lies in the entrance plane of the diffuser? Assume that the isentropic efficiency of the subsonic portion of the diffuser is still 85 percent and the kinetic energy at the compressor face can be neglected. What is the overall isentropic efficiency of this diffuser?

7

Thermodynamics of Heat Engines

OBJECTIVES

After studying this chapter, the student should be able to

(a) discuss the Otto, Carnot, Diesel, Stirling, Rankine and Brayton power cycles including their advantages and limitations;

(2) define the terms compression ratio, mean effective pressure, indicated and brake thermal efficiencies, and air-standard cycle;

(3) analyze and describe the purpose of interstage cooling, a regenerator, a turbo-supercharger, a condenser, a reheater and a feed water heater;

(4) describe the basic refrigeration cycles and determine the coefficient of performance for a reversible refrigeration system.

INTRODUCTION

Three important qualities of energy have been discussed in the previous chapters. The first of these is that energy can be stored in various forms--for instance: (1) as internal energy in fuels, (2) as kinetic energy in a revolving flywheel, (3) as potential energy in an elevated water reservoir. The second is that heat and work are forms of energy. The third quality is that energy can be changed from one form to another. This chapter will describe some of the processes currently used or being developed for heat engines which transfer energy from the less useful state of internal energy to the more useful form of mechanical energy. Refrigeration systems and heat pumps will also be discussed since they are reverse heat engines.

Man first learned to utilize the transformation of energy for his own benefit in burning wood for heat, that is in converting internal energy into heat. Man himself and domestic animals provided the initial sources for mechanical power, but he slowly discovered how to use sails, wind mills and waterwheels to convert the kinetic energy contained in the wind and falling water into useful power. These primitive devices were only useful where and when falling water and winds were available. It was not until the development of heat engines in the eighteenth century that man finally had the means to supply large quantities of power on demand.

Reciprocating engines which presently account for nearly 95 percent of the nation's prime mover[†] capacity were the first successful heat engines developed. These devices were an outgrowth of the piston-and-cylinder air pump which was introduced in the 17th century. In fact, the reciprocating engine must operate as an efficient pump over much of its working cycle. For this reason we will begin our discussion of power cycles with an introduction to reciprocating compressors.

RECIPROCATING ENGINES

Gas Compressors

Figure 7-1 depicts a reciprocating gas compressor attached to a device which

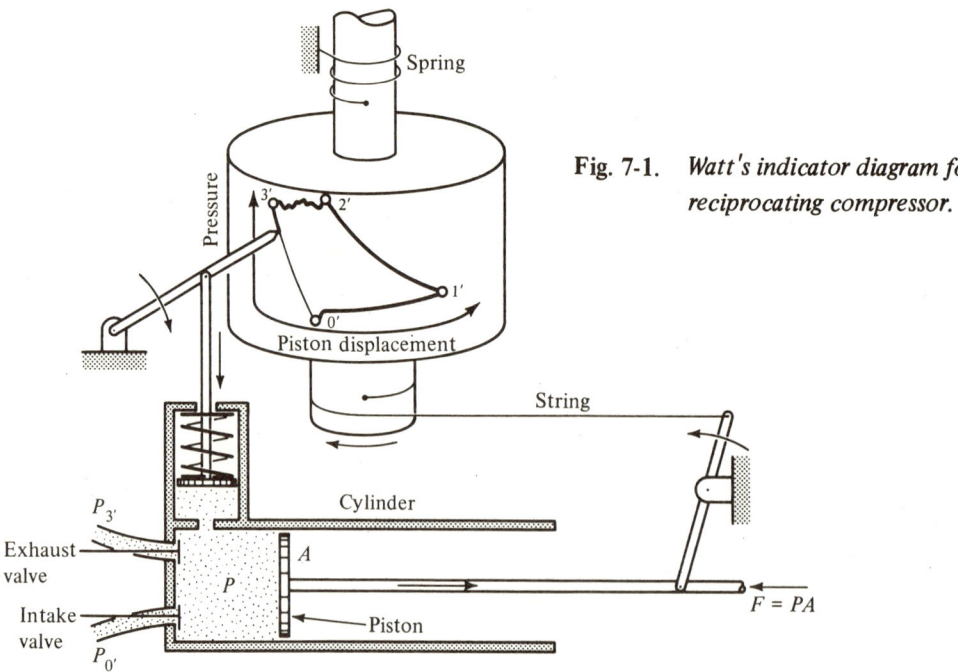

Fig. 7-1. *Watt's indicator diagram for a reciprocating compressor.*

[†] Defined as engines for converting fuels to mechanical energy.

plots the cylinder pressure versus the piston displacement. We see that this plot, which is called an indicator diagram, is really just a *PV* diagram. The enclosed area on this diagram would, therefore, represent the work per cycle that the piston does on the gas and is called the "indicated work." The work that the compressor itself absorbs per cycle would be greater than the indicated work due to friction and other mechanical losses. The indicator diagram has played an important part in the development of reciprocating engines even though it was kept a trade secret for many years by its inventor, James Watt, during his early work on steam engines.

The compressor's cycle begins with the intake stroke $0' - 1'$ as shown in Figs. 7-1 and 7-2. Low pressure gas is drawn through the intake valve into the

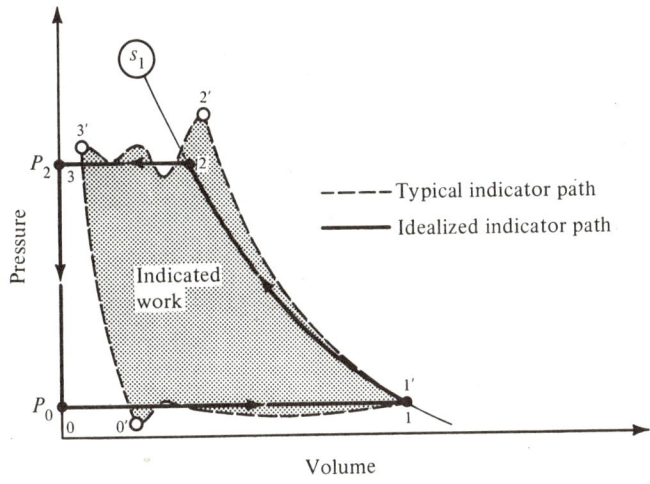

Fig. 7-2. *Typical indicator diagram for a reciprocating compressor and the corresponding reversible cycle.*

cylinder as the piston moves between the displacements $0'$ and $1'$. The intake valve then closes and the gas is compressed along path $1' - 2'$ to the discharge pressure during the first segment of the piston's compression stroke. Next the exhaust valve opens and most of the gas is discharged as the piston moves between displacement positions $2'$ and $3'$. The exhaust valve then closes and the piston again moves outward until the chamber pressure is slightly below the inlet pressure, whereupon the intake valve opens to repeat the cycle. The volume in the cylinder at the point where the intake valve opens is called the clearance volume.

To analyze the basic performance characteristics of reciprocating compressors we will consider the idealized compressor cycle which is superimposed on the actual cycle in Fig. 7-2. This idealized reversible cycle assumes that the intake (0—1) and discharge (2—3) processes are isobaric and that all the gas is discharged at 3. Unless there is some special cooling provision made, the actual compression process is essentially adiabatic. This will be idealized as the adiabatic reversible or isentropic process 1—2 in our model. In this chapter, processes in which a state variable remains constant will be represented by a line on the process diagram ending with a circle in which the constant property is identified.

Our idealized cycle obviously does not faithfully reproduce the actual compressor cycle, but it does qualitatively characterize the basic processes. We, therefore, have a simple model from which approximate quantitative estimates can be made and some general conclusions drawn without knowing the details of the compressor itself. Engineers commonly use highly idealized models like this one not only to render the analysis of actual systems tractable, but also to provide convenient standards to which the performance of actual systems can be compared. For instance, an idealized isentropic compressor will typically absorb only 75 to 90 percent of the work used by an actual compressor to compress the same amount of gas. A compressor's isentropic efficiency, η_{comp}, is therefore defined as

$$\eta_{comp} = \frac{W_{adiab,\,rev}}{W_{act}} \qquad (3\text{-}20)$$

where W_{act} represents the work absorbed by the actual compressor and $W_{adiab,\,rev}$ is the work absorbed by the idealized isentropic compressor operating between the same inlet and discharge pressures. A designer would be able to estimate the power requirement of an actual compressor by calculating the power absorbed by an ideal compressor and dividing the result by the measured efficiency of comparable compressors.

EXAMPLE PROBLEM 7-1. Calculate the power needed to compress 1,000 pounds of air per hour (454 kg/hr) adiabatically from 1 atmosphere and 80 °F (27 °C) to 7 atmospheres. Assume air to be an ideal gas and the compressor's isentropic efficiency to be 90 percent.

We will take the compressor as our system and base our calculations on one complete cycle of the pump. After neglecting any kinetic and potential energy changes, the generalized mass and energy balance equations reduce to

$$0 = m(h_1 - h_2) - m w_{adiab}$$

where m is the mass of air that enters and leaves the pump per cycle, h_1 and h_2 are the specific enthalpies that the air enters and leaves with respectively, and w_{adiab} is the specific pump work. Thus

$$w_{adiab} = h_1 - h_2 = -\int_1^2 dh = -\int_1^2 (T\,ds + v\,dP)$$

where Eq. (4-18) was utilized in the last step. For a reversible adiabatic pump, the compression process is isentropic. That is,

$$ds = 0$$

which implies that

$$w_{adiab,rev} = -\int_{P_1}^{P_2} v\, dP$$

The enclosed area 0–1–2–3–0 in Fig. 7-2 therefore represents the work performed on the pump during one cycle. The use of the isentropic relationship for an ideal gas with constant heat capacities

$$Pv^\gamma = P_1 v_1^\gamma = \text{constant} \tag{5-11}$$

reduces the above integral to

$$w_{adiab,rev} = -P_1^{1/\gamma} v_1 \int_{P_1}^{P_2} P^{-1/\gamma}\, dP = \frac{-\gamma}{\gamma-1} P_1 v_1 \left[\left(\frac{P_2}{P_1}\right)^{(\gamma-1)/\gamma} - 1\right]$$

$$= \frac{-\gamma}{\gamma-1} RT_1 \left(r_P^{(\gamma-1)/\gamma} - 1\right) \tag{7-1}$$

where r_P is the pressure ratio, P_2/P_1. Numerical evaluation of this expression gives

$$w_{adiab,rev} = -\frac{1.4}{0.4} \times \frac{1.986\ \text{BTU}}{\text{lb mole}\ ^\circ\text{R}} \times \frac{539.7\ ^\circ\text{R}}{28.96\ \text{lb}_m/\text{lb mole}}$$

$$\times \left[\left(\frac{7\ \text{atm}}{1\ \text{atm}}\right)^{0.4/1.4} - 1\right] = -96.3\ \frac{\text{BTU}}{\text{lb}_m\ \text{air}}$$

The actual specific work needed is

$$w_{act} = \frac{w_{adiab,rev}}{\eta_{comp}} = (-96.3\ \text{BTU/lb}_m)/0.9 = -107.0\ \text{BTU/lb}_m\ \text{air}$$

resulting in a power requirement of

$$\dot{W} = \dot{m} w_{act} = -1{,}000\ \frac{\text{lb}_m}{\text{hr}} \times 107.0\ \frac{\text{BTU}}{\text{lb}_m} = -1.07 \times 10^5\ \frac{\text{BTU}}{\text{hr}}\ (-31.3\ \text{kW})$$

The minus sign indicates that work is done on the system.

We can reduce the amount of work required to compress this gas if the compression process is carried out isothermally. In this case, the energy balance taken over one cycle reduces to the form

$$h_2 - h_1 = q - w$$

or

$$w = q - (h_2 - h_1) = q - \int_1^2 (T\,ds + v\,dP)$$

For a reversible isothermal compression process, this equation becomes

$$w_{T,\text{rev}} = q_{\text{rev}} - T(s_2 - s_1) - \int_{P_1}^{P_2} v_T\,dP$$

The entropy balance for a reversible process is given by Eq. (3-4). For an isothermal process in a closed system, this equation simplifies to

$$\frac{q_{\text{rev}}}{T} = s_2 - s_1$$

After combining the above two equations, we have

$$w_{T,\text{rev}} = -\int_{P_1}^{P_2} v_T\,dP$$

For instance, the isothermal compression work in the above example is

$$w_{T,\text{rev}} = -\int_{P_1}^{P_2} v_T\,dP = -RT_1 \int_{P_1}^{P_2} dP/P = -RT_1 \ln P_2/P_1$$

$$= -\frac{1.986 \times 539.7 \times \ln 7}{28.96} = -72.0 \text{ BTU/lb}_m \text{ air}$$

which is 25 percent less than the work required by the isentropic compressor in Example Prob. 7-1. This result is shown graphically in Fig. 7-3 with path 1–2″ depicting the isothermal process.

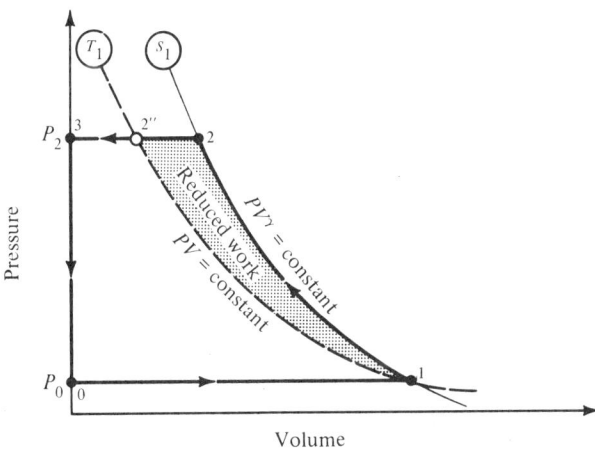

Fig. 7-3. *The influence of heat transfer in reducing the compression work.*

Obtaining an isothermal compression in an actual system is very difficult, but various cooling techniques can cause the actual compression process to fall between the isentropic process Pv^γ = constant and the isothermal process Pv = constant. These processes are called polytropic processes and are denoted by Pv^k = constant where k is called the polytropic coefficient and has a value between 1 and γ. For instance, cooling of a low speed compressor by the use of cooling fins or a water-jacket can drop the polytropic coefficient from 1.4 (the value of γ for a diatomic gas) to around 1.2 while direct cooling by spraying water into the cylinder itself can be used to obtain the same reduction at higher compressor speeds.

When comparatively large pressure changes are desired, multi-stage compression is used with each stage usually having a pressure ratio $r_p = P_2/P_1$ between 3 and 7. In this case, cooling the gas between the stages can be quite effective. Figure 7-4 shows an example of interstage cooling where the gas is compressed

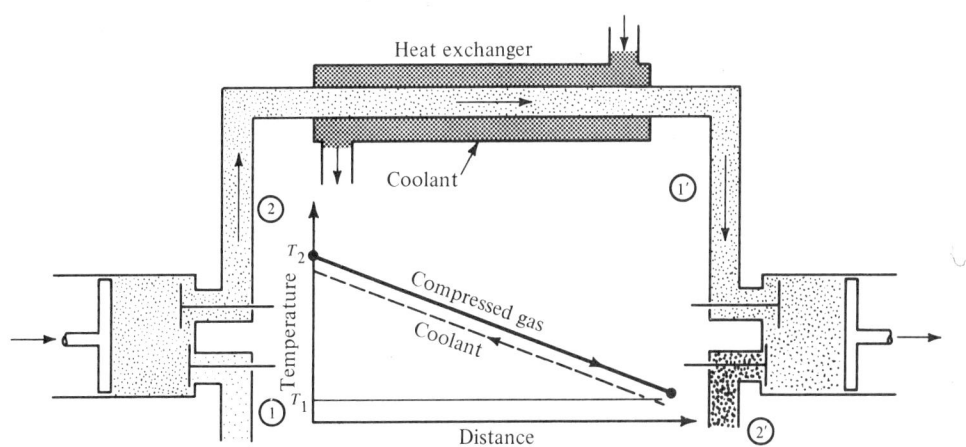

Fig. 7-4. *Multistage compression with interstage cooling.*

from state 1 to state 2 in the first compressor; passes through a counterflow heat exchanger 2–1' in which the compressed gas is cooled to its initial temperature T_1; and then enters a second compressor to be compressed from state 1' to 2'. The corresponding idealized PV and TS diagrams for this system are shown in Fig. 7-5 where the compression processes are assumed to be isentropic and any

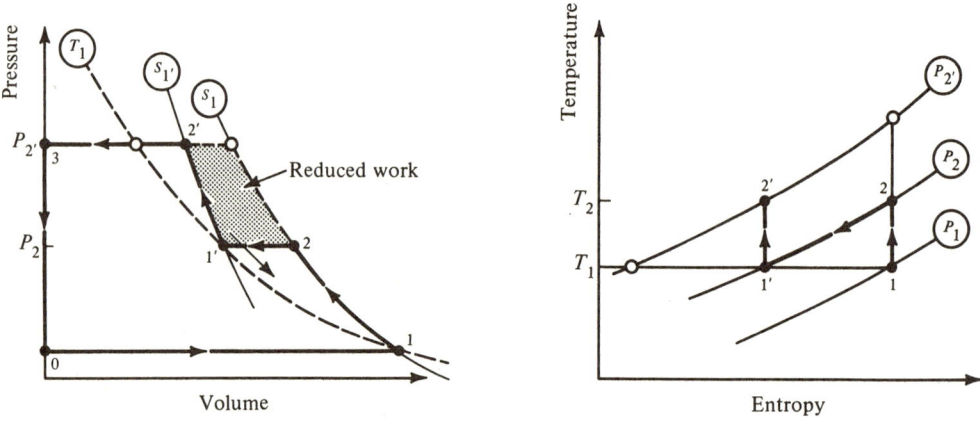

Fig. 7-5. *PV and TS diagram for multistage compression with interstage cooling.*

pressure drop in the heat exchanger is neglected so that $P_{1'} = P_2$. The hatched area on the PV diagram represents the work saved by the interstage cooling.

The question arises as to what the pressure ratios across the two compressors P_2/P_1 and $P_{2'}/P_2$, should be to minimize the total compression work,

$$w_{total} = w_{stage_1} + w_{stage_2}$$

If the fluid is an ideal gas, the work absorbed by an isentropic compressor can be calculated from Eq. (7-1),

$$w_{total} = -\frac{\gamma R T_1}{\gamma - 1}\left[\left(\frac{P_2}{P_1}\right)^{(\gamma-1)/\gamma} - 1\right] - \frac{\gamma R T_1}{\gamma - 1}\left[\left(\frac{P_{2'}}{P_{1'}}\right)^{(\gamma-1)/\gamma} - 1\right]$$

Since $P_{1'} = P_2$

$$w_{total} = -\frac{\gamma R T_1}{\gamma - 1}\left[\left(\frac{P_2}{P_1}\right)^{(\gamma-1)/\gamma} + \left(\frac{P_{2'}}{P_2}\right)^{(\gamma-1)/\gamma} - 2\right]$$

To find the interstage pressure which will minimize the total work, we differentiate the total work with respect to the unknown interstage pressure P_2 and equate the result to zero to obtain the answer

$$P_2 = (P_1 P_{2'})^{1/2}$$

or

$$\frac{P_2}{P_1} = \sqrt{\frac{P_{2'}}{P_1}} = \frac{P_{2'}}{P_{1'}}$$

If the compression process in Example Prob. 7-1 had used two compressors with interstage cooling, the pressure ratio of each compressor should be

$$r_P = \frac{P_2}{P_1} = \frac{P_{2'}}{P_{1'}} = \sqrt{\frac{P_{2'}}{P_1}} = \sqrt{\frac{7}{1}} = 2.65$$

The total work absorbed by these ideal compressors is

$$w_{\text{total}} = \frac{-2\gamma R T_1}{\gamma - 1}\left[\left(\frac{P_2}{P_1}\right)^{(\gamma-1)/\gamma} - 1\right]$$

$$= \frac{-2 \times 1.4 \times 1.986 \times 539.7}{0.4 \times 28.96}\left(2.65^{0.4/1.4} - 1\right)$$

$$= -83.2 \text{ BTU/lb}_m \text{ air}$$

which is 14 percent less than the work required by the isentropic compressor in Example Prob. 7-1.

This general conclusion that the pressure ratios should be equal for multistage compression with interstage cooling applies with good accuracy to real compressors even though the result itself was obtained from a very elementary analysis. In the examination of various power cycles to follow, we will continue to idealize the cycles to obtain their qualitative characteristics. Accurate quantitative information would have to come from a detailed analysis which, for the additional work involved, may not lead to a much better insight.

The piston-and-cylinder air pump mechanism was converted into a practical "atmospheric steam engine" by Newcomen in 1712. Steam at atmospheric pressure instead of air was drawn into the cylinder from a boiler during the intake stroke. The intake valve was then closed and the steam was condensed by injecting cold water into the cylinder, thereby causing the cylinder pressure to drop below the atmospheric pressure. The resulting pressure difference across the piston forced the piston back down the cylinder doing useful work as it moved.

The water was then drained from the cylinder and the cycle was repeated. Since the pressure difference across the piston would necessarily be less than one atmosphere, the specific power obtained from a system of this type was quite low.

Instead of condensing atmospheric steam to form a partial vacuum in the cylinder, the next generation of steam engines used high pressure steam to obtain higher specific power outputs by essentially running the pump cycle shown in Fig. 7-2 backwards. An idealized cycle would have the high pressure steam entering the cylinder between piston displacements 3 and 2 whereupon the valve is closed. The steam next expands adiabatically and reversibly between displacements 2 and 1 and is then exhausted along path 1—0. The enclosed area 0—1—2—3—0 in Fig. 7-2 now represents the work that the steam does on the piston.

The steam engine proved to be a heavy and expensive engine with a scandalously low thermal efficiency. For example, the steam locomotive which had its last stronghold in the United States had only a 10 percent thermal efficiency when it was finally phased out in the 1950's. Some of the reasons for the steam engine's poor performance will be discussed later in conjunction with the steam turbine system.

It is interesting to note here that the first and second laws of thermodynamics evolved principally from the development of steam engines. It was Sadi Carnot's desire to understand the production of power by steam engines that led him to publish his first and only scientific work in 1824 which contained the general conclusions that were later developed into the second law. Similarly, the first law of thermodynamics which merges the mechanical and thermal forms of energy owes much of its conceptual basis to the energy conversion studies performed on steam engines. Historically, the second law preceded the first law of thermodynamics.

The Otto Cycle

The high capital and operating costs of the steam engine induced many inventors to search for viable alternatives. One idea that received much attention was to replace steam as the working fluid with air to permit the use of internal combustion in the cylinder itself. This would avoid the heavy firebox, boiler and complex system of valves and pipes which the steam engine needed for generating and handling steam. Many of the first experimental internal combustion engines were essentially converted steam engines that employed a two stroke cycle called the Lenoir cycle. The first half of the initial stroke was used to draw an air-fuel mixture into the cylinder through the open intake valve. The mixture was then ignited by an electric spark after the intake valve was closed. The resulting explosion increased the cylinder pressure and the subsequent expansion which completed the first stroke provided the power. The exhaust valve was then opened and the return stroke discharged the spent mixture. Some storage device such as a large flywheel had to be provided to smooth out the large impulses generated by the periodic explosions which also proved to be destructive to the engine. These engines turned out to be no more efficient than the steam engine and therefore had a short history.

Nicolaus Otto demonstrated an internal combustion engine in 1876 which produced a thermal efficiency that was around three times as great as any steam engine of its day. Otto had discovered that the compression of the fuel-air mixture before combustion would increase the engine's efficiency. He, therefore, developed the four stroke cycle depicted in Figs. 7-6 and 7-7. A metered mixture

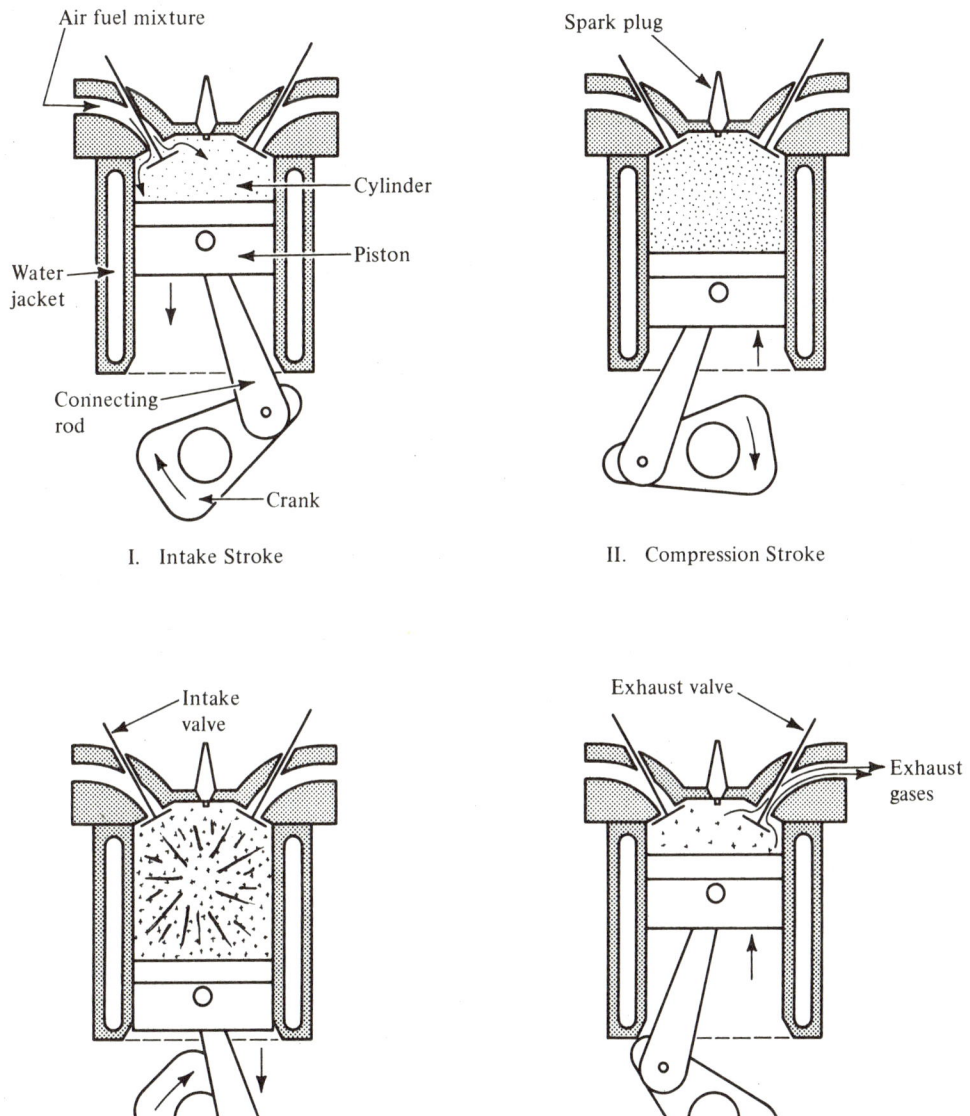

Fig. 7-6. *Otto cycle engine.*

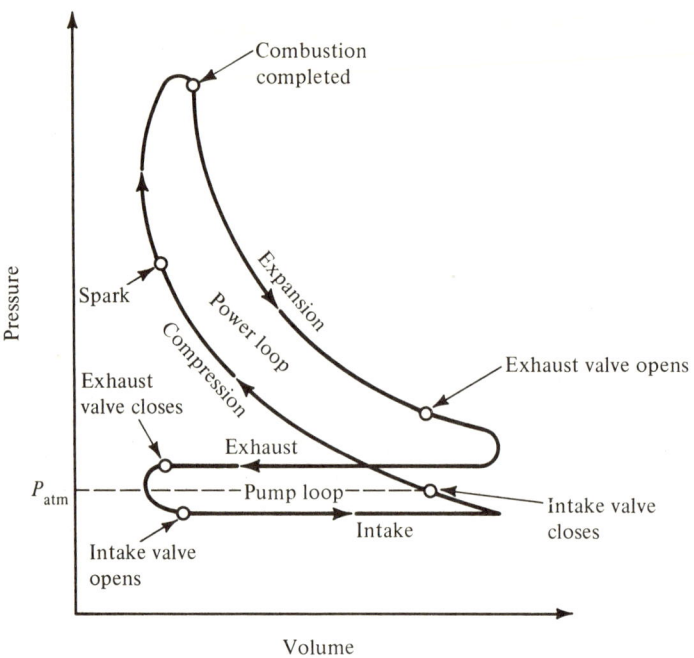

Fig. 7-7. *Typical indicator diagram for an Otto cycle.*

of fuel and air is drawn into the cylinder during the intake stroke, but now that mixture is compressed to a moderate pressure by a compression stroke before ignition occurs. The compression stroke is followed by the expansion or power stroke and finally by the exhaust stroke. Otto was also able to obtain relatively smooth combustion by experimenting with the fuel-air mixture ratio and ignition timing.

The first generation of internal combustion engines weighed about a ton for each horsepower produced and used mainly illuminating gas for the fuel. Gasoline was originally considered to be extremely dangerous and it was not until the carburetor was invented in the 1890's that its use was possible. Many other notable improvements were made to the Otto engine; some of which reduced the weight of the engine to such an extent that it was used to power the early airplanes and many of the first automobiles. Present aircraft Otto cycle engines weigh between 1.1 and 2.6 pounds for each horsepower produced. Automobiles, tractors and small airplanes currently use the Otto cycle almost exclusively whether in the conventional or rotary (Wankel) form because of its high specific power output, smooth and quiet operation and rapid response to load changes. It also serves well in a variety of other applications since it can be manufactured inexpensively in a range of sizes under 500 horsepower (373 kW). Today it is the most common internal combustion engine.

To analyze why the addition of a compression stroke had such a positive influence on the engine's performance, we will again formulate a very simple model. First of all, we note that the air-fuel mixture ratio for internal combustion engines is usually quite large. For instance, the stoichiometric amount of air required to burn 1.0 lb_m of gasoline is 15.1 lb_m which is essentially the ratio of air to fuel used for the Otto cycle. Other engines such as gas turbine systems use mixture ratios as high as 50:1. Because of the large amount of air involved, we will assume that both the original fuel-air mixture and the resulting products of combustion have the properties of air which we will then treat as an ideal gas. Next we approximate the actual open cycle with a reversible closed cycle as shown in Fig. 7-8. This is accomplished by replacing the combustion process with a reversible

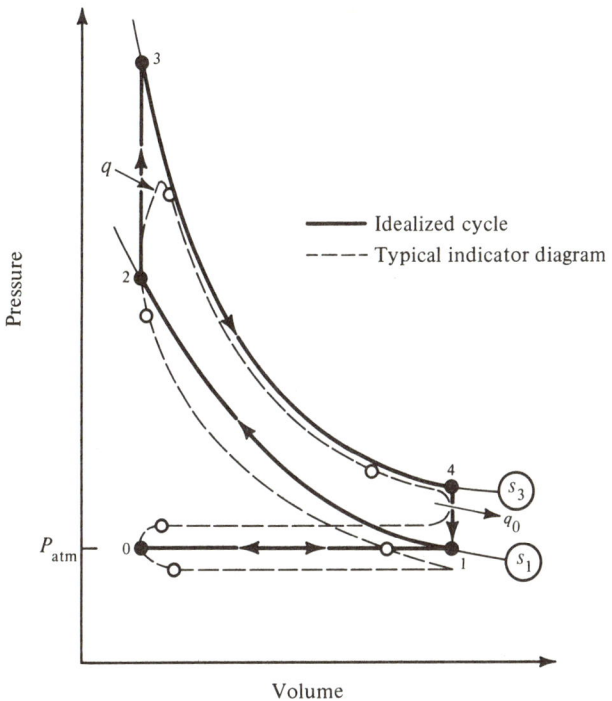

Fig. 7-8. *Indicator diagram for an air-standard Otto cycle.*

heat-transfer process from an external source, and by closing the cycle with a constant volume process instead of the exhaust process used by an actual engine. The combination of all of these assumptions is called the "air-standard cycle." These simplifying assumptions will also be used to facilitate the analysis of other gas power cycles examined in this chapter.

The idealized Otto cycle shown in Fig. 7-8 approximates the compression stroke up to ignition with the adiabatic reversible or isentropic process 1—2; assumes that the combustion process is rapid enough to be represented by the reversible constant volume process 2—3; and approximates the power stroke with the isentropic expansion 3—4. The cycle is then closed with the reversible, constant volume process 4—1. The pump loop, shown in Fig. 7-7 and superimposed on the idealized Otto cycle in Fig. 7-8, typically takes 3 to 7 percent of the engine's power. This pump loop will be neglected by the closed air-standard cycle, but it can also be thought of as having been replaced by an idealized intake, 0—1, and exhaust, 1—0, processes which require no net work. This idealized cycle may again deviate significantly from actual cycles, but the model does incorporate the basic characteristics of the Otto cycle. Therefore, the influence that compression has on the engine's performance can be easily determined from this model.

For instance, we see from the corresponding *TS* diagram, Fig. 7-9, that heat

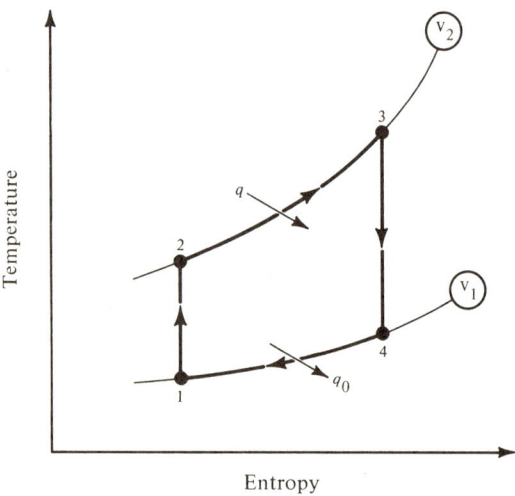

Fig. 7-9. *TS diagram of idealized Otto cycle.*

addition is initiated at state 2. State 2 is at a higher temperature than the inlet temperature T_1 due to the isentropic compression process 1—2. It was pointed out in Chap. 3 that the Carnot cycle has the highest thermal efficiency possible for any heat engine cycle operating between the same two temperature extremes. This is due to the fact that all the heat is added at the cycle's maximum temperature and exhausted at the cycle's minimum temperature. Compression, therefore, improved the performance of the Otto cycle by increasing the average temperature at which heat is added to the cycle.

To quantitatively determine the influence compression has on the idealized Otto cycle, we start with the generalized energy balance equation (2-25). Since there is no mass entering or leaving the idealized engine and the gas returns to its initial state at the end of the cycle, the work performed per unit mass during the cycle must equal the net heat supplied per unit mass,

$$w = q + q_0$$

In the idealized Otto cycle, heat transfer only occurs during the reversible constant volume processes 2–3 and 4–1. Since the working medium is assumed to be an ideal gas with constant heat capacity,

$$w = c_V(T_3 - T_2) + c_V(T_1 - T_4)$$

The thermal efficiency for this cycle is therefore

$$\eta_{therm} = w/q = \frac{c_V(T_3 - T_2) - c_V(T_4 - T_1)}{c_V(T_3 - T_2)}$$

$$= 1 - \frac{T_4 - T_1}{T_3 - T_2} = 1 - \left(\frac{T_1}{T_2}\right)\left[\frac{(T_4/T_1) - 1}{(T_3/T_2) - 1}\right] \quad (7\text{-}2)$$

The above expression can be simplified by using the isentropic relationship between the temperature and volume ratios given by Eq. (5-12). That is,

$$\frac{T_1}{T_2} = \left(\frac{v_2}{v_1}\right)^{\gamma - 1}$$

and

$$\frac{T_3}{T_4} = \left(\frac{v_4}{v_3}\right)^{\gamma - 1} \quad (5\text{-}12)$$

As shown in Fig. 7-8 for the idealized cycle, $v_2 = v_3$ and $v_1 = v_4$. Substitution of these equalities into Eq. (5-12) shows that

$$\frac{T_1}{T_2} = \frac{T_4}{T_3}$$

or

$$\frac{T_4}{T_1} = \frac{T_3}{T_2}$$

Combining these results with Eq. (7-2) will give the air-standard Otto cycle thermal efficiency

$$\eta_{therm} = 1 - \left(\frac{v_2}{v_1}\right)^{\gamma - 1}\left[\frac{(T_4/T_1) - 1}{(T_4/T_1) - 1}\right]$$

or

$$\eta_{\text{therm}} = 1 - \frac{1}{r^{\gamma-1}} \qquad (7\text{-}3)$$

where r is defined as the compression ratio, v_1/v_2.

A plot of Eq. (7-3) shown in Fig. 7-10 indicates that the thermal efficiency

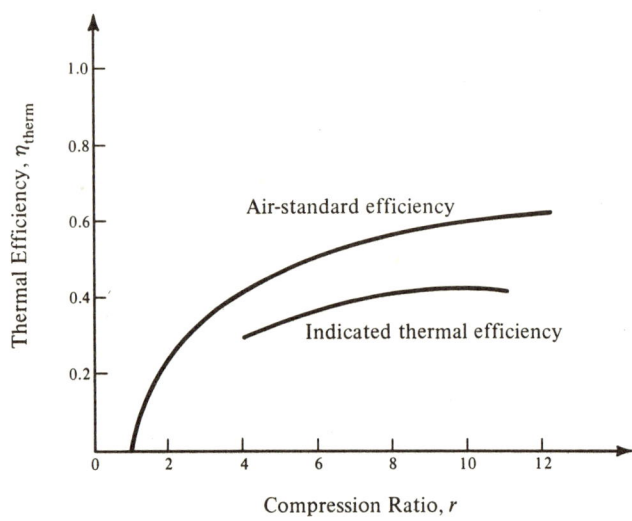

Fig. 7-10. *Thermal efficiency of an Otto cycle versus compression ratio.*

of the idealized Otto cycle increases with the compression ratio. For instance the thermal efficiency increases from 24 to 60 percent when the compression ratio increases from 2 to 10. The model also indicates that the advantage of operating at increasing compression ratios lessens rapidly due to the decreasing slope, $d\eta_{\text{therm}}/dr$. To compare these results to the actual "indicated thermal efficiency" of an engine, we would measure the work performed during the power loop on the indicator diagram shown in Fig. 7-7 and divide it by the heat of combustion of the fuel used during the cycle. The heat of combustion is a measure of the amount of heat liberated when the fuel is completely burned. The indicated thermal efficiency of a typical Otto engine which is also presented in Fig. 7-10 does, in fact, increase with the compression ratio, but peaks at an efficiency of about 40 percent.

The compression ratios in the early Otto engines were approximately 2, but these ratios are now running between 4 and 12. As the compression ratio is increased, the temperatures and pressures during the compression stroke also increase. This causes the fuel to burn faster and finally leads to detonation and spontaneous combustion which manifests itself in the familiar pinging or knocking of the engine. The "octane number" of a fuel is a measure of its ability to resist spontaneous combustion. This resistance can be improved with additives—tetraethyl lead being the most common additive for gasoline.

The actual work delivered by the engine will, of course, be less than the work developed during the power loop due to the pump loop, friction and other mechanical losses. The "brake thermal efficiency" is therefore defined as the actual work delivered to the output shaft, the so-called "brake work," divided by the heat of combustion of the fuel used. The brake thermal efficiency for automobile engines averages around 25 percent and has increased very little in the last fifty years.

There are many other criteria besides thermal efficiency that are used to evaluate the performance and suitability of an engine. For instance, a high specific power based on engine weight would be desirable in transportation applications, but this index of performance is really not too satisfactory since it is a strong function of engine speed. Increasing the engine speed to increase the power has the detrimental effect of increasing both the mechanical friction and the inertia loads which are both functions of the piston velocity. A more meaningful index of performance which is frequently used to compare engine outputs is the "mean effective pressure," mep, which is defined as the work performed during the cycle divided by the piston displacement volume. In the case of the idealized Otto cycle, the indicated work performed during the cycle is represented by the enclosed area 1—2—3—4—1 shown in Fig. 7-8 and the displacement volume is $V_1 - V_2$. Then,

$$\text{mep} = \frac{W_{12341}}{V_1 - V_2} = \frac{w_{12341}}{v_1 - v_2} \tag{7-4}$$

A high mep would indicate that the work output per cycle is large compared to the engine size (displacement). This is an important mechanical consideration since the piston displacement would have to increase as the mep decreases in order to maintain a given work/cycle output. The larger displacement would have higher frictional losses which, in turn, would decrease the mechanical efficiency.

In practice, the brake mean effective pressure (bmep) and the indicated mean effective pressure (imep) are defined by using the brake work or the indicated work respectively for the work term in Eq. (7-4). The imep for our air-standard cycle can be calculated as follows:

$$\text{imep} = \frac{w_{12341}}{v_1 - v_2} = \frac{\eta_{\text{therm}} q}{v_1(1 - v_2/v_1)} = \frac{\eta_{\text{therm}} q}{\frac{RT_1}{P_1}(1 - 1/r)}$$

where we have used the ideal gas equation of state,

$$P_1 v_1 = RT_1 \tag{5-1}$$

The relationship between the thermal efficiency and the compression ratio is given by Eq. (7-3). We can also substitute P_0/T_0 for P_1/T_1 since the pressure and temperature remain constant during the idealized intake process. After making these substitutions, we find

$$\text{imep} = \frac{P_0 q}{RT_0} \frac{1 - r^{(1-\gamma)}}{1 - 1/r} \qquad (7\text{-}5)$$

From Eq. (7-5), we see that the indicated mean effective pressure increases linearly with the intake pressure and heat addition, slowly with the compression ratio and inversely with the intake temperature. In order to facilitate the comparisons between cycles that are to be made in this chapter, we will put this equation in a more convenient form. We note again that

$$q = c_V(T_3 - T_2)$$

so that

$$\text{imep} = \frac{P_0 c_V (T_3/T_0 - T_2/T_0)(1 - r^{(1-\gamma)})}{R(1 - 1/r)}$$

Remembering that the inlet temperature T_0 is equal to T_1, and that

$$\frac{T_2}{T_1} = \left(\frac{v_1}{v_2}\right)^{\gamma-1} = r^{\gamma-1}$$

we obtain the relationship that

$$\text{imep} = \frac{P_0 r (T_3/T_1 - r^{\gamma-1})(1 - r^{1-\gamma})}{(\gamma - 1)(r - 1)}$$

This relationship gives the mean effective pressure in terms of the inlet pressure P_0, compression ratio r, and the ratio of the maximum to minimum cycle temperatures T_3/T_1. This equation, which is plotted in Fig. 7-11, indicates that a reversible Otto Engine with an inlet pressure of 1 atm and a compression ratio and temperature ratio of 10 will have a mep of 12.5 atm. This means that a pressure of 12.5 atm acting on the piston through one power stroke will produce the same net work that our example Otto engine will in one cycle. The brake mean effective pressure of a corresponding actual engine will, of course, not be quite as high. Looking back to the first steam engine, we note that Newcomen's atmospheric steam engine had a bmep of less than one atmosphere.

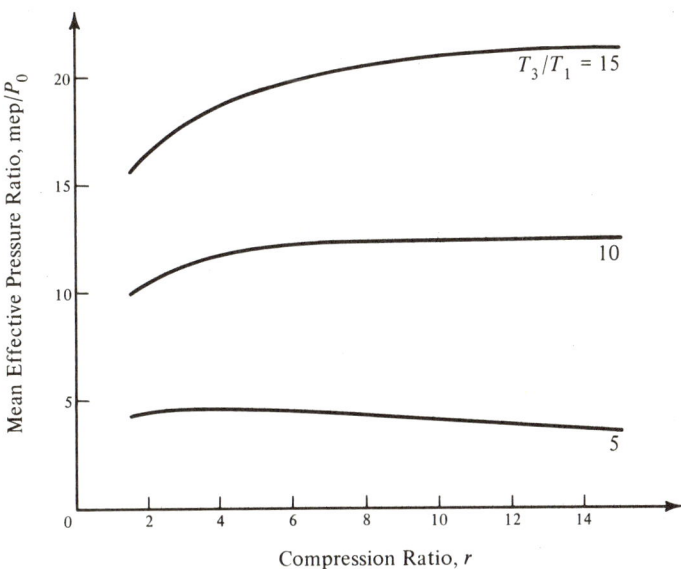

Fig. 7-11. *Mean effective pressure ratio vs compression ratio for Otto cycle.*

As an application of our Otto cycle analysis, consider the following example.

EXAMPLE PROBLEM 7-2. A World War I airplane engine with a compression ratio of 5 is operating in an environment where the temperature is 60 °F (16 °C) and the pressure is one atmosphere. If a stoichiometric air-gasoline mixture is used, determine the theoretical thermal efficiency, net work per pound of air, indicated mean effective pressure, and the pressures and temperatures at states 2, 3 and 4 in Fig. 7-8. The nominal heat of combustion of gasoline is 19,000 BTU/lb_m (44,164 joule/gram).

The thermal efficiency of an engine operating on the air standard Otto cycle is given by Eq. (7-3)

$$\eta_{therm} = 1 - \frac{1}{r^{\gamma-1}} = 1 - \frac{1}{5^{(1.4-1)}} = 0.47$$

The specific work is

$$w = \eta_{therm} q = (\eta_{therm}) \times (\text{fuel-air mixture ratio}) \times (\text{heat of combustion})$$

$$= \frac{0.47 \times 19{,}000 \text{ BTU/}lb_m \text{ gasoline}}{15.1 \; lb_m \text{ air/}lb_m \text{ gasoline}}$$

$$= 591.4 \text{ BTU/}lb_m \text{ air } (1375 \text{ joule/gram})$$

The pressure at state 2 is calculated from the isentropic equation

$$P_1 v_1^\gamma = P_2 v_2^\gamma \tag{5-11}$$

or

$$P_2 = P_1 \left(\frac{v_1}{v_2}\right)^\gamma = P_1 r^\gamma = (14.7 \text{ psia})(5)^{1.4}$$

$$= 139.9 \text{ psia } (9.6 \text{ bars})$$

Similarly, from Eq. (5-12), we find

$$\frac{T_2}{T_1} = \left(\frac{v_1}{v_2}\right)^{\gamma-1}$$

or

$$T_2 = T_1 r^{\gamma-1} = 520\,°\text{R} \times (5)^{(1.4-1)} = 990\,°\text{R} = 530\,°\text{F } (277\,°\text{C})$$

For heat addition at constant volume,

$$q = c_V(T_3 - T_2)$$

therefore,

$$T_3 = T_2 + \frac{q}{c_V} = T_2 + \frac{q}{(5/2)R}$$

$$= 990\,°\text{R} + \frac{19{,}000 \text{ BTU/lb}_m \text{ gasoline} \times 28.96 \text{ lb}_m \text{ air/lb mole}}{\frac{5}{2} \times 1.986 \frac{\text{BTU}}{\text{lb mole }°\text{R}} \times 15.1 \frac{\text{lb}_m \text{ air}}{\text{lb}_m \text{ gasoline}}}$$

$$= 8329\,°\text{R} = 7869\,°\text{F } (4354\,°\text{C})$$

The maximum temperature encountered in actual engines is around 5,000 °F, which is much lower than the value just calculated. This difference is mainly caused by our assumption that the specific heat remains constant while, in fact, the specific heat is found to increase significantly for this large a temperature increase. For large temperature variations, charts or tables[10,12,13] giving the thermodynamic properties of air-fuel mixtures should be consulted. It should also be noted that Fig. 7-7 indicates that some expansion occurs before the combustion process has been completed which would also decrease the temperature of an actual cycle. Even so, 5000 °F is a very high temperature and is the essential reason why the Otto engine is so much more efficient than the steam engine. Steam engines are limited by their continuous combustion process to boiler temperatures around 1000 °F. The reciprocating nature of the Otto cycle means that the metal surfaces are only momentarily exposed to combustion temperatures of 5000 °F and can, therefore, be protected with a cooling system.

The pressure corresponding to this maximum temperature T_3 is predicted from the constant volume relation developed from the ideal gas equation of state, Eq. (5-1),

$$P_3 = P_2\left(\frac{T_3}{T_2}\right) = 139.9 \text{ psia} \times \frac{8329}{990} = 1177 \text{ psia } (81.1 \text{ bars})$$

The exhaust temperature T_4 is calculated from the isentropic relation, Eq. (5-12)

$$T_4 = T_3\left(\frac{V_3}{V_4}\right)^{\gamma-1} = T_3\left(\frac{1}{r}\right)^{\gamma-1} = 8329 \text{ °R} \times \left(\frac{1}{5}\right)^{1.4-1}$$

$$= 4375 \text{ °R } (2431 \text{ K})$$

and the corresponding pressure is

$$P_4 = P_3\left(\frac{V_3}{V_1}\right)^{\gamma} = P_3\left(\frac{1}{r}\right)^{\gamma} = 1177 \text{ psia} \times \left(\frac{1}{5}\right)^{1.4}$$

$$= 123.7 \text{ psia } (8.5 \text{ bars})$$

This high exhaust pressure P_4 and temperature T_4 is, of course, very wasteful in that additional work can be extracted from this exhaust.

To obtain the indicated mean effective pressure, we will rewrite Eq. (7-5) in the form

$$\text{imep} = \frac{P_0 w}{RT_0(1 - 1/r)}$$

where $w = \eta_{\text{therm}} q$. The evaluation of this expression gives

$$\text{imep} = \frac{14.7 \text{ psia} \times 591.4 \, \frac{\text{BTU}}{\text{lb}_m \text{ air}} \times 28.96 \, \frac{\text{lb}_m \text{ air}}{\text{lb-mole}}}{1.986 \, \frac{\text{BTU}}{\text{lb}_m \text{ mole °R}} \times 520 \text{ °R} \times \left(1 - \frac{1}{5}\right)}$$

$$= 305 \text{ psia } (21.0 \text{ bars})$$

For comparison purposes, we note that a Carnot cycle operating between the same two temperature extremes would have a thermal efficiency of

$$\eta_{therm,c} = 1 - T_0/T_3 \qquad (3\text{-}8)$$

$$= 1 - 520/8329 = 0.94$$

which is twice as efficient as this ideal Otto cycle.

Equation (7-5) indicates that the imep varies linearly with the intake pressure P_0 and inversely with the absolute intake temperature T_0. Therefore, as an airplane flies higher, there will be a loss of power due to the ambient pressure decreasing more rapidly than the ambient temperature. For instance, we see from Example Prob. 5-2 that if this airplane was flying at 15,000 feet instead of sea level, the pressure and temperature would be approximately 8 psia (0.6 bars) and 6 °F (−14 °C) respectively. The resulting imep would be

$$\text{imep}_{15,000} = \text{imep}_0 \frac{P_{15,000'}}{P_0} \frac{T_0}{T_{15,000'}}$$

$$= 305 \text{ psia} \times \frac{8}{14.7} \times \frac{520}{466}$$

$$= 185 \text{ psia } (12.7 \text{ bars})$$

which represents a 39 percent reduction in power. This serious power loss could not be alleviated by increasing the heat addition q since the typical fuel-air mixture was near the stoichiometric value. On the other hand, the air pressure at the engine's intake could be increased by precompression. Thus, the supercharger, which is nothing more than a small centrifugal compressor, was developed during World War I.

EXAMPLE PROBLEM 7-3. Repeat Example Prob. 7-2 for a supercharged reversible engine at 15,000 feet altitude. Assume that the supercharger shown in Fig. 7-12 has an isentropic efficiency of 60 percent and discharge pressure P_1 of one atmosphere. The TS diagram for this process is also presented in Fig. 7-12.

We will base our calculations on one lb_m of air flowing through our system which is first defined to be the supercharger. If the compression process in the supercharger is assumed to be steady-state and adiabatic and the kinetic and potential energy changes are neglected, the energy balance equation reduces to

$$h_0 - h_1 - w_{comp} = 0$$

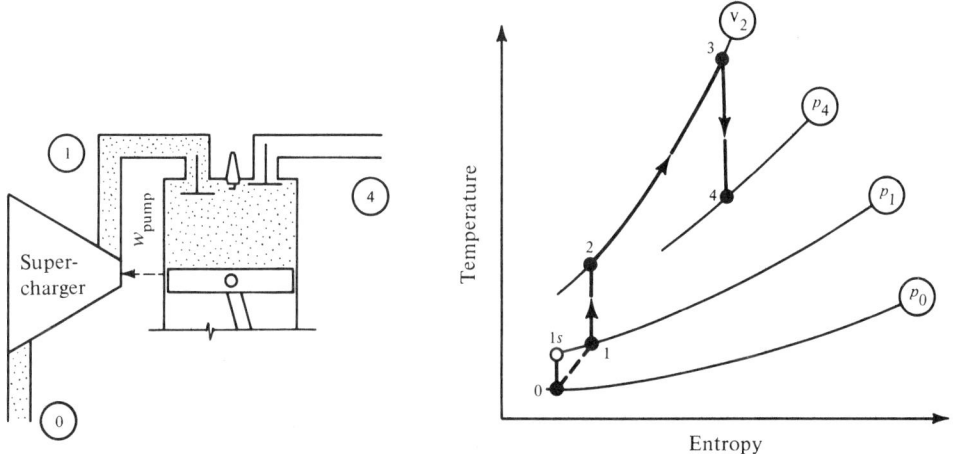

Fig. 7-12. *TS diagram of supercharged Otto cycle.*

If the supercharger were isentropic, we could use Eq. (5-12) to estimate its exhaust temperature T_{1s}.

$$T_{1s} = T_0 \left(\frac{P_1}{P_0}\right)^{(\gamma-1)/\gamma} = 466\,°\text{R} \times \left(\frac{14.7}{8.0}\right)^{0.4/1.4} = 554\,°\text{R (308 K)}$$

The supercharger's inlet pressure P_0 and temperature T_0 were taken from Example Prob. 5-2. The efficiency of the supercharger is 60 percent. Thus,

$$\eta_{\text{comp}} = \frac{w_{\text{adiab, rev}}}{w_{\text{act}}} = 0.6 = \frac{h_0 - h_{1s}}{h_0 - h_1} = \frac{c_P(T_0 - T_{1s})}{c_P(T_0 - T_1)}$$

Then

$$T_1 = T_0 + \frac{T_{1s} - T_0}{\eta_{\text{comp}}} = 466\,°\text{R} + \frac{554 - 466}{0.6} = 613\,°\text{R (340 K)}$$

The work that is absorbed by the compressor is

$$w_{\text{comp}} = h_0 - h_1 = c_P(T_0 - T_1) = \frac{7}{2} R (T_0 - T_1)$$

$$= \frac{7/2 \times 1.986 \text{ BTU/lb mole }°\text{R} \times (466 - 613)\,°\text{R}}{28.96 \text{ lb}_m/\text{lb mole}}$$

$$= -35.3 \frac{\text{BTU}}{\text{lb}_m \text{ air}} = -82.0 \frac{\text{joule}}{\text{gram air}}$$

which is only 6 percent of the work delivered by the engine. Repeating the other calculations presented in Example Prob. 7-2 gives

$$T_2 = 1167 \text{ R } (648 \text{ K})$$
$$T_3 = 8506 \text{ R } (4726 \text{ K})$$
$$T_4 = 4468 \text{ R } (2482 \text{ K})$$
$$P_2 = 140 \text{ psia } (9.6 \text{ bars})$$
$$P_3 = 1020 \text{ psia } (70.3 \text{ bars})$$
$$P_4 = 107 \text{ psia } (7.4 \text{ bars})$$

and imep = 259 psia (17.8 bars) which is a 40 percent increase over the unsupercharged engine.

Modern supercharges can, in fact, increase the power output of an engine by as much as 30 percent without much of a weight penalty. It should be noted that this is not due to an increase in efficiency but is due to the engine handling more air and fuel per cycle. The power needed to drive the supercharger was originally supplied mechanically by the engine and would, therefore, reduce the above engine's net output by 6 percent. This power loss was eliminated with the development of the turbosupercharger which uses the engine's high pressure exhaust gas, P_4, to power a gas turbine which in turn drives the supercharger. This device will be discussed later in this chapter.

As mentioned previously, the Otto cycle serves the nation as its main prime mover; but it is also responsible for a large portion of the nation's air pollution. Oxides of nitrogen which are primarily responsible for the brownish haze over many American cities are produced along with CO during the combustion process. The amount of these pollutants produced increases rapidly with the combustion temperature. These gases would normally undergo further reactions to produce less harmful products but they are cooled so fast in the Otto cycle that these constituents are "frozen" in the exhaust. Raising the compression ratio has, therefore, increased the pollution problem since this has increased the maximum temperature in the Otto cycle. Also, the lead compounds produced from the anti-knock lead additives that must be added to the fuel of high compression engines are considered by the government to be environmentally dangerous. As a result of these factors, automobile manufacturers have reduced the average compression ratio of an automobile engine from a high of 9.2 in 1958 to a present value around 8 even though this implies a reduction in thermal efficiency and power.

The fuel-air mixture for automobiles has actually been fuel rich in the past for well-controlled ignition and combustion (more like 13 lb_m of air per lb_m of fuel instead of the stoichiometric 15.1 lb_m of air per lb_m of fuel). The pollution contributed by the unburned hydrocarbons which result from rich fuel mixtures is being cured by developing engines that can operate effectively on lean mixtures. One method that has been proposed to burn lean mixtures is to burn the fuel in two stages using a stratified charge. A fuel rich mixture is first ignited in a precombustion chamber and the resulting flame is used to ignite a mixture in the main combustion chamber that is so lean that it ordinarily would not ignite. This procedure lowers the combustion temperature and thereby reduces the formation of nitrogen oxides.

Otto had the mistaken impression that his engine obtained its superior combustion from a stratified charge effect in the combustion chamber itself. He wrote this explanation into his patent and it eventually caused him to lose his patent protection when his competitors were able to persuade the authorities that the relatively smooth combustion was achieved by proper mixture and timing.

Diesel's Carnot Engine

In 1824 Sadi Carnot first introduced his theoretical cycle which was discussed in Chap. 3. In Example Prob. 7-2 we found that the Otto cycle efficiency is around one-half that of the Carnot cycle since heat is added and exhausted over a wide temperature range. It would, therefore, appear natural to try to adopt the Carnot cycle to the reciprocating engine. An indicator diagram for a simulated Carnot cycle engine is shown in Fig. 7-13. Air would again be drawn into the

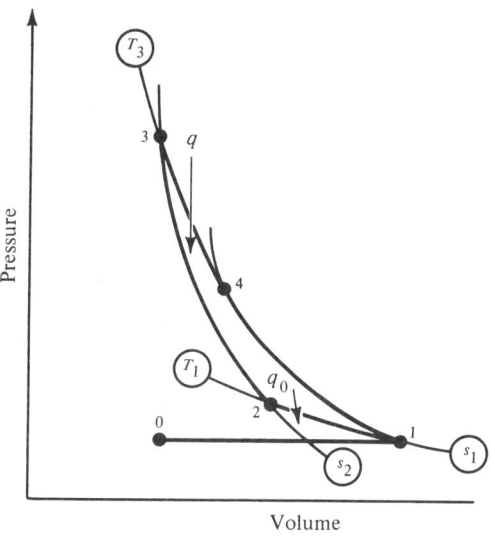

Fig. 7-13. *Indicator diagram for a Carnot cycle.*

idealized engine during the intake stroke 0–1, but now it would have to be compressed isothermally during the first part of the compression stroke 1–2 and isentropically during the last part of this stroke 2–3. During the expansion stroke, heat is added at a rate which would maintain a constant temperature during the first part of the stroke 3–4, while the final expansion 4–1 is isentropic. The expansion stroke is followed by the exhaust stroke 1–0.

To accomplish the isothermal compression process mechanically at a reasonable rate is very difficult as was pointed out in the section on reciprocating compressors but such a process can be approached by proper design of the cooling system. The more challenging process to obtain would be the isothermal combustion process—how can fuel be burned and the temperature not increase?

In 1892 a German engineer, Rudolf Diesel, thought he had found the answer and patented the idea. Air by itself would be compressed to such a high pressure P_3 that the corresponding high temperature T_3 would spontaneously ignite the fuel as it was injected into the high temperature air. This type of ignition is called compression ignition as compared to the spark ignition of the Otto cycle. The isothermal expansion (3–4) could then be maintained by injecting the fuel into the cylinder at the proper rate.

To get an idea of the actual performance that this proposed engine would have, we will calculate its mean effective pressure as a function of compression ratio, v_1/v_3, using the relationship

$$\text{imep} = \frac{w_{12341}}{v_1 - v_3} = \frac{q + q_0}{v_1(1 - 1/r)} \tag{7-6}$$

Referring to Fig. 7-13 and using the generalized energy balance equation, (2-25), it is seen that the heat added during the isothermal process 3–4 per pound of fluid is

$$q = u_4 - u_3 + w_{34} = u_4 - u_3 + \int_{v_3}^{v_4} P\,dv$$

if the kinetic and potential energy changes are neglected. Assuming the fluid to be an ideal gas implies that the internal energy does not vary during the isothermal process

$$u_3 = u_4$$

Also
$$P = RT_3/v$$
Therefore,

$$q = RT_3 \int_{v_3}^{v_4} dv/v = RT_3 \ln(v_4/v_3) \tag{7-7}$$

Since the path between states 1 and 4 is isentropic, we find from Eq. (5-12) that

$$T_1 v_1^{\gamma-1} = T_4 v_4^{\gamma-1} \tag{5-12}$$

or

$$v_4/v_1 = (T_1/T_4)^{1/(\gamma-1)} = (v_4/v_3)(v_3/v_1)$$

Solving this last expression for v_4/v_3, substituting this term into Eq. (7-7), and noting that $T_3 = T_4$ gives

$$q = RT_3 \ln [(T_1/T_4)^{1/(\gamma-1)} (v_1/v_3)] = RT_3 \left[\ln r - \frac{1}{\gamma-1} \ln(T_3/T_1)\right]$$

By a similar procedure, the heat exhausted by the cycle is found to be

$$q_0 = -RT_1 \left(\ln r - \frac{1}{\gamma-1} \ln T_3/T_1\right) \tag{7-8}$$

Combining these two equations with Eq. (7-6) and simplifying gives the final result

$$\text{imep} = \frac{rP_1}{r-1} \left(\frac{T_3}{T_1} - 1\right)\left(\ln r - \frac{1}{\gamma-1} \ln T_3/T_1\right) \tag{7-9}$$

We see that the imep for the Carnot cycle is a function of the compression and temperature ratios and is, as in the Otto cycle, a linear function of the intake pressure.

A plot of Eq. (7-9) presented in Fig. 7-14 indicates that the imep increases

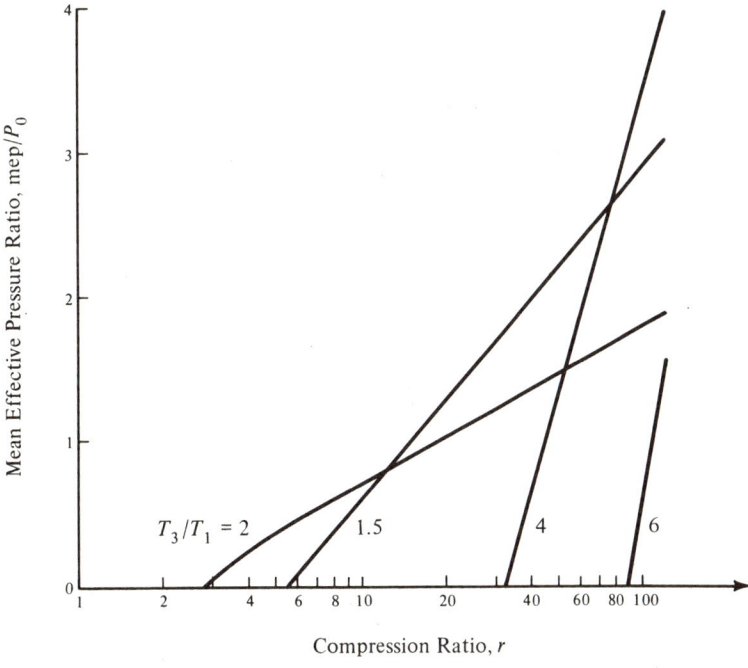

Fig. 7-14. *Mean effective pressure ratio vs compression ratio and temperature ratio for Carnot cycle.*

with the compression ratio for a given temperature ratio. However, close examination of this graph shows that the imep is extremely low even for a ridiculously high compression ratio. For example, consider a simulated Carnot engine that intakes air at 1 atmosphere pressure and a temperature of 500 °R, and compresses the air to a temperature of 2000 °R ($T_3/T_1 = 4$). In this case, using a compression ratio of 60 will only develop an imep of 2 atmospheres according to Fig. 7-14. If we compare the mean effective pressures produced by reversible Otto and Carnot cycles, we note from Figs. 7-11 and 7-14 that the Carnot cycle cannot produce an acceptable mep with a practical compression ratio while the reversible Otto cycle can. This characteristic of Carnot engines confines their usefulness to theory alone.

The simulated Carnot cycle has a low imep due to the small specific work that is done per cycle. The compression and expansion indicator paths for this cycle would be barely separated from each other making the enclosed indicated work area very small as schematically indicated in Fig. 7-13. In fact, an actual reciprocating Carnot engine would not develop enough power to overcome its own friction let alone develop any useful power. Diesel's theoretical and experimental studies both resulted in this same conclusion forcing him to abandon isothermal combustion as a practical process.

The Diesel Cycle

Even though the use of an isothermal combustion process was not practical, the concept of employing compression ignition and fuel injection to control the combustion process was still an intriguing idea. The combustion process could easily be changed from the proposed isothermal combustion to a constant pressure process, for example, by injecting fuel at the proper increased rate. Diesel essentially did this and in 1897 announced the first successful ignition compression engine. Figure 7-15 depicts a typical indicator diagram for a modern Diesel engine. Air is drawn into the cylinder and compressed adiabatically to a sufficiently high pressure and temperature to cause immediate ignition when the fuel is injected. The fuel is then injected at a controlled rate to produce the indicated combustion process. The figure shows that the actual combustion process is a compromise between a constant pressure and a constant volume process. After the fuel injection is terminated, the products of combustion are then expanded adiabatically and finally exhausted to the atmosphere.

The idealized air-standard approximation for this cycle is shown in Fig. 7-16, where the expansion and compression processes are assumed to be isentropic as was done in the Otto cycle and the combustion process is now approximated by a reversible isobaric process.

The thermal efficiency of this air-standard Diesel cycle is given by

$$\eta_{\text{therm}} = \frac{w}{q} = \frac{q + q_0}{q} = 1 + \frac{q_0}{q}$$

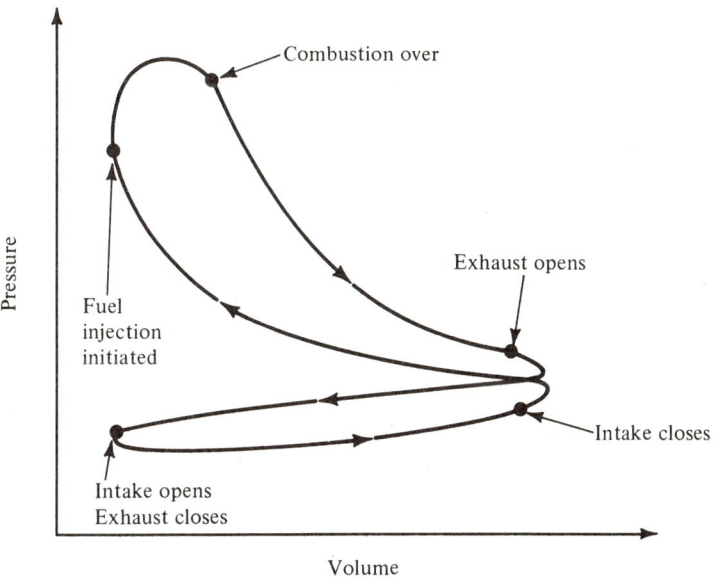

Fig. 7-15. *Indicator diagram for a Diesel engine.*

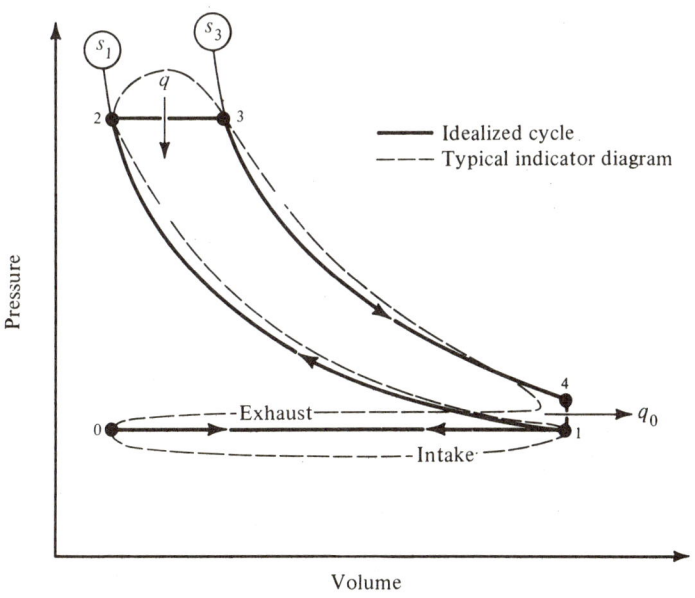

Fig. 7-16. *Indicator diagram for an air-standard and actual Diesel cycle.*

Since heat is added during the constant pressure process and exhausted during the constant volume process,

$$\eta_{\text{therm}} = 1 + \frac{c_V(T_1 - T_4)}{c_P(T_3 - T_2)} = 1 - \frac{(T_4/T_1) - 1}{\gamma(T_2/T_1)(T_3/T_2 - 1)} \tag{7-10}$$

Combining the isentropic relations from Eq. (5-12)

$$\frac{T_1}{T_2} = \left(\frac{v_2}{v_1}\right)^{\gamma-1} \quad \text{and} \quad \frac{T_4}{T_3} = \left(\frac{v_3}{v_4}\right)^{\gamma-1}$$

we obtain

$$\frac{T_4}{T_1} = \frac{T_3}{T_2}\left(\frac{v_1}{v_2}\frac{v_3}{v_4}\right)^{\gamma-1}$$

Since $v_1 = v_4$ and $Pv = RT$,

$$\frac{T_4}{T_1} = \frac{T_3}{T_2}\left(\frac{v_3}{v_2}\right)^{\gamma-1} = \frac{P_3 v_3}{P_2 v_2}\left(\frac{v_3}{v_2}\right)^{\gamma-1}$$

However, $P_2 = P_3$ which simplifies the above expression to

$$\frac{T_4}{T_1} = \left(\frac{v_3}{v_2}\right)^{\gamma} = r_c^{\gamma}$$

where r_c is called the (fuel) cutoff ratio or load ratio. Inserting this expression into Eq. (7-10) and simplifying gives

$$\eta_{\text{therm}} = 1 - \frac{1}{r^{\gamma-1}}\left[\frac{r_c^{\gamma} - 1}{\gamma(r_c - 1)}\right] \tag{7-11}$$

The thermal efficiency of the Diesel cycle differs from the thermal efficiency of an Otto cycle, Eq. (7-3), by the bracketed term which is greater than unity. As shown in Fig. 7-17, the thermal efficiency of the Diesel cycle is, therefore, less than that of an Otto cycle when compared at the same compression ratio. Figure 7-17 also indicates that a cutoff ratio near one is desirable to obtain a maximum thermal efficiency, but the indicated work would obviously approach

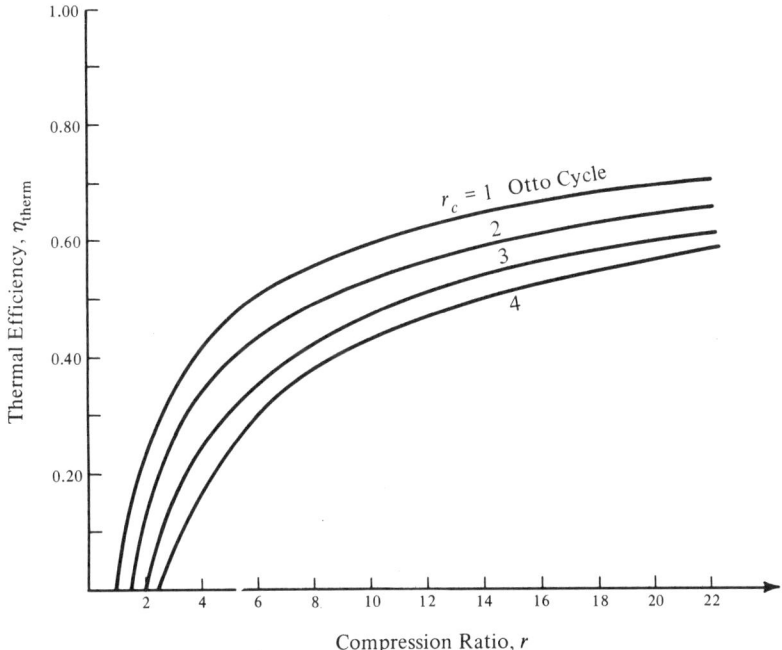

Fig. 7-17. *Thermal efficiency of Diesel cycle.*

zero as r_c approached 1 since no heat would be added. We can also demonstrate this point by calculating the indicated mean effective pressure. From Eq. (7-4),

$$\text{imep} = \frac{w_{12341}}{v_1 - v_2} = \frac{q + q_0}{v_1 - v_2}$$

$$= \frac{c_P(T_3 - T_2) + c_V(T_1 - T_4)}{v_1 - v_2}$$

which after some manipulation reduces to

$$\text{imep} = \frac{rP_0}{(r-1)(\gamma-1)} [\gamma r^{\gamma-1}(r_c - 1) + 1 - r_c^\gamma] \qquad (7\text{-}12)$$

This relationship is plotted in Fig. 7-18 and indicates that a high r_c, which is accomplished by burning more fuel, is desirable from a power aspect as would be expected. Thus, a design compromise between power and efficiency must be made. The imep is also plotted as a function of the temperature ratio T_3/T_1 instead of r_c in Fig. 7-18 so that these results can be compared with the corresponding imep's for the Otto and Carnot cycles (Figs. 7-11 and 7-14).

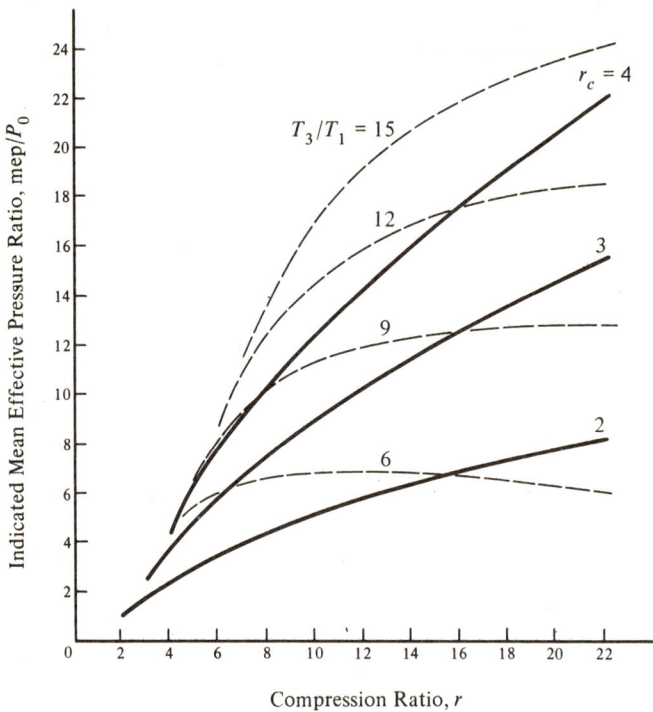

Fig. 7-18. *Indicated mean effective pressure of a Diesel cycle.*

Since the compression ratio of a Diesel cycle is not limited by preignition problems, it can utilize higher compression ratios than can the Otto cycle to increase both its efficiency and power. Compression ratios typically range from 11 to 22 or around twice that of the spark-ignition engines. We also see from Eq. (7-12) that the imep can again be increased by increasing the intake pressure. Superchargers are much more commonly used on Diesel engines than on Otto engines. Some Diesel engine superchargers increase the intake pressure by as much as a factor of four. The bmep of Diesel engines are currently running from 70 to 270 psi (4.8 to 18.6 bars) and have exceeded 400 psi (27.6 bars) experimentally.

The Diesel engine has the advantage of being able to burn the comparatively low cost fuel oils which are extracted from crude oil after the lower boiling range gasoline and kerosene have been removed. At the same time, some large modern engines can achieve brake thermal efficiencies as high as 42 percent. The Diesel engine has, therefore, one of the highest thermal efficiencies of all power cycles that are commercially available. It is interesting to note that Diesel's original engines had a very respectable thermal efficiency of 26 percent.

The Diesel engine is used in a broad range of heavy-duty applications because of its high efficiency, mep and reliability. These range from 40,000 hp (30 megawatt) versions for electric power generation and marine applications to the 4,000 hp diesel-electric locomotives which dramatically increased the thermal efficiency of locomotives from 10 to 35 percent when the conversion from steam was made. Diesels are also used in a few low weight applications including some automobiles. The smaller engines generally use the Otto cycle because it results in a lighter and less expensive engine than the Diesel cycle.

Besides having a higher thermal efficiency, the Diesel engine also has an advantage over the spark-ignition engine of being generally less polluting. The leaner mixtures and the combustion process produces much less CO and around half the hydrocarbons of a typical unmodified spark ignition engine. However, all high-temperature combustion engines have problems in forming oxides of nitrogen.

EXAMPLE PROBLEM 7-4. Calculate the specific work, the thermal efficiency, imep, and the maximum temperature for an air-standard Diesel cycle with a compression ratio of 15 and a cutoff ratio of 2. The inlet conditions are 1 atmosphere and 70 °F (21 °C).

The thermal efficiency of an idealized Diesel cycle is given by Eq. (7-11). The numerical evaluation of this expression gives

$$\eta_{therm} = 1 - \frac{1}{r^{\gamma-1}} \left[\frac{r_c^\gamma - 1}{\gamma(r_c - 1)} \right]$$

$$= 1 - \frac{1}{15^{(1.4-1)}} \frac{2^{1.4} - 1}{1.4(2-1)} = 0.60$$

which agrees with the value given by Fig. 7-17.

As the air-standard cycle in Fig. 7-16 indicates, the heat addition occurs during a constant pressure process. Thus, the specific work can be calculated from the expression

$$w = \eta_{therm} q = \eta_{therm} c_P (T_3 - T_2)$$

once the temperatures T_2 and T_3 are determined. For the isentropic compression process, Eq. (5-12) is used to give

$$T_2 = T_1 (v_1/v_2)^{\gamma-1} = T_1 r^{\gamma-1} = 530 \, °R \times 15^{(1.4-1)}$$

$$= 1566 \, °R \, (870 \, K)$$

For the constant pressure process, the ideal gas equation, Eq. (5-1), indicates that

$$T_3 = T_2 \left(\frac{v_3}{v_2}\right) = T_2 r_c = 1566 \, °R \times 2$$

$$= 3132 \, °R \, (1740 \, K)$$

which is the maximum temperature of the cycle. Referring back to Example Prob. 7-2, we see that this maximum temperature is much lower than the maximum temperature predicted for the Otto cycle which is some indication of why the Diesel cycle is less polluting.

The specific work can now be determined. From the definition of thermal efficiency,

$$w = \eta_{\text{therm}} q \tag{3-7}$$

$$= \eta_{\text{therm}} c_P (T_3 - T_2) = 0.60 \times 0.24 \frac{\text{BTU}}{\text{lb}_m \, °\text{R}} \times (3132 - 1566) \, °\text{R}$$

$$= 226 \text{ BTU/lb}_m \text{ (525 joule/gram)}$$

which is approximately one third the value predicted for the Otto cycle in Example Prob. 7-2. This value can now be used to calculate the indicated mean effective pressure ratio, imep/P_0.

$$\text{imep} = \frac{w}{v_1 - v_2} = \frac{w}{\dfrac{RT_1}{P_1}(1 - 1/r)} \tag{7-4}$$

so that

$$\frac{\text{imep}}{P_0} = \frac{w}{RT_0(1 - 1/r)} = \frac{226 \text{ BTU/lb} \times 28.96 \text{ lb/lb mole}}{1.986 \dfrac{\text{BTU}}{\text{lb mole} \, °\text{R}} \times 530 \, °\text{R} \, (1 - 1/15)}$$

$$= 6.7$$

This value agrees with the value which may be obtained from Fig. 7-18.

The Stirling Cycle

The Diesel and Otto engines are presently the world's principal prime movers because of their versatility, reliability, reasonable efficiency and relatively low manufacturing cost. These characteristics have evolved after vast investments in research, development and plants for mass production. In the past, there has been little impetus in general to seek and initiate large scale development of alternatives to these engines. However, the recent pollution laws and the escalating costs and uncertainty in the supplies of petroleum have changed this picture.

One of the alternate cycles presently under serious development is the Stirling cycle depicted in Fig. 7-19. This idealized cycle is composed of a reversible

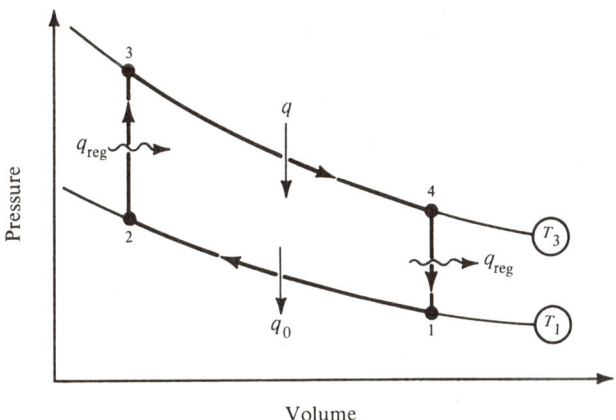

Fig. 7-19. *Idealized Stirling cycle.*

isothermal compression process, 1—2; a reversible constant volume heating process, 2—3; a reversible isothermal expansion process, 3—4; and a reversible constant volume cooling process, 4—1. A closed cycle engine that approximated this theoretical cycle was first patented in 1816 by a Scottish minister, Robert Stirling. Engines of this type were used up until the 1920's in small power plants and on farms for pumping water, but were then replaced with internal combustion engines.

We note that all four processes of this reversible cycle transfer heat. In the case of the constant volume processes, there is no work performed. If the working medium is an ideal gas, the internal energy changes $u_3 - u_2$ and $u_4 - u_1$ are equal since the constant volume processes are between the same two temperatures. The application of the energy balance equation to these two processes would, therefore, imply that the heat added during the reversible constant volume process 2—3 is equivalent to the heat rejected during the constant volume process 4—1. If the heat rejected during process 4—1 could be internally stored and reversibly transferred to the heat addition process 1—2, the only external heat transfer would occur during the isothermal processes 1—2 and 3—4. The Stirling cycle could then achieve the same theoretical efficiency as the Carnot cycle since all the external heat transfers occur reversibly at the maximum and minimum temperatures of the cycle. We can also demonstrate this attribute by calculating the thermal efficiency of the Stirling cycle,

$$\eta_{therm} = \frac{w_{12341}}{q} = \frac{w_{12} + w_{34}}{q} \tag{7-13}$$

We note again that processes 1—2 and 3—4 are both isothermal. For an ideal gas, the generalized energy balance tells us that the work performed by a unit mass of gas during an isothermal process is equal to the heat added, i.e.,

$$w_{34} = q$$

since Δu is zero for an ideal gas at constant temperature. For a closed system, the work performed per unit mass is

$$w = \int P\,dv$$

The isothermal work terms are obtained by integrating this equation for an ideal gas undergoing an isothermal process. Substituting these terms into Eq. (7-13) gives

$$\eta_{therm} = \frac{RT_3 \ln v_4/v_3 - RT_1 \ln v_1/v_2}{RT_3 \ln v_4/v_3}$$

or

$$\eta_{therm} = \frac{T_3 - T_1}{T_3} = \eta_{therm,c}$$

since $v_4/v_3 = v_1/v_2$ (Fig. 7-19).

In terms of specific output, we would expect the Stirling cycle to be superior to the Carnot cycle since the replacement of the isentropic processes with constant volume processes greatly increases the enclosed PV area as can be seen by comparing Figs. 7-13 and 7-19. To make a quantitative comparison, the Stirling cycle's theoretical imep is calculated using the relationship

$$imep = \frac{w_{12341}}{v_1 - v_2} = \frac{R(T_3 - T_1)\ln v_1/v_2}{v_1(1 - v_2/v_1)}$$

or, since $P_1 v_1 = RT_1$,

$$imep = \frac{P_1 r(T_3/T_1 - 1)\ln r}{r - 1} \tag{7-14}$$

We see from this expression which is plotted on Fig. 7-20 that the Stirling cycle can achieve theoretical imep's comparable if not superior to those we predicted for the Otto and Diesel cycles while operating over physically realistic compression and temperature ratios—something the Carnot cycle cannot do.

A simple mechanical system that would approximate the Stirling cycle is schematically depicted in Fig. 7-21. The engine consists of a cylinder with a

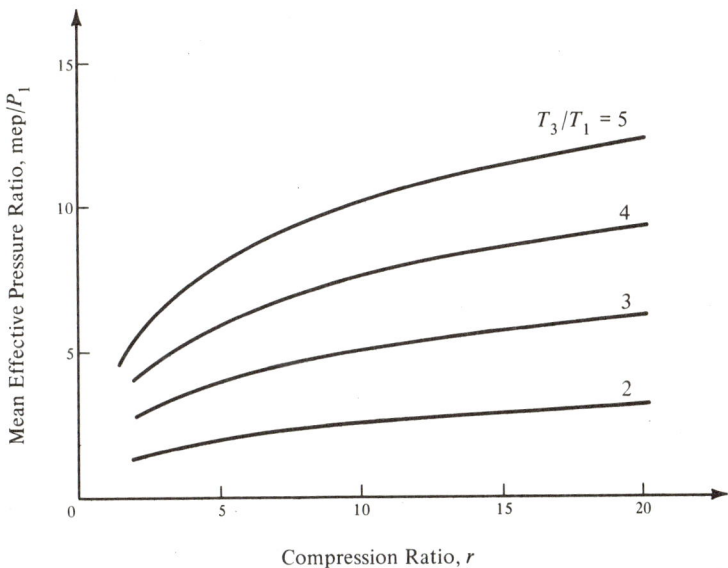

Fig. 7-20. *Mean effective pressure ratio vs compression ratio and temperature ratio for Stirling cycle.*

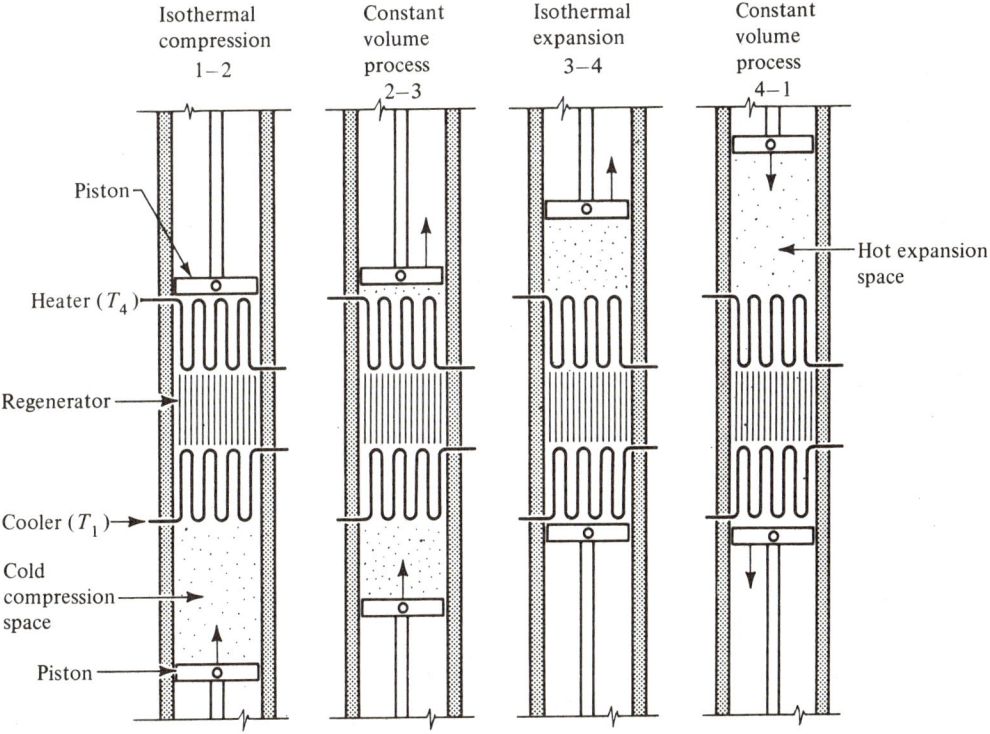

Fig. 7-21. *Schematic of the apparatus for a Stirling cycle.*

piston at each end and three heat exchangers (heater, regenerator, and cooler) between the pistons. The regenerator is the energy storage device or economizer as Stirling referred to it. It consists of a porous metallic or ceramic matrix that has a large internal surface area for good heat transfer and a large heat capacity for energy storage. The temperature through the idealized regenerator is assumed to vary smoothly from the cooler temperature T_1 to the heater temperature T_3, as indicated in Fig. 7-22 so that the working medium is heated reversibly as it passes

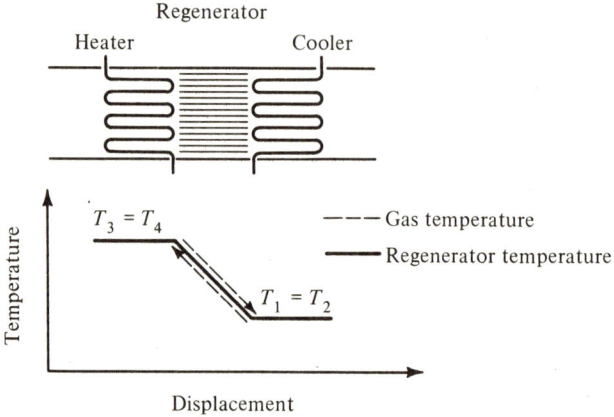

Fig. 7-22. *Regenerator's temperature distribution.*

through the regenerator. Of course, reversible heat transfer would require an infinitesimal temperature difference between the regenerator and the gas throughout the regenerator which is not practical if the engine is to run at any reasonable speed.

The cycle begins with the bulk of the gas in the cold compression space shown in Fig. 7-21. The bottom piston moves upward to compress the gas, process 1—2, while the cooler essentially maintains a constant gas temperature T_1. The bulk of the gas is then transferred to the hot expansion space via the constant volume process 2—3 by having both pistons move upward at the same speed. As the gas passes through the regenerator during this process, enough heat q_{reg} is transferred from the regenerator to the gas to heat the gas to the heater temperature T_3.

After the bulk of the gas has been transferred to the hot expansion space, the bottom piston stops and the top piston continues to move upward, process 3—4. The heater maintains a constant temperature T_3 during this expansion. Upon completion of the expansion, both pistons move downward at the same velocity to return the gas to the cold compression space via the constant volume process 4—1. The regenerator absorbs an amount of energy q_{reg} from the gas during this return passage such that the regenerator's net heat transfer will be zero during the cycle. We see that Stirling's regenerator provides a practical means of eliminating low temperature external heat addition and high temperature exhaust; both of which lead to inefficiencies. The use of internal heat transfer in place of external heat transfer during low temperature heating and high temperature cooling is an important feature of many other modern cycles. These cycles are referred to as regenerative cycles.

One reason for the renewed interest in the Stirling cycle is that the cycle utilizes external-combustion which can be controlled more effectively than the internal-combustion to produce substantially less air pollution. External combustion is also not dependent upon a particular fuel. Wood, coal and corn cobs, as well as petroleum derivatives, were burned in the fire box of this engine in the past. Solar radiation and radioisotopes may be utilized as energy sources in the future.

External combustion has the severe disadvantage that some surfaces are continuously exposed to the cycle's maximum working temperature. Therefore, the maximum cycle temperatures for external combustion are much lower than the temperatures that can be tolerated in the Otto and Diesel engines during the periodic internal combustion. Of course, lower temperatures imply lower efficiencies. Gas temperatures of 1150 to 1250 °F (621 to 677 °C) are currently possible with modern heat resistant materials appropriate for this application.

Actual Stirling cycles achieve only a fraction of the Carnot efficiency due to the difficulties in the attainment of isothermal compression and expansion in machines operating at reasonable speeds; the pressure drops in the heat exchangers and the finite temperature difference between the regenerator and the passing gas. Brake thermal efficiencies in the range of 30 to 40 percent are presently being obtained.

Equation (7-14) indicates that the imep is proportional to the initial pressure P_1 and can, therefore, be increased by pressurizing the whole system. In practice, very high mean pressures reaching 2000 psia (138 bars) and higher are used to obtain adequate mep's while using small compression ratios.

The present Stirling engines are comparatively expensive and heavy and must run at moderate speeds since large aerodynamic losses will occur in the heat exchangers at high gas velocities. The performance of the Stirling engine can be increased by using hydrogen or helium instead of air as the working fluid. The thermophysical characteristics of these gases permit much higher heat transfer rates than air. Since hydrogen and helium are also much lighter than air, their aerodynamic losses in the heat exchanger will be less than air's for the same flow conditions. In terms of the near future, it is conceivable that Stirling engines matching the thermal efficiency of the Diesel and the lightness of the Otto engine will start to appear.

TURBINE SYSTEMS

Turbines

After the reciprocating engine, the next major class of prime movers in terms of total national capacity involves systems built around the turbine. Prime movers producing power outputs exceeding about 40,000 hp (30 megawatts) are presently the exclusive domain of gas and steam turbines of which the steam turbine reigns supreme for power outputs exceeding 100 megawatts.

The turbine is similar to the windmill and water wheel in that they all convert the kinetic energy of a flowing fluid into the kinetic energy of rotary motion or torque applied to a rotating shaft. The steam turbine was first introduced by a British engineer named Charles Parsons in 1884. In a steam turbine, the fluid kinetic energy is obtained by expanding high pressure steam to a low pressure under nearly adiabatic conditions. Water turbines which were in use in Parsons' time were capable of extracting 70 to 80 percent of the fluid's kinetic energy. Parsons reasoned that similar efficiencies would be obtained from a steam turbine if the flow across the turbine blades remained basically incompressible. He therefore used many rows of turbine blades as shown in Fig. 7-23 instead of the single

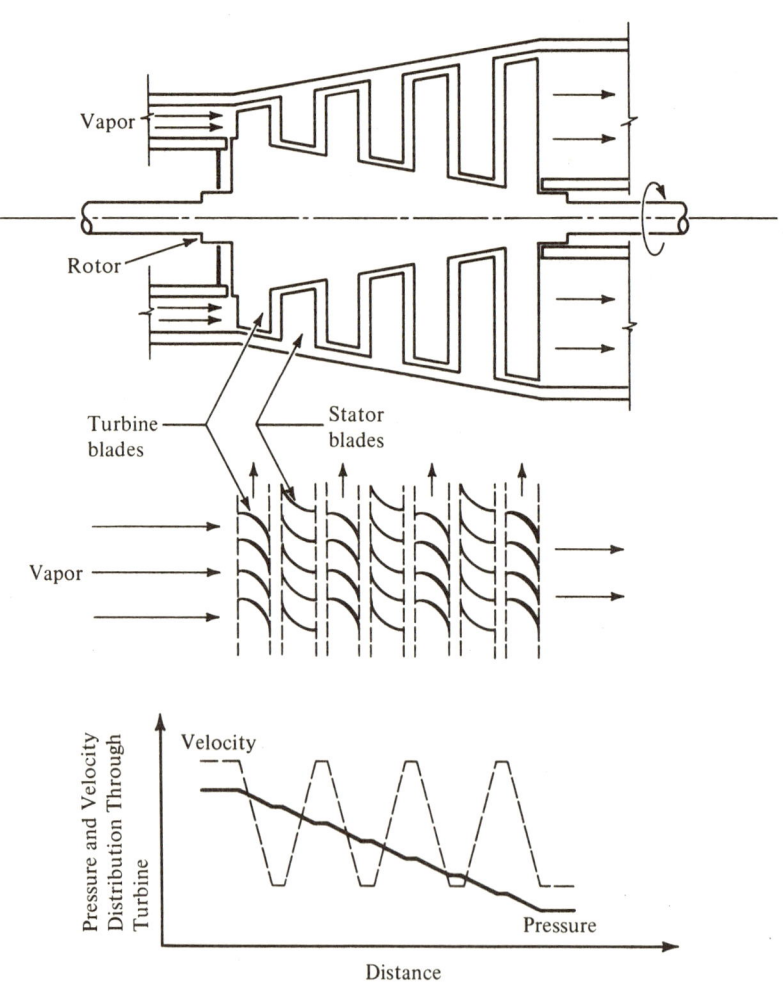

Fig. 7-23. *Schematic of a steam turbine.*

row used in a water turbine to make the pressure drop across each row small. Between each row of rotating turbine blades is a row of fixed blades called stators that are used to direct the flow against the moving blades. A row of stator blades and the adjoining row of rotating blades are termed a stage. As indicated in Fig. 7-23, each successive stage must be larger than the previous stage to accommodate the increased specific volume of the steam as it expands to a lower pressure through the turbine.

Since the expansion is essentially adiabatic, the efficiency of the overall turbine or of an individual stage of the turbine is defined by comparing the actual work output w_{act} with that obtained from a reversible adiabatic turbine (or stage) $w_{adiab, rev}$, operating between the same inlet state and outlet pressure. The work output from an adiabatic, reversible or isentropic turbine is used as the standard of comparison since it can be theoretically calculated and, of course, represents the maximum output obtainable from an adiabatic turbine.

$$\eta_{turb} = \frac{w_{act}}{w_{adiab, rev}} \qquad (3\text{-}19)$$

Turbines are usually designed to convert almost immediately any increase in the fluid's kinetic energy obtained during its expansion through the turbine into torque applied to the rotating rotor. As is indicated by Fig. 7-23, the velocity change and the corresponding kinetic energy change across an individual turbine stage or across the whole turbine would be small in this case. Thus, the change in the fluids kinetic energy across a turbine stage or across the whole turbine itself can be neglected along with the change in potential energy and heat transfer. Under these conditions, the generalized energy balance equation indicates that the work performed is equal to the enthalpy drop. Equation (3-19) can, therefore, be written in the more convenient form

$$\eta_{turb} = \frac{\Delta h}{\Delta h_s} \qquad (7\text{-}15)$$

where Δh indicates the actual enthalpy drop and Δh_s is the enthalpy drop for an adiabatic reversible or isentropic turbine or turbine stage.

Figure 7-24 presents the hs diagram for a flow expanding through two stages of a turbine. The thermal efficiencies of the individual stages are determined from Eq. (7-15) to be

$$\eta_{turb, 1-2} = \frac{h_1 - h_2}{h_1 - h_{2s}}$$

and

$$\eta_{turb, 2-3} = \frac{h_2 - h_3}{h_2 - h_{3's}}$$

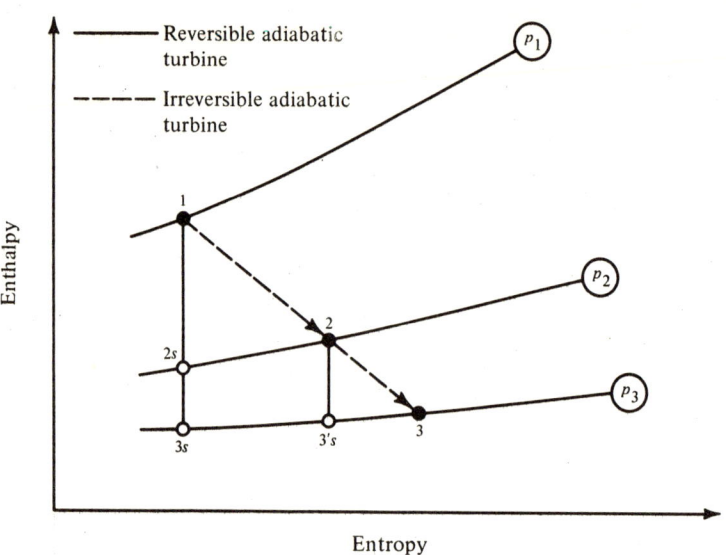

Fig. 7-24. *Expansion of steam through two turbine stages.*

while the overall thermal efficiency of the two stages is

$$\eta_{\text{turb, 1-3}} = \frac{h_1 - h_3}{h_1 - h_{3s}} = \frac{[(h_1 - h_2) + (h_2 - h_3)]}{h_1 - h_{3s}}$$

This last expression can be rewritten in terms of the stage thermal efficiencies

$$\eta_{\text{turb, 1-3}} = \frac{\eta_{\text{turb,1-2}}(h_1 - h_{2s}) + \eta_{\text{turb, 2-3}}(h_2 - h_{3's})}{h_1 - h_{3s}} \qquad (7\text{-}16)$$

If the stage efficiencies are equal, that is,

$$\eta_{\text{turb, 1-2}} = \eta_{\text{turb, 2-3}} = \eta$$

Eq. (7-16) simplifies to the form

$$\eta_{\text{turb, 1-3}} = \frac{\eta[(h_1 - h_{2s}) + (h_2 - h_{3's})]}{h_1 - h_{3s}} \qquad (7\text{-}17)$$

Figure 7-24 indicates that the pressure curves diverge with increasing entropy on the *hs* diagram. This causes the sum of the Δh_s values for the individual stages

$(h_1 - h_{2s}) + (h_2 - h_{3's})$ to be greater than the enthalpy drop $h_1 - h_{3s}$ obtained from the continuous isentropic expansion, 1—3s. The ratio

$$\frac{(h_1 - h_{2s}) + (h_2 - h_{3s})}{h_1 - h_{3s}}$$

in Eq. (7-17) is therefore somewhat greater than one. As a consequence, the overall thermal efficiency $\eta_{\text{turb}, 1-3}$ is greater than the individual stage thermal efficiency η. That is,

$$\eta_{\text{turb}, 1-3} > \eta$$

This surprising result means that a portion of the work lost in the first irreversible turbine stage is recovered in the next stage. Graphically, this is a consequence of the diverging pressure curves on the $h-s$ diagram. In other words, the enthalpy and therefore the corresponding available energy[†] are greater after the irreversible expansion 1—2 than after the corresponding isentropic expansion 1—2s. This conclusion is, of course, not limited to a two stage turbine with equal stage efficiencies but can be extended to a typical turbine having an arbitrary number of stages with varying efficiencies.

EXAMPLE PROBLEM 7-5. Referring to Fig. 7-24, saturated steam enters a two stage turbine at a pressure P_1 of 150 psia (10.3 bars) and leaves the turbine at a pressure P_3 of one atmosphere. If the efficiency of each stage is 39 percent and the steam pressure leaving the first stage P_2 is 70 psia (4.8 bars), calculate the overall turbine efficiency.

Taking the turbine as our system and basing our calculations on one pound of steam flowing through the system, we find that the mass and energy balance equations indicate that the work produced by an isentropic turbine would be

$$w_{\text{turb}, s} = h_1 - h_{3s}$$

From Table A-2,

$$h_1 = 1194.1 \frac{\text{BTU}}{\text{lb}_m}$$

$$s_1 = 1.5695 \frac{\text{BTU}}{\text{lb}_m \, °R} = s_{3s}$$

[†] See Chap. 3.

The isentropic expansion of the saturated vapor to one atmosphere pressure will put state 3s in the liquid-vapor region. Thus,

$$x_{3s} = \frac{s_1 - s_f}{s_{fg}} = \frac{1.5695 - 0.3121}{1.4447} = 0.8704$$

and

$$h_{3s} = h_f + x_{3s} h_{fg} = 180.2 + 0.8704 \times 970.3 = 1024.8 \; \frac{\text{BTU}}{\text{lb}_m}$$

We can now calculate the isentropic work,

$$w_{\text{turb}, s} = h_1 - h_{3s} = 1194.1 - 1024.8 = 169.3 \; \frac{\text{BTU}}{\text{lb}_m}$$

The quantities h_{2s}, h_2, and $h_{3's}$ that appear in Eq. (7-16) will now be calculated. In the case of the first stage, the exit pressure is 70 psia. The quality after an isentropic expansion is

$$x_{2s} = \frac{s_1 - s_f}{s_{fg}} = \frac{1.5695 - 0.4411}{1.1905} = 0.9478$$

and the corresponding enthalpy is

$$h_{2s} = h_f + x_{2s} h_{fg} = 272.7 + 0.9478 \times 907.8 = 1133.1 \; \frac{\text{BTU}}{\text{lb}_m}$$

From Eq. (7-15),

$$h_1 - h_2 = \eta(h_1 - h_{2s})$$

or

$$h_2 = h_1 - \eta(h_1 - h_{2s}) = 1194.1 - 0.39 \times (1194.1 - 1133.1)$$

$$= 1170.3 \; \frac{\text{BTU}}{\text{lb}_m}$$

The quality leaving the first stage is

$$x_2 = \frac{h_2 - h_f}{h_{fg}} = \frac{1170.3 - 272.7}{907.8} = 0.9888$$

and the corresponding entropy is

$$s_2 = s_f + x_2 s_{fg} = 0.4411 + 0.9888 \times 1.1905 = 1.618 \frac{\text{BTU}}{\text{lb}_m {}^\circ\text{R}}$$

Expanding the vapor isentropically from state 2 to 14.7 psia results in a quality of

$$x_{3's} = \frac{s_2 - s_f}{s_{fg}} = \frac{1.618 - 0.312}{1.4447} = 0.9040$$

and an enthalpy of

$$h_{3's} = h_f + x_{3s} h_{fg} = 180.2 + 0.9040 \times 970.3 = 1057.4 \frac{\text{BTU}}{\text{lb}_m}$$

From Eq. (7-17), the overall turbine thermal efficiency is

$$\eta_{\text{turb}} = 0.39 \times \frac{(1194.1 - 1133.1) + (1170.3 - 1057.4)}{1194.1 - 1024.8}$$

$$= 0.40$$

which is slightly larger than the individual stage thermal efficiency of 0.39.

We might note here that the turbine depicted in Fig. 7-23 can be converted into an axial flow compressor by reversing the flow and the rotor's rotation. The pressure of the fluid now increases as it is compressed into a decreasing cross-sectional area. Experience indicates that the expansion of flows can be performed more efficiently than the corresponding compression process. That is, the efficiency of a compressor stage will be less than a corresponding turbine stage with the same pressure difference across it. Also, an analysis of a multistage compressor will show that the overall efficiency of the compressor decreases with each additional stage. The flow through a turbine and a compressor are therefore quite different. For these reasons, the development of the axial flow compressor lagged behind the development of the reciprocating and centrifugal compressors.

Steam Turbine System

As depicted in Fig. 7-25, the simplest mechanical system utilizing a steam turbine would also require a pump and a boiler to supply the turbine with high pressure steam. In contrast to the reciprocating internal-combustion engine that uses essentially a single device (the piston-cylinder) to accomplish its complete cycle, we see that turbine systems consist of several basic components that are individually dedicated to only a segment of the cycle. Larger systems will employ additional components which can be combined in a large variety of mechanical

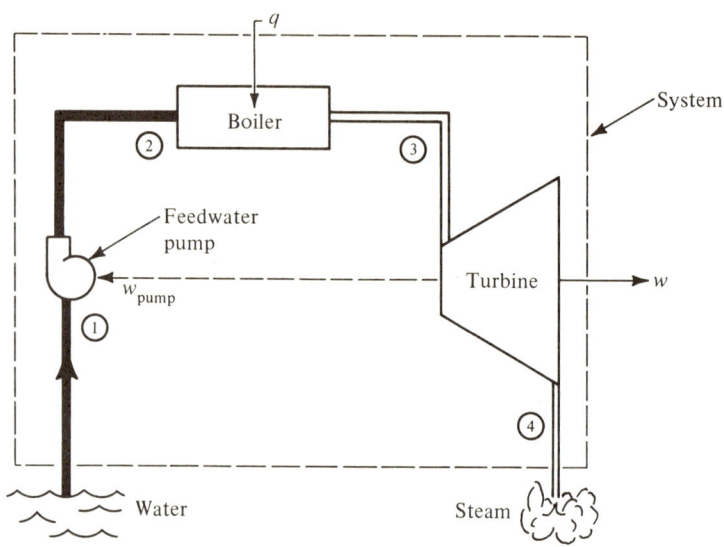

Fig. 7-25. *Simple steam-turbine system.*

arrangements. The performance of the total system can be predicted by repeated application of the generalized balance equations with the estimated efficiencies of the individual components themselves. Our calculation of the overall isentropic turbine efficiency in Example Prob. 7-5 from knowledge of the individual stage efficiencies is a simple illustration of this point.

EXAMPLE PROBLEM 7-6. As another example of this method of analysis, we will assume the following states and component efficiencies for the open system depicted in Fig. 7-25 which are characteristic of Parsons' early steam turbine system:

State	T(°F)	P(psia)	x(%)	Component Efficiency
1	80	14.7		
				$\eta_{pump} = \dfrac{w_{adiab,\,rev}}{w_{act}} = 0.30$
2		150		
				$\eta_{boiler} = \dfrac{\text{heat transferred to the system}}{\text{heat liberated by fuel}}$
				$= \dfrac{q}{\Delta h_c} = 0.50$
3		150	100	
				$\eta_{turb} = \dfrac{w_{act}}{w_{adiab,\,rev}} = 0.40$
4		14.7		

Calculate the overall thermal efficiency of this system based on the fraction of the total energy in the fuel which can be converted to heat, Δh_c.

To calculate the thermal efficiency of the total system, we need to calculate the work performed by the turbine w_{turb}, the pump work w_{pump}, and the heat transferred to the system q.

$$\eta_{therm} = \frac{w}{q} = \frac{w_{turb} + w_{pump}}{q} \qquad (3\text{-}7)$$

Neglecting any changes in kinetic and potential energy again, we can use the results of Example Prob. 7-5 to obtain the turbine output,

$$w_{turb} = \eta_{turb} w_{turb,s} = 0.40 \times 169.3 \, \frac{\text{BTU}}{\text{lb}_m}$$

$$= 67.7 \, \frac{\text{BTU}}{\text{lb}_m} \, (157.5 \text{ joule/gram})$$

The pump work is

$$w_{pump} = h_1 - h_2 = w_{pump,s}/\eta_{pump}$$

$$= (h_1 - h_{2s})/\eta_{pump}$$

As shown on Fig. 7-26, states 1 and 2 are in the subcooled liquid region. A tabula-

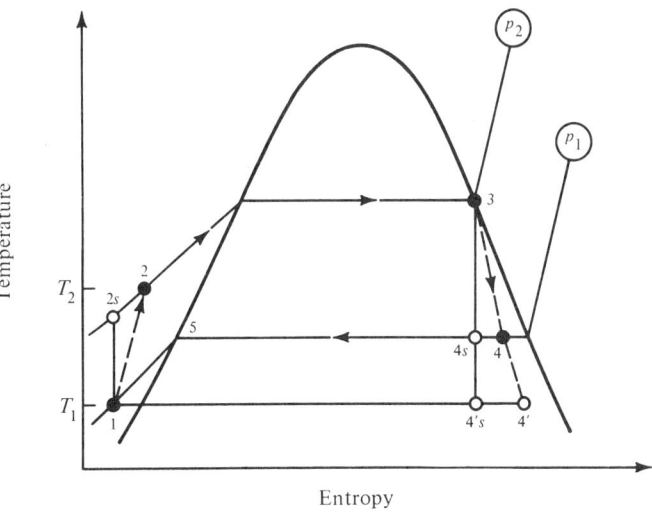

Fig. 7-26. *T-s diagrams of simple steam turbine system.*

tion of the state properties in the liquid region are not given in this text but they can be found in other sources such as the steam tables by Keenan and Keyes[2] The required enthalpy difference can also be estimated from the property relationship

$$dh = T ds + v\, dP \tag{4-18}$$

by integrating this equation along an isentropic path between state 1 and pressure P_2. If water is assumed to be incompressible in this integration,

$$v_1(T_1, P_1) = v_f(T_1) = v_2$$

we obtain the result

$$h_{2s} - h_1 = v_1(P_2 - P_1)$$

From Table A-2, we find that $v_f = 0.0161$ ft^3/lb$_m$. Thus,

$$h_{2s} - h_1 = 0.0161\, \frac{\text{ft}^3}{\text{lb}_m} \times (150 - 14.7)\, \frac{\text{lb}_f}{\text{in}^2} \times 144\, \frac{\text{in}^2}{\text{ft}^2} / 778.2\, \frac{\text{ft} \cdot \text{lb}_f}{\text{BTU}}$$

$$= 0.40\, \frac{\text{BTU}}{\text{lb}_m}\; (0.94\ \text{joule/gram})$$

which is equal to the work absorbed by the adiabatic reversible pump $w_{pump,s}$. The actual pump work, therefore, is equal to

$$w_{pump} = \frac{w_{pump,s}}{\eta_{pump}} = \frac{-0.40\ \text{BTU/lb}_m}{0.30} = -1.3\, \frac{\text{BTU}}{\text{lb}_m}\; (-3.1\ \text{joule/gram})$$

To calculate h_1 and h_2, we start with the expression for the enthalpy differential

$$dh = \left(\frac{\partial h}{\partial T}\right)_P dT + \left(\frac{\partial h}{\partial P}\right)_T dP = c_P dT + \left(\frac{\partial h}{\partial P}\right)_T dP$$

and integrate it along a constant pressure path from the known saturated liquid state $h_f(P_1)$ to $h_1(P_1, T_1)$. If c_P is assumed to be constant, the integration results in the relationship

$$h_1(P_1, T_1) - h_f(P_1) = c_P\, [T_1 - T_f(P_1)]$$

where the saturated liquid values h_f and T_f are obtained from the steam tables. Then,

$$h_1 = h_f(P_1) + c_p [T_1 - T_f(P_1)]$$

$$= 180.2 \frac{BTU}{lb_m} + 1.0 \frac{BTU}{lb_m \, °F} \times (80\,°F - 212\,°F)$$

$$= 48.2 \frac{BTU}{lb_m} \, (112.0 \text{ joule/gram})$$

and

$$h_2 = h_1 - w_{pump} = 48.2 + 1.3$$

$$= 49.5 \text{ BTU/lb}_m \, (115.1 \text{ joule/gram})$$

The heat transferred to the water from the boiler is

$$q_{boiler} = h_3 - h_2 = h_g(P_2) - h_2$$

$$= 1194.1 \frac{BTU}{lb_m} - 49.5 \frac{BTU}{lb_m} = 1144.6 \frac{BTU}{lb_m} \, (2662 \text{ joule/gram})$$

but only 50 percent of the heat liberated when the fuel is completely burned Δh_c is transferred to the water. The remaining energy is lost in the stack gases. The available energy of the fuel consumed by the system is therefore

$$q = \Delta h_c = \frac{q_{boiler}}{\eta_{boiler}} = \frac{1144.6}{0.5} = 2289 \frac{BTU}{lb_m} \, (5321 \text{ joule/gram})$$

The following table summarizes the results, some of which were obtained in Example Problem 7-5.

State	T(°F)	P(psia)	h (BTU/lb$_m$)	s (BTU/°R lb$_m$)	x(%)	q or w (BTU/lb$_m$)	Component Efficiency
1	80	14.7	48.2				
2s		150	48.6			−0.4	
2		150	49.5			−1.3	$\eta_{pump} = 0.30$
3	358	150	11.94.1	1.5695	100	1144.6	$\eta_{boiler} = 1.00$
						2289.2	$\eta_{boiler} = 0.50$
4s	212	14.7	1024.8	1.5695	87	169.3	
4	212	14.7	1126.4		98	67.7	$\eta_{turb} = 0.40$

The thermal efficiency of this early turbine system is

$$\eta_{therm} = \frac{w_{turb} + w_{pump}}{q} = \frac{67.7 - 1.3}{2289.2} = 0.03$$

which is, as expected, quite low. The low component efficiencies are not the only cause of this system's poor performance since the thermal efficiency of the reversible system is not too impressive either,

$$\eta_{therm, rev} = \frac{w_{turb,s} + w_{pump,s}}{h_3 - h_{2s}} = \frac{169.3 - 0.4}{1194.1 - 48.6}$$

$$= 0.15$$

We note that the heat addition occurs between 80 °F and 358 °F, while the steam is exhausted at 212 °F. Heat is, therefore, added to the system at a relatively low average temperature and is exhausted at a relatively high temperature. We know from our discussion of the Carnot heat engine that this combination of heat transfer characteristics will lead to a thermally inefficient system.

The Rankine Cycle

Many improvements have been made to Parsons' original turbine system over the years to improve its thermal efficiency. The first refinement which was made by Parsons himself involved the addition of a condenser between the pump and the turbine to form the closed system depicted in Fig. 7-27. As shown in Fig.

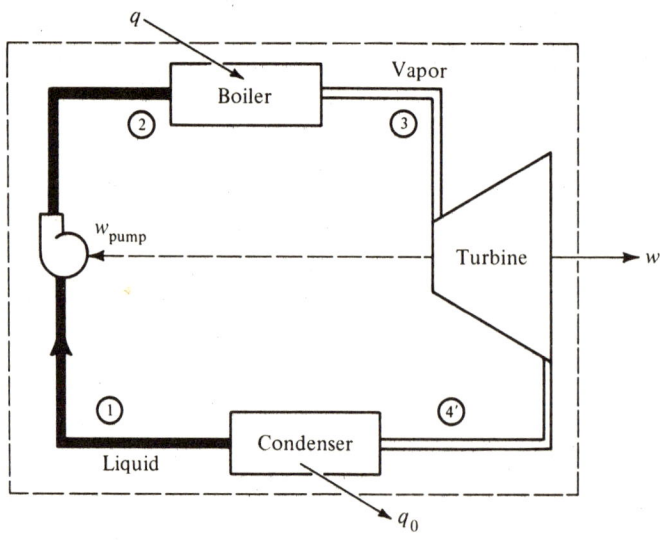

Fig. 7-27. *Condensing steam turbine system.*

7-26, the closed system permits the turbine exhaust pressure $P_{4'}$ to be below the atmospheric pressure $P_1 = P_4$, used in the previous example. The work performed by the reversible closed system is given by the enclosed area 1–2s–3–4's–1 in Fig. 7-26 while the enclosed area 1–2s–3–4s–5–1 represents the work performed by the reversible open system. Process 4s–5–1 in the open system depicts the isobaric cooling of the exhausted steam by the atmosphere to its original subcooled liquid state. The enclosed area 1–5–4s–4's–1 therefore represents the increased output that is obtained when the turbine exhaust pressure P_4 is decreased. The heat transferred to both the reversible open and closed systems is given by the area under the curve 1–2s–3. Thus, the efficiency of the reversible turbine system

$$\eta = \frac{w}{q} = \frac{\text{area } 1-2s-3-4's-1}{\text{area under } 2s-3}$$

will increase as the turbine exhaust pressure is decreased.

The steam tables indicate that the back pressure $P_{4'}$ in the condenser would only be 0.51 psia if the condenser temperature were maintained at 80 °F. This would make the pressure ratio across the turbine 29 times larger than the noncondensing system described in Example Prob. 7-6. If we inserted this condenser into this system, the work obtained from the turbine is found by the same procedure used in Example Prob. 7-6. For the isentropic expansion between states 3 and 4's shown in Fig. 7-26, we have

$$s_{4's} = s_3 = 1.5695 \text{ BTU/lb}_m \, °R$$

From Table 7-1. This implies that the quality at state 4's is

$$x_{4's} = \frac{s_3 - s_f}{s_{fg}} = \frac{1.5695 - 0.0932}{1.9426} = 0.76$$

The work obtained from an isentropic turbine operating between states 3 and 4's is, therefore,

$$w_{\text{turb}, s} = h_3 - h_{4's} = h_3 - (h_f + x_{4's} h_{fg}^*)$$

$$= 1194.1 - (48.04 + 0.76 \times 1048.4) = 349.3 \text{ BTU/lb}_m$$

If we again assume that the thermal efficiency of the actual turbine is only 40 percent, then the turbine output would be 139.7 BTU/lb$_m$ which represents a 106 percent increase over the specific work obtained from the open cycle turbine in Example Prob. 7-6.

In a closed system, there is no reason to subcool the water at state 1 significantly; so we will assume state 1 to be a saturated liquid as shown in Fig. 7-28. If

238   Chap. 7   Thermodynamics of Heat Engines

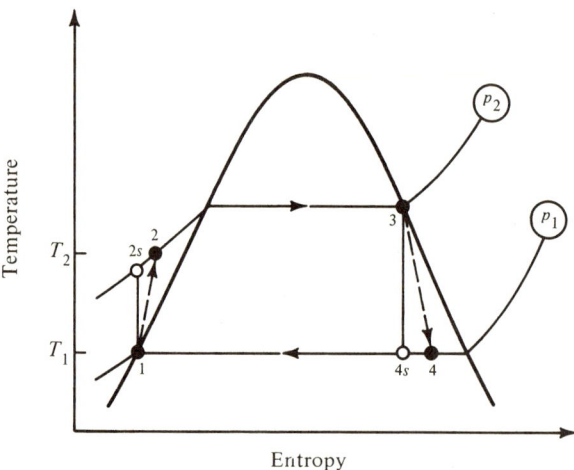

Fig. 7-28. *Rankine cycle.*

the analysis of this system was continued, the following cycle characteristics would be obtained:

State	T(°F)	P(psia)	h (BTU/lb$_m$)	s (BTU/°R lb$_m$)	x(%)	q or w (BTU/lb$_m$)	Component Efficiency
1	80	0.51	48.0		0		
2s		150	48.4			−0.44	
2		150	49.5			−1.5	$\eta_{pump} = 0.30$
3	358	150	1194.1	1.5695	100	1144.6	$\eta_{boiler} = 1.00$
						2289.2	$\eta_{boiler} = 0.50$
4s	80	0.51	844.8	1.5695	76	349.3	
4	80	0.51	1054.4		96	139.7	$\eta_{turb} = 0.40$

The thermal efficiency of this system is

$$\eta = \frac{w_{turb} + w_{pump}}{q} = \frac{139.7 - 1.5}{2289.2} = 0.06$$

which is double the efficiency obtained by the open cycle.

The corresponding reversible cycle shown in Fig. 7-28 is composed of an isentropic compression in the liquid phase 1—2s; a constant pressure heat addition 2s—3; an isentropic expansion 3—4s; and a constant pressure condensing process 4—1. The thermal efficiency of this reversible cycle is

$$\eta_{rev} = \frac{w_{turb,s} + w_{pump,s}}{q} = \frac{349.3 - 0.4}{1145.7} = 0.30$$

which is almost the efficiency that would be obtained from a Carnot cycle ($\eta_c = 0.34$) operating between the same two temperature extremes. This reversible cycle was proposed independently by Rankine and Clausius but is generally referred to as the Rankine cycle.

The reason that the Rankine cycle approaches the Carnot efficiency in this case is that all the heat is exhausted at the minimum cycle temperature T_1 during the phase change 4—1; and almost all of the heat (75 percent) is added at the cycle's maximum temperature T_3 during the vaporization. Another salient feature of Rankine's vapor cycle is that the pump absorbs a small fraction (≈ 1 percent) of the turbine's actual output since the compression process (1—2 in Fig. 7-28) is confined to the liquid phase. We are, therefore, able to obtain power from the Rankine cycle even with very poor component performance. This characteristic can be compared to the gas reciprocating cycles where the compression process was found to absorb a large fraction of the output of the power stroke.

The condensing steam turbine system was rapidly adopted at the turn of the century for use in large ships and in electrical power plants where the turbine was the natural mate for the generator. Figure 7-29 indicates that the thermal effi-

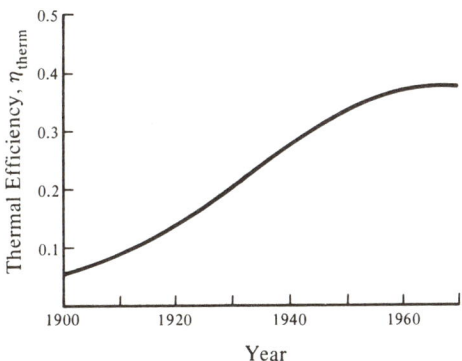

Fig. 7-29. *Historical thermal efficiency of state of the art power plants.*

ciency obtained by an average power plant at the turn of the century is comparable to the 6 percent we obtained for the condensing system example. These early power plants had an output of only a few hundred kilowatts. To calculate the overall thermal efficiency of an electrical generating plant, we should multiply the efficiency of the turbine system by the generator's efficiency, but generators have always been very efficient with modern units being able to convert 99 percent of the mechanical power supplied to them into electrical power.

Increasing the thermal efficiency of a turbine system by further decreasing the condenser temperature below the 80 °F assumed here is not a realistic alternative for a large system since condensers typically depend upon some natural body of water or air for cooling. In addition, the condenser temperature usually runs 20 to 40 °F above the temperature of the cooling medium to insure adequate heat transfer rates.

One avenue that was obviously open for improving the thermal efficiencies of the early turbine systems was to improve the performance of the individual components. Thanks to the advancements made in fluid mechanics, isentropic efficiencies of the order of 96 percent are obtained in large modern turbines while smaller turbines generally achieve efficiencies between 60 and 80 percent. It should be noted here that the efficiency of a turbine is obviously dependent upon the rotor's angular velocity, since there would be no power delivered if the rotor is stationary or if the rotor moves at such a speed that there is no normal velocity component between the turbine blades and the gas. The maximum efficiency must occur somewhere between these two extreme cases, implying that a turbine must operate near its design load to obtain the high efficiencies quoted above.

Large modern feedwater pumps achieve isentropic efficiencies around 85 percent while smaller centrifugal pumps now have isentropic efficiencies in the range of 40 to 60 percent. As for boiler efficiencies, modern units can be expected to transfer around 90 percent of the available chemical energy in the fuel to the water in contrast to the 50 percent assumed in Example Prob. 7-6.

If these modern components had been available in Parsons' time, his small condensing turbine system (Fig. 7-27) would be expected to operate at a thermal efficiency near

$$\eta = \frac{w_{turb} + w_{pump}}{q} \approx \frac{\eta_{turb} w_{turb,s}}{q_{boiler}/\eta_{boiler}} = \frac{0.80 \times 349.3}{1144.6/0.90} = 0.22$$

which is comparable to the modern Otto cycle.

It is interesting to note here that the reciprocating steam engine was also based on the Rankine cycle, where the expansion process was against a piston instead of through a turbine. The operating pressures and temperatures of these engines were basically limited to values close to the ones used in the above example, but the efficiency of these engines was under 15 percent as compared to the 22 percent that we estimated for a comparable modern steam turbine system. We would expect that the reciprocating engine is mechanically less efficient than a turbine system but this accounts for only part of the difference. The fact that it was impractical to expand the steam completely and the heating and cooling of the metal cylinder during intake and expansion processes were also major contributors in making the reciprocating engine less efficient. The reciprocating steam engine's very low thermal efficiency has, therefore, relegated this engine to the museums.

Besides increasing the component efficiencies to the modern levels, engineers have made numerous modifications to Parsons' system to increase its efficiency. One obvious improvement immediately pursued was to increase the average temperature at which heat was absorbed in the cycle from an external energy source. A superheater was therefore added after the boiler as indicated in Fig. 7-30 with the corresponding reversible cycle depicted in Fig. 7-31. The superheater increases

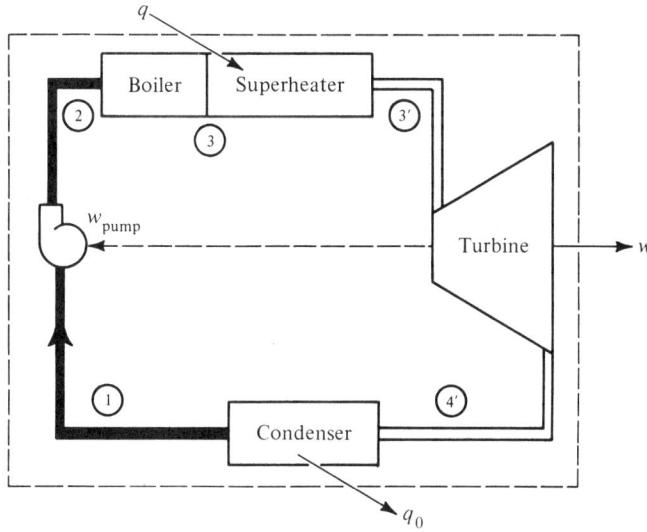

Fig. 7-30. *Condensing steam turbine system with superheater.*

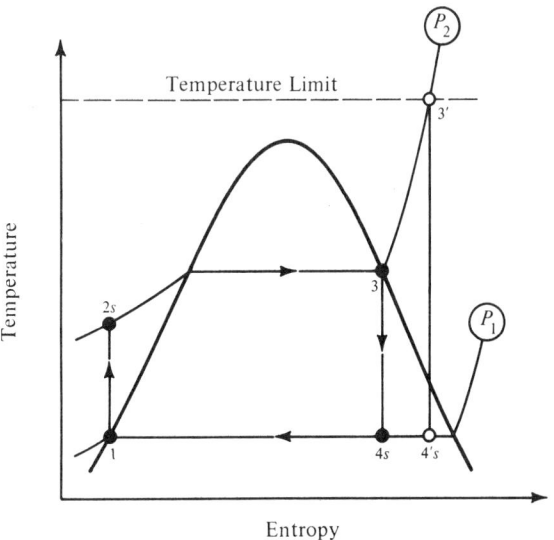

Fig. 7-31. *T-s diagram of a Rankine cycle with a superheated region.*

the temperature from the saturated value T_3 at the boiler exit to a superheated value $T_{3'}$. The enclosed area $4s-3-3'-4's-4s$ shown in Fig. 7-31 represents the additional work that can be obtained from a reversible turbine by superheating while the area under path $3-3'$ represents the additional heat that must be added.

The maximum steam temperature $T_{3'}$ is limited by material problems involving corrosion and allowable stresses at elevated temperatures. Steam temperatures around 1000 °F are presently economic, but this temperature level was obtained only after many years of research on high temperature metals. The maximum operating temperature of the steam Rankine cycle is, therefore, not expected to increase much in the near future.

If we now add a superheater to our previous example of a condensing turbine system and heat the steam to 1000 °F the following changes in the characteristics of the reversible cycle are found:

State	T(°F)	P(psia)	h (BTU/lb$_m$)	s (BTU/lb$_m$ °R)	x(%)	q or w (BTU/lb$_m$)
3'	1000	150	1530.5	1.8751		1482.1
4's	80	0.51	1012.5	1.8751	92	518.0

The resulting efficiency of this reversible Rankine cycle is

$$\eta_{\text{rev}} = \frac{w_{\text{turb},s} + w_{\text{pump},s}}{q} = \frac{518.0 - 0.44}{1482.0} = 0.35$$

which represents a 17 percent increase in efficiency over the reversible system without superheat. However, a Carnot cycle operating between the same two temperature extremes would have a much higher thermal efficiency (63 percent) since most of the heat in this superheated Rankine cycle is no longer added at or near the maximum temperature of the cycle.

Figure 7-32 shows the effect that increasing the maximum pressure P_2 has on the Rankine cycle if the temperature extremes T_1 and T_3 are held constant. As the pressure increases ($P_2 < P_{2'} < P_{2''}$), we note that the average temperature at which heat is added to the cycle along the paths marked $2-3$ appears to also increase. The efficiency of the Rankine cycle does in fact increase with the pressure ratio P_2/P_1, which is also the case for the reciprocating gas cycles that were previously discussed.

In the 1930's, the steam pressure in a typical power plant was increased from below 300 psi to over 1200 psi, and modern plants now utilize pressures at least double this later value. Cycle $1-2''-3''-4''-1$ in Fig. 7-32 depicts the case where the maximum pressure $P_{2''}$ has exceeded the critical pressure ($P_{2''} > 3200$ psi); and there is, therefore, no phase change between the liquid and vapor states. Plants utilizing pressures up to 5000 psia have been built, but the American So-

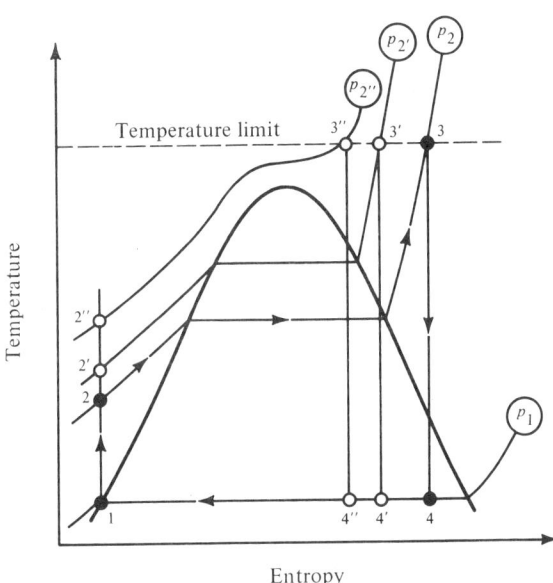

Fig. 7-32. *Effect of pressure on the Rankine cycle.*

ciety of Mechanical Engineers Boiler Codes that set the guidelines for safe boiler construction require the use of very heavy structures in this case. The thermodynamic gains obtained by operating at high supercritical pressures are, therefore, generally offset by the much higher capital costs and also by the severe maintenance problems encountered. The practical pressure and temperature limits for conventional Rankine steam cycles are generally considered to be around 3500 psia and 1000 °F respectively.

Figure 7-32 indicates that the quality x_4 of the steam that leaves the turbine decreases as the maximum pressure P_2 increases if the temperature extremes of the cycle remain constant. The water droplets in steam having qualities below 90 percent were found to cause severe erosion of the turbine and stator blades and to decrease efficiency in the final low pressure stages. The practice of reheating the steam as shown in Figs. 7-33 and 7-34 was, therefore, introduced in the 1930's along with the use of high pressure steam. In this cycle, the steam is expanded to some intermediate pressure $P_{4'}$ usually near the saturation line where it is reheated and then expanded in a second turbine to the terminal pressure P_4. The corresponding quality x_4 is higher than the quality $x_{4''}$ that would have resulted from an expansion process without reheat.

Figure 7-34 also shows that reheating will increase the total turbine output, but the efficiency of the cycle may or may not be improved since the average temperature at which heat is added is not necessarily increased by reheating. Proper use of reheat will increase both the quality of the steam leaving the turbine and the system's efficiency. Some large supercritical power stations, in fact, use double reheat.

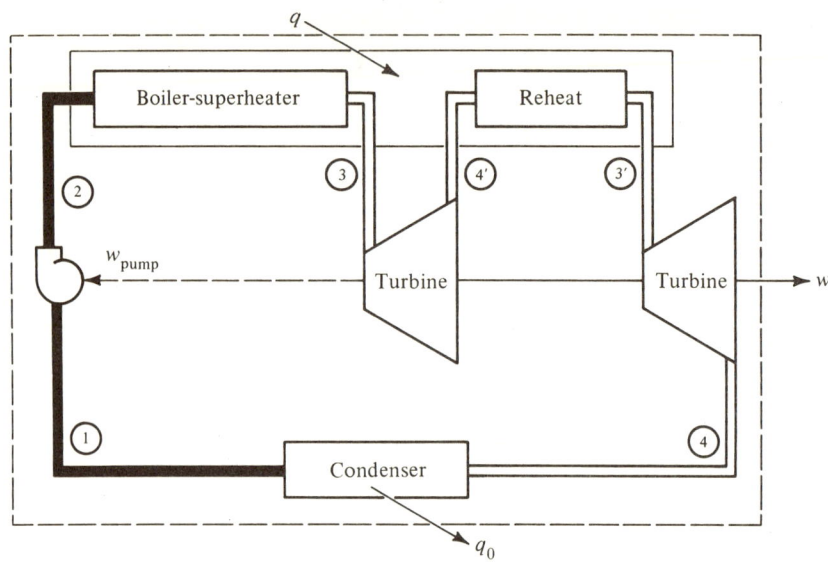

Fig. 7-33. *Condensing Rankine cycle with reheat.*

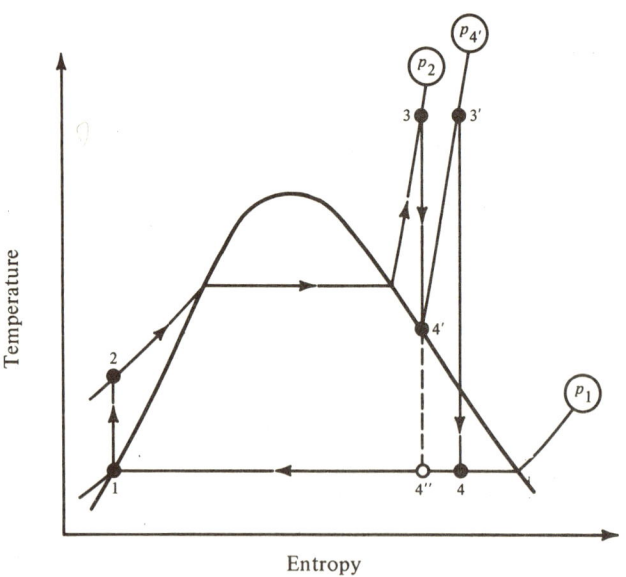

Fig. 7-34. *T-s diagram of a Rankine cycle with reheat.*

Regenerative heating, which was introduced with the Stirling cycle, will also improve the performance of the steam turbine system. This involves the use of internal heating during the lower temperature portion of the cycle's heat addition process. The average temperature at which heat is added to the cycle from an external source is thereby increased which in turn increases the cycle's thermal efficiency. The use of a regenerative feedwater heater to preheat the water before it enters the boiler-superheater is depicted in Figs. 7-35 and 7-36. In this example,

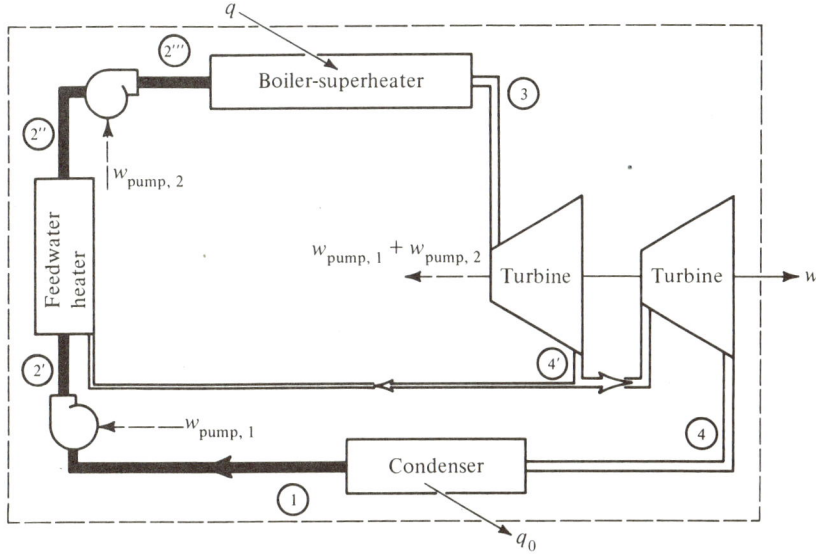

Fig. 7-35. *Condensing steam turbine system with open feedwater heater.*

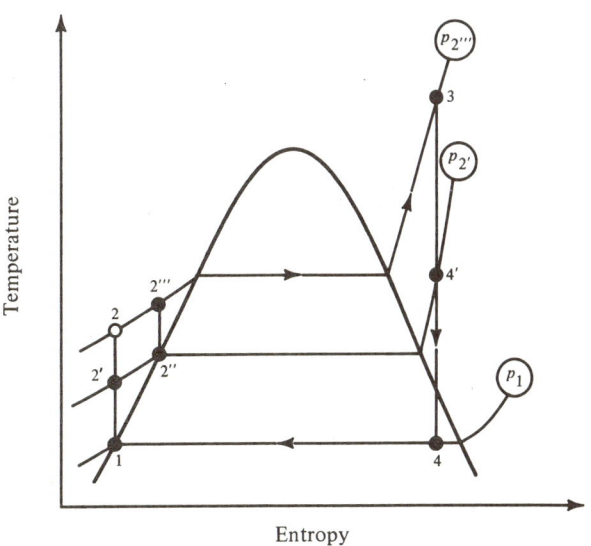

Fig. 7-36. *T-s diagram of a Rankine cycle with feedwater heater.*

high temperature steam is extracted from the turbine at state 4' to preheat the water (2'—2") before it enters the boiler-superheater where it is heated by an external source (2'''—3). The regenerative feedwater heater not only improves the cycle efficiency but also reduces the mass flow rate in the final turbine stages 4'—4. This is advantageous since the low pressure turbine must be large and, therefore, expensive in order to handle the large volume flow rate.

Two types of feedwater heaters are used. The open feedwater heater depicted in Fig. 7-35 mixes the steam extracted from the turbine directly with the feedwater. In a closed feedwater heater, the steam extracted from the turbine is not mixed with the feedwater inside the heater but instead condenses on the outside of the feedwater tubes. A closed feedwater heater has the advantage of allowing the steam and water pressures to be quite different. In theory, an infinite number of regeneration stages, with the entrance temperatures of the steam and

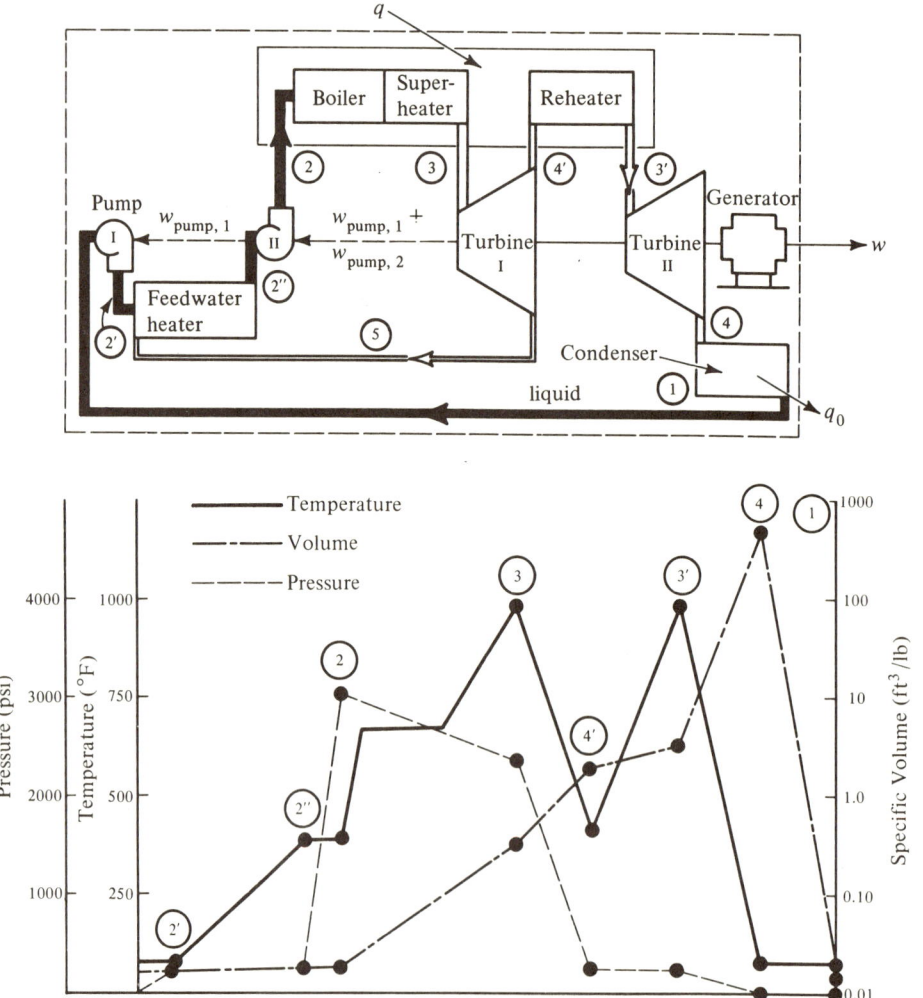

Fig. 7-37. *Temperature, pressure and specific volume of steam as it passes through the above steam turbine system.*

water at each stage being different by only an infinitesimal amount, could be assembled into a reversible regeneration system. In practice, as many as eight successive regeneration stages are employed to approximate the ideal regenerator.

Figure 7-37 depicts a system which combines all of the components that have been introduced individually—condenser, superheater, reheater and open feedwater heater. The indicated pressure and temperature variation is representative of what might be expected in a modern facility, except that additional feedwater heaters and possibly another reheat stage might be included. Also, three separate turbines (a high pressure, an intermediate, and a low pressure turbine) would be employed to handle the indicated three orders of magnitude change in the steam's specific volume. Note that the pressure drop in the boiler-superheater has been included, but the pressure drops in the feedwater heater, reheater and condenser have been neglected again since the actual variations are much smaller. The performance of this turbine cycle is calculated in the following example.

EXAMPLE PROBLEM 7-7. The system depicted in Fig. 7-37 has the following characteristics:

State	T(°F)	P(psia)	x(%)	Component Efficiency
1	80		0.0	
				$\eta_{pump,1} = 0.80$
2'		220		
2''		220	0.0	
				$\eta_{pump,2} = 0.80$
2		3100		
				$\eta_{boiler} = 0.88$
3	1000	2400		
				$\eta_{turb,1} = 0.90$
4'		220		
				$\eta_{reheat} = 0.88$
3'	1000	220		
				$\eta_{turb,2} = 0.90$
4	80			

Calculate the thermal efficiency of this system.

The thermal efficiency of this system is

$$\eta_{therm} = \frac{w}{q} = \frac{w_{turb,1} + w_{pump,2} + \frac{\dot{m}_1}{\dot{m}_2}(w_{turb,2} + w_{pump,1})}{q_{2-3}/\eta_{boiler} + \frac{\dot{m}_1}{\dot{m}_2} q_{4'-3'}/\eta_{reheat}} \quad (7\text{-}18)$$

where $\dot{m}_1/\dot{m}_2$ represents the fraction of steam that passes through all the turbines and pumps. We will follow the same analytical procedure used in Example Prob. 7-6. The specific work of the first pump is

$$w_{pump,1} = h_1 - h_{2'} \cong v_1(P_1 - P_2)/\eta_{pump,1}$$

$$= v_f(T_1)(P_1 - P_2)/\eta_{pump,1}$$

$$= 0.0161 \frac{ft^3}{lb_m} \times (0.51 - 220) \frac{lb_f}{in^2} \times 144 \frac{in^2}{ft^2} / (778.2 \frac{ft \cdot lb_f}{BTU} \times 0.8)$$

$$= -0.82 \text{ BTU}/lb_m$$

and

$$h_{2'} = h_1 - w_{pump,1} = 48.04 + 0.82 = 48.86 \text{ BTU}/lb_m$$

The feedwater is mixed with enough steam in the feedwater heater to produce a saturated liquid which is then compressed to the boiler pressure P_2. The specific work of this second pump is evaluated in the same manner as the first pump,

$$w_{pump,2} = v_f(P_{2''})(P_{2''} - P_2)/\eta_{pump,2}$$

$$= 0.0185 \times (220 - 3100) \times 144/(778.2 \times 0.8)$$

$$= -12.32 \frac{BTU}{lb_m}$$

Assuming water to be incompressible in the above calculation for so large a pressure increase is not a good approximation, and steam tables[2] containing enthalpies for subcooled water should be consulted for a more accurate result. Since the work absorbed by the pump will be a small percent of the turbine's output, this approximation will not significantly influence our performance calculation.

The enthalpy of the water entering the boiler is

$$h_2 = h_{2''} - w_{\text{pump},2} = h_f(P_{2''}) - w_{\text{pump},2}$$

$$= 364.2 + 12.3 = 376.5 \, \frac{\text{BTU}}{\text{lb}_m}$$

and the heat transferred to the water by the boiler-superheater is

$$q_{2-3} = h_3 - h_2 = 1460.9 - 376.5 = 1084.4 \, \frac{\text{BTU}}{\text{lb}_m}$$

An isentropic expansion from state 3 ($s_3 = 1.5332$ BTU/lb$_m$ °R) to $P_{4'} = 220$ psia would result in state $4's$ being in the liquid-vapor region. Thus,

$$x_{4's} = \frac{s_{4's} - s_f}{s_{fg}} = \frac{1.5332 - 0.5540}{0.9834} = 0.9957$$

The corresponding enthalpy is

$$h_{4's} = h_f + x_{4's} h_{fg} = 364.2 + 0.9957 \times 835.4 = 1196.0 \, \frac{\text{BTU}}{\text{lb}_m}$$

The output of the first turbine is, therefore,

$$w_{\text{turb},1} = \eta_{\text{turb},1}(h_3 - h_{4's}) = 0.9 \times (1460.9 - 1196.0) = 238.4 \, \frac{\text{BTU}}{\text{lb}_m}$$

and the enthalpy of the steam leaving this turbine is

$$h_{4'} = h_3 - w_{\text{turb},1} = 1460.9 - 238.4 = 1222.5 \, \frac{\text{BTU}}{\text{lb}_m}$$

We can calculate the mass fraction of steam that will pass through both turbines $\dot{m}_1/\dot{m}_2$ by performing a mass and energy balance on the feedwater heater.

$$\dot{m}_{2'} + \dot{m}_5 = \dot{m}_{2''}$$

$$\dot{m}_{2'} h_{2'} + \dot{m}_5 h_5 = \dot{m}_{2''} h_{2''}$$

But Fig. 7-37 indicates that $\dot{m}_{2'} = \dot{m}_1$, $\dot{m}_{2''} = \dot{m}_2$ and $h_5 = h_{4'}$. Combining all of these relationships results in the expression

$$\frac{\dot{m}_1}{\dot{m}_2} = \frac{h_{2''} - h_{4'}}{h_{2'} - h_{4'}}$$

250 Chap. 7   Thermodynamics of Heat Engines

The numerical evaluation of this expression gives

$$\frac{\dot{m}_1}{\dot{m}_2} = \frac{364.2 - 1222.5}{48.9 - 1222.5} = 0.73$$

Thus, 27 percent of the flow must be extracted after the first turbine at state 4' and mixed with the feedwater at state 2' to obtain a saturated liquid at state 2".
The heat added by the reheater is

$$q_{4'-3} = h_{3'} - h_{4'} = 1528.5 - 1222.5 = 306.0 \frac{\text{BTU}}{\text{lb}_m}$$

and the quality of the steam that would leave an isentropic turbine after expanding to the condenser pressure P_4 is

$$x_{4s} = \frac{s_{3'} - s_f}{s_{fg}} = \frac{1.8318 - 0.0932}{1.9426} = 0.8950$$

The corresponding enthalpy is

$$h_{4s} = h_f + x_{4s}h_{4s} = 48.0 + 0.8950 \times 1048.4 = 986.3 \frac{\text{BTU}}{\text{lb}_m}$$

The performance of the second turbine is

$$w_{\text{turb},2} = \eta_{\text{turb},2}(h_{3'} - h_{4s}) = 0.9 \times (1528.5 - 986.3) = 488.0 \frac{\text{BTU}}{\text{lb}_m}$$

In summary,

State	T(°F)	P(psia)	h (BTU/lb$_m$)	x(%)	q or w (BTU/lb$_m$)	Component Efficiency
1	80	0.51	48.0	0.0		
2'		220	48.9		−0.82	$\eta_{\text{pump}} = 0.80$
2"	390	220	364.2	0.0		
2		3100	376.5		−12.3	$\eta_{\text{pump}} = 0.80$
3	1000	2400	1460.9		1084.4	$\eta_{\text{boiler}} = 0.88$
4'	426	220	1222.5		238.4	$\eta_{\text{turb}} = 0.90$
3'	1000	220	1528.5		306.0	$\eta_{\text{reheat}} = 0.88$
4	80	0.51	1040.5	95	488.0	$\eta_{\text{turb}} = 0.90$

The thermal efficiency of this system is

$$\eta_{therm} = \frac{w}{q} = \frac{w_{turb,1} + w_{pump,2} + \frac{\dot{m}_1}{\dot{m}_2}(w_{turb,2} + w_{pump,1})}{q_{2-3}/\eta_{boiler} + \frac{\dot{m}_1}{\dot{m}_2} q_{4'-3'}/\eta_{reheat}} \qquad (7\text{-}18)$$

$$= \frac{238.4 - 12.3 + 0.73 \times (488.0 - 0.8)}{1084.4/0.88 + 0.73 \times 306.0/0.88}$$

$$= 0.39$$

In our simple analysis, we have not included all of the losses that would occur in an actual system and we did not account for the power needed to drive the peripheral equipment such as fuel handling machinery. Thus, the overall thermal efficiency of an actual plant would be less than the 39 percent predicted here. Very large and complex fossil fuel power plants can obtain overall thermal efficiencies as high as 41 percent; but, as shown in Fig. 7-29, the rate of improvement in system efficiencies is diminishing rapidly. Since the component efficiencies are presently quite high and a plateau in operating pressures and temperatures appears to have been reached, only slight efficiency gains can be expected for the steam turbine system in the foreseeable future. In fact, the average thermal efficiency of utilities may actually drop for a while because of the increasing reliance on nuclear power to supply the heat for the steam turbine system. The boiling water reactor, which is the current nuclear workhorse, can only obtain thermal efficiencies near 34 percent because of the temperature (less than 700 °F) and pressure (less than 1000 psia) limitations required by the reactor. These nuclear power plants are, therefore, producing approximately 10 to 35 percent more waste heat than comparable size modern fossil fuel plants. An advanced reactor called the high temperature gas reactor may significantly change this situation since it can operate at temperatures that are comparable to the fossil fuel plants.

Electrical generating plants are presently supplying about 25 percent of the nation's energy needs, and this percentage is expected to increase to 50 percent by the year 2000. Because the total cost of producing power decreases with power plant size, the size of the units have steadily increased and plants are now exceeding 1000 megawatts. The enormity of this figure may be appreciated if we calculate the rate that coal must be supplied to a plant of this magnitude.

$$\dot{m}_{coal} = \frac{\dot{W}}{\eta_{thermal}(\Delta h_c)_{coal}}$$

where $(\Delta h_c)_{coal}$ is the specific heat of combustion for coal. If we take Δh_c to be 12,000 BTU/lb_m and the total power plant thermal efficiency η_{therm} to be 0.39, the rate that coal must be supplied is

$$\dot{m}_{coal} = \frac{10^9 \text{ watts} \times 3.41 \text{ BTU/hr watt}}{0.39 \times 12000 \text{ BTU/lb}_m \times 2000 \text{ lb}_m/\text{ton}} = 364 \text{ tons/hr}$$

This would correspond to around 100 railroad coal cars per day. We see that the fuel saved by a slight increase in efficiency of a large system will offset substantial capital investments in reheaters, feedwater heaters and other devices to increase the plants thermal efficiency.

The waste heat that is exhausted from this plant into the environment is

$$\dot{Q}_0 = \frac{1 - \eta_{thermal}}{\eta_{thermal}} \dot{W}$$

$$= \frac{(1 - 0.39) \times 10^9 \text{ watts}}{0.39} = 1.56 \times 10^9 \text{ watts}$$

In addition, many tons of stack effluent in the form of gases and particulates are also discharged into the atmosphere. The magnitude of this thermal pollution and stack gas pollution is so large that it can adversely affect the local environment and must, therefore, be carefully controlled.

The Working Medium and Binary Systems

Water has been used as the working medium in all of the vapor cycle examples since it is the dominant fluid used in the large power plants. These examples have illustrated the dependence of the cycle efficiency on both the operating conditions of the cycle and on the properties of water. Some of the advantages and disadvantages of using water that have been indicated in the examples would lead us to seek a fluid with the following desirable characteristics that are depicted in Fig. 7-38.

(1) High critical temperature (T_C) – this will permit vaporization and therefore isothermal heat addition at the cycle's maximum temperature.

(2) Large latent heat of vaporization (2'–3) – this would help to maximize the average temperature at which heat is added to the system and would tend to minimize the equipment size needed for a given power output by reducing the required flow rate.

(3) Low vapor pressures at high temperature (P_2) – this would reduce the material costs and safety hazards incurred in handling high pressure fluids.

(4) Steep saturated liquid line — this would minimize the external low temperature heat addition (2—2′) and, therefore, reduce the need for regenerators.

(5) Steep sloped or positive sloped saturated vapor line — this would reduce or eliminate the problem of producing a wet vapor during the expansion process (3—4) and thereby reduce the need for reheat.

(6) Condensing pressure slightly above one atmosphere (P_1) — this would eliminate the troublesome contamination due to leaks in systems operating below atmospheric pressures.

(7) Noncorrosive and chemically stable.

(8) Easily and safely handled.

(9) Readily available and inexpensive.

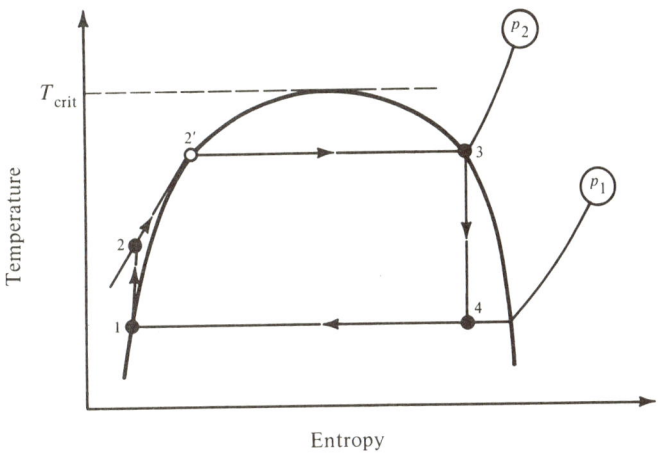

Fig. 7-38. *Idealized phase diagram for Rankine cycle.*

No single fluid, of course, possesses all of these characteristics. Water rates high on only items 2, 8, and 9; but these are very important properties which eliminate many fluids with other desirable characteristics from being viable alternatives.

The wide range of thermal properties that are possessed by the available working fluids does permit, though, the tailoring of a Rankine cycle for a specific application. In space power systems, for instance, a fluid with a low vapor pressure at high temperatures (mercury, sodium, potassium, rubidium, etc.) is required since lightweight condensers must operate at relatively high temperatures to provide an adequate heat transfer to space which is the only available heat sink. Working fluids of this type are also being considered for the high temperature portion of binary systems for large power plants where two separate Rankine cycles are combined as shown in Fig. 7-39. The high temperature cycle would be

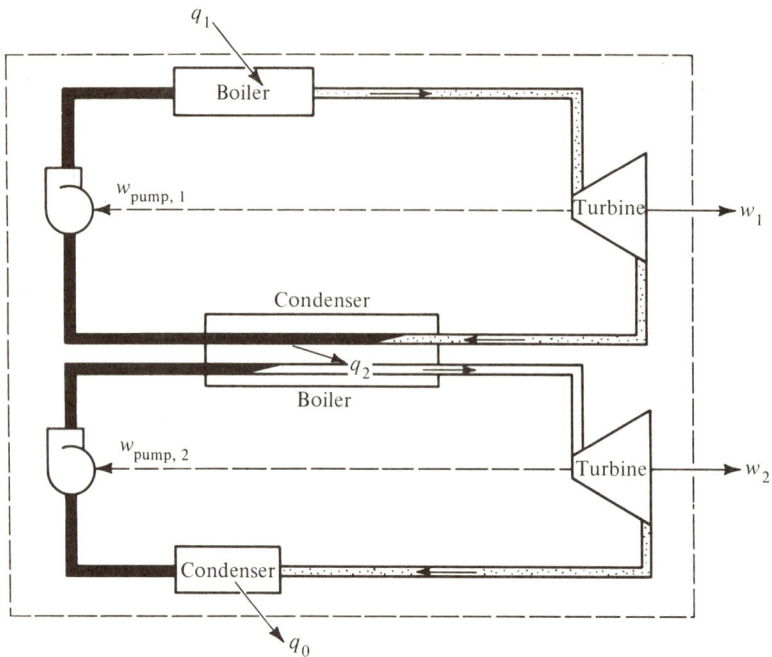

Fig. 7-39. *Binary Rankine cycle.*

designed to optimize the high temperature external heat addition (items 1—3). The heat removed from the high temperature condenser would provide the input for the boiler of the low temperature cycle which would probably be a conventional system. Binary systems of this type could offer an overall thermal efficiency on the order of 50 percent, but the total cost of producing electricity is estimated to be higher than for conventional steam systems due to the significantly higher capital costs.

Another example where the flexible characteristics of the Rankine cycle are being adopted to a special application involves the use of the cycle to partly recover the relatively high temperature waste energy from the exhaust of Diesel and Otto engines. In this combined cycle application, studies have indicated that the use of some organic compounds as the working medium would be superior to steam.

Gas Turbine Systems and the Brayton Cycle

Around the turn of the century, serious development of turbine systems using air instead of water vapor as the working fluid was initiated since this would allow the mating of the turbine with internal combustion. A simple gas turbine system would consist of a compressor, combustor and a turbine as shown in Fig. 7-40. The corresponding T–s diagram for this system is also given by Fig. 7-40.

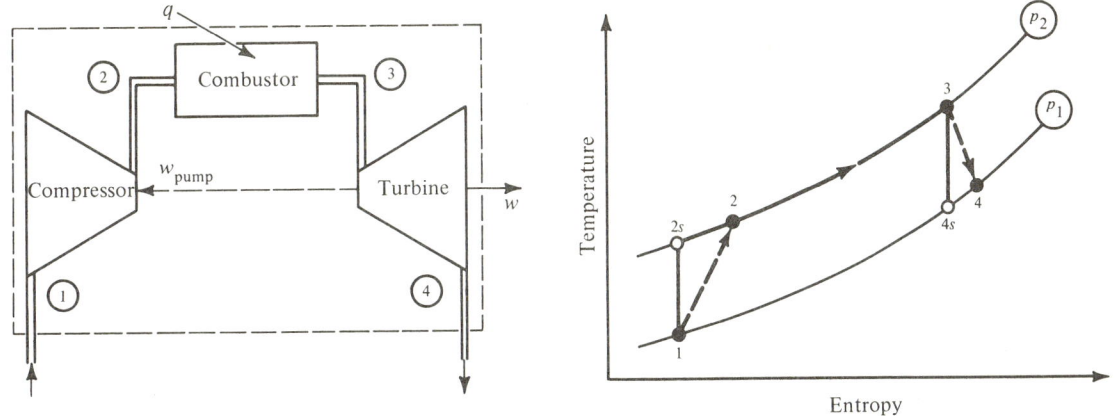

Fig. 7-40. *Simple gas turbine system.*

We note that the internal combustion process has been replaced again with an external heat addition process, and that the pressure drop in the combustor has been neglected in this idealization of the cycle. As in the case of the reciprocating engines, the compression process 1—2 shown in Fig. 7-40 absorbs a large percent of the turbine output (3—4). The compressors that were used in the early experiments had such low efficiencies that they required more power than could be delivered by the turbine. With the development of blade theory, the isentropic efficiencies of gas compressors were increased to around 60 percent, and modern compressors obtain efficiencies in the range of 80 to 90 percent. The performance of early gas turbine systems was also curtailed by low turbine inlet temperatures which had to be imposed due to the high temperature strength characteristics of the available materials. The following example illustrates the performance that was obtained from a typical early gas turbine system.

EXAMPLE PROBLEM 7-8. Determine the thermal efficiency of the simple gas turbine system depicted in Fig. 7-40 having the following properties:

State	P(psia)	T(°F)	Component Efficiency
1	15	60	$\eta_{comp} = 0.60$
2	45		
3	45	1000	$\eta_{turb} = 0.85$

Assume that air and the products of combustion can both be treated as an ideal gas with the properties of air.

The thermal efficiency of this system, based on a unit mass of air passing through the system, is

$$\eta_{therm} = \frac{w_{turb} + w_{comp}}{q} = \frac{(h_3 - h_4) + (h_1 - h_2)}{h_3 - h_2}$$

where we have applied the energy balance equation to the pump, the turbine and the combustor. If the gas acts as an ideal gas with constant specific heat capacities, this equation simplifies to

$$\eta_{therm} = \frac{c_P(T_3 - T_4) + c_P(T_1 - T_2)}{c_P(T_3 - T_2)} = 1 - \frac{T_4 - T_1}{T_3 - T_2}$$

$$= 1 - \frac{(T_4/T_3)(T_3/T_1) - 1}{(T_3/T_1) - (T_2/T_1)} \qquad (7\text{-}19)$$

By noting that the isentropic compressor efficiency for this ideal gas medium can be rewritten in the form

$$\eta_{comp} = \frac{h_{2s} - h_1}{h_2 - h_1} = \frac{T_{2s} - T_1}{T_2 - T_1}$$

we can obtain an expression for the temperature ratio T_2/T_1

$$T_2/T_1 = 1 + (T_{2s}/T_1 - 1)/\eta_{comp} \qquad (7\text{-}20)$$

A similar expression for the isentropic turbine efficiency

$$\eta_{turb} = \frac{h_3 - h_4}{h_3 - h_{4s}} = \frac{T_3 - T_4}{T_3 - T_{4s}}$$

can be used to find the temperature ratio T_4/T_3,

$$T_4/T_3 = 1 - \eta_{turb}(1 - T_{4s}/T_3) \qquad (7\text{-}21)$$

Since the isentropic expansion processes 3—4s and the isentropic compression process 1—2s are between the same two pressure limits,

$$\frac{T_{2s}}{T_1} = \frac{T_3}{T_{4s}} = \left(\frac{P_2}{P_1}\right)^{(\gamma-1)/\gamma} = r_P^{(\gamma-1)/\gamma} \qquad (7\text{-}22)$$

where r_P is defined as the pressure ratio P_2/P_1. Substitution of this result into Eqs. (7-20) and (7-21) gives

$$T_2/T_1 = 1 + (r_P^{(\gamma-1)/\gamma} - 1)/\eta_{comp} \tag{7-23}$$

and

$$T_4/T_3 = 1 + \eta_{turb}(r_P^{(1-\gamma)/\gamma} - 1) \tag{7-24}$$

When these two terms are inserted into the system's thermal efficiency equation, Eq. (7-19), we obtain

$$\eta_{therm} = 1 - \frac{[1 + \eta_{turb}(r_P^{(1-\gamma)/\gamma} - 1)](T_3/T_1) - 1}{T_3/T_1 - [1 + (r_P^{(\gamma-1)/\gamma} - 1)/\eta_{comp}]} \tag{7-25}$$

In our case, the temperature ratio is

$$\frac{T_3}{T_1} = \frac{1460\,°R}{520\,°R} = 2.81$$

and the pressure ratio r_P is 3. The resulting thermal efficiency for this gas turbine system is only

$$\eta_{therm} = 1 - \frac{[1 + 0.85 \times (3^{-(0.4/1.4)} - 1)] \times 2.81 - 1}{2.81 - [1 + (3^{0.4/1.4} - 1)/0.6]}$$

$$= 0.02$$

The corresponding reversible gas turbine cycle 1–2s–3–4s–1 in Fig. 7-40 is called the Brayton cycle. The cycle's heat rejection process 4s–1 represents the cooling of the exhaust gas in the open system by the atmosphere to its initial state. The thermal efficiency of the reversible gas turbine system is

$$\eta_{therm,\,rev} = \frac{w_{turb} + w_{comp}}{q} = \frac{c_P(T_3 - T_{4s}) + c_P(T_1 - T_{2s})}{c_P(T_3 - T_{2s})}$$

$$= 1 - \frac{T_{4s} - T_1}{T_3 - T_{2s}} = 1 - \frac{T_1}{T_{2s}} \frac{(T_{4s}/T_1 - 1)}{(T_3/T_{2s} - 1)} \tag{7-26}$$

Equation (7-22) implies that

$$T_{4s}/T_1 = T_3/T_{2s}$$

so that Eq. (7-26) simplifies to

$$\eta_{\text{therm, rev}} = 1 - \frac{T_1}{T_{2s}} = 1 - r_P^{(1-\gamma)/\gamma} = 1 - \frac{1}{r^{\gamma-1}} \qquad (7\text{-}27)$$

which is the same expression that we obtained for an Otto cycle (Eq. 7-3). For our example where the pressure ratio is 3, the Brayton cycle has a thermal efficiency of

$$\eta_{\text{therm, rev}} = 1 - 3^{-(0.4/1.4)} = 0.27$$

This value is also not very impressive since a Carnot cycle operating between the same two temperature extremes would be 64 percent efficient. One reason for the Brayton cycle's poor performance is its high exhaust temperature T_4 which is

$$T_{4s} = T_3 r_P^{(1-\gamma)/\gamma} = 1460\,°\text{R} \times 3^{-(0.4/1.4)} \qquad (7\text{-}22)$$

$$= 1067\,°\text{R} = 607\,°\text{F}$$

for the above example.

As mentioned previously, the performance of gas cycles generally suffer from the fact that the compression process absorbs a large fraction of the work output from the power process. For the reversible case, the ratio of work absorbed by the compressor $-w_{\text{comp}}$ to the turbine's output w_{turb} is

$$\left(\frac{-w_{\text{comp}}}{w_{\text{turb}}}\right)_{\text{rev}} = \frac{-c_p(T_1 - T_{2s})}{c_p(T_3 - T_{4s})} = \frac{(T_{2s}/T_1) - 1}{(T_3/T_1) - (T_{4s}/T_3)(T_3/T_1)}$$

$$= \frac{r_P^{(\gamma-1)/\gamma} - 1}{T_3/T_1 \left(1 - r_P^{(1-\gamma)/\gamma}\right)} = \frac{T_1}{T_3} r_P^{(\gamma-1)/\gamma} \qquad (7\text{-}28)$$

For the values of r_P and T_3/T_1 of the foregoing example, the back work ratio is

$$\left(\frac{-w_{\text{comp}}}{w_{\text{turb}}}\right)_{\text{rev}} = \frac{3^{0.4/1.4}}{2.81} = 0.49$$

The back work ratio in our actual cycle is much larger due to irreversibilities that increase the work absorbed by the compressor and decrease the turbine's output.

$$\left(\frac{-w_{comp}}{w_{turb}}\right)_{act} = \frac{-c_p(T_1 - T_2)}{c_p(T_3 - T_4)} = \frac{(T_2/T_1) - 1}{(T_3/T_1)(1 - T_4/T_3)}$$

$$= \frac{(r_P^{(\gamma-1)/\gamma} - 1)/\eta_{comp}}{(\bar{T}_3/T_1)\eta_{turb}\left(1 - r_P^{(1-\gamma)/\gamma}\right)} \quad (7\text{-}29)$$

$$= \frac{(3^{0.4/1.4} - 1)/0.6}{2.81 \times 0.85 \times (1 - 3^{-(0.4/1.4)})} = 0.96$$

The result clearly illustrates the advantage of a vapor turbine cycle over a gas turbine cycle. We found that the pump in a steam turbine system absorbed only around 1 percent of the turbine's output even with component efficiencies lower than the values used here.

The low or negative efficiencies of the gas turbine engine in its early stages of development prevented it from being a viable engine, but the basic components found an immediate application in the form of the turbo-supercharger to boost the power capability of Otto and Diesel engines. As an example of the use of a turbo-supercharger, we will return to Example Prob. 7-3 and place a turbine in the exhaust of the Otto engine to drive the compressor as depicted in Fig. 7-41. The

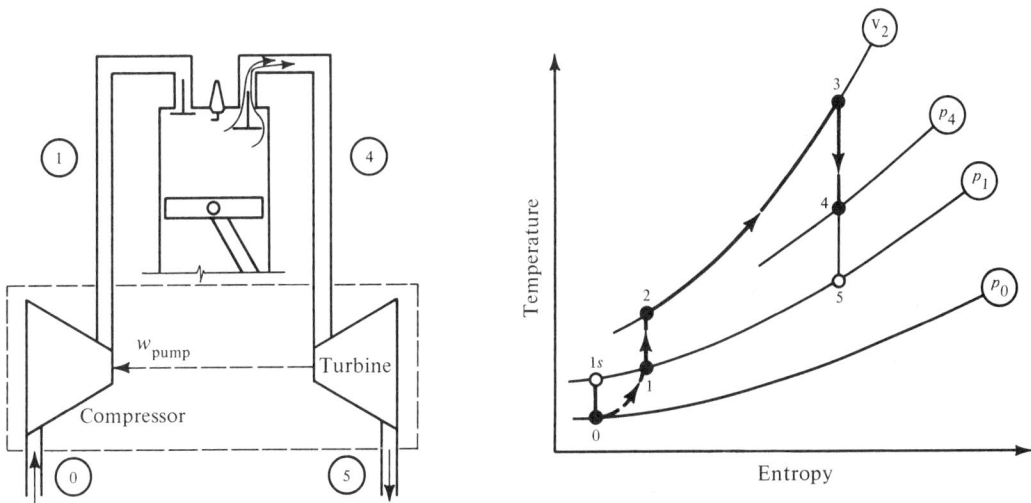

Fig. 7-41. *Turbo-supercharged Otto cycle.*

air in this case is compressed from the atmospheric state 0 to state 1 by an irreversible compressor. The air then enters a reversible Otto engine to be further compressed by the isentropic process 1—2, heated by the constant volume process 2—3 and is expanded isentropically during the power stroke 3—4. The analysis of this portion of the process was performed in Example Prob. 7-3. The exhaust valve now opens at state 4 and the gas is now free to expand through the "blow down" turbine.

EXAMPLE PROBLEM 7-9. Calculate the work that would be produced by an isentropic turbine placed in the exhaust of Example Prob. 7-3 if the turbine exhaust pressure P_5 is 1 atmosphere. Assume that the piston is stationary at the bottom of its stroke when the exhaust valve opens and during the isentropic blow down process as the gas expands through the turbine.

This nonsteady flow problem will be analyzed by looking at the open system in Fig. 7-42 just before the exhaust valve opens, state 4, and just after the expansion has been completed but before the exhaust stroke begins, state 5. By apply-

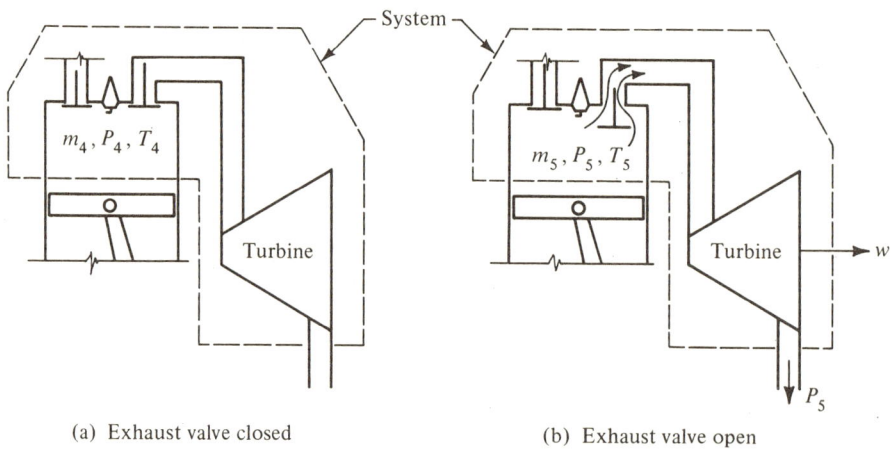

Fig. 7-42. *Turbo-charged system before the exhaust valve opens (a) and after exhaust valve opens and the gas has expanded from state 4 to a final state 5 with the piston stationary (b).*

ing the generalized energy balance to this nonsteady state and open system, we obtain the relationship

$$m_5 u_5 - m_4 u_4 = -(m_4 - m_5) h_5 - W$$

m_4 is the mass in the cylinder before the exhaust valve opens and m_5 is the mass left in the system after the blow down process has ended. The heat transferred to the system is zero because the blow down process is assumed to be isentropic. The gas that remains in the system after blow down m_5 and the gas that left the system $m_4 - m_5$ will both have the same state 5 since the gas is expanded isentropically to the same pressure P_5 in both cases. The kinetic energy in the system before and after blow down is zero and we have neglected the kinetic energy of the mass leaving along with any potential energy changes. The specific work obtained from the turbine during the blow down is, therefore,

$$\frac{W}{m_4} = w = u_4 - \frac{m_5}{m_4} u_5 - (1 - m_5/m_4) h_5 \tag{7-30}$$

If the system's volume outside the cylinder itself is neglected, we find that

$$m_4 v_4 = m_5 v_5 = \text{cylinder volume}$$

or

$$\frac{m_5}{m_4} = \frac{v_4}{v_5} = \frac{T_4}{T_5} \frac{P_5}{P_4}$$

In the case of an isentropic expansion of an ideal gas to 1 atmosphere P_5, the exhaust temperature and final chamber temperature is

$$T_5 = T_4 \left(\frac{P_5}{P_4}\right)^{(\gamma-1)/\gamma}$$

$$= 4468\,°\text{R} \left(\frac{14.7}{107}\right)^{0.4/1.4} = 2534\,°\text{R}$$

where P_4 and T_4 were obtained from Example Prob. 7-3. Thus

$$\frac{m_5}{m_4} = \frac{T_4}{T_5} \frac{P_5}{P_4} = \frac{4468}{2534} \frac{14.7}{107} = 0.24$$

The work performed by the isentropic turbine during the blow down period for an ideal gas medium can now be obtained from Eq. (7-30),

$$w = u_4 - h_5 + \frac{m_5}{m_4}(h_5 - u_5) \tag{7-30}$$

$$= u_4 - u_5 - P_5 v_5 + \frac{m_5}{m_4}(u_5 + P_5 v_5 - u_5)$$

$$= c_V(T_4 - T_5) + \left(\frac{m_5}{m_4} - 1\right) RT_5$$

$$= c_V\left[T_4 - T_5 + \left(\frac{m_5}{m_4} - 1\right)(\gamma - 1) T_5\right]$$

$$= \frac{\frac{5}{2} \times 1.986 \frac{\text{BTU}}{\text{lb mole °R}}}{28.96 \text{ lb}_m/\text{lb mole}} [4468 - 2534 + (0.24 - 1) \times (1.4 - 1) \times 2534] \text{ °R}$$

$$= 200 \frac{\text{BTU}}{\text{lb}_m \text{ air}}$$

This is about one third of what the reversible Otto engine itself is putting out in Example Problems 7-2 and 7-3. Since only 35 BTU/lb_m must be supplied to the compressor, we can afford to have a turbine system that is only 18 percent as efficient as our reversible system and still have enough power to run the compressor which is just about what happens with actual turbo-superchargers.

With the development of efficient compressors and satisfactory high temperature materials and cooling techniques for turbine blades, the gas turbine has become one of the most important current prime movers and its influence is expanding. Typical modern turbine inlet temperatures range between 1600 and 2200 °F, but the development efforts underway should make temperatures in the 2500 to 3000 °F range practical in the 1980's.

The efficiency of a Brayton cycle was found to be identical to that of an Otto cycle which increases with the pressure ratio r_p. Figure 7-43 indicates that

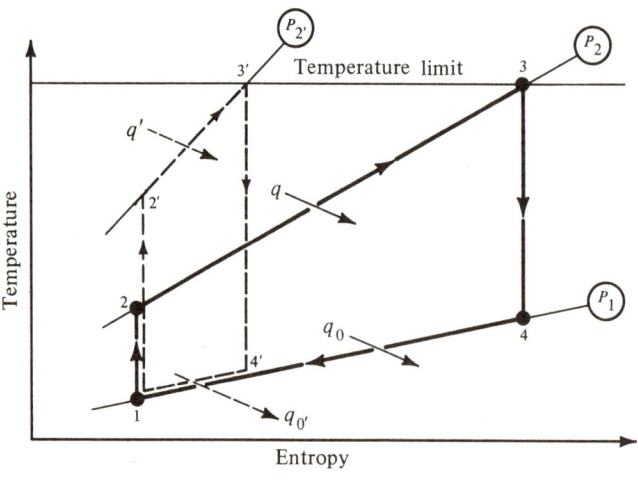

Fig. 7-43. *Effect of pressure ratio on the thermal efficiency of a Brayton cycle.*

the specific work performed during a cycle, which is equal to the enclosed area 1—2—3—4—1, will decrease as the pressure ratio increases if the cycle's minimum and maximum temperatures T_1 and T_3 remain constant. In other words, the back work ratio which can be calculated from Eq. (7-28) increases with the pressure ratio. The thermal efficiency of the reversible system increases with the pressure ratio because the average temperature at which heat is added along path 2—3 increases and the average temperature at which heat is exhausted 4—1 decreases. For an actual engine, there obviously exists an optimum pressure ratio that will maximize the system's efficiency since the efficiency is zero for a pressure ratio of one and also approaches zero as the back work ratio approaches 1 for increasing pressure ratios. One of the main problems that design engineers face is in matching the compressor with the other system components to obtain the maximum thermal efficiency. The following example problem illustrates this point.

EXAMPLE PROBLEM 7-10. Find the pressure ratio that will give the maximum thermal efficiency for a simple gas turbine system operating with an inlet air temperature of 70 °F and an inlet turbine temperature of 1600 °F. The compressor and turbine isentropic efficiencies are 0.80 and 0.85 respectively.

The thermal efficiency of the system is given by Eq. (7-25). To find the pressure ratio that produces the maximum cycle efficiency, we could differentiate this equation with respect to r_p and equate the result to zero. This pressure ratio could also be determined by plotting the cycle efficiency against the pressure ratio as is done in Fig. 7-44. The temperature ratio for this gas turbine system is

$$\frac{T_3}{T_1} = \frac{1600 + 460}{70 + 460} = 3.89$$

Figure 7-44 indicates that the maximum thermal efficiency is 25 percent and occurs at a pressure ratio around 9.

The temperatures at states 2 and 4 can now be calculated from Eqs. (7-23) and (7-24). If the inlet pressure for this example is taken as one atmosphere, the following cycle states are found:

State	P(psia)	T(°F)	Component Efficiency
1	14.7	70	
			$\eta_{comp} = 0.80$
2	132.3	649	
3	132.3	1600	
			$\eta_{turb} = 0.85$
4	14.7	784	

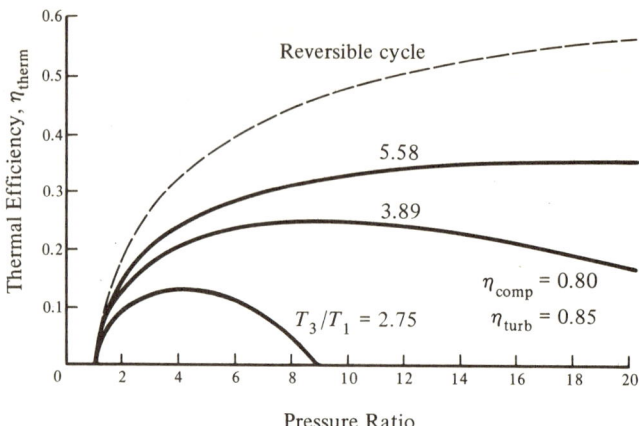

Fig. 7-44. *Effect of pressure and temperature ratios on the thermal efficiency of a Brayton cycle.*

We again note the high exhaust temperature T_4 which is a major cause of the system's low efficiency despite a relatively high maximum temperature T_3.

For comparison purposes, the thermal efficiencies for a temperature ratio of 2.75 ($T_3 = 1000\,°F$) and 5.58 ($T_3 = 2500\,°F$) are also plotted on Fig. 7-44. These three curves indicate the effect that improved high temperature materials (1000 to 1600 °F) and cooling techniques (1600 to 2500 °F) have had on the performance of the gas turbine system.

Pressure ratios for stationary power facilities typically range between 4 and 6, while jet engines operate at pressure ratios greater than 8 because of their higher turbine inlet temperatures. The jet engine first flew in 1943 and, with the help of massive development programs, quickly came to dominate both commercial and military aviation. This is due to the jet engine's high power to weight ratio and to the speed ceiling of around 500 mph on aircraft driven by propellers.

The thermal efficiency of the gas turbine system can be greatly improved by the addition of a regenerator as depicted in Fig. 7-45 to partly recover the waste

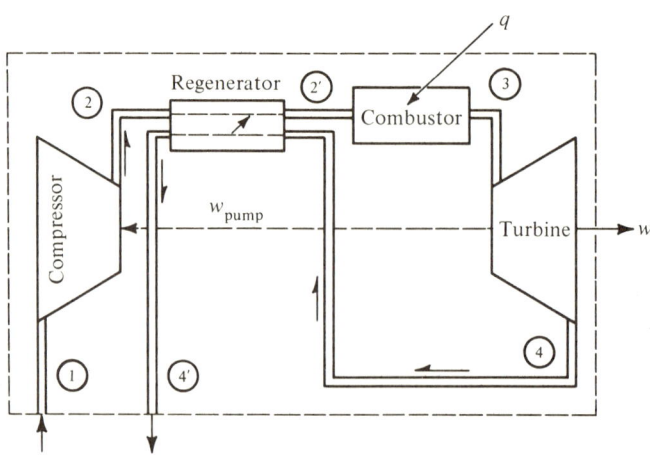

Fig. 7-45. *Regenerative Brayton cycle.*

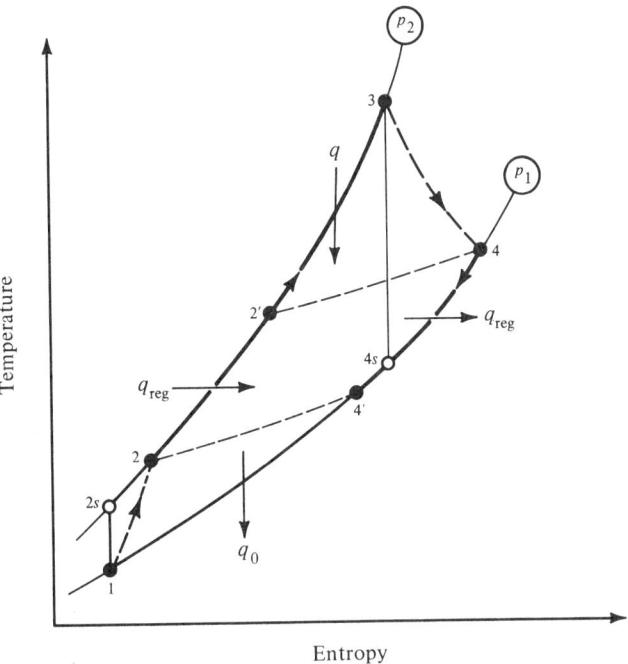

Fig. 7-45. (*Continued*)

energy that is contained in the high temperature turbine exhaust. The corresponding Ts diagram for this system is also presented in Fig. 7-45. It consists of the compression process 1—2; the internal heating of the compressed gas 2—2' by the exhaust turbine gas; the heating of the compressed gas in the combustor which is represented as the external heat addition process 2'—3; the expansion process 3—4 in the turbine; and the cooling of the turbine exhaust gas 4—4' in the regenerator by the compressed gas. We note that pressure drops in the regenerator and combustor have been neglected in Fig. 7-45.

If we perform an energy balance on the regenerator, we would find that the enthalpy lost by the turbine exhaust gas $h_4 - h_{4'}$ is gained by the compressed gas $h_{2'} - h_2$. With the assumption that the products of combustion have the properties of air which is assumed to be an ideal gas with constant heat capacity, we find that

$$q_{reg} = c_P (T_{2'} - T_2) = c_P (T_4 - T_{4'})$$

where q_{reg} is the heat transferred internally in the regenerator. In the case of a reversible regenerator, the compressed gas would leave the regenerator at the exhaust turbine gas temperature T_4 and the turbine gas leaves at the compressed gas inlet temperature T_2. The reversible heat transfer is, therefore,

$$q_{reg, rev} = c_P (T_4 - T_2)$$

Actual regenerators only exchange about 80 percent of this amount. That is

$$\eta_{reg} = \frac{(\text{actual heat transferred})_{reg}}{(\text{reversible heat transferred})_{reg}} \cong 0.80$$

Since the working fluid is an ideal gas with constant heat capacity, the regenerator efficiency becomes

$$\eta_{reg} = \frac{c_p(T_{2'} - T_2)}{c_p(T_4 - T_2)} = \frac{T_{2'} - T_2}{T_4 - T_2}$$

The thermal efficiency of our regenerative gas turbine system is

$$\eta_{therm} = \frac{w_{turb} + w_{comp}}{q} = \frac{c_p(T_3 - T_4) + c_p(T_1 - T_2)}{c_p(T_3 - T_{2'})}$$

$$= \frac{(T_3 - T_4) - (T_2 - T_1)}{(T_3 - T_2) - \eta_{reg}(T_4 - T_2)}$$

$$= \frac{\dfrac{T_3}{T_1} - \dfrac{T_4}{T_3}\dfrac{T_3}{T_1} - \dfrac{T_2}{T_1} + 1}{\dfrac{T_3}{T_1} - \dfrac{T_2}{T_1} - \eta_{reg}\left(\dfrac{T_4}{T_3}\dfrac{T_3}{T_1} - \dfrac{T_2}{T_1}\right)} \qquad (7\text{-}31)$$

As presented in Example Prob. 7-8, the temperature ratio T_2/T_1 can be expressed in terms of the pressure ratio and the compressor efficiency

$$T_2/T_1 = 1 + (r_P^{(\gamma-1)/\gamma} - 1/\eta_{comp}) \qquad (7\text{-}23)$$

and similar results were obtained for the temperature ratio T_4/T_3

$$T_4/T_3 = 1 + \eta_{turb}(r_P^{(1-\gamma)/\gamma} - 1) \qquad (7\text{-}24)$$

Substitution of these ratios into the above thermal efficiency equation, Eq. (7-31), will express the thermal efficiency of the regenerative cycle in terms of the compressor, turbine and regenerator efficiencies; the pressure ratio P_2/P_1 and the temperature ratio T_3/T_1.

In the case of a reversible system,

$$\eta_{comp} = \eta_{turb} = \eta_{reg} = 1$$

The closed cycle 1–2s–3–4s–1 depicts this reversible system on the Ts diagram given in Fig. 7-45. The cycle is closed with the reversible cooling of the turbine gas leaving the regenerator from temperature T_{2s} to T_1. This replaces an actual exhaust process where the exhaust gas is cooled by the atmosphere to its initial state. The thermal efficiency for this reversible cycle can be calculated from Eq. (7-31); or more directly, we find from Fig. 7-45 that

$$\eta_{therm, rev} = \frac{w_{turb, rev} + w_{comp, rev}}{q_{rev}}$$

$$= \frac{c_P(T_3 - T_{4s}) - c_P(T_{2s} - T_1)}{c_P(T_3 - T_{4s})}$$

$$= 1 - \frac{T_{2s} - T_1}{T_3 - T_{4s}} = 1 - \frac{T_1}{T_3}\left(\frac{T_{2s}/T_1 - 1}{1 - T_{4s}/T_3}\right)$$

$$= 1 - \frac{T_1}{T_3}\frac{(r_P^{(\gamma-1)/\gamma} - 1)}{(1 - r_P^{(1-\gamma)/\gamma})}$$

$$= 1 - \frac{T_1}{T_3} r_P^{(\gamma-1)/\gamma} \qquad (7\text{-}32)$$

Figure 7-46 is a plot of both Eqs. (7-31) and (7-32) for two temperature ratios. Unlike the simple Brayton cycle, we see that the thermal efficiency of a reversible regenerative cycle decreases with pressure ratio. Likewise, the irreversible regenerative systems are found to have lower optimum pressure ratios than the corresponding simple systems presented in Fig. 7-44. A lower pressure ratio implies a lower specific work output. Thus, a trade off between thermal efficiency and specific power must again be made.

Figure 7-46 indicates that regenerative systems with inlet turbine temperatures around 1600 °F ($T_3/T_1 = 3.89$) can currently be designed to obtain efficiencies as high as 36 percent. The projected advances in gas turbine technology indicate that efficiencies exceeding the Rankine cycle's 41 percent will be practical in the near future due to both increases in the turbine inlet temperature and component efficiencies. Even though the currently available gas turbine systems with reasonable life-spans cannot match the efficiencies obtained from steam

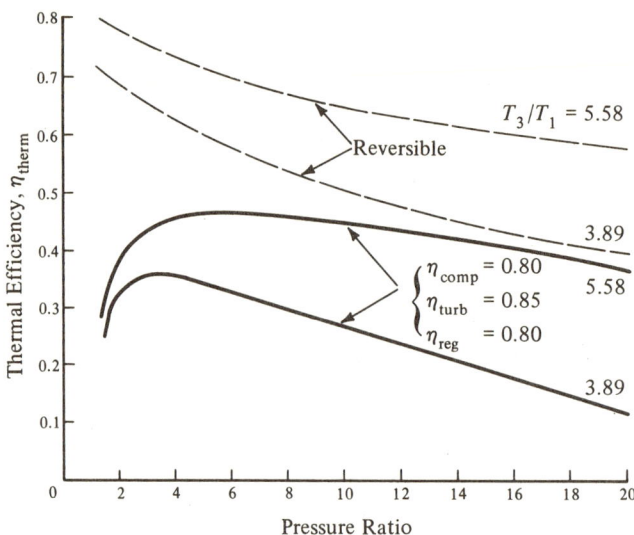

Fig. 7-46. *Effect of pressure and temperature ratios on the thermal efficiency of a regenerative gas turbine cycle.*

turbine systems, it has still become an important generating unit for the electric utility industry. Its simplicity results in substantially lower capital costs and construction time than an equivalent steam power plant. The gas turbine system is also not dependent upon water for cooling and it can be started much more rapidly than a steam turbine system. Unfortunately, its lower thermal efficiency and high quality, high cost fuel requirements prevents it from presently being used for general base load applications. The intermittent use of gas turbines to help supply up to around 20 percent of the peak load does result in a reduction in the total power costs, and it is in this capacity that gas turbines are presently used by the utilities.

Regenerators are usually made relatively large in order to keep the pressure drop across them as small as possible. As a consequence, they are expensive and were normally employed in only large gas turbine systems in the past. However, the recent development of compact and efficient regenerators for small gas turbine systems has made these systems more competitive in the small engine market.

Several other modifications can be made to increase the thermal efficiency of gas turbine systems. Interstage cooling for systems employing large compression ratios may be used to reduce the compression work. Figure 7-47 illustrates the addition of an interstage cooler $2'-1'$ to the Brayton cycle $1-2-3-4-1$. The analysis of the intercooler is similar to the one made for the reciprocating compressor discussed at the beginning of the chapter. In both cases, the compression work is minimized if the pressure ratios across the compressor stages are identical. Reheat $4'-3'$ can also be utilized in a gas turbine system. Figure 7-47 indicates

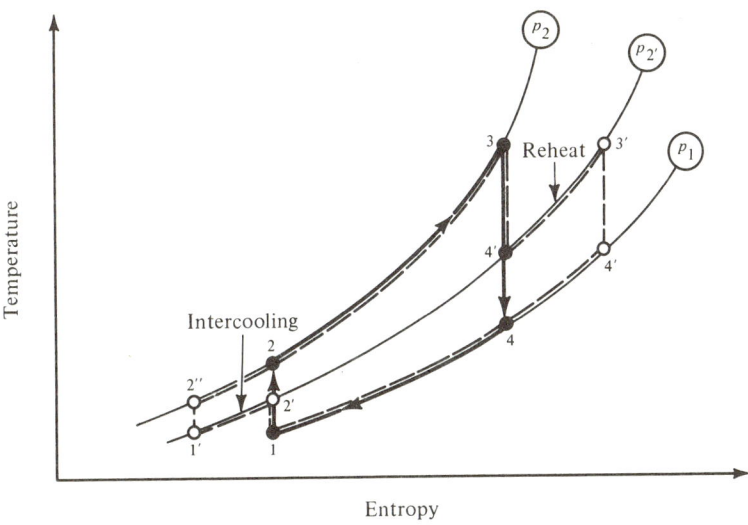

Fig. 7-47. *Brayton cycle with reheat and interstage cooling.*

that both these modifications will increase the enclosed area and, therefore, the net specific work. However, the heat addition is also increased by an amount represented by the area under path 2″—2 in Fig. 7-47 for intercooling and by the area under curve 4′—3′ for reheat. The net thermal efficiency of this turbine system would actually decrease with the use of either reheat or intercooling or both unless used in conjunction with a regenerator, whereupon the reversible system's efficiency increases.

The direct coupling of a closed-cycle gas turbine system with a high-temperature gas-cooled reactor (HTGR) is also being developed. In this case, helium will be utilized as the working fluid. It was pointed out in the discussion of the Stirling cycle that helium is an excellent working medium for heat exchangers. In addition, the use of a monatomic gas in place of air in a turbine system will tend to move the point of maximum thermal efficiency to a lower pressure ratio. For example, the regenerative system described by Figs. 7-45 and 7-46 has an optimum pressure ratio of 3.4 for a temperature ratio of 3.89. If helium were used in this system, essentially the same thermal efficiency would be obtained but at a pressure ratio of only 2.4. The specific power output is, of course, reduced along with the pressure ratio. Some representative values for a proposed system of this type are given in the following example.

EXAMPLE PROBLEM 7-11. Calculate the thermal efficiency of the HTGR shown in Fig. 7-48 where the corresponding state values are given in the following table.

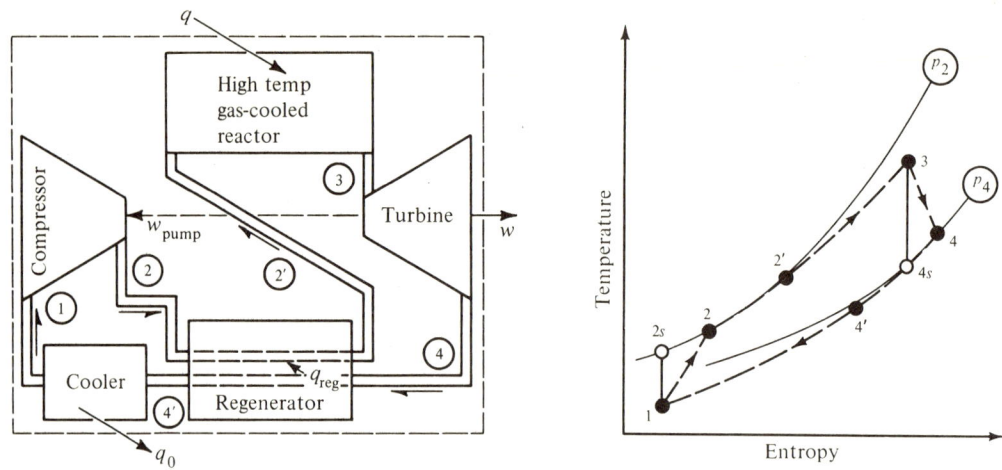

Fig. 7-48. *High-temperature helium gas turbine cycle.*

State	P(psia)	T(°F)	Component Efficiency
1	445	105	
2	1000		$\eta_{comp} = 0.90$
2'	980		$\eta_{reg} = 0.82$
3	970	1500	
4	465		$\eta_{turb} = 0.90$
4'	450		

Note that pressure drops in the cooler, regenerator and reactor have not been neglected. The working fluid is helium.

We will base our calculations on one pound of helium flowing through the system. The temperature of the gas leaving the compressor is given by Eq. (7-23) if we assume helium to be a monoatomic ideal gas with constant heat capacities.

$$T_2 = T_1 \left\{ 1 + [(P_2/P_1)^{(\gamma-1)/\gamma} - 1]/\eta_{comp} \right\} \qquad (7\text{-}23)$$

$$= (565\,°\text{R}) \left\{ 1 + \left[\left(\frac{1000}{445}\right)^{(1.67-1)/1.67} - 1 \right]/0.90 \right\}$$

$$= 806\,°\text{R} = 346\,°\text{F}$$

and the turbine outlet temperature by Eq. (7-24)

$$T_4 = T_3 \left\{ 1 + \eta_{turb} \left[(P_3/P_4)^{(1-\gamma)/\gamma} - 1 \right] \right\}$$

$$= (1960\,°R) \left\{ 1 + 0.90 \left[\left(\frac{465}{970} \right)^{0.67/1.67} - 1 \right] \right\}$$

$$= 1509\,°R = 1049\,°F$$

The inlet temperature to the nuclear reactor is obtained from the definition of regeneration efficiency.

$$T_{2'} = T_2 + \eta_{reg}(T_4 - T_2) = 806 + (0.82)(1509 - 806)$$

$$= 1382\,°R = 922\,°F$$

The thermal efficiency for this system is

$$\eta_{therm} = \frac{w_{turb} + w_{comp}}{q} = \frac{c_p(T_3 - T_4) + c_p(T_1 - T_2)}{c_p(T_3 - T_{2'})}$$

$$= \frac{(1500 - 1049) + (105 - 346)}{1500 - 922} = 0.36$$

This efficiency should increase substantially in the future with the utilization of higher core temperatures. Note the simplicity of this system and the low pressure ratio, of the order 2:1, as compared to modern steam turbines which have pressure ratios of the order of 3000:1. The pressures in the system, 450 to 1000 psia, are kept high in order to reduce the size of the plant. The relatively high temperature at the cooler inlet $T_{4'}$ which can be found by making an energy balance on the regenerator

$$c_p(T_4 - T_{4'}) = c_p(T_{2'} - T_2)$$

or

$$T_{4'} = T_4 + T_2 - T_{2'} = 1049 + 346 - 922 = 473\,°F$$

also makes dry air cooling practical. This plant would not, therefore, be dependent upon a water source for cooling.

272 Chap. 7   Thermodynamics of Heat Engines

The Combined Gas and Steam Cycle

As previously mentioned, large gas turbines having inlet temperatures exceeding 2800 °F (1538 °C) and reasonable lifetimes are expected to be available in the future. Figure 7-46, which summarizes our analysis of the regenerative gas turbine system, indicates that thermal efficiencies of the order of 50 percent can be expected with turbine temperatures of this magnitude. Unfortunately, the same figure also shows that the specific power output of these systems will be small due to the low optimum pressure ratio for maximum thermal efficiency. In comparing the simple gas turbine system (Fig. 7-44) to a regenerative system having the same temperature ratio (Fig. 7-46), we find that the regenerative system has the higher thermal efficiency but its optimum pressure ratio occurs at a lower value. This indicates that the specific power output of the simple gas turbine system is larger than the regenerative gas turbine system.

Instead of using a regenerator and trading specific power for thermal efficiency, simple gas turbine systems are being combined with the steam turbine cycle. In this combination, the high temperature exhaust from the gas turbine is used to heat the steam-generator of a conventional steam turbine cycle. Thus the high temperature attributes of the simple gas turbine system are combined with the low-temperature heat rejection capabilities of the steam Rankine cycle. In the following example, we will form a combined gas and steam system by placing a simple advanced gas turbine system on top of the steam cycle analyzed in Example Prob. 7-7.

EXAMPLE PROBLEM 7-12. The components of an advanced gas turbine system are shown in Fig. 7-49. This gas turbine system is assumed to operate under the following conditions:

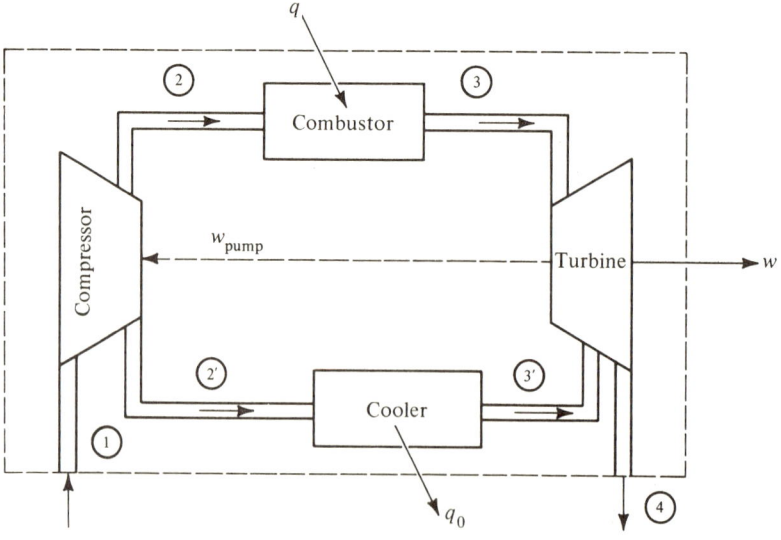

Fig. 7-49. *Advanced gas turbine system.*

State	P(psia)	T(°F)	Component Efficiency
1	15	70	
2	185		$\eta_{comp} = 0.90$
2'	185		
3	185	2800	
3'	185	200	
4	16	1200	

In this system, 8 percent of the intake air is bled from the compressor (1–2'), cooled to 200 °F (2'–3'), and then used to cool some of the turbine components in order to protect them from the high gas temperature at the turbine entrance. The pressure drops through the combustor and cooler have been neglected and the internal combustion process has again been replaced with external heat addition.

Calculate the thermal efficiency of this system and the combined thermal efficiency when the steam cycle in Example Prob. 7-7 is used as a bottoming cycle to recover part of the high temperature waste energy from the gas turbine exhaust. Assume that the exhaust gas enters both the reheater and the boiler-superheater at 1200 °F (649 °C) and leaves them at 500 °F (260 °C).

Air will be assumed to be an ideal gas with constant heat capacities and we will base our calculations on one pound of air flowing through the gas turbine system. The air temperature after compression is

$$T_2 = T_1 [1 + (r_P^{(\gamma-1)/\gamma} - 1)/\eta_{comp}] \tag{7-23}$$

$$= 530\,°R \times \left\{1 + \left[\left(\frac{185}{15}\right)^{0.4/1.4} - 1\right]/0.9\right\}$$

$$= 1148\,°R\,(638\,K)$$

and the compression work is

$$w_{comp} = h_1 - h_2 = c_P(T_1 - T_2)$$

$$= 0.24\,\frac{BTU}{lb_m\,°R} \times (530 - 1148)\,°R$$

$$= -148\,BTU/lb_m\,(-344\,joule/gram)$$

The output of the gas turbine w_{turb} is found by performing a mass and an energy balance on the turbine system shown in Fig. 7-50.

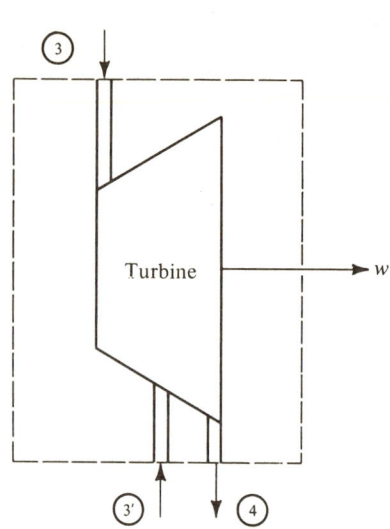

Fig. 7-50. *Gas turbine with bleed air cooling at 3'.*

$$\dot{m}_3 + \dot{m}_{3'} = \dot{m}_4$$

and

$$\dot{m}_3 h_3 + \dot{m}_{3'} h_{3'} - \dot{m}_4 h_4 - \dot{m}_4 w_{turb} = 0$$

where $\dot{m}_{3'}$ represents the air bled from the compressor to cool the turbine blades and stators. This bled air is forced through pores in the surface of the blades and causes a thin film of relatively cool air to form around them. This film of cool air insulates the blades from the hot gas stream before it mixes with the primary stream.

After rewriting the energy balance equation

$$\frac{\dot{m}_3}{\dot{m}_4} h_3 + \frac{\dot{m}_{3'}}{\dot{m}_4} h_{3'} - h_4 - w_{turb} = 0$$

and noting that the compressor bleed is 8 percent, we find that

$$w_{turb} = 0.92 \, h_3 + 0.08 \, h_{3'} - h_4$$

If the products of combustion are also assumed to be an ideal gas with constant heat capacities and the properties of air, then the above equation reduces to

$$w_{turb} = c_P(0.92 \times T_3 + 0.08 \times T_{3'} - T_4)$$

$$= 0.24 \frac{\text{BTU}}{\text{lb}_m \, °\text{R}} \times (0.92 \times 3260 + 0.08 \times 660 - 1660)$$

$$= 334 \text{ BTU/lb}_m \text{ air } (776 \text{ joule/gram})$$

Similarly, the heat transferred to the combustor per pound of air through the combustor is

$$q_{2-3} = h_3 - h_2 = c_P(T_3 - T_2)$$

$$= 0.24 (3260 - 1148)$$

$$= 507 \text{ BTU/lb}_m \text{ air } (1178 \text{ joule/gram})$$

The heat transfer to the system per pound of air through the system is then

$$q = \frac{\dot{m}_2}{\dot{m}_4} q_{2-3} = 0.92 \times 507 = 466 \text{ BTU/lb}_m \text{ air } (1084 \text{ joule/gram})$$

which gives a net thermal efficiency for the simple Brayton cycle of

$$\eta_{therm} = \frac{w_{turb} + w_{pump}}{q} = \frac{334 - 148}{466} = \frac{186}{466} = 0.40$$

We see that the thermal efficiency of only the gas turbine part of the combined system in this example matches the thermal efficiency obtained by modern steam cycles. However, the above analysis neglected pressure drops in the combustor and cooler, and treated the working medium as a perfect gas with constant heat capacities. Gas tables[10] would have to be consulted and the pressure drops included if a more comprehensive analysis is to be made.

The analysis of the steam part of the system was performed in Example Prob. 7-7 where it was found that the steam system had a thermal efficiency of 39 percent. A heat exchanger configuration for the steam generator and reheater is shown in Fig. 7-51. The turbine exhaust enters the bottom of the heat ex-

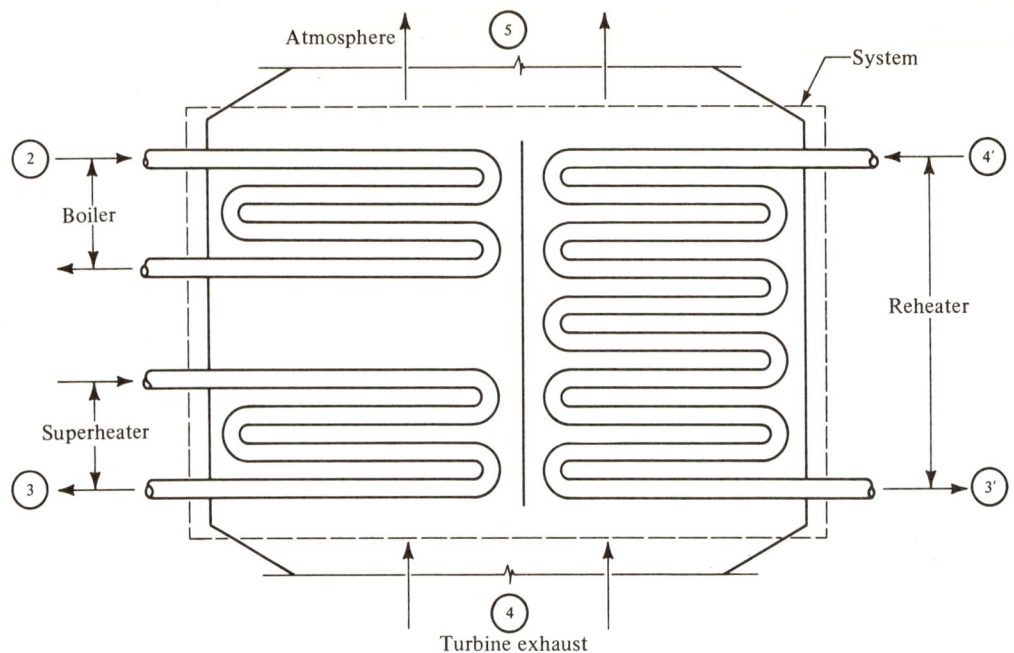

Fig. 7-51. *Heat exchanger between gas and steam systems.*

changer at state 4 (16 psia and 1200 °F) and is then exhausted to the atmosphere at 500 °F (260 °C). States 2, 3, 3' and 4' correspond to states in the steam cycle of Example Prob. 7-7. The temperature differentials between the water and gas are very low when compared to conventional steam generator designs which will make this heat exchanger large. An energy balance performed on the heat exchanger in the exhaust shows that the heat transferred to the water by the gas is

$$q_{H_2O} = (h_4 - h_5)_{air}$$

$$= [c_p(T_4 - T_5)]_{air}$$

$$= 0.24 \frac{BTU}{°F\, lb_m\, air} (1200 - 500)\, °F$$

$$= 168 \frac{BTU}{lb_m\, air} \quad (390\, joule/gram)$$

Since the bottom steam cycle is 39 percent efficient, the work it does per pound of air through the combined system is

$$w_{H_2O} = \eta_{therm} q_{H_2O} = 0.39 \times 168 \frac{BTU}{lb_m \text{ air}}$$

$$= 65.5 \frac{BTU}{lb_m \text{ air}} \text{ (152 joule/gram)}$$

The net specific work produced by the combined system is

$$w = w_{gas} + w_{H_2O} = 186 + 66 = 252 \frac{BTU}{lb_m \text{ air}} \text{ (586 joule/gram)}$$

and its net thermal efficiency is

$$\eta_{therm} = \frac{w}{q} = \frac{252}{466} = 0.54$$

This would represent a significant increase in the thermal efficiency that can be achieved from practical heat engines.

This combined cycle may also be very important for environmental reasons. The thermal pollution of cooling water would be reduced both because of the increased efficiency and because a significant amount of the waste heat is injected directly into the atmosphere. Also, shortages of natural gas and crude oil have forced the utilities to turn to coal where there are ample reserves and to nuclear power. Restrictions on sulfur and particulate emissions from coal burning plants often preclude the direct use of coal in the boiler furnace. However, it may be economically feasible to produce a fuel gas of adequate heating value for gas turbines from the coal in which the sulfur and ash have been removed. Current coal gasification processes have thermal efficiencies ranging from 70 to 75 percent while the predicted thermal efficiencies of advanced coal gasifiers ranges from 85 to 95 percent. With the use of advanced combined cycles, overall thermal efficiencies of around 45 percent are projected in some studies for the coal processing plant and power plant. It is also interesting to note that underground coal gasification is also being considered as a means of extracting the energy from coal that cannot be mined by conventional techniques for use in advanced combined cycles.

REVERSE HEAT ENGINES: REFRIGERATORS AND HEAT PUMPS

Reverse Carnot Cycle

Besides being able to transform various forms of energy into work, man also wants to be able to take a given space (which many times includes himself) and heat it above or cool it below the temperature of its surroundings. Both early and modern man have accomplished these effects by such methods as burning fuel for heat and melting ice for cooling.

It was noted in Chap. 3 that the same effects could be accomplished using a cyclic process by reversing the Carnot heat engine as shown in Fig. 7-52. That is,

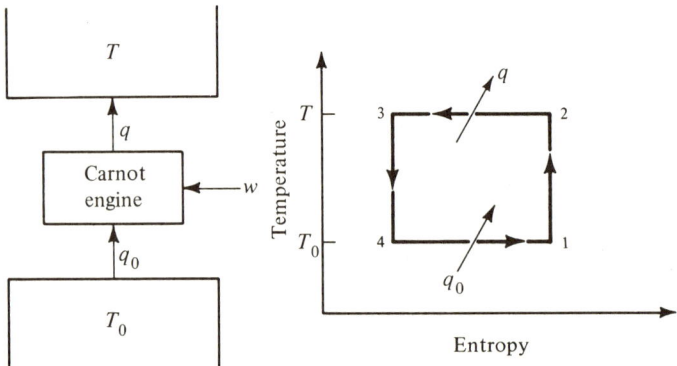

Fig. 7-52. *Reverse Carnot cycle.*

heat is transferred from a low temperature region T_0 to a high temperature region T but work must now be supplied to the reverse Carnot engine. If the desired effect of such a system is to remove heat from the cold temperature region, the system is called a refrigerator. On the other hand, if the desired effect is to add heat to the high temperature region, the system is called a heat pump.

The coefficient of performance for a refrigerator was defined in Chap. 3 to be

$$\beta = \frac{\text{heat removed from low temperature source}}{\text{work required by refrigerator}} = -\frac{q_0}{w} \qquad (3\text{-}11)$$

which is sometimes called COP. For the reverse Carnot cycle, the coefficient of performance was found to be

$$\beta_c = -\frac{q_0}{w} = \frac{T_0}{T - T_0} \qquad (3\text{-}12)$$

which can range from zero to infinity depending upon the two reservoir temperatures T and T_0. For instance, the coefficient of performance for a reverse Carnot cycle that is used to cool a 0 °C refrigerator compartment in a 25 °C room is

$$\beta_c = \frac{T_0}{T - T_0} = \frac{273}{298 - 273} = 10.9$$

This means that 10.9 joules of heat q_0 will be removed from the refrigerator compartment for every joule of work w used to operate the reverse Carnot engine.

Similarly, a coefficient of a performance for the heat pump is defined as

$$\text{COP} = \frac{\text{heat added to high temperature region}}{\text{work required by heat pump}}$$

For the reverse Carnot cycle shown in Fig. 7-52, this becomes

$$\text{COP}_c = \frac{q}{w} = \frac{q}{q + q_0} = \frac{T(s_3 - s_2)}{T(s_3 - s_2) + T_0(s_1 - s_4)}$$

$$= \frac{T}{T - T_0}$$

We note that the COP_c is the inverse of the thermal efficiency for a Carnot engine. If we now want to use the reverse Carnot engine to transfer heat to our 25 °C room from 0 °C outside atmosphere, the coefficient of performance for this system is

$$\text{COP}_c = \frac{T}{T - T_0} = \frac{298}{298 - 273} = 11.9$$

11.9 joules of heat are, therefore, transferred to the room for every joule of work used to operate the reverse Carnot engine. We see that 10.9 joules of this heat come from the cold region and the remaining one joule is the heat equivalent of the work consumed by the cycle. This means that we would get almost 12 times as much energy into the room by using our electrical power to run a heat pump than by converting the same amount of electrical energy directly into heat through the use of electrical heaters.

Of course the coefficient of performance obtained from a reverse Carnot cycle represents the theoretical maximum that can be obtained from any system operating between the same two temperatures. As was the case for the power cycle, the Carnot refrigeration and heat pump cycle is not a practical cycle but serves as a useful standard to compare other cycles against.

Vapor-Compression Refrigeration

Figure 7-53 illustrates the refrigeration cycle that is used for most applications. The idealized cycle consists of the isobaric vaporization of a liquid-vapor mixture 1—2 in the evaporator; the reversible adiabatic compression of the dry vapor 2—3; the isobaric condensing of the superheated vapor to a saturated liquid 3—4; and the expansion of the liquid through a throttling valve 4—1, which is a

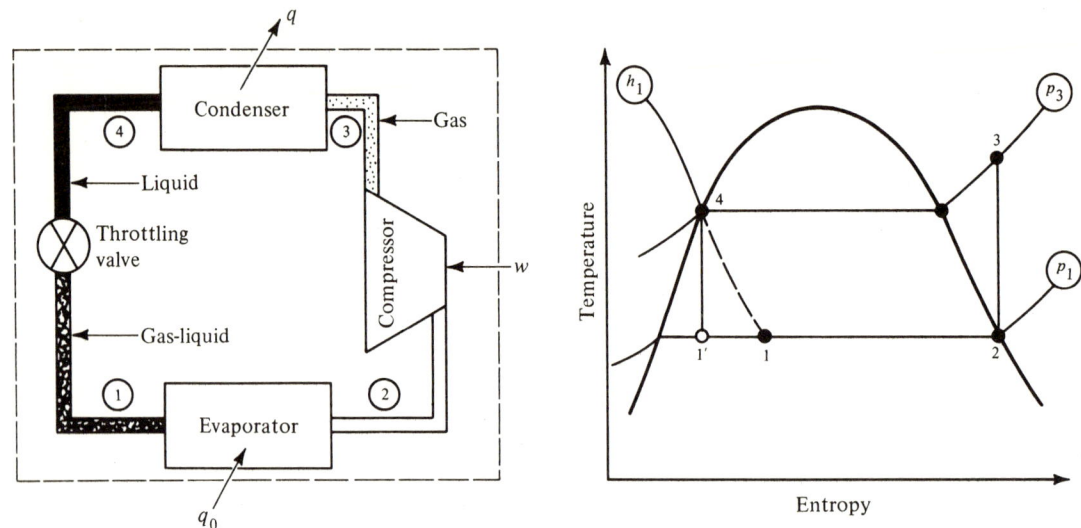

Fig. 7-53. *Vapor-compression refrigeration cycle.*

constant enthalpy process for small kinetic energy changes. We note that this system would have the same components as the reversible steam turbine system shown in Fig. 7-27 if the throttling valve were replaced with an isentropic turbine, process 4–1'. In this case, cycle 1'–2–3–4–1' would be a reverse Rankine cycle and the turbine could supply part of the work needed to operate the compressor. As a rule, the simplicity of design obtained with the throttling valve outweighs the benefit that would be derived with the use of a turbine as the expansion device.

EXAMPLE PROBLEM 7-13. Calculate the coefficient of performance for a vapor-compression refrigeration cycle using water as the working medium to cool a 0 °C refrigeration compartment in a 25 °C room

From Eq. (3-11),

$$\beta = -\frac{q_0}{w} = -\frac{h_2 - h_1}{h_2 - h_3}$$

where the generalized energy balance equation has been applied to both the evaporator and the compressor to find q_0 and w. From Table A-2, we find that

$$h_1 = h_4 = h_f(25\,°C) = 104.6 \frac{\text{joule}}{\text{gram}} \left(45.0 \frac{\text{BTU}}{\text{lb}_m}\right)$$

$$P_4 = P_3 = P_f(25\,°C) = 0.032 \text{ bars } (0.459 \text{ psia})$$

$$h_2 = h_g(0\,°C) = 2501.6 \frac{\text{joule}}{\text{gram}} \left(1075.5 \frac{\text{BTU}}{\text{lb}_m}\right)$$

$$s_2 = s_3 = s_g(0\,°C) = 9.15\,\frac{\text{joule}}{\text{gram K}}\left(2.1873\,\frac{\text{BTU}}{\text{lb}_m\,°R}\right)$$

Since P_3 and s_3 are known, h_3 can be found by consulting a more complete steam table than Table A-2 or a Mollier chart[1].

$$h_3(P_3, s_3) = 2759\,\frac{\text{joule}}{\text{gram}}\left(1186\,\frac{\text{BTU}}{\text{lb}_m}\right)$$

$$T_3 = 137\,°C\,(278\,°F)$$

In summary,

State	T(°C)	P(bars)	h(joule/gram)	v(cm³/gram)
1	0	0.006	104.6	
2	0	0.006	2502	2.06×10^5
3	137	0.032	2759	
4	25	0.032		

The states in the above table refer to Fig. 7-53.

The coefficient of performance of this ideal vapor-compression refrigeration cycle is then

$$\beta = \frac{h_2 - h_1}{h_3 - h_2} = \frac{2502 - 105}{2759 - 2502} = 9.3$$

as compared to the 10.9 we calculated for the corresponding Carnot refrigeration cycle. The coefficient of performance of an actual system will of course be only a fraction of the above calculated value due to irreversibilities.

In selecting a refrigerant, many of the desirable characteristics that were listed for a Rankine cycle working fluid are still applicable, but the low temperature properties of the fluid are now of paramount importance.

(1) High critical and low freezing temperatures—this will permit vaporization and, therefore, refrigeration over a broad range of temperatures.
(2) Large latent heat of vaporization—this permits a large amount of heat to be transferred per unit mass of refrigerant circulated.
(3) Convenient evaporation and condensation pressures—evaporator pressures above atmospheric pressure are desirable to avoid leakage of air into the system. In order to avoid heavy equipment, moderate pressures are desirable.

(4) Low specific vapor volume—a high specific vapor volume requires larger piping and may not permit the use of a reciprocating compressor.

(5) Noncorrosive and chemically stable.

(6) Easily and safely handled.

(7) Inexpensive.

We see that water is not a very good refrigerant. For one thing, it can not be used below 0 °C. In fact, water could not have been used in the above example problem since the evaporator temperature in an actual system must be significantly colder than the region to be cooled in order to obtain sufficient heat transfer rates. The specific volume of saturated water vapor is excessively large under normal temperatures as was found for state 2 in Example Prob. 7-13. Pressures below atmospheric will also be experienced in the evaporator for temperatures below 100 °C when water is used as the working medium.

There are many commercially important refrigerants in use today. Ammonia (NH_3), which was one of the first refrigerants used, is still used extensively in large commercial installations. If ammonia is used in our vapor-compression refrigeration example instead of water, the following data can be developed from ammonia tables that are available in the literature[14]:

State	T(°C)	P(bars)	h(joule/gram)	v(cm³/gram)
1	0	4.29	298.7	
2	0	4.29	1443.5	289.5
3	58	10.02	1559.0	
4	25	10.02	298.7	

The states in the above table refer again to Fig. 7-53. We see that the saturated vapor specific volume and the cycle pressures are far superior to those obtained with water. The coefficient of performance for this system is

$$\beta = \frac{h_2 - h_1}{h_3 - h_2} = \frac{1443.5 - 298.7}{1559.0 - 1443.5}$$

$$= 9.9$$

or slightly better than what was obtained with water. The coefficient of performance of an actual ammonia system would be around 4.8 instead of the 9.9 calculated for this ideal system.

Ammonia is toxic and corrosive to copper-bearing alloys and can, therefore, only be used in installations where reasonable precautions are taken. For air conditioning applications and domestic refrigeration, the halogenated hydrocarbons which are marketed under the trade name of Freon are widely used because they are nontoxic, inert and economical.

The unit commonly used in practice to describe the capacity of refrigeration is the ton which is approximately equivalent to the rate of heat removal from a ton (2000 lb) of ice at 0 °C in order to melt it in 24 hours, or equivalently,

$$1 \text{ ton} = 12{,}000 \text{ BTU/hr} = 3.516 \text{ kilowatts}$$

EXAMPLE PROBLEM 7-14. Size the compressor for the ideal ammonia system discussed above by determining the mass flow rate, the maximum volume flow rate and the power required for a one half ton unit. This is the cooling rating for an average household refrigerator.

The heat transferred from the refrigerator compartment per unit mass of ammonia circulated q_0 is equal to the enthalpy difference $h_2 - h_1$. For one half ton of cooling Q_0 the mass flow rate must be

$$\dot{m} = \frac{\dot{Q}_0}{h_2 - h_1} = \frac{0.5 \text{ ton} \times 3.516 \frac{\text{kW}}{\text{ton}} \times 10^3 \frac{\text{joule/sec}}{\text{kW}}}{(1443.5 - 298.7)\frac{\text{joule}}{\text{gram}}}$$

$$= 1.54 \text{ gram/sec}$$

The ammonia enters the compressor with the maximum specific volume of the cycle v_2. Thus,

$$\dot{V}_2 = \dot{m}v_2 = 1.54 \frac{\text{gram}}{\text{sec}} \times 289.5 \frac{\text{cm}^3}{\text{gram}}$$

$$= 445.8 \text{ cm}^3/\text{sec}$$

The power required by the adiabatic reversible compressor is

$$\dot{W} = \frac{\dot{Q}_0}{\beta} = \frac{0.5 \text{ ton} \times 3.516 \frac{\text{kW}}{\text{ton}}}{9.9} \quad (3\text{-}11)$$

$$= 0.18 \text{ kilowatts}$$

Gas Refrigeration Cycles

We have noted that the Stirling cycle is a practical cycle that has the same theoretical performance as the Carnot cycle. As a refrigeration cycle, the reverse Stirling cycle has proven to be very suitable for extremely low temperatures (cryogenic) ranging between 75 and 200 K. Figure 7-54 depicts the ideal Stirling

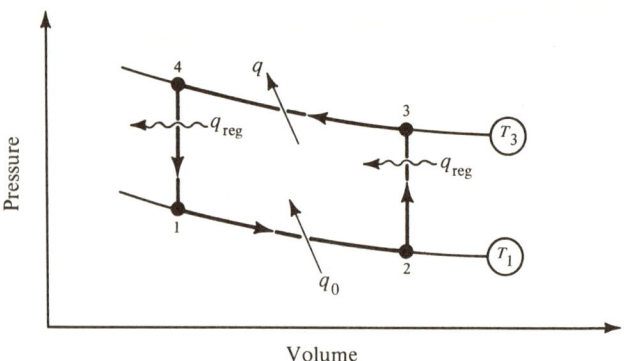

Fig. 7-54. *PV diagram for Stirling refrigeration cycle.*

refrigeration cycle which is just the reverse of Fig. 7-19. The cycle begins with the gas absorbing heat q_0 from the cold environment as it expands isothermally 1—2 in the cold space shown in Fig. 7-21. The gas is then transferred through the regenerator 2—3 which heats it to the upper working temperature T_3. Next the gas discharges heat q to the warm environment as it is isothermally compressed along path 3—4. The cycle is completed with the transfer of the working gas to the cold space through the regenerator 4—1 which cools the gas to the lower working temperature T_1. The energy absorbed by the regenerator in cooling the working gas q_{reg} is temporarily stored by the regenerator until the gas passes through it again during process 2—3.

We see from Eq. (3-12) that the coefficient of performance for a Stirling cycle operating between room temperatures and cryogenic temperatures will be quite low due to the large temperature range. For instance, a unit operating at the liquid-air temperature 83 K while exhausting to a 300 K environment will have a theoretical coefficient of performance of only 0.38. The coefficient of performance of actual machine is about 40 percent of this value. That is, for every joule of heat removed from this cold environment, approximately 6.6 joules of mechanical work must be supplied.

The reverse Brayton cycle depicted in Fig. 7-55 is also employed as a gas refrigeration cycle. The cycle starts with the working gas absorbing heat q_0 from the cold environment during the constant pressure process 1—2. The working gas is then compressed isentropically to state 3. From state 3 the working gas discharges heat q to the warm environment as it is cooled during a constant pressure process 3—4. The cycle ends with an isentropic expansion 4—1, to the cycle's initial state. We note that the temperature of the warm environment must be T_4 or less and the temperature of the cold environment T_2 or more to obtain the proper heat transfer.

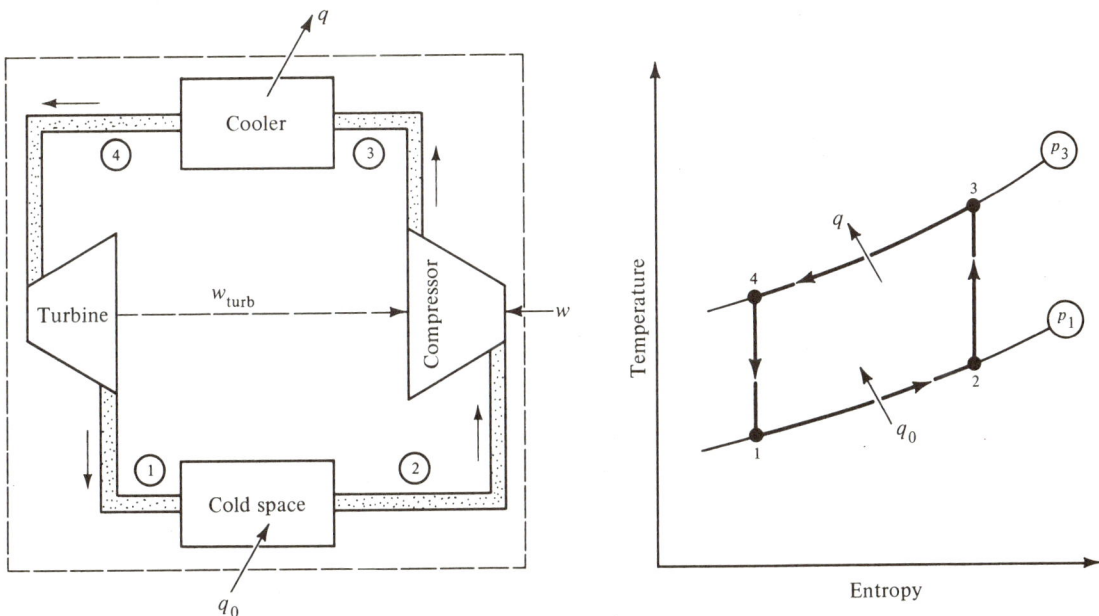

Fig. 7-55. *Brayton refrigeration cycle.*

EXAMPLE PROBLEM 7-15. The simple Brayton refrigeration cycle shown in Fig. 7-55 uses air as the working medium and has a pressure ratio of 5. The ambient temperature is 25 °C and the refrigeration space temperature is 0 °C. Determine the coefficient of performance, the mass flow rate and the power required by an idealized system having a one-half ton capacity.

We will base our calculation on one second of operation. If air is assumed to be an ideal gas with constant heat capacities and the generalized energy balance equation is applied to the components in the cycle, we find that

$$\beta = -\frac{q_0}{w} = -\frac{c_p(T_2 - T_1)}{c_p(T_4 - T_3) + c_p(T_2 - T_1)}$$

$$= \frac{T_1 - T_2}{(T_2 + T_4) - (T_1 + T_3)}$$

For the isentropic expansion processes, we know from Eq. (5-12) that

$$T_1 = T_4/r_P^{(\gamma-1)/\gamma} = (25 + 273.2)/5^{(1.4-1)/1.4}$$

$$= 188.3 \text{ K}$$

and

$$T_3 = T_2 r_P^{(\gamma-1)/\gamma} = (0 + 273.2) \text{ K} \times 5^{(1.4-1)/1.4}$$

$$= 432.7 \text{ K}$$

Then

$$\beta = \frac{188.3 - 273.2}{273.2 + 298.2 - 188.3 - 432.7} = 1.7$$

The mass flow rate for one-half ton of refrigeration is

$$\dot{m} = \frac{\dot{Q}_0}{q_0} = \frac{\dot{Q}_0}{h_2 - h_1} = \frac{\dot{Q}_0}{c_P(T_2 - T_1)}$$

$$= \frac{0.5 \text{ ton} \times 3.516 \frac{\text{kW}}{\text{ton}} \times 10^3 \frac{\text{joule/sec}}{\text{kW}} \times 28.96 \frac{\text{gram}}{\text{gram mole}}}{\frac{7}{2} \times 8.314 \frac{\text{joules}}{\text{gram mole K}} (273.2 - 188.3) \text{ K}}$$

$$= 20.6 \text{ gram/sec}$$

The required power for this ideal system is

$$\dot{W} = -\frac{\dot{Q}_0}{\beta} = \frac{-1/2 \text{ ton} \times 3.516 \frac{\text{kW}}{\text{ton}}}{1.7} = -1.03 \text{ kW}$$

We see that this reverse Brayton cycle example has a low coefficient of performance in comparison to the corresponding vapor-compression examples 7-13 and 7-14. Actual systems based on the reverse Brayton cycle have coefficients of performance that are less than one. Even so, early air conditioners especially aboard ships used this reverse Brayton cycle with air as the working gas since there was no danger of toxic leaks. Open cycles of this type are still important in the air conditioning of jet aircraft where a nontoxic working medium and small lightweight equipment must be used. In this case, high pressure air is bled from one of the first compressor stages of a turbojet engine, state 3 in Fig. 7-55; passed through a cooler 3—4; and then expanded through a turbine 4—1 before being passed through the cabin 1—2 and exhausted to the outside atmosphere. The power from the turbine is commonly used to power the circulation fans.

The reverse Brayton cycle with regeneration is also used to obtain the cryogenic temperature needed in the liquefaction of air. In this case, the gas at state 2 in Fig. 7-55 is used to further cool the gas at state 4 before it is expanded. The resulting cycle is shown in Fig. 7-56.

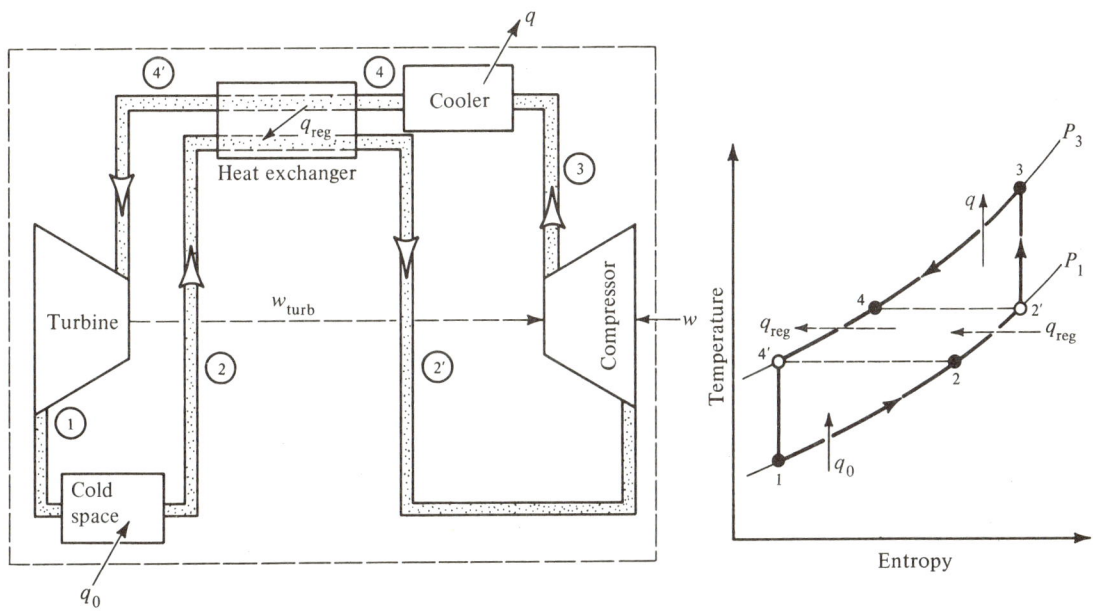

Fig. 7-56. *Reverse Brayton cycle refrigerator with regeneration.*

Absorption Refrigeration Cycle

In the previous refrigeration systems discussed, vapor or gas compression was employed. The absorption refrigeration cycle is a scheme to replace the gas compression process with a liquid compression process and thereby significantly reduce the power required for pumping. As depicted in Fig. 7-57, this is accomplished by absorbing a soluble refrigerant vapor in a liquid solvent 2–2′ as it leaves the evaporator and compressing the resulting strong solution in a liquid pump 2′–3. The strong solution is then transferred to the generator where it is heated to boil off part of the dissolved refrigerant vapor. The resulting weak solution is then returned to the absorber to repeat its cycle and the refrigerant vapor passes through the condenser, throttling valve and evaporator to complete its cycle. We see that the only difference between the absorption refrigeration system and the conventional vapor-compression refrigeration system in Fig. 7-53 is that the vapor compressor is replaced with an absorber, liquid pump and generator. Many combinations of refrigerant and absorbing liquid have been proposed but the ammonia-water combination is by far the most common.

We note that even though the pump work will be quite small in this system, heat must be supplied to the generator q_{gen} to boil off the refrigerant vapor. Unless a high temperature waste heat source is available, the savings in pumping costs do not usually offset the extra equipment cost and the cost to heat the generator. In many plants, there is a suitable waste heat source that the absorption refrigeration cycle can utilize. It is often more efficient in terms of a plant's total

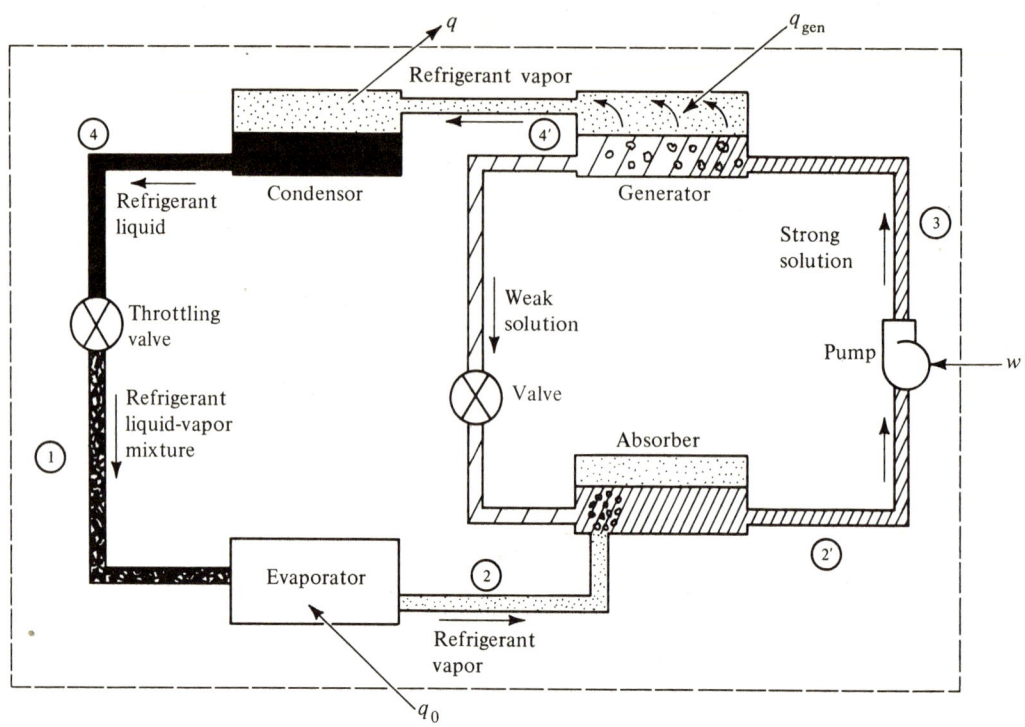

Fig. 7-57. *Elements of an absorption refrigeration system.*

energy usage to operate a waste heat producing system below its maximum possible thermal efficiency to increase the waste heat production for use in heating or refrigeration.

PROBLEMS

7-1. Estimate the maximum power that a windmill with a 125 foot diameter rotor can put out as a function of wind velocity. Real windmills will deliver only around 10 percent of this value.

7-2. Nitrogen is to be compressed from 1 atmosphere pressure and 25 °C to 64 atmospheres in an ideal multistage compressor with interstage cooling. The compressors are water jacketed such that the compression process can be represented by the polytropic process having an exponent k equal to 1.30. If the temperature leaving each interstage cooler is 25 °C,

(a) How many stages would you recommend?
(b) Determine the work of compression per gram of N_2 and compare this with the work that would be done by a single stage using the same polytropic process.

7-3. An air compressor-plant takes in 50,000 cubic feet of air at 70 °F and 1 atmosphere per hour and compresses it to 200 psia. Calculate the horsepower that would be required by a reversible compressor if the compression process is (a) isentropic, (b) isothermal, or (c) isentropic with optimum interstage cooling between two compressor stages.

7-4. An early steam engine took water in at 1 atmosphere pressure and 80 °F, compressed it adiabatically to 100 psia, formed saturated steam in a constant pressure boiler, and then used the reverse pump cycle 3—2—1—0 shown in Fig. 7-2 to obtain mechanical work. If the discharge pressure P_0 for this system is 1 atmosphere and the engine is assumed to be reversible, what is the engine's (a) specific work, (b) thermal efficiency, and (c) indicated mean effective pressure.

7-5. The first experimental reciprocating engines to use internal combustion did not employ a compression stroke. The fuel-air mixture was drawn into the cylinder, ignited and the combustion products expanded against the piston before being exhausted to the atmosphere. An idealization of these processes would consist of a constant pressure intake 0—1; constant volume heating 1—2; an isentropic expansion back to atmospheric pressure 2—3; and the constant pressure exhaust stroke 3—0. The air standard cycle 1—2—3—1 is called the Lenoir cycle after the engine's inventor.

    (a) Derive the thermal efficiency of this air standard cycle in terms of γ, T_1, T_2, and T_3 assuming constant specific heat capacities.

    (b) If the intake pressure and temperature are 1 atmosphere and 25 °C respectively, what is the indicated mean effective pressure of this idealized engine if a stoichiometric air-gasoline mixture is used.

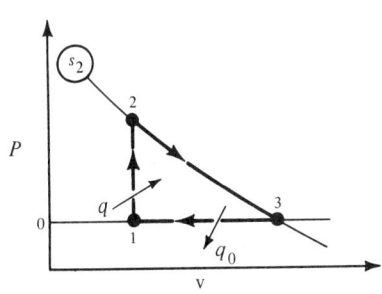

Fig. Problem 7-5.

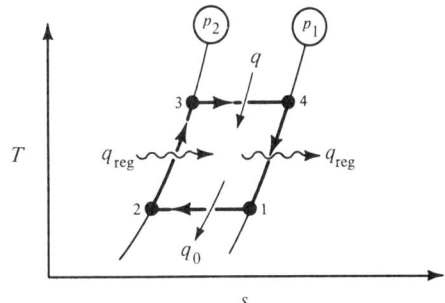

Fig. Problem 7-6.

7-6. A regenerative cycle similar to the Stirling cycle was proposed by a Swedish inventor, John Ericsson. His cycle consists of an isothermal compression process 1—2; an isobaric heating of this gas by a regenerator 2—3; an isothermal expansion process 3—4; and an isobaric cooling of the gas in a regenerator 4—1. This cycle differs from the Stirling cycle in that isobaric processes replace the Stirling cycle's constant volume regenerative processes. Show that the thermal efficiency of this reversible engine is equal to that of the Carnot engine and determine its mean effective pressure in terms of P_1, r, T_1/T_4 and v_1/v_2.

7-7. An air standard Diesel engine intakes air at a pressure of 1 atmosphere and a temperature of 30 °C. The initial volume V_1 is 40 liters. The compression ratio of the cycle is 14 and the cut off ratio is 2.5. Tabulate the pressure, temperature and volume at states 2, 3 and 4. Determine the total work, mean effective pressure and thermal efficiency of this Diesel cycle. Assume the ratio of specific heat capacities to be 1.36.

7-8. Actual indicator diagrams from modern Otto and Diesel engines show that the combustion process is neither a constant volume nor a constant pressure process but can be better represented as a combination of the two. Calculate the thermal efficiency of the dual combustion cycle shown below in terms of the cycle temperatures and the ratio of specific heat capacities.

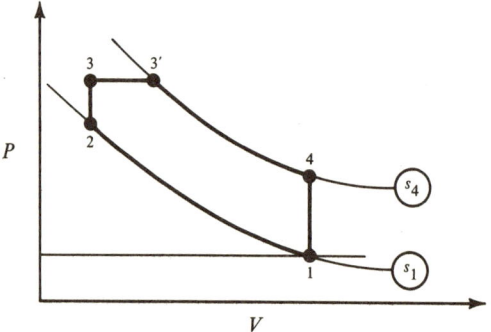

7-9. A six cylinder Otto engine with a compression ratio of 8 develops 80 brake horsepower at 2000 rpm. What is the brake mean effective pressure of this engine if the bore is 3.625 inches and the stroke is 3.6 inches? How does this compare to the theoretically predicted value?

7-10. An air standard Diesel engine with a compression ratio of 14 intakes air at 15 psia and 70 °F. If 250 BTU/lb_m of heat are liberated during the combustion process, determine the thermal efficiency of this reversible Diesel engine and compare it to the thermal efficiency of an Otto cycle operating with the same compression ratio.

7-11. In the case that the isentropic efficiencies of all the compressor stages are equal, show that the overall isentropic efficiency of a multistage compressor must be less than the individual stage efficiencies.

7-12. For the reversible Rankine cycle shown in Fig. 7-32, the temperature limit T_3 is 1000 °F and the condensing temperature T_1 is 80 °F. What is the maximum vaporization pressure P_3 permitted if the quality of the steam leaving the turbine must not fall below 90 percent?

7-13. The simple steam turbine system depicted in Figs. 7-27 and 7-28 has turbine inlet conditions of 900 °F and 400 psia and an exhaust pressure of 1 psia. The turbine has an isentropic efficiency of 80 percent and the pump is 75 percent efficient.

Determine the thermal efficiency of this system and compare it to a Carnot cycle operating between the same two temperature extremes.

7-14. A reversible Rankine cycle with reheat similar to the cycle depicted in Fig. 7-34 has a condensing temperature T_1 of 80 °F, a superheat temperature T_3 and a reheat temperature $T_{3'}$ of 600 °F and a vaporization pressure P_2 of 900 psia. Instead of expanding the steam through the first turbine to saturated vapor as depicted in Fig. 7-34, the steam is expanded until it has a 90 percent steam quality at state 4' and then reheated.

    (a)   Draw the new cycle diagram.

    (b)   Calculate the thermal efficiency of this cycle and compare it to the thermal efficiency that a Carnot cycle would obtain when operating between the same two temperature extremes.

7-15. In the Rankine cycle with a feedwater heater depicted in Figs. 7-35 and 7-36, the condensing temperature T_1 is 80 °F; the superheat temperature T_3 is 700 °F and the vaporization pressure P_3 is 470 psia. Enough steam is bled from the turbine at 100 psia $P_{2'}$ to heat the feed water 2' to a saturated liquid 2''. Determine the thermal efficiency of this reversible cycle and compare it to a Carnot cycle operating between the same two temperature extremes. Also compute the net work output per pound of steam.

7-16. An industrial plant needs 20,000 pounds of saturated steam at 280 °F per hour for heating purposes and one megawatt of electrical power. A small turbine system is to be designed to handle both of these needs. Steam is supplied to a turbine at 400 psia pressure from a boiler and expanded until 280 °F saturated steam is obtained whereupon the steam needed for heating is extracted. The rest of the steam is expanded in a second turbine to a condensor pressure of 5 psia, where it is condensed, compressed to 400 psia and returned to the boiler. The heating steam returns as a liquid at 200 °F and 1 atmosphere pressure where it is compressed to 400 psia and returned to the boiler. Both turbines have an isentropic efficiency of 72 percent and both feedwater pumps have an efficiency of 50 percent. The power from the turbines is used to drive the feed water pumps and an electrical generator.

    (a)   At what temperature must the steam be generated in the boiler to supply the 280 °F saturated heating steam?

    (b)   How much steam must the boiler supply?

    (c)   If one pound of coal delivers 11,000 BTU of heat to the water in the boiler, how many pounds of coal are consumed per hour?

    (d)   How many pounds of coal per hour would be needed if a separate system were used to supply the heating steam?

7-17. A reversible air standard Brayton cycle with a pressure ratio of 5 has a 100 °F compressor inlet temperature and a 1500 °F turbine inlet temperature. Find

    (a) the specific compressor work.
    (b) the specific turbine work.
    (c) the back work ratio.
    (d) the thermal efficiency.

7-18. The air standard Brayton cycle with a pressure ratio of 3 has a 30 °F compressor inlet temperature and a 1500 °F turbine inlet temperature. The isentropic turbine efficiency is 85 percent and the compressor efficiency is 75 percent. Find

    (a) the thermal efficiency of this system.
    (b) the thermal efficiency of the system if a regenerator that is 80 percent efficient is also employed.

7-19. If the working gas in the High Temperature Gas Reactor in Example Prob. 7-11 is air instead of helium, what would the system's thermal efficiency be now assuming the same pressure and temperatures. Compare the specific work output of the helium and air systems.

7-20. Small-gas turbine regenerator systems applicable for automobiles have pressure ratios around 3.5, maximum temperatures around 1800 °F, and power outputs near 300 hp. Assuming isentropic efficiencies of 90 and 85 percent for the turbine and compressor respectively and 70 percent for the regenerator, estimate the gasoline consumption rate at maximum power if the total energy in the fuel which can be converted to heat is 110,000 BTU/gal.

7-21. The vapor-compression system depicted in Fig. 7-53 is used as a heat pump with water as the working fluid. The water is evaporated at 35 °F by the outside atmosphere 1–2, isentropically compressed to 2 psia 2–3, and introduced into radiators in the house 3–4. The vapor condenses to a saturated liquid in the radiators and leaves the radiators at a pressure of 1 psia. The saturated water then expands through a throttle valve into the evaporator to complete the cycle. Calculate the coefficient of performance of this system.

7-22. Air from a compressor of a jet engine is bled off at 1 atmosphere pressure and cooled isobarically to 38 °C. It is then expanded isentropically through a turbine to the cabin pressure of 0.7 atmosphere. This air is mixed with the cabin air and eventually exhausted to the outside. 3000 watts of heat must be removed from the cabin to maintain its temperature at 24 °C.

    (a) At what rate must air be bled from the compressor?
    (b) How much work will the turbine do?
    (c) What would happen if we replaced the turbine with a throttle valve as was used in the vapor-compression cycle?

8

Equations of State

OBJECTIVES

After studying this chapter, the student should be able to

(1) predict numerical values of the critical properties for a single component system;
(2) evaluate coefficients in the van der Waals equation of state, Eq. (8-3), the Redlich Kwong equation of state, Eq. (8-22), and the virial equation of state, Eq. (8-26), from a limited amount of physical property data; and
(3) calculate values of pressure, temperature or specific volume for a single component system at equilibrium given any two of these three variables, using either an analytical or graphical form of an equation of state.

INTRODUCTION

As we have shown in Chap. 5, the ideal gas equation of state,

$$Pv = RT \tag{5-1}$$

is an analytical relationship which represents the experimentally observed phase behavior of gases at low pressures and temperatures well above the critical point. You will recall that in Chap. 1, we defined the critical point as the state represented by the highest temperature and the highest pressure at which liquid and vapor can coexist in equilibrium. The ideal gas equation has proven to be extremely useful in evaluating changes of thermodynamic properties of systems to which it is applicable, such as air at normal atmospheric conditions.

Unfortunately, it has been found experimentally that real gas behavior deviates from ideal gas behavior, in some cases even at low pressure. For example, if we plot the product of Pv as a function of P at various fixed temperatures, as shown in Fig. 8-1, we do not obtain a series of horizontal lines but rather a series

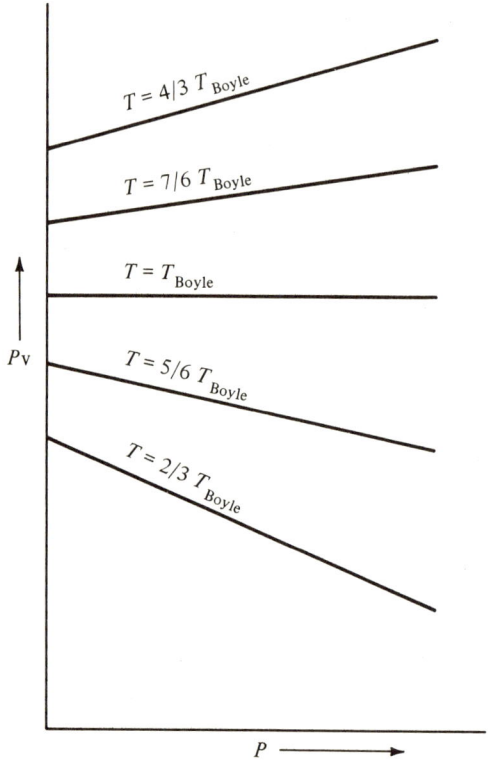

Fig. 8-1. Pv product as a function of P.

of straight lines which can be represented by the equation

$$Pv = RT - \alpha_R P$$

where α_R is the slope of the line and is a function of temperature. Only at a temperature called the Boyle temperature is α_R equal to zero so that Boyle's Law is obeyed. Further, since

$$\alpha_R = \frac{RT}{P} - v \qquad (8\text{-}1)$$

we see that α_R is the difference between the volume occupied by the gas if it were ideal RT/P and the actual volume occupied by the gas v. For this reason α_R is called the residual volume. The generalized two-constant Callendar equation of state,

$$Pv = RT\left[1 - \frac{1}{3}\left(\frac{P}{P_c}\right)\left(\frac{T_c}{T}\right)^4\right] \qquad (8\text{-}2)$$

is based on the observation that below the Boyle temperature,

$$\alpha_R = \frac{RT_c^4}{3P_c T^3}$$

Note that the subscript c indicates a property of the critical state. Not only do real gases deviate from ideal gas behavior as pressure increases, but also liquids and solids do not follow ideal gas behavior. As a result, other equations of state have been developed to more accurately represent experimental PVT behavior of gases over a wider range of pressures as well as the experimental PVT behavior of liquids and solids.

EQUATIONS OF STATE FOR GASES

van der Waals' Equation

In 1873, J. D. van der Waals gave a simple and elegant description of the behavior of real gases. He observed that the ideal gas law is correct only when the volume of and the interaction between molecules can be neglected. Neither of these approximations are rigorous for real gases. The volume occupied by the molecules is approximately equal to the total volume of the gas if it were in the liquid state at the same temperature. Further, the heat released during the condensation is evidence of the attractive energy between particles.

Van der Waals corrected the ideal gas law by subtracting a constant b from the volume v to account for the volume occupied by molecules and not available to other molecules. He also added a correction term a/v^2 to the pressure P in the ideal gas law to account for the attractive forces between molecules. With these additions to the ideal gas law, we have

$$\left(P + \frac{a}{v^2}\right)(v - b) = RT$$

or

$$P = \frac{RT}{v - b} - \frac{a}{v^2} \qquad (8\text{-}3)$$

Equation (8-3) is known as van der Waals' equation of state. Note that at very low pressures where v increases toward infinity that Eq. (8-3) will reduce to the ideal gas equation of state Eq. (5-1).

The constants in the van der Waals equation can be selected in such a manner that the equation will represent observed phase behavior. Typical single component phase behavior is shown in Fig. 8-2 which is a plot of pressure versus specific

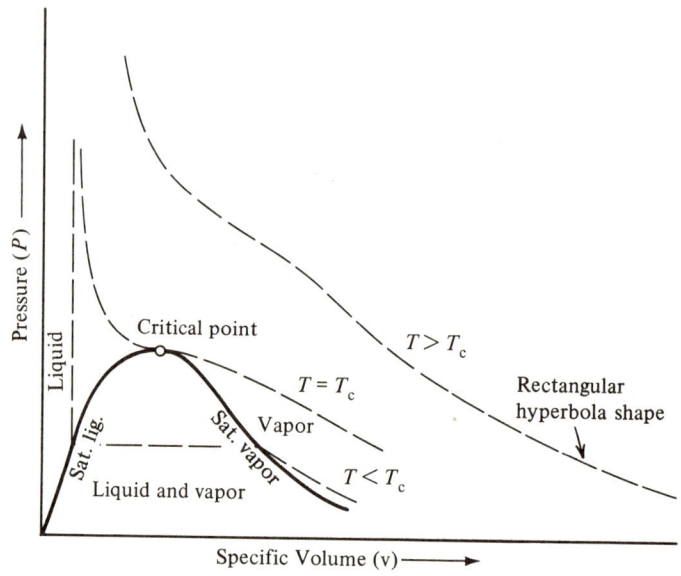

Fig. 8-2. *Pressure volume diagram.*

volume with lines of constant temperature called isotherms.

The dome shaped region on this diagram represents mixtures of liquid and vapor. As we have noted previously, the isotherms inside this coexistence region are perfectly horizontal. The isotherms outside the coexistence region have a variety of shapes. The fact that the low-pressure vapor isotherms approximate rectangular hyperbolas was used in developing the equation of state $Pv = RT$ for ideal gases. The liquid isotherms are very steep and almost straight at the lower temperatures. At higher temperatures, the isotherms curve concave upward near the saturated liquid line. The isotherms at temperatures well above the critical temperature have a very definite bump which distinguishes them from hyperbolas.

The critical isotherm itself possesses two inflection points just as do some of the curves immediately above it. However, there is an inflection at the critical point which is different from that anywhere else in the whole diagram. At this point, the critical isotherm becomes perfectly horizontal, as well as undergoing an inflection.

If the van der Waals equation of state is to represent the observed behavior of the critical isotherm at the critical point, we can write

$$P_c = \frac{RT_c}{v_c - b} - \frac{a}{v_c^2}$$

Further, because of the existence of a horizontal inflection point at the critical point,

$$\left(\frac{\partial P}{\partial v}\right)_c = \frac{-RT_c}{(v_c - b)^2} + \frac{2a}{v_c^3} = 0$$

and

$$\left(\frac{\partial^2 P}{\partial v^2}\right)_c = \frac{2RT_c}{(v_c - b)^3} - \frac{6a}{v_c^4} = 0$$

Solving these three equations simultaneously, we find that

$$a = \frac{27}{64} R^2 \frac{T_c^2}{P_c} \tag{8-4}$$

and

$$b = \frac{RT_c}{8P_c} \tag{8-5}$$

These results may be combined to show that for all gases obeying van der Waals' equation of state,

$$v_c = \frac{3RT_c}{8P_c} \tag{8-6}$$

Equations (8-4), (8-5), and (8-6) led to the theorem of corresponding states, our next topic.

Theorem of Corresponding States

The Compressibility Factor. The theorem of corresponding states was first proposed by van der Waals when he observed that the constants in his equation could be estimated in terms of the critical properties of a component. This led to the concept that at the same reduced temperature T_r and reduced pressure P_r where

$$T_r = \frac{T}{T_c} \tag{8-7}$$

and

$$P_r = \frac{P}{P_c} \tag{8-8}$$

all substances which followed van der Waals' equation of state would show the same deviation from the ideal gas law. Sidney Young introduced the term "corresponding state" to gases at the same T_r and P_r in 1899.

The next step in the development of this theory occurred in 1931, when generalized charts of the compressibility factor z where

$$z = \frac{(PV)_{actual}}{(PV)_{ideal\ gas}} = \frac{PV}{nRT}$$

or

$$PV = znRT \tag{8-9}$$

appeared.

Experimental data based primarily on hydrocarbon gases was used as the basis for the original plots of z as a function of reduced pressure with reduced temperature as a parameter. Although this correlation appeared to give satisfactory results for the compounds on which it was based, it was not satisfactory for either compounds which boil at very low temperatures such as hydrogen, helium or neon or for polar compounds.

In 1935, R. H. Newton suggested that the compressibility factor plot could be used for hydrogen, helium, and neon if pseudo-critical pressures and temperatures were used to calculate their respective reduced pressures and temperatures. For these three gases, he recommended using

$$P_r = \frac{P}{P_c + 8}$$

where P is in atmospheres, and

$$T_r = \frac{T}{T_c + 8}$$

where T is in K.

Further problems arose when the two parameter compressibility factor correlation was used to predict the PVT behavior of compounds which differed substantially in shape and polarity from the hydrocarbons upon which the correlation was based. In 1951, H. P. Meissner and R. Seferian suggested the use of a third parameter, the critical compressibility factor, where

$$z_c = \frac{P_c v_c}{RT_c} \tag{8-10}$$

to account for shape factors and polarity.

The theorem of corresponding states was extended to say that any two substances having the same critical compressibility factors have the same z at the same reduced temperature and pressure. Such substances are said to be conformal. However, the value of z_c depends upon a measured value of v_c, a parameter which is very difficult to measure. As vapor pressures are more easily measured, in 1955 K. S. Pitzer suggested the use of an alternate third parameter defined in terms of the vapor pressure. This parameter is called the accentric factor ω and is defined by the relationship

$$\omega = -\log(P_r^\square)_{T_r = 0.7} - 1.000$$

where the reduced vapor pressure $P_r^\square$ is measured at a reduced temperature of 0.7[15]. The accentric factor is closely related to the slope of the vapor pressure curve. This slope, in turn, is related through the relationship

$$\left(\frac{dP^\square}{dT}\right)_V = \left(\frac{dS^\square}{dV}\right)_T \tag{4-37}$$

to the entropy of vaporization, which depends upon the interactions between molecules. The accentric factors of helium, hydrogen and neon are 0.0, while other compounds have accentric factors which differ from 0.0. If we define a simple fluid to be a substance which has an accentric factor of 0.0, the accentric factor of a real substance may be regarded as a measure of the increase in the entropy of vaporization over that of a simple fluid.

Now, the compressibility factor of any substance can be closely approximated by the equation

$$z = z^{(0)} + \omega z^{(1)} \tag{8-11}$$

where $z^{(0)}$ is the compressibility factor of a simple fluid at the P_r and T_r of the system and $z^{(1)}$ is a compressibility factor correction term, also measured at the P_r and T_r of the system.

Values of $z^{(0)}$ can be read directly from Fig. A-1A and values of $z^{(1)}$ can be read directly from Fig. A-1B in the Appendix. Critical pressures, temperatures and specific volumes as well as accentric factors of some representative compounds are presented in Table A-4 in the Appendix[16,17].

The substances for which this correlation is valid are called normal fluids. Unfortunately, highly polar or hydrogen bonding molecules such as H_2O and NH_3 do not form normal fluids. Errors of several percent can be expected when applying the accentric factor correlation to non-normal fluids. Further, molecules with a mass equal to or less than the mass of neon exhibit significant quantum effects at extremely low temperatures which are not taken into account by these correlations.

EXAMPLE PROBLEM 8-1. Estimate the specific volume of ethane (MW = 30) at 500 lb_f/in^2 and 200 °F using Figs. A-1A and A-1B. Compare the result with a value of 0.4048 ft^3/lb_m reported in the literature[4].

From Table A-4, we find for ethane

$$T_c = 305.4 \text{ K}$$

$$P_c = 48.2 \text{ atm}$$

and

$$\omega = 0.0980$$

At the conditions of the problem from Eqs. (8-7) and (8-8),

$$T_r = \frac{200 + 460}{(305.4)(1.8)} = 1.2$$

$$P_r = \frac{500}{(48.2)(14.7)} = 0.706$$

Entering Fig. A-1A at P_r = 0.706 and reading vertically to T_r = 1.2, we find that

$$z^{(0)} = 0.857$$

and similarly, from Fig. A-1B, we find

$$z^{(1)} = 0.034$$

Then

$$z = z^{(0)} + \omega z^{(1)} \tag{8-11}$$

$$z = 0.857 + (0.0980)(0.034)$$

or

$$z = 0.860$$

Then, from Eq. (8-9),

$$v = \frac{zRT}{P}$$

$$= \frac{(0.860)(10.73)(660)}{(30)(500)}$$

$$= 0.4060 \text{ ft}^3/\text{lb}_m$$

The percent error, then, is

$$\text{error} = \frac{0.4048 - 0.4060}{0.4048} \times 100$$

$$= -0.3 \text{ percent}$$

Estimation of Critical Constants. The equation

$$PV = znRT \tag{8-9}$$

where

$$z(P_r, T_r, \omega)$$

is an extremely useful equation if the critical properties of a component are known. We can find much of this information in tables. However, sometimes critical constants must be estimated. Methods of doing this are summarized conveniently by R. C. Reid and T. K. Sherwood[18].

Fortunately, there is a certain regularity in critical constants of different compounds. C. M. Guldberg first noted that to a good approximation, the critical temperature T_c can be estimated from the atmospheric boiling point temperature T_b using the relationship

$$\frac{T_b}{T_c} = \frac{2}{3}$$

or from the atmospheric freezing point temperature T_f using the relationship

$$\frac{T_f}{T_c} = \frac{1}{3}$$

Reid and Sherwood recommend that A. L. Lydersen's modification of Guldberg's ratio,

$$\frac{T_b}{T_c} = k_L \tag{8-12}$$

where

$$k_L = 0.567 + \Sigma\Delta_T - (\Sigma\Delta_T)^2$$

be used. Values of Δ_T for atomic increments which are added to form compounds are presented in Table 8-1[19].

TABLE 8-1. Lydersen's Critical Property Increments[19]

	Δ_T	Δ_P	Δ_V
Nonring increments:			
—CH_3	0.020	0.227	55
—CH_2—	0.020	0.227	55
—CH—	0.012	0.210	51
—C—	0.00	0.210	41
=CH_2	0.018	0.198	45
=CH—	0.018	0.198	45
=C—	0.0	0.198	36
=C=	0.0	0.198	36
≡CH	0.005	0.153	(36)
≡C—	0.005	0.153	(36)
Ring increments:			
—CH_2—	0.013	0.184	44.5
—CH—	0.012	0.192	46
—C—	(−0.007)	(0.154)	(31)
=CH—	0.011	0.154	37
=C—	0.011	0.154	36
=C=	0.011	0.154	36
Halogen increments:			
—F	0.018	0.224	18
—Cl	0.017	0.320	49
—Br	0.010	(0.50)	(70)
—I	0.012	(0.83)	(95)

TABLE 8-1. *(continued)*

	Δ_T	Δ_P	Δ_V
Oxygen increments:			
—OH (alcohols)	0.082	0.06	(18)
—OH (phenols)	(0.035)	(−0.02)	(3)
—O— (nonring)	0.021	0.16	20
—O—(ring)	(0.014)	(0.12)	(8)
—C=O (nonring)	0.040	0.29	60
—C=O (ring)	(0.033)	(0.2)	(50)
HC=O (aldehyde)	0.048	0.33	73
—COOH (acid)	0.085	(0.4)	80
—COO— (ester)	0.047	0.47	80
=O (except for combinations above)	(0.02)	(0.12)	(11)
Nitrogen increments:			
—NH$_2$	0.031	0.095	28
—NH (nonring)	0.031	0.135	(37)
—NH (ring)	(0.024)	(0.09)	(27)
—N— (nonring)	0.014	0.17	(42)
—N— (ring)	(0.007)	(0.13)	(32)
—CN	(0.060)	(0.36)	(80)
—NO$_2$	(0.055)	(0.42)	(78)
Sulfur increments:			
—SH	0.015	0.27	55
—S— (nonring)	0.015	0.27	55
—S— (ring)	(0.008)	(0.24)	(45)
=S	(0.003)	(0.24)	(47)
Miscellaneous:			
—Si—	0.03	(0.54)	
—B—	(0.03)		

Notes
1. There are no increments for hydrogen.
2. All bonds shown as free are connected with atoms other than hydrogen.
3. Values in parentheses are based upon too few experimental values to be reliable.
4. It has been suggested that the C—H ring increment common to two condensed saturated rings be given the value of $\Delta_T = 0.064$.

Reprinted by permission of the University of Wisconsin, Engineering Experimental Station.

In order to estimate critical pressures and volumes, Reid and Sherwood suggest that one could use Lydersen's parameters in L. Riedel's correlation,

$$P_c(\text{atm}) = \frac{MW}{(\Sigma\Delta_P + 0.34)^2} \tag{8-13}$$

to predict critical pressures of compounds and Lydersen's correlation,

$$v_c(\text{cc/gm mole}) = 40 + \Sigma\Delta_V \tag{8-14}$$

to predict critical volumes. Values of Δ_P and Δ_V for atomic increments which are added to form compounds are also presented in Table 8-1.

In addition, Reid and Sherwood suggest that Edmister's equation,

$$\omega + 1 = 3/7[k_L/(1 - k_L)] \log P_c \tag{8-15}$$

where

$$k_L = \frac{T_b}{T_c}$$

and

P_c has units of atmospheres

be used to estimate accentric factors when accurate vapor pressure data is not available. Values of k_L can be estimated using Eq. (8-12) and values of P_c can be estimated using Eq. (8-13).

The methods shown here are not the only methods for evaluating critical constants. In some cases, other methods give better results. In using empirical correlations to predict critical properties, however, one should compare the predictions with experimental critical properties of similar compounds in order to avoid any gross computational errors.

EXAMPLE PROBLEM 8-2. Estimate the critical properties of ethyl acetate ($MW = 88.10$),

$$\begin{array}{c} \text{O} \\ \text{H} \quad \| \quad \text{H} \quad \text{H} \\ \text{HC} - \text{C} - \text{O} - \text{C} - \text{CH} \\ \text{H} \quad \quad \text{H} \quad \text{H} \end{array}$$

using Eqs. (8-12), (8-13), and (8-14), and compare the results with the following data from the literature:

$$T_c = 523.3 \text{ K}$$

$$P_c = 37.8 \text{ atm}$$

$$v_c = 286 \text{ cc/gm mole}$$

The atmospheric boiling point of ethyl acetate is 350.3 K.
In order to estimate T_c, we find, using Table 8-1,

$$\Sigma \Delta_T = 2(\text{HC}\overset{\text{H}}{\underset{\text{H}}{-}}) + (-\overset{\text{H}}{\underset{\text{H}}{\text{C}}}-) + (-\text{O}-) + (-\overset{\overset{\text{O}}{\|}}{\text{C}}-)$$

$$= 2(0.02) + 0.02 + 0.021 + 0.04$$

$$= 0.121$$

Then, from Eq. (8-12),

$$k_L = 0.567 + 0.121 - 0.121^2$$

$$= 0.673$$

and

$$T_c = \frac{350.3}{0.673}$$

$$= 520.5 \text{ K}$$

The percent error between this value and the experimental value is

$$\text{error} = \frac{523.3 - 520.5}{523.3} \times 100$$

$$= 0.5 \text{ percent}$$

Now, in order to estimate P_c, we find, using Table 8-1,

$$\Sigma\Delta_P = 2(HC-\underset{H}{\overset{H}{|}}) + (-\underset{H}{\overset{H}{\underset{|}{C}}}-) + (-O-) + (-\overset{\overset{O}{\|}}{C}-)$$

$$= 2(0.227) + 0.227 + 0.16 + 0.29$$

$$= 1.131$$

Then, from Eq. (8-13),

$$P_c = \frac{88.10}{(1.131 + 0.34)^2}$$

$$= 40.7 \text{ atm}$$

The percent error between this value and the experimental value is

$$\text{error} = \frac{37.8 - 40.7}{37.8} \times 100$$

$$= -7.7 \text{ percent}$$

Finally, in order to estimate v_c, we find, using Table 8-1,

$$\Sigma\Delta_V = 2(HC-\underset{H}{\overset{H}{|}}) + (-\underset{H}{\overset{H}{\underset{|}{C}}}-) + (-O-) + (-\overset{\overset{O}{\|}}{C}-)$$

$$= 2(55) + 55 + 20 + 60$$

$$= 245$$

Then, from Eq. (8-14),

$$v_c = 40 + 245$$

$$= 285 \text{ cc/gm mole}$$

The percent error between this value and the experimental value is

$$\text{error} = \frac{286 - 285}{286} \times 100$$

$$= 0.3 \text{ percent}$$

Other Equations of State

When we showed that the constants a and b in van der Waals' equation were functions of P_c and T_c, we assumed that not only was van der Waals' equation valid at the critical point, but also that

$$\left(\frac{\partial P}{\partial v}\right)_c = 0$$

and

$$\left(\frac{\partial^2 P}{\partial v^2}\right)_c = 0$$

at the critical point and we eliminated v_c from the resulting equations. We might just as easily have eliminated T_c and shown that

$$a = 3P_c v_c^2 \tag{8-16}$$

and

$$b = v_c/3 \tag{8-17}$$

Had we eliminated P_c, we would have found that

$$a = 9RT_c v_c/8 \tag{8-18}$$

and

$$b = v_c/3 \tag{8-19}$$

Unfortunately, the values of a and b calculated using different critical properties do not necessarily agree. Consider argon, for example. The critical temperature of argon is 150.9 K, the critical pressure is 48.3 atm, and the critical volume is 75.2 cc/gm mole. The van der Waals constants may then be calculated with the following results:

TABLE 8-2. van der Waals' Constants for Argon

Basis	$a\left[\left(\frac{cm^3}{gm\text{-}mole}\right)^2 atm\right]$	$b\left(\frac{cm^3}{gm\text{-}mole}\right)$
Eqs. (8-4), (8-5)	1.3392×10^6	32.05
Eqs. (8-16), (8-17)	0.8194×10^6	25.07
Eqs. (8-18), (8-19)	1.0475×10^6	25.07

Obviously, the result we would obtain from the van der Waals equation would depend upon how the constants were obtained. Even different values of a and b would be found if we evaluate these constants using experimental PVT data points taken at non-critical conditions.

Another problem with the van der Waals equation is that the predicted value of z_c is

$$z_c = \frac{P_c v_c}{RT_c} = \frac{P_c}{RT_c} \cdot \frac{3RT_c}{8P_c} = 0.375$$

for all gases while experimentally, values of z_c are much less than this.

J. J. Martin[20] has observed that if the constants for the van der Waals equation are based on experimental P_c and T_c, the predicted v_c will be too large by a factor of $0.375/z_c$. That is, since

$$v_c \text{ (van der Waals)} = \frac{3RT_c}{8P_c} = \frac{0.375 RT_c}{P_c}$$

and, from the definition of the critical compressibility factor,

$$v_c \text{ (actual)} = \frac{z_c RT_c}{P_c}$$

we find

$$v_c \text{ (van der Waals)} = \left(\frac{0.375}{z_c}\right) v_c \text{ (actual)}$$

Similarly, if the van der Waals constants are based on experimental v_c and T_c, the predicted P_c will be too large by a factor of $0.375/z_c$; and if the van der Waals constants are based on experimental P_c and v_c, the predicted T_c will be too low by a factor of $z_c/0.375$.

In order to fit the experimental data on P_c, T_c and v_c, Martin[20] suggests that we might shift the critical isotherm predicted by van der Waals' equation so that it would pass through the actual v_c using the linear transformation

$$v = v + c$$

Then, with these modifications to van der Waals' equation, we obtain the relationship

$$P = \frac{RT}{(v+c)-b} - \frac{a}{(v+c)^2}$$

Next we might examine how this modified van der Waals equation represents the variation of pressure with temperature at constant specific volume. A single component phase diagram of pressure as a function of temperature with volume as a parameter is shown in Fig. 8-3. The vapor-pressure curve separating the liquid and vapor regions, shown in Fig. 8-3, rises smoothly with increasing slope all the

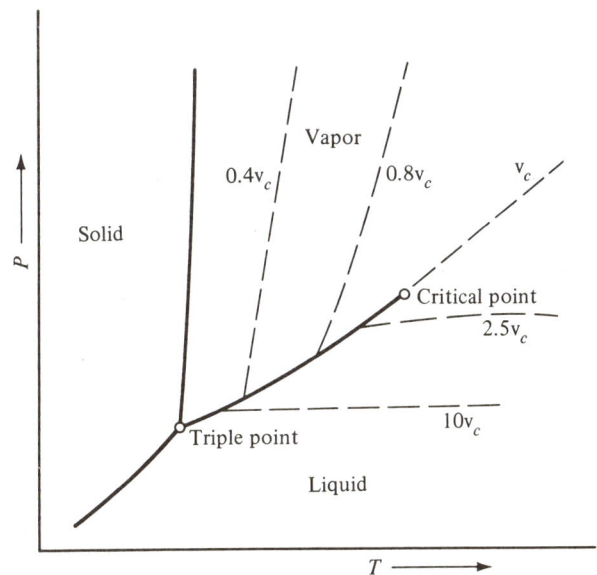

Fig. 8-3. *Pressure temperature diagram.*

way to the critical point. Within a few degrees of the critical point, the slope of the vapor pressure curve changes rapidly. Saturated liquid-vapor mixtures may exist anywhere on the vapor-pressure curve in all proportions from pure liquid to pure vapor. The dashed lines of constant volume shown in Fig. 8-3 are almost straight. In fact, Martin has stated that extremely close inspection of the data for most compounds shows these constant volume lines, or isochores as they are often called, are perfectly straight at very low pressures or very large volumes, and at all volumes at high temperatures.[21] The critical isochore is almost straight and its slope is the same as the slope of the vapor pressure curve at the critical point.

Isochores lying between the critical volume and extremely large volumes tend to curve slightly downward. This effect is especially pronounced in the region of about twice the critical volume. Isochores lying between the critical volume and approximately one-half to one-third the critical volume tend to curve upward just a little. This effect is most pronounced for volumes just a little smaller than the critical volume. Isochores for volumes less than one-third the critical volume tend to curve slightly downward. The curvature of the isochores is greatest at the low-temperature end of the vapor-pressure curve. At higher temperatures, all isochores tend to become perfectly straight.

Now, we find that the slope of the isochores predicted from the modified van der Waals equation of state is given by the relationship

$$\left(\frac{\partial P}{\partial T}\right)_V = \frac{R}{(v+c)-b}$$

Curves with a constant slope are straight lines. In order to account for curvature in the isochores, shown in Fig. 8-3, we need to further modify van der Waals' equation of state by assuming that a is a function of temperature. Martin has suggested that a vary linearly with temperature. If we use Martin's suggestion and apply the restriction that at the critical point, the first and second derivatives of pressure with respect to volume are zero, we obtain the reduced two constant Martin equation of state[20],

$$P_r = \frac{T_r}{z_c\left(v_r - \frac{0.085}{z_c}\right)} - \frac{9.0(4.0 - T_r)}{64 z_c^2 \left(v_r + \frac{0.04}{z_c}\right)^2} \tag{8-20}$$

A similar two constant equation of state is the Redlich-Kwong equation, which Martin shows has the reduced form,

$$P_r = \frac{T_r}{z_c\left(v_r - \frac{0.08664}{z_c}\right)} - \frac{0.42748}{T_r^{0.5} v_r z_c^2 \left(v_r + \frac{0.08664}{z_c}\right)} \tag{8-21}$$

The Redlich-Kwong equation of state is used widely in the following form:

$$P = \frac{RT}{v-b} - \frac{a}{T^{1/2} v(v+b)} \tag{8-22}$$

Using procedures similar to those used in developing Eqs. (8-4) and (8-5), Redlich and Kwong found that

$$a = 0.4278 \frac{R^2 T_c^{2.5}}{P_c} = A^2 R^2 T^{2.5} \tag{8-23}$$

and

$$b = 0.0867 \frac{RT_c}{P_c} = BRT \tag{8-24}$$

The Redlich and Kwong values of a and b differ from the van der Waals constants.

Another widely used equation of state is the Benedict-Webb-Rubin equation of state,

$$P = \frac{RT}{v} + \frac{(B_o RT - A_o - C_o/T^2)}{v^2} + \frac{bRT - a}{v^3} + \frac{a\alpha}{v^6}$$

$$+ \frac{c}{T^2 v^3}\left(1 + \frac{\gamma}{v^2}\right)\exp\left(\frac{\gamma}{v^2}\right) \tag{8-25}$$

This equation represents experimental PVT data from infinite volume down to a density of 1.8 times the critical density. Values of the constants for this equation for 39 different compounds may be found in Table 8-3A[22,23]. These constants have been obtained by fitting the Benedict-Webb-Rubin equation to experimental data. The magnitude of the individual constants depends not only on the particular analytical and statistical techniques used to reduce the data but also on the range of densities covered by the data. This is the reason that more than one set of constants have been tabulated for the same compound. When solving a problem requiring the use of an equation of state, it is generally advisable to use that set of constants that was derived from experimental data taken under conditions most nearly like the conditions of the problem. Table 8-3B may be used as a guide in selecting which constants to use for any particular problem involving the use of the Benedict-Webb-Rubin equation of state.

EXAMPLE PROBLEM 8-3. Repeat Example Prob. 8-1 using the van der Waals and the Redlich-Kwong equations.

Equations of state that are implicit in volume are frequently solved for volume on a digital computer by an iterative technique using the compressibility factor.

van der Waals' Equation

The van der Waals equation, Eq. (8-3), can be written in the form

$$z = \frac{v}{v - b} - \frac{a}{RTv}$$

where, From Eq. (8-9),

$$v = \frac{zRT}{P} \tag{8-9}$$

If we assume a value for z, we can calculate v from Eq. (8-9). This value may then be substituted into the first equation to give an improved estimate of z. For this case, from Eq. (8-4), we find

$$a = 20{,}727 \, \frac{(\text{lb}_f/\text{in.}^2)(\text{ft}^6)}{(\text{lb mole})^2}$$

TABLE 8-3A. Benedict-Webb-Rubin Equation of State Coefficients[22, 23]

	a	A_o	b	B_o	$c \times 10^{-6}$	$C_o \times 10^{-6}$	α	γ
Methane	.0494000	1.85500	.00338004	.0426000	.00254500	.0225700	.000124359	.0060000
Methane	.0435200	1.79894	.00252033	.0454625	.00358780	.0318382	.000330000	.0105000
Ethane	.345160	4.15556	.0111220	.0627724	.0327670	.179592	.000243389	.0118000
Propane	.947700	6.87225	.0225000	.0973130	.129000	.508256	.000607175	.0220000
n-Butane	1.88231	10.0847	.0399983	.124361	.316400	.992830	.00110132	.0340000
i-Butane	1.93763	10.23264	.0424352	.137544	.286010	.849943	.00107408	.0340000
n-Pentane	4.07480	12.1794	.0668120	.156751	.824170	2.12121	.00181000	.0475000
i-Pentane*	3.75620	12.7959	.0668120	.160053	.695000	1.74632	.00170000	.0463000
2,2 Dimethyl Propane	5.1429	13.470	.08333	.1584	.8333	1.4833	.001700	.0480
n-Hexane	3.4905	12.9635	.0668120	.170530	.546	1.273	.002	.05
n-Hexane†	7.11671	14.4373	.109131	.177813	1.51276	3.31935	.00281086	.0666849
n-Hexane†	5.36245	22.9802	.0964342	.338519	1.64757	1.64757	.00229048	.0546777
i-Hexane†	7.4286	14.930	.1215	.1729	.78015O410	2.8500	.0235	.0620
3-methyl pentane**	5.9716	12.203	.11224	.081505	.95556	2.2125	.00225	.062890
2,2 dimethyl butane**	10.108	11.842	.1400	.19214	1.7483	3.3595	.002189	.056500
2,3 dimethyl butane**	4.6956	16.430	.07900	.19000	1.1346	2.5534	.0035948	.075
n-heptane	10.36475	17.5206	.151954	.19900	2.47000	4.74574	.00435611	.0900000
n-heptane†	7.71375	1.74107	.164879	−.0510148	.360400	3.57721	.00226059	0
3 methyl hexane**	7.5854	14.310	.14321	.091423	1.3252	3.1564	.0028155	.07446
2,2 dimethyl pentane**	11.786	12.423	.17721	.20246	.2586	5.1237	.002764	.06799
n-nonane†	55.1599	−131.560	.856466	−2.32091	.781821	−3.20417	.00232899	0
n-decane†	125.122	−358.180	1.96701	−6.23189	.00442954	.131900	.00214459	0
methyl chloride	.521422468	2.20450849	.0108436	.00967625	.0921988450	.739067438	.00016866659	.0093
Perfluorocyclobutane	4.632380	6.869300	.095225	.109250	.5081500	1.403000	.0017570	.0530000
Tetrafluoromethane	.398530	1.86000	.0121147	.0552000	.0243300	.1545000	.0002500089	.0125000
Pentafluoromonochloroethane	3.3978459	1.9495299	.07351784	.01736622	.28297636	.96936284	.0073924	.02911417
Ethylene	.259000	3.33958	.0086000	.0556833	.021120	.131140	.000178000	.00093000
Ethylene	.2936449	2.1186242	.0100240	.24261342	.0910866	1.628585	.00013338	.0076587
Propylene	.774056	6.11220	.0187059	.0850647	.102611	.439182	.00455696	.0182900
Propylene	1.065493	4.01365	.0240732	.0467375	.116990	.556093	.00349627	.0162000
i-Butylene	1.69270	8.95325	.0348156	.116025	.274920	.927280	.000910889	.0295945
i-Butylene	1.682	9.060	.0348156	.116025	.274920	.927280	.000910889	.0295945
Propyne	.6970948	5.1079342	.01482999	.06946403	.10984375	.64062824	.00027363248	.01245167
Benzene	5.570	6.509772	.07663	.05030055	1.176418	3.42997	.0007001	.02930
Ammonia	.10354029	3.7892819	.00071958516	.051646121	.0015753298	.17857089	.000000465217790	.019805156
Argon	.000061797763	2.0259528	.0028513822	.0033649627	.0017232401	.060409764	.00016797770	.0066251015
Carbon Dioxide	.0288358	.823417	.00215289	.022282597	.0007982437	.01314125	.000003558895	.0023382711
Carbon Dioxide	.136814	2.73742	.00721045	.0499101	.0149180	.138567	.0000847	.005394
Carbon Dioxide	.136814	2.51606	.0041239	.0448849	.0149180	.147443	.00008464663197	.0052533158
Carbon Monoxide	.2091177	1.5348410	.0060337	2.0637953	.0158231	1.742216	.000044352	.0039018
Carbon Monoxide	.03665	1.34122	.00263158	.0545425	.001040	.00856209	.000135	.006
Carbon Monoxide	.03665	1.03115	.00263158	.0400	.001040	.011240	.000135	.006
Diborane	.08404260	8.041757	.01352699	.1395514	1.32183 × 10⁻¹⁰	.03089757	.0008614493	.0300
Diborane	2.816671	2.973888	.068512733	.01958304	−.06975970	.2019797	.003232267	.0300
Helium	.00057339	.040962	−.000000019727	.023661	−.000000005521	.00000016227	.0000002673	.00077942
Nitrogen	.025102	1.053642	.0023377	.0407426	.00072841	.00805900	.0001272	.005300
Nitrogen	.0312319	.872086	.0032351	.0281066	.000547364	.00781375	.0000709232	.0045
Nitric Oxide	.350821484	2.19573852	.00753154391	.060455814	.0115237289	.01795570089	.00001563696033	.002
Nitric Oxide	.234879752	4.344693	.0086235211388	.00976280227	.00467829761	.02744415181826	.00001246080662	.00195
Oxygen	.16774177	2.5441140	.0051001644	.0394584521	.015946	.14793968	.00026435918	.0048129414
Oxygen	.162689940	.950851963	.0035883473614	.0000003328505	.0128273741	.0326435918	−3.927058894	.0301
Sulfur Dioxide	.84468	2.12044	.014653	.026182	.11335	.79384	.000071955	.059236
Hydrogen	4.85937 × 10⁻⁵	.231182	6.14400 × 10⁻⁴	.026434	3.4200 × 10⁻⁶	37.6754 × 10⁻⁶	.002	.003

Metric Units

a (liters/g-mole)³ atm
A_o (liters/g-mole)² atm
b (liters/g-mole)²
B_o (liters/g-mole)
c (liters/g-mole)³ (K)² atm
C_o (liters/g-mole)² (K)² atm
α (liters/g-mole)³
γ (liters/g-mole)²

*These constants were derived from P-V-T data but are not necessarily the "best" fit. The constants were modified to fit a generalized equation in order to predict extrapolated constants for 2-methyl heptane and 2-methyl hexane.

†These constants were derived from P-V-T data by means of a conventional least squares technique. However, the constants were only evaluated for the following values of γ/V_c^2: 0, 0.4, and ∞. The value of γ/V_c^2 which produced the smallest relative error in predicting density is shown in the table.

**These constants were derived from P-V-T data but are not necessarily the "best" fit. The constants were used to produce a generalized correlation for paraffinic isomers.

Reproduced by permission from *Hydrocarbon Processing*, copyright 1967, Gulf Publishing Co., Houston, Texas

TABLE 8-3B. Range of Conditions[22, 23]

Substance	VAPOR PHASE				LIQUID PHASE			
	Density $\left(\frac{gm\ moles}{liter}\right)$		Temperature (°C)		Pressure, (atm)		Temperature, (°C)	
	From	To	From	To	From	To	From	To
Methane	2.0	18.0	−70	200	4.5	41.8	−140	−85
Methane	.75	12.5	0	350	Not Studied		Not Studied	
Ethane	.5	10.0	25	275	5.5	41.4	−50	25
Propane	1.0	9.0	96.8	275	2.0	28.0	−25	75
n-Butane	.5	7.0	150	300	1.2	22.5	4.0	121
i-Butane	.5	7.0	104.4	237.8	1.0	27.2	−12	119
n-Pentane	.46	4.8	140	280	2.1	25.5	60	180
i-Pentane	.5	5.0	130	280	.34	21.5	0	160
i-Pentane	1.5	5.5	200	300	Not recommended for liquid region			
2,2 Dimethyl Propane	1.0	6.0	160	275	Not Studied		Not Studied	
n-Hexane	2.5	5.0	225	275	1.0	23.8	70	220
n-Hexane		Not recommended for gas phase			0	670	38	238
i-Hexane	1.5	5.5	250	275	Not recommended for liquid region			
3-methane pentane	3.2	6.0	250	275	Not recommended for liquid region			
2,2 dimethyl butane	1.8	5.0	225	275	Not recommended for liquid region			
2,3 dimethyl butane	1.5	5.0	250	275	Not recommended for liquid region			
n-heptane	1.0	4.0	275	350	1.0	13.6	100	221
n-heptane		Not recommended for gas phase			0	27	177	267
3 methyl hexane	1.5	5.0	250	275	Not recommended for liquid region			
2,2 dimethyl pentane	1.8	5.0	225	275	Not recommended for liquid region			
n-nonane		Not recommended for gas phase			0	6.5	204	238
n-decane		Not recommended for gas phase			0	5.3	38	238
methyl chloride	.0847	4.0	35	225	8.013	102.71	35	140
Perfluorocyclobutane	.75	7.0	30	350	1.0	27.412	−6	115
Tetrafluoromethane	.75	12.5	0	350	Not Studied		Not Studied	
Pentafluoromonochloroethane	.64	10.5	−68	174	.03	27.	−76	67
Ethylene	1.0	12.8	0	198.5	2.1	31.9	−90	−10
Ethylene	.15	18.9	50	125	Not recommended for liquid region			
Propylene	.5	8.0	25	300	2.1	24.9	−30	60
Propylene	.5	9.5	100	250	Not Studied		Not Studied	
i-Butylene	1.0	7.0	150	275	1.0	27.2	−7	123
i-Butylene	1.0	8.0	150	250	Not given		Not given	
Propyne	.3	11.4	50	200	Not given		Not given	
Benzene	.6	8.1	240	355	Not given		Not given	
Ammonia	0	2.6	27	307	Not recommended for liquid region			
Ammonia		Not recommended for gas phase			20	800	37	127
Argon	.02	29.8	−111	327	Not recommended for liquid region			
Carbon Dioxide	0	14.5	Up to	138	0	66	−23	31
Carbon Dioxide	0	14.5	138	238	Not recommended for liquid region			
Carbon Dioxide	.15	18.9	50	125	Not recommended for liquid region			
Carbon Monoxide	.15	9.0	−140.2	−25	3.13	34.53	−180	−25
Carbon Monoxide	.09	1.8	−25	200	Not recommended for liquid region			
Diborane	.04	15.7	−93	27	1	39.5	−93	9
Diborane	.04	8.35	−93	27	Not recommended for liquid region			
Helium	2.0	50.0	−270	−253	0	100	−270	−267
Nitrogen	.02	24.34	−170	93	Not recommended for liquid region			
Nitrogen	0	22.2	−173	100	Not given		Not given	
Nitric Oxide	.04	1.6	5	105	Not given		Not given	
Nitric Oxide	0	10.0	27	105	Not given		Not given	
Nitrous Oxide	0	25.0	−30	150	Not recommended for liquid region			
Oxygen	0	2.4	27	727	Not given		Not given	
Sulfur Dioxide	0	22.8	10	250	Not recommended for liquid region			
Hydrogen		Not given			Not given			

Reproduced by permission from *Hydrocarbon Processing*, copyright 1967, Gulf Publishing Co., Houston, Texas

and from Eq. (8-5), we find

$$b = 1.04 \text{ ft}^3/\text{lb mole}$$

Then,

$$v = \frac{(10.73)(660)}{500} z = 14.164 z$$

and

$$z(v) = \frac{v}{v - 1.04} - \frac{20{,}727}{(10.73)(660)v}$$

$$= \frac{v}{v - 1.04} - \frac{2.927}{v}$$

The results of an iterative calculation are shown in the table that follows.

Trial	Assumed z	v = 14.164z	Calc. z(v)
1	1.0	14.164	0.872
2	0.872	12.35	0.855
3	0.855	12.11	0.852
4	0.852	12.06	0.852

An assumed initial value of 1.0 for z is usually satisfactory for gases. Other techniques are generally required to solve for liquid volumes. Even though an equation of state can be solved for liquid volumes, it should be pointed out that the liquid state is not in the range of validity for most equations of state.

Using the z factor calculated using the van der Waals equation, we find, from Eq. (8-9),

$$v = \frac{(0.852)(10.73)(660)}{(30)(500)}$$

$$= 0.4022 \text{ ft}^3/\text{lb}_m$$

The percent error, then is

$$\text{error} = \frac{0.4048 - 0.4022}{0.4048} \times 100$$

$$= 0.6 \text{ percent}$$

Redlich-Kwong equation

The Redlich-Kwong equation, Eq. (8-22), may be arranged in the form:

$$\frac{Pv}{RT} = \frac{v}{v - BRT} - \frac{A^2 RT}{v + BRT}$$

where A is defined by Eq. (8-23) and B is defined by Eq. (8-24). If we further define

$$h = \frac{BRT}{v} = \frac{BP}{z}$$

The Redlich-Kwong equation becomes

$$z = \frac{1}{1-h} - \frac{A^2}{B}\left(\frac{h}{1+h}\right)$$

Now,

$$A^2 = \frac{a}{R^2 T^{2.5}} \tag{8-23}$$

where

$$a = 0.4278 \frac{R^2 T_c^{2.5}}{P_c}$$

and

$$B = \frac{b}{RT} \tag{8-24}$$

where

$$b = 0.0867 \frac{RT_c}{P_c}$$

Then,

$$A^2 = 5.62 \times 10^{-3} \text{ atm}^{-1}$$

and

$$B = 1.50 \times 10^{-3} \text{ atm}^{-1}$$

so that

$$h = \frac{(1.50) \times 10^{-3}(500/14.7)}{z} = \frac{0.051}{z}$$

and

$$z(h) = \frac{1}{1-h} - 3.747 \frac{h}{1+h}$$

Again, solving for z iteratively, as would be done on a digital computer, we find on tabulating our results,

Trial	Assumed z	h	Calc. z(h)
1	1.0	0.051	0.872
2	0.872	0.058	0.856
3	0.856	0.0596	0.852
4	0.852	0.0599	0.852

Then, we find from Eq. (8-9),

$$v = \frac{zRT}{P}$$

$$= \frac{(0.852)(10.73)(660)}{(30)(500)}$$

$$= 0.4022 \text{ ft}^3/\text{lb}_m$$

The percent error using the Redlich-Kwong equation is

$$\text{error} = \frac{0.4048 - 0.4022}{0.4048} \times 100$$

$$= 0.6 \text{ percent}$$

An equation of state which is based on molecular concepts is the virial equation of Clausius. The virial equation of state may be written

$$\frac{Pv}{RT} = 1 + \frac{B(T)}{v} + \frac{C(T)}{v^2} + \ldots \quad (8\text{-}26)$$

where

> $B(T)$ = second virial coefficient and accounts for interactions between pairs of molecules

and

> $C(T)$ = third virial coefficient and accounts for interactions between triplets of molecules.

Although the form of this equation is based on molecular considerations, the constants may be determined from experimental data. Multiplying both sides of Eq. (8-26) by RT/P, we find

$$v = \frac{RT}{P} + \frac{RTB}{Pv} + \frac{RTC}{Pv}\rho$$

where the molar density ρ is

$$\rho = \frac{1}{v}$$

Then, since

$$\alpha_R = \frac{RT}{P} - v \qquad (8\text{-}1)$$

and

$$\frac{Pv}{RT} = z \qquad (8\text{-}9)$$

we find

$$-\alpha_R = \frac{B}{z} + \frac{C}{z}\rho \qquad (8\text{-}27)$$

At low pressure, the value of z approaches unity, and

$$-\alpha_R = B + C\rho \qquad (8\text{-}28)$$

In this case, a plot of the negative of the residual volume α_R as a function of density will be a straight line having a slope of C and an intercept of B. Values of these virial coefficients are available in the literature.

M. L. McGlashan[24] examined the data for methane, argon, krypton, and xenon and found the second virial coefficient of these gases could be represented as a function of temperature using reduced temperature and volume as parameters in the form

$$\frac{B(T)}{v_c} = 0.430 - 0.886\left(\frac{T_c}{T}\right) - 0.694\left(\frac{T_c}{T}\right)^2 \qquad (8\text{-}29)$$

A similar reduced correlation for argon, krypton, and xenon was prepared by K. S. Pitzer[25]

$$\frac{B(T)P_c}{RT_c} = 0.1445 - \frac{0.330}{T_r} - \frac{0.1385}{T_r^2} - \frac{0.0121}{T_r^2} \qquad (8\text{-}30)$$

Recently, similar correlations for the third virial coefficients of several components have been presented.[26]

At low densities, deviations from ideality are adequately described by the second virial coefficient, whereas at higher densities, more virial coefficients must be used. At about the density of liquid, the virial series expansion diverges. Hence, the primary application of the virial equation of state is in the study of gases at low to moderate densities.

EXAMPLE PROBLEM 8-4. Repeat Example Prob. 8-1 using the virial equation of state through the second virial coefficient. At 200 °F, the second virial coefficient of ethane is -1.89 ft^3/lb mole.

The virial equation through the second virial,

$$\frac{Pv}{RT} = 1 + \frac{B(T)}{v}$$

can be rearranged to give

$$z = 1 - \frac{1.89}{v}$$

But

$$v = \frac{zRT}{P}$$

$$= \frac{(10.73)(660)}{500} z$$

$$= 14.164 z$$

Then,

$$z = 1 - \frac{1.89}{14.164z}$$

Solving for z, we find

$$z^2 - z + 0.1334 = 0$$

so that

$$z = \frac{1 \pm \sqrt{1 - 0.5336}}{2}$$

$$= 0.8415$$

Then,

$$v = \frac{(14.164)(0.8415)}{(30)}$$

$$= 0.3973 \text{ ft}^3/\text{lb}_m$$

The percent error, then, is

$$\text{error} = \frac{0.4048 - 0.3973}{0.4048} \times 100$$

$$= 1.9 \text{ percent}$$

Somewhat better results would be obtained if we had included the third virial coefficient.

This problem brings out an interesting point, however. There are two values of z to satisfy our requirements: $z = 0.8415$, and $z = 0.1585$. The unfortunate fact is that all equations of state for real gases have multiple roots. The method of successive substitution used in Example Prob. 8-3 will generally converge to the root nearest the initial assumed value of z. Thus, the engineer must use judgement in selecting the proper root. One possible procedure might be to check the final answer from any equation of state using Fig. A-1.

EQUATIONS OF STATE FOR LIQUIDS AND SOLIDS

As attempts have been made to develop equations of state to represent PVT behavior within experimental precision over wider and wider ranges, the complexity of the equations has increased markedly. Whereas the ideal gas law satisfactorily represents phase behavior at very low densities, Benedict, Webb and Rubin found it necessary to use eight constants to represent the behavior of hydrocarbon gases over a range extending from low densities to a density of 1.8 times the critical density.

Fortunately, it seems that the problem is not so much that of representing *PVT* behavior of a liquid or solid component at high densities. Rather, the problem lies in attempting to develop a single equation that is valid for all densities. Thus, rather than attempting to develop an analytical expression which simultaneously represents both the behavior of low density and high density materials, we will find it convenient when studying liquids and solids to consider an equation of state developed for use primarily in limited regions of the phase diagram.

Since specific volume is a function of temperature and pressure, we may write

$$dv = \left(\frac{\partial v}{\partial P}\right)_T dP + \left(\frac{\partial v}{\partial T}\right)_P dT \tag{4-9}$$

The partial derivatives in this equation can be evaluated from experimental *PVT* measurements.

Two parameters have been found useful in the integration of Eq. (4-9). The first is the coefficient of isothermal compressibility κ where

$$\kappa_T = -\frac{1}{v}\left(\frac{\partial v}{\partial P}\right)_T \tag{8-31}$$

and the second is the coefficient of thermal expansion β where

$$\beta = \frac{1}{v}\left(\frac{\partial v}{\partial T}\right)_P \tag{8-32}$$

Then, the equation of state for a liquid or solid takes the form

$$\frac{dv}{v} = \beta \, dT - \kappa_T dP \tag{8-33}$$

Experimental data shows us that the coefficients of isothermal compressibility and thermal expansion of liquids and solids are relatively small quantities which vary both with temperature and pressure. In general, the effects of temperature are somewhat greater than the effects of pressure.

The effect of temperature on the coefficient of isothermal compressibility of liquid water at one atmosphere is shown in Table 8-4.

TABLE 8-4. Effect of Temperature on the Coefficient of Isothermal Compressibility of Liquid Water

$T(°C)$	$\kappa_T \times 10^6 (atm^{-1})$
0	51.1
20	46.4
40	44.7
60	44.9
80	46.3

The effect of pressure on the coefficient of isothermal compressibility of liquid water at 20 °C is shown in Table 8-5.

TABLE 8-5. Effect of Pressure on the Coefficient of Isothermal Compressibility of Liquid Water

$P(atm)$	$\kappa_T \times 10^6 (atm^{-1})$
1	46.4
200	43.5
400	41.5
500	39.5

Over a wide range of conditions, the coefficient of isothermal compressibility for a liquid is well approximated by the Tait equation,

$$\kappa_T = \frac{C}{L + P} \qquad (8\text{-}34)$$

where C is a dimensionless constant for a given substance, and L is a function of temperature. For liquid water, C is 0.1368, and L is approximately 3000 atmospheres.

The effect of temperature on the coefficient of thermal expansion of liquid water is shown in Table 8-6.

TABLE 8-6. Effect of Temperature on the Coefficient of Thermal Expansion of Liquid Water

$T(°C)$	$\beta \times 10^6 (°C^{-1})$
0	−67
20	208
40	390
60	522
80	643

Although κ and β are functions of pressure and temperature, they may be assumed to be constants over limited ranges. In fact, it is common practice to evaluate the effect of pressure and temperature on the specific volumes of high density materials by using average values of κ and β as constants. The average value of κ is defined by the relationship

$$\kappa_{avg} = -\frac{1}{v^\circ}\left(\frac{v - v^\circ}{P - P^\circ}\right) \quad (8\text{-}35)$$

and the average value of β is defined by the relationship

$$\beta_{avg} = \frac{1}{v^\circ}\left(\frac{v - v^\circ}{T - T^\circ}\right) \quad (8\text{-}36)$$

where v° is the specific volume at a reference pressure P° and a reference temperature T°.

EXAMPLE PROBLEM 8-5. The specific volume of saturated liquid water at 20 °C is 0.01603 ft^3/lb$_m$. Estimate the specific volume of liquid water at 20 °C and 100 atm, and compare the result with a volume of 0.01599 ft^3/lb$_m$ reported in the literature[2].

We will use the relationship

$$v = v^\circ [1 - \kappa_{avg}(P - P^\circ)] \quad (8\text{-}35)$$

From Table 8-5, we find

$$\kappa_T = 46.4 \times 10^{-6} \text{ atm}^{-1} \text{ at 1 atm}$$

$$\kappa_T = 44.95 \times 10^{-6} \text{ atm}^{-1} \text{ at 100 atm}$$

since κ varies with pressure, we must use an average value over the pressure range of interest. Thus,

$$\kappa_{avg} = 45.68 \times 10^{-6} \text{ atm}^{-1}$$

Substituting this value into Eq. (8-35) and noting that the vapor pressure of water at 20 °C is essentially zero, we find

$$v = 0.01603 [1 - 45.68 \times 10^{-6}(100 - 0)]$$

$$= 0.01596 \text{ ft}^3/\text{lb}_m$$

The percent error between this value and the value reported in the literature[2] is

$$\text{error} = \frac{0.01596 - 0.01599}{0.01599} \times 100$$

$$= -0.2 \text{ percent}$$

Frequently, we encounter problems in which the change of length of a material with temperature is of more interest than the change in volume with temperature. Suppose we have a rectangular solid with three independent dimensions L_x, L_y, and L_z. Then, the volume of the solid is

$$V = L_x L_y L_z$$

From this relationship, we find that

$$\left(\frac{\partial V}{\partial T}\right)_P = L_y L_z \left(\frac{\partial L_x}{\partial T}\right)_P + L_x L_z \left(\frac{\partial L_y}{\partial T}\right)_P + L_x L_y \left(\frac{\partial L_z}{\partial T}\right)_P$$

or

$$\frac{1}{V}\left(\frac{\partial V}{\partial T}\right)_P = \frac{1}{L_x}\left(\frac{\partial L_x}{\partial T}\right)_P + \frac{1}{L_y}\left(\frac{\partial L_y}{\partial T}\right)_P + \frac{1}{L_z}\left(\frac{\partial L_z}{\partial T}\right)_P \quad (8\text{-}37)$$

If we define a quantity α_τ called the coefficient of linear expansion such that

$$\alpha_\tau = \frac{1}{L}\left(\frac{\partial L}{\partial T}\right)_\tau \quad (8\text{-}38)$$

where we have replaced the restriction of constant pressure, a three-dimensional concept, with the restriction of constant tension, a one-dimensional concept, Eq. (8-37) becomes

$$\beta = \alpha_{\tau_x} + \alpha_{\tau_y} + \alpha_{\tau_z} \quad (8\text{-}39)$$

If the properties of the solid are independent of direction,

$$\beta = 3\alpha_\tau \quad (8\text{-}40)$$

Representative values of the coefficient of linear expansion for solids at 20 °C are presented in Table 8-7.

TABLE 8-7. Coefficients of Linear Expansion

Compound	$\alpha_T \times 10^6 \, (°C^{-1})$
Copper	16.2
Iron, cast	11.8
Iron, wrought	11.9
Steel	11.4
Tin	21.4

PROBLEMS

8-1. Show that for the van der Waals equation

$$a = \frac{27}{64} R^2 \frac{T_c^2}{P_c}$$

and

$$b = \frac{RT_c}{8P_c}$$

8-2. Using constants in terms of P_c and T_c, show that for any gas which obeys the van der Waals equation

$$z = 1 + \frac{1}{8} \frac{P_r}{T_r} - \frac{27}{64} \frac{P_r}{T_r^2 z} + \frac{27}{512} \frac{P_r^2}{T_r^3 z^2}$$

This equation is an analytic representation of the Theorem of Corresponding States.

8-3. The following data is available from the literature:

Compound	Molecular Weight	Boiling Point, (°R)	Critical Pressure ($lb_f/in.^2$)	Critical Temperature (°R)	Critical Volume (ft^3/lb_m)
n-Butane	58	490.7	551.1	765.4	0.0704
Benzene	78	635.6	714.2	1011.8	0.0530
Ethyl alcohol	46	632.3	929.3	927.3	0.0582
Freon-12	121	438.3	596.9	693.6	0.0287

Calculate the critical properties of these compounds using Lydersen's parameters and compare your results with these data from the literature. Also, calculate the critical compressibility factor for these compounds using the data from the literature.

8-4. Using the van der Waals equation and the Redlich-Kwong equation, estimate the pressure required to compress two gram moles of nitrogen into a volume of 0.1 ft^3 at 70°F.

8-5. Using the reduced two constant Martin equation, estimate the specific volume of ethane ($MW = 30$) at 500 lb$_f$/in.^{2} and 200°F. Compare the results with a value of 0.4048 ft^3/lb$_m$ reported in the literature[4].

8-6. The volume occupied by 1 pound of n-octane ($MW = 114$) at 27 atm is 0.20 cu ft. Using the equation

$$PV = znRT$$

and the generalized compressibility factor chart, find the temperature.

8-7. One pound of carbon dioxide ($MW = 44$) occupies 0.6 ft^3 at a pressure of 200 lb$_f$/in.^{2} Using the generalized two constant Callendar equation, estimate the temperature of the system.

8-8. Representative values of the volumetric properties of saturated methane are as follows[4]:

P(lb$_f$/in.^{2})	T(°F)	v$_f$(ft^3/lb$_m$)	v$_g$(ft^3/lb$_m$)
14.7	−258.7	0.03775	8.7050
101.0	−205.0	0.04248	1.4500
214.9	−175.0	0.04565	0.6818
297.0	−160.0	0.04944	0.4814
401.0	−145.0	0.05347	0.3382
482.0	−135.0	0.05735	0.2672
575.0	−125.0	0.06362	0.1989
673.1	−116.5	0.09901	0.0990

On a graph representing the two phase volumetric behavior of methane, plot isotherms at $T_r = 0.8$, $T_r = 1.0$, and $T_r = 1.2$ from a specific volume of 0.01 ft^3/lb$_m$ to a specific volume of 1.0 ft^3/lb$_m$ obtained from

(a) The van der Waals equation.

(b) The Redlich-Kwong equation.

Hint: The use of the digital computer may be of value in the solution of this problem.

8-9. Using the Benedict-Webb-Rubin equation of state, estimate the specific volume of ethane ($MW = 30$) at 500 lb_f/in^2 and 200 °F. Compare the result with a value of 0.4048 ft^3/lb_m reported in the literature[4].

Hint: This problem might be more easily solved if the digital computer is used.

8-10. The specific heat at constant pressure c_p of saturated liquid water at 60 °F is 1.0 BTU/lb_m °R. What is the specific heat at constant volume c_V of saturated liquid water at 60 °F in BTU/lb_m °R?

8-11. The volume of one gram of mercury at one atmosphere is given as a function of temperature in the following table:

T(°C)	v(cc/gm)
−10	0.0734205
0	0.0735540
10	0.0736877
20	0.0738215
30	0.0739552
40	0.0740891

Estimate the coefficient of thermal expansion of mercury in °C^{-1} as a function of temperature.

8-12. The pressure on a cube of copper ($MW = 63.5$) having a mass of 1.0 kg is increased reversibly and isothermally at 300 K from 1.0 to 1000 atm. Determine the work done and the heat transferred from the copper, assuming that over the range of pressures involved in this problem,

$$v = 7.062 \text{ cm}^3/\text{gm mole}$$

$$\beta = 49.2 \times 10^{-6} \text{ K}^{-1}$$

$$\kappa_T = 77.6 \times 10^{-14} \text{ cm}^2/\text{dyne}$$

9

Evaluation of Thermodynamic Functions

OBJECTIVES

After studying this chapter, the student should be able to

(1) apply equations of state in either analytical, tabular or graphical form for the solution of new problems involving the application of the principle of conservation of mass, the principle of conservation of energy as well as the second law of thermodynamics to single component systems; and

(2) define and estimate numerical values of the fugacity for a single component system.

GRAPHICAL TECHNIQUES

Differentiation

We have indicated that thermodynamic functions such as u, h, and s have been introduced for convenience in analyzing processes. Unfortunately, these functions can not be measured directly, but must be estimated from other measurable properties of the system. For example, we have shown that the differential relationship between u, T and v is

$$du = \left[T\left(\frac{\partial P}{\partial T}\right)_V - P\right]dv + c_V\, dT \qquad (4\text{-}39)$$

This equation can be integrated to yield a value of u relative to the value of u° in some reference state. However, before we can integrate we must evaluate the coefficient of the derivative dv as a function of v and the coefficient of the derivative dT as a function of T.

If we are satisfied that one of the equations of state discussed in the last chapter adequately represents the PVT behavior of the fluid in the specific volume and temperature range of interest, we can use it to evaluate these coefficients. For example, if the behavior of a gas is satisfactorily represented by van der Waals' equation of state,

$$P = \frac{RT}{v-b} - \frac{a}{v^2} \tag{8-3}$$

then

$$T\left(\frac{\partial P}{\partial T}\right)_V = \frac{RT}{v-b}$$

$$= P + \frac{a}{v^2}$$

and

$$T\left(\frac{\partial P}{\partial T}\right)_V - P = \frac{a}{v^2}$$

Substituting this result into Eq. (4-39), we find that at constant temperature,

$$\Delta u_T = \int_{v_1}^{v_2} \frac{a}{v^2} dv$$

On the other hand, it is also possible to evaluate functions such as

$$T\left(\frac{\partial P}{\partial T}\right)_V - P$$

directly from experimental measurements. As we have seen it is possible to measure T with a thermometer, P with a pressure gage, and V with a yardstick. However, we cannot directly measure $(\partial P/\partial T)_V$, but we must estimate it some other way.

One method we might use is to find an analytical relationship between the variables of interest. Let's assume that we have measured P over a limited range at various values of v at constant temperature. From our knowledge of *PVT* behavior in general, we anticipate that an equation of state such as

$$Pv^m = A$$

where m and A are constants, will adequately represent the data. In order to check this assumption, we can linearize the assumed relationship by writing

$$\ln P = -m\ln v + \ln A$$

Here, we have rewritten the assumed equation in a form which can be plotted as a straight line. If a plot of ln *P* as a function of ln v falls on a straight line on regular coordinate paper, with a slope of $-m$ and an intercept of $\ln A$, then the equation of state we assumed may be satisfactory.

At this point, we might ask: What are the "best" values of the slope m, and the intercept $\ln A$? Frequently, these can be determined with sufficient accuracy by using a straight edge to draw a line by eye through the points. However, a more rigorous way to determine the "best" values of the slope and intercept would be to find that line from which the difference or deviations between the points and the line are a minimum. Now the general equation for a straight line is

$$y = mx + b$$

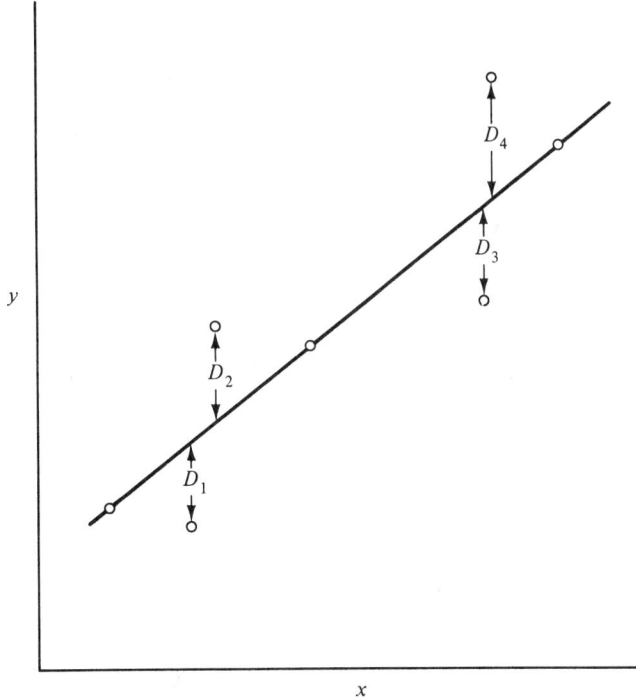

Fig. 9-1. *y as a function of x.*

If y is a measured value, and $mx + b$ is the value of y predicted by the fitted equation, then the deviation D between the measured and predicted values of y, as shown in Fig. 9-1, is

$$D_i = y_i - mx_i - b$$

Then, the problem of finding the "best" values of m and b becomes one of minimizing the sum of the deviations for all points. Now, some deviations will be positive and some deviations will be negative. If we added positive and negative deviations, they might cancel each other.

It has been found better first to square all the deviations and then minimize the sum of the squares of the deviations,

$$\Sigma D_i^2 = \Sigma(y_i - mx_i - b)^2$$

Now, the only variables that affect D^2 are m and b. In order that D^2 be a minimum, we require that

$$\left(\frac{\partial \Sigma D_i^2}{\partial m}\right)_b = 0$$

and

$$\left(\frac{\partial \Sigma D_i^2}{\partial b}\right)_m = 0$$

That is,

$$m\Sigma x_i + nb = \Sigma y_i \tag{9-1}$$

and

$$m\Sigma x_i^2 + b\Sigma x_i = \Sigma x_i y_i \tag{9-2}$$

where n is the number of data points. By using the experimental data to evaluate Σx_i, Σy_i, Σx_i^2, and $\Sigma x_i y_i$, we can solve these equations for the best values of m and b. This method is called the method of least squares. Once we have the desired analytical relationship between x and y, we can then evaluate the desired derivative.

EXAMPLE PROBLEM 9-1. Find the derivative of P with respect to T from the following experimental data taken at constant specific volume.

P	T
0	0
2	1
4	2
8	4

A plot of P versus T on regular coordinate paper suggests that

$$P = mT + b$$

will adequately represent the data. Thus, we must solve the equations

$$m\Sigma T_i + nb = \Sigma P_i$$

and

$$m\Sigma T_i^2 + b\Sigma T_i = \Sigma T_i P_i$$

for m and b. Now, from the data given, we find

	P	T	PT	T^2
	0	0	0	0
	2	1	2	1
	4	2	8	4
	8	4	32	16
Totals	14	7	42	21

Thus, Eq. (9-1) gives us $7m + 4b = 14$ where the number of data points n is 4; and Eq. (9-2) gives us $21m + 7b = 42$. Solving these equations simultaneously, we find $m = 2$ and $b = 0$. Thus, the desired analytical relationship is

$$P = 2T$$

Finally, the desired derivative is

$$\frac{dP}{dT} = 2$$

A problem which arises when we fit experimental data with an analytical expression by the method of least squares is that we have no way of telling whether or not the analytic function we have used is the correct expression. Another problem is that this method, if used without due caution, gives equal weight to all points and does not take any bias in the data into account.

In general, it is better to obtain the required derivatives directly from the data. As a first approximation, one might plot y versus x on regular coordinate graph paper. Then, by laying a straight edge along the data points, we can estimate the slope of the curve. This method, however, is not very reliable.

A more exact method of finding a derivative is called the chord area method of differentiation. This method might best be understood by consideration of the following example problem.

EXAMPLE PROBLEM 9-2. Find the value of the derivative $(\partial P/\partial v)_T$ for steam at a temperature of 1000 °F and a specific volume of 14.0 ft³/lb$_m$ using data from the steam tables in Table A-2.

Values of specific volume as a function of pressure at 1000 °F are given in Table A-2 in the Appendix. From these values, we must first find the average rate of change of pressure between intervals of volume $\Delta P/\Delta v$ as follows:

$P(lb_f/in.^2)$	$v(ft^3/lb_m)$	ΔP	Δv	$(\Delta P/\Delta v)$
50	17.35			
		5	−1.58	−3.16
55	15.77			
		5	−1.32	−3.79
60	14.45			
		5	−1.11	−4.50
65	13.34			
		5	−0.96	−5.21
70	12.38			
		5	−0.83	−6.02
75	11.55			

If we now plot $\Delta P/\Delta v$ on the ordinate and v on the abscissa on Fig. 9-2, we will obtain a series of rectangles and the area under the rectangles will be ΔP.

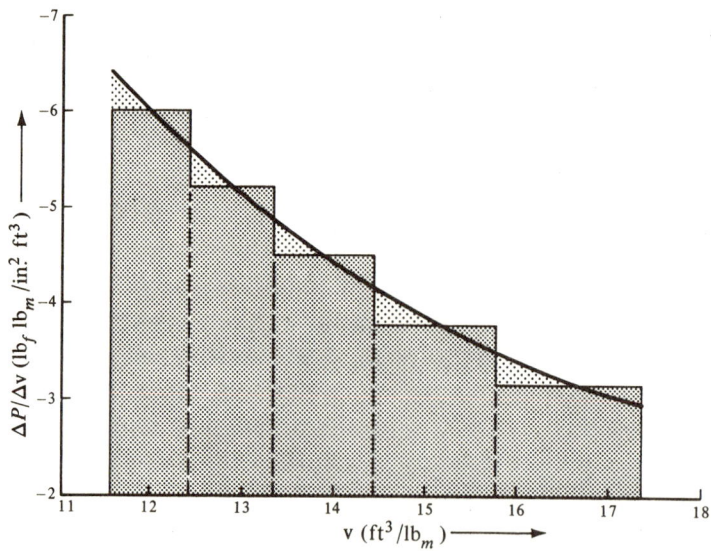

Fig. 9-2. *Chord area method of graphical differentiation.*

In order to obtain the local rate of change of pressure with respect to specific volume, we must now draw a smooth curve through the rectangles such that the area under the smooth curve is equal to the area under the rectangles. Then, the desired derivative, $(\partial P/\partial v)_T$ at $v = 14$ ft^3/lb$_m$ can be read directly from the smooth curve and is equal to -4.4 (lb$_f$ lb$_m$)/(in.^{2} ft^3).

The location of the smoothed curve plotted in this manner can be adjusted such that the area under a segment of the curve connecting any two points on the curve must be equal to the change in P associated with that particular change in v. We also could have plotted $(\partial v/\partial P)_T$ as a function of P. The area under any segment on this curve is Δv. However, if we had plotted $(\partial P/\partial v)_T$ as a function of P or $(\partial v/\partial P)_T$ as a function of v, the area under the curve would have had no physical counterpart.

Integration

Once we have evaluated the coefficients of the derivatives in an expression such as

$$du = \left[T\left(\frac{\partial P}{\partial T}\right)_V - P\right]dv + c_V\,dT \qquad (4\text{-}39)$$

using experimental data, the next step required to find Δu is integration. We have demonstrated in Chap. 4 how to integrate Eq. (4-39) for the case in which the coefficients of dv and dT were in the form of analytical expressions obtained by differentiating the ideal gas equation of state. However, if the coefficients are given as a plotted curve or in tabular form, we must integrate Eq. (4-39) graphically.

Many methods are available for graphical integration. Consider the problem of evaluating the integral,

$$\int_{x_1}^{x_2} y\,dx$$

The simplest method of evaluating this integral graphically is called the trapezoidal method. In this case, a series of trapezoids are superimposed on a graphical representation of the function that is to be integrated as shown in Fig. 9-3(a). Then, the value of the integral, which is numerically equal to the area under the curve, is assumed to be equal to the sum of the areas of the trapezoids. The smaller the base of the trapezoid, the closer the answer obtained in this manner will be to the true value of the integral.

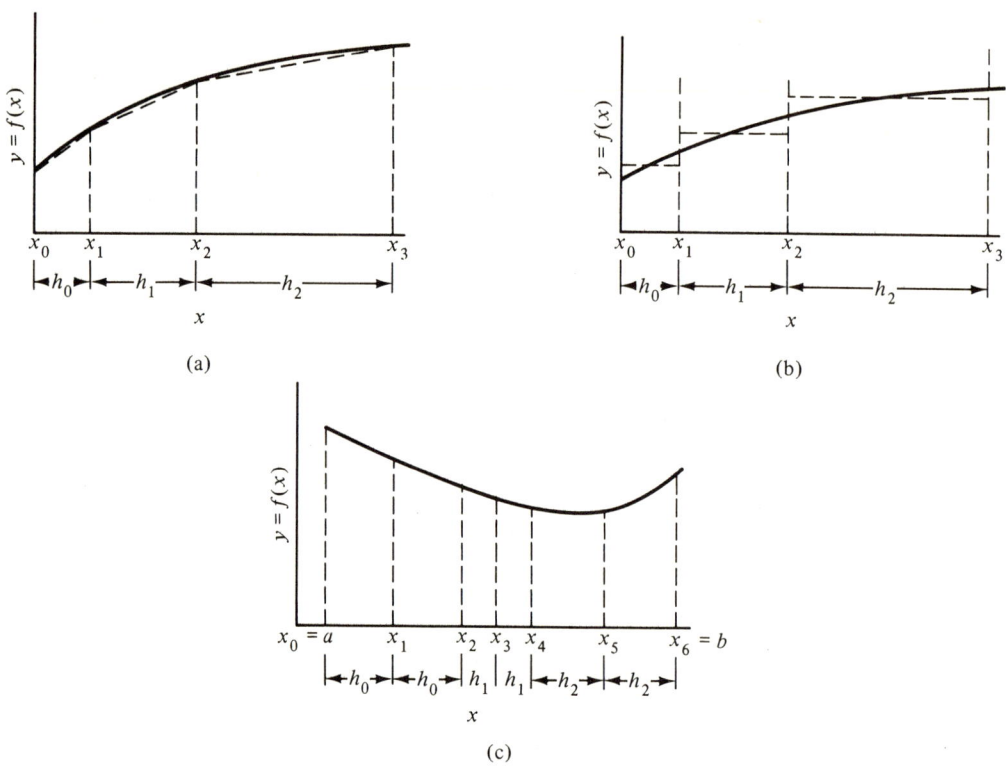

Fig. 9-3. *Graphical integration.*

In superimposing trapezoids over the area whose value it is desired to determine, the curve between any two adjacent points is replaced by the straight line, $y = mx + b$. Further, $x_2 = x_1 + h_1$ where h_1 is the width of the trapezoid between x_1 and x_2. Now, $y(x)$ has been replaced by an analytically integrable function. Proceeding formally, we find

$$\int_{x_1}^{x_2} y(x)\, dx = \tfrac{1}{2}m(x_2^2 - x_1^2) + b(x_2 - x_1)$$

$$= \tfrac{1}{2}m(x_1^2 + 2x_1 h_1 + h_1^2 - x_1^2) + b(x_1 + h_1 - x_1)$$

$$= bh_1 + \tfrac{1}{2}m\,(2x_1 h_1 + h_1^2)$$

Now, from the relationships

$$y_1 = mx_1 + b$$

and

$$y_2 = mx_1 + mh_1 + b$$

we find

$$m = \frac{y_2 - y_1}{h_1}$$

and

$$b = y_1 - \left(\frac{y_2 - y_1}{h_1}\right) x_1$$

Substituting these values into the integrated equation, the value of the integral between x_1 and x_2 is

$$\int_{x_1}^{x_2} f(x)\, dx = \frac{1}{2} h_1 (y_1 + y_2)$$

This is the equation for the area of the trapezoid between x_1 and x_2.

Finally, the total value of the integral from x_0 to x_n is found by adding the areas of all the trapezoids,

$$\int_{x_0}^{x_n} f(x)\, dx = \frac{1}{2} \sum_i h_i (y_i + y_{i+1}) \qquad (9\text{-}3)$$

If the bases of all trapezoids are of equal width,

$$\int_{x_0}^{x_n} f(x)\, dx = \frac{h}{2}(y_0 + 2y_1 + 2y_2 + \ldots + 2y_{n-1} + y_n) \qquad (9\text{-}4)$$

The same result can be obtained by superimposing rectangles over the graphical representation of the function that is to be integrated, as shown in Fig. 9-3(b), since the area of each rectangle is

$$A_i = y_{\text{avg}} h_i$$

where

$$y_{\text{avg}} = \frac{y_i + y_{i+1}}{2}$$

Frequently, more reliable results can be obtained by using Simpson's rule to numerically integrate the data. Let us divide the interval $a - b$, shown in Fig. 9-3(c), into an even number $2n$ of subintervals at

$$x_0 = a, x_1, x_2, x_3, \ldots, x_{2n} = b$$

such that the first two subintervals are equal, the next two are equal, ... Now, if the curve $y(x)$ between a and x_2 is replaced by the equation

$$y = Ax^2 + Bx + C$$

we can estimate the area under the curve between a and x_2 using a technique similar to that used in establishing the trapezoidal rule. If this procedure is repeated over a set of subintervals in which all subintervals are of equal width and the areas added, we find

$$\int_a^b f(x)\, dx = \frac{h}{3}(y_0 + 4y_1 + 2y_2 + 4y_3 + \ldots$$

$$+ 2y_{2n-2} + 4y_{2n-1} + y_{2n}) \qquad (9\text{-}5)$$

where

$$h = \frac{b-a}{2n}$$

EXAMPLE PROBLEM 9-3. Find the value of the integral

$$\int_{17.35}^{11.55} \left(\frac{\partial P}{\partial v}\right)_T dv$$

for steam at 1000 °F using data from Example Prob. 9-2 and Simpson's rule.

Let us divide the interval of integration into four intervals so that $2n = 4$ or $n = 2$.

Then

$$h = \Delta v = \frac{b-a}{2n}$$

$$= \frac{11.55 - 17.35}{2 \times 2}$$

$$= -1.45$$

From Fig. 9-2, we obtain the following data:

$v(ft^3/lb_m)$	$\left(\dfrac{\partial P}{\partial v}\right)_T \left(\dfrac{lb_f \cdot lb_m}{ft^3/in.^2}\right)$
11.55	−6.50
13.00	−5.08
14.45	−4.14
15.90	−3.42
17.35	−2.95

Then, from Simpson's rule,

$$\int_{17.35}^{11.55} \left(\frac{\partial P}{\partial v}\right)_T dv_T = \frac{h}{3}(y_0 + 4y_1 + 2y_2 + 4y_3 + y_4)$$

$$= \frac{-1.45}{3}(-6.50 + 4(-5.08) + 2(-4.14)$$

$$+ 4(-3.42) - 2.95)$$

$$= \frac{-1.45 \times -51.73}{3}$$

$$= 25.00 \; lb_f/in.^2$$

We can check this result by noting that

$$\int_{17.35}^{11.55} \left(\frac{\partial P}{\partial v}\right)_T dv_T = \int_{50}^{75} dP_T$$

Since P is a point function,

$$\int_{50}^{75} dP_T = 75 - 50$$

$$= 25 \; lb_f/in.^2$$

Ordinarily, we wouldn't expect such an exact agreement.

GENERALIZED CHARTS

The procedure for evaluating a relationship such as

$$dh = \left[v - T\left(\frac{\partial v}{\partial T}\right)_P\right]dP + c_P\, dT$$

can be greatly facilitated by using an equation of state. For example, let us assume

$$Pv = zRT$$

where

$$z = z(P_r, T_r, \omega)$$

adequately represents the *PVT* behavior of the gas. Any other equation of state that adequately represents the phase behavior of the gas could just as easily be used. Then,

$$\left(\frac{\partial v}{\partial T}\right)_P = \frac{zR}{P} + \frac{RT}{P}\left(\frac{\partial z}{\partial T}\right)_P$$

since z is a function of P and T. Thus,

$$dh_T = \left[\frac{zRT}{P} - \frac{zRT}{P} - \frac{RT^2}{P}\left(\frac{\partial z}{\partial T}\right)_P\right]dP_T$$

or

$$dh_T = -\frac{RT^2}{P}\left(\frac{\partial z}{\partial T}\right)_P dP_T \tag{9-6}$$

Since $P = P_c P_r$ and $T = T_c T_r$, we find that $dP = P_c dP_r$ and $dT = T_c dT_r$. Then,

$$dh_T = -\frac{RT_c^2 T_r^2}{P_c P_r}\left(\frac{\partial z}{T_c \partial T_r}\right)_{P_r} P_c\, dP_r$$

or

$$\frac{h - h^*}{RT_c} = -T_r^2 \int_0^{P_r} \left(\frac{\partial z}{\partial T_r}\right)_{P_r} \frac{dP_r}{P_r}$$

where h is the specific enthalpy of the real gas at P and T and h^* is the specific enthalpy of the real gas at zero pressure and T. The asterisk indicates that at zero pressure, a real gas behaves as if it were an ideal gas.

Now, since

$$z = z^{(0)} + \omega z^{(1)} \tag{8-11}$$

then

$$\left(\frac{\partial z}{\partial T_r}\right)_{P_r} = \left(\frac{\partial z^{(0)}}{\partial T_r}\right)_{P_r} + \omega \left(\frac{\partial z^{(1)}}{\partial T_r}\right)_{P_r}$$

and

$$\frac{h - h^*}{RT_c} = -T_r^2 \int_0^{P_r} \left(\frac{\partial z^{(0)}}{\partial T_r}\right)_{P_r} \frac{dP_r}{P_r} - \omega T_r^2 \int_0^{P_r} \left(\frac{\partial z^{(1)}}{\partial T_r}\right)_{P_r} \frac{dP_r}{P_r}$$

or

$$\frac{h^* - h}{RT_c} = \left(\frac{h^* - h}{RT_c}\right)^{(0)} + \omega \left(\frac{h^* - h}{RT_c}\right)^{(1)} \tag{9-7}$$

Now, if we had an analytical expression for

$$z = z(P_r, T_r, \omega)$$

we could integrate Eq. (9-7) directly. However, in Figs. A-1A and A-1B, we have a graphical representation of

$$z = z(P_r, T_r, \omega)$$

so we must take derivatives and integrate graphically.

The results of the graphical differentiation and integration of Eq. (9-7) are presented in the Appendix[27]. At a given value of reduced pressure and reduced temperature, values of the enthalpy function for a simple fluid $[(h^* - h)/RT_c]^{(0)}$, can be read directly from Fig. A-2A while values of the enthalpy function correction term $[(h^* - h)/RT_c]^{(1)}$, can be read directly from Fig. A-2B.

In order to find the effect of pressure at constant temperature on internal energy, we use the definition of enthalpy in the following form:

$$\Delta u = \Delta h - \Delta(Pv)$$

where Δh is obtained from Figs. A-2A and A-2B. Then, at constant temperature,

$$\Delta(Pv) = RT(z_2 - z_1) \tag{9-8}$$

The compressibility factors are obtained from Figs. A-1A and A-1B.

Another topic of interest is the effect of pressure on entropy. It can be shown that

$$ds = \left(\frac{\partial v}{\partial T}\right)_P dP + \frac{c_P}{T} dT$$

At constant temperature,

$$(s_P - s_0^*)_T = \int_0^P -\left(\frac{\partial v}{\partial T}\right)_P dP_T$$

where s_P is the specific entropy of the real gas at P and T and s_0^* is the specific entropy of the real gas at zero pressure and T. The asterisk indicates that at zero pressure, a real gas behaves as if it were an ideal gas. Now,

$$\lim_{P \to 0} -\left(\frac{\partial v}{\partial T}\right)_P = \infty$$

Thus, our equation for Δs_T is not integrable in the limit as pressure approaches zero. In order to circumvent this problem, we might select an isothermal path of integration in which the gas is first expanded from system pressure P to the ideal gas state of zero pressure and then the ideal gas is recompressed to some reference pressure P°. By doing this, the low pressure effects on entropy tend to cancel.

If the gas obeys the equation of state

$$Pv = zRT \tag{8-9}$$

we find that at constant T,

$$s_0^* - s_P = -\int_P^0 \frac{zR}{P} dP - \int_P^0 \frac{RT}{P}\left(\frac{\partial z}{\partial T}\right)_P dP$$

For an ideal gas, we find

$$s_{P^0}^* - s_0^* = \int_0^{P^0} -\frac{R}{P} dP$$

$$= \int_0^P -\frac{R}{P} dP + \int_P^{P^0} -\frac{R}{P} dP$$

Adding the last two equations,

$$s_{P^0}^* - s_P = R\int_0^P (z-1)\frac{dP}{P} + R\ln\frac{P}{P^0} + RT\int_0^P \left(\frac{\partial z}{\partial T}\right)_P \frac{dP}{P}$$

or, in reduced form

$$\frac{s_{P^0}^* - s_P}{R} = \int_0^{P_r} (z-1)\frac{dP_r}{P_r} + \ln\frac{P_r}{P_r^0} + T_r\int_0^{P_r}\left(\frac{\partial z}{\partial T_r}\right)_{P_r}\frac{dP_r}{P_r} \qquad (9\text{-}9)$$

Again, we have an equation that must be evaluated graphically. Two common reference states for P^0 are one atmosphere and the actual system pressure. If we select the system pressure as the reference, Eq. (9-9) becomes

$$\frac{s_P^* - s_P}{R} = \int_0^{P_r} (z-1)\frac{dP_r}{P_r} + T_r\int_0^{P_r}\left(\frac{\partial z}{\partial T_r}\right)_{P_r}\frac{dP_r}{P_r}$$

Now, since

$$z = z^{(0)} + \omega z^{(1)} \qquad (8\text{-}11)$$

$$\frac{s_P^* - s_P}{R} = \int_0^{P_r} (z^{(0)} - 1) \frac{dP_r}{P_r} + T_r \int_0^{P_r} \left(\frac{\partial z^{(0)}}{\partial T_r}\right)_{P_r} \frac{dP_r}{P_r}$$

$$+ \omega \int_0^{P_r} z^{(1)} \frac{dP_r}{P_r} + \omega T_r \int_0^{P_r} \left(\frac{\partial z^{(1)}}{\partial T_r}\right)_{P_r} \frac{dP_r}{P_r}$$

$$= \left(\frac{s_P^* - s_P}{R}\right)^{(0)} + \omega \left(\frac{s_P^* - s_P}{R}\right)^{(1)} \tag{9-10}$$

where $s_P^* - s_P$ is the difference in the specific entropy of an ideal gas and a real gas at the same pressure and temperature. The results of the evaluation of this equation are presented in Figs. A-3A and A-3B in the Appendix[27].

In order to determine the change in entropy of a real gas resulting from an isothermal change in pressure using Figs. A-3A and A-3B, we must take into account the effect of pressure on the entropy of an ideal gas. That is,

$$s_{P_2} - s_{P_1} = \left(s_{P_1}^* - s_{P_1}\right) - \left(s_{P_2}^* - s_{P_2}\right) - R \ln \frac{P_2}{P_1} \tag{9-11}$$

where the last term is equal to $s_{P_2}^* - s_{P_1}^*$ at constant temperature.

EXAMPLE PROBLEM 9-4. Ethane ($MW = 30$) is flowing irreversibly through a horizontal section of pipe. At the inlet of the section, the pressure is 1,420 lb$_f$/in.^{2}, the temperature is 365 °F, and the velocity of the ethane is 200 ft/sec. At the outlet of the section, the pressure is 1,060 lb$_f$/in.^{2} and the temperature is 365 °F.

Assuming that ethane follows the equation of state

$$Pv = zRT \tag{8-9}$$

and using the section of the pipe as a system,

(1) Find the quantity of heat exchanged between the system and the surroundings in BTU/lb-mole of ethane flowing.
(2) Find the lost work for this process in BTU/lb-mole of ethane flowing.

Taking the pipe as the system and basing our calculations on one mole of ethane flowing, the mass balance equation, Eq. (2-1), reduces to

$$m_{in} = m_{out}$$

The combined mass and energy balances, Eqs. (2-1) and (2-25), reduce to

$$\left(h_{in} + \frac{v_{in}^2}{2}\right) - \left(h_{out} + \frac{v_{out}^2}{2}\right) + q = 0$$

and the combined mass and entropy balances, Eqs. (2-1) and (3-16), reduce to

$$s_{in} - s_{out} + \frac{q}{T} + \frac{lw}{T} = 0$$

Using critical constants given in Table A-4, we find for ethane ($\omega = 0.0980$) at the inlet of the pipe, $P_r = 2.0$ and $T_r = 1.5$, while for ethane at the outlet of the pipe, $P_r = 1.5$ and $T_r = 1.5$. Then, from the mass balance,

$$v_{out} = v_{in}\left(\frac{P_{in}}{P_{out}}\right)\left(\frac{z_{out}}{z_{in}}\right)$$

Using compressibility factors from Figs. A-1A and A-1B, where

$$z_{in} = 0.83 + 0.0980 \times 0.18$$

$$= 0.848$$

and

$$z_{out} = 0.87 + 0.0980 \times 0.14$$

$$= 0.884$$

we find

$$v_{out} = (200)\left(\frac{1420}{1060}\right)\left(\frac{0.884}{0.848}\right)$$

$$= 279.3 \text{ ft/sec}$$

and

$$\frac{\Delta v^2}{2} = \frac{(279.3^2 - 200^2)(30)}{(2)(32.2)(778)}$$

$$= 22.8 \text{ BTU/lb-mole}$$

344 Chap. 9   *Evaluation of Functions*

Further,

$$\Delta h = RT_c \left[\left(\frac{h^* - h}{RT_c} \right)_{in} - \left(\frac{h^* - h}{RT_c} \right)_{out} \right]$$

From Figs. A-2A and A-2B,

$$\left(\frac{h^* - h}{RT_c} \right)_{in} = 1.04 + 0.0980 \times 0.04 = 1.044$$

and

$$\left(\frac{h^* - h}{RT_c} \right)_{out} = 0.75 + 0.0980 \times 0.04 = 0.754$$

so

$$\Delta h = (1.986)(549.7)(1.044 - 0.754)$$

$$= 316.6 \text{ BTU/lb-mole}$$

so that from the reduced energy balance,

$$q = 316.6 + 22.8$$

$$= 339.4 \text{ BTU/lb-mole of ethane flowing}$$

Now to find the lost work, we will use the relationship

$$lw = w_{rev} - w_{act} \qquad (3\text{-}13)$$

where w_{rev} is the reversible work that could have been obtained from this flow process. If the process were an isothermal, reversible process, the combined mass and entropy balances would reduce to

$$q_{rev} = T\Delta s$$

From Eq. (9-11),

$$s_{out} - s_{in} = (s^*_{in} - s_{in}) - (s^*_{out} - s_{out}) - R \ln \frac{P_{out}}{P_{in}}$$

From Figs. A-3A and A-3B,

$$\left(\frac{s^* - s}{R}\right)_{in} = 0.51 + 0.0980 \times 0.10 = 0.520$$

and

$$\left(\frac{s^* - s}{R}\right)_{out} = 0.36 + 0.0980 \times 0.09 = 0.369$$

Then

$$s_{out} - s_{in} = 1.986\left(0.520 - 0.369 + \ln\frac{1420}{1060}\right)$$

$$= 0.881 \text{ BTU/lb-mole } °R$$

Thus,

$$q_{rev} = (825)(0.881) = 726.8 \text{ BTU/lb-mole of ethane flowing}$$

Now, from the reduced mass and energy balance

$$w_{rev} = q_{rev} - \Delta h - \frac{\Delta v^2}{2}$$

$$= 726.8 - 316.6 - 22.8$$

$$= 387.4 \text{ BTU/lb-mole of ethane flowing}$$

Then, since w_{act} is zero,

$$lw = 387.4 \text{ BTU/lb-mole of ethane flowing}$$

FUGACITIES

A thermodynamic function which we previously found to be of use in developing relationships between derived and measurable thermodynamic properties and which we will later find to be useful in our studies of phase equilibrium is the Gibbs free energy G defined by the relationship

$$G = H - TS \tag{4-19}$$

The combination of the thermodynamic variables H and TS arises naturally in studies of isothermal flow processes. Consider, for example, the problem of estimating the work associated with a turbine operating reversibly and isothermally in a flow process.

If we consider our system to include only the turbine, the general mass balance reduces to

$$m_{in} = m_{out}$$

Neglecting changes in velocity and elevation, the general mass and energy balances reduce to

$$H_{in} - H_{out} + Q - W = 0$$

and the generalized mass and entropy balances reduce to

$$S_{in} - S_{out} + \frac{Q}{T} + \frac{LW}{T} = 0$$

If the turbine operates reversibly,

$$LW = 0$$

and

$$Q = T\Delta S$$

Combining this result with the reduced energy balance, we find that the reversible work which we can obtain from a turbine is

$$-W_{rev} = \Delta H_T - T\Delta S_T$$

But at constant temperature,

$$\Delta G_T = \Delta H_T - T\Delta S_T$$

Thus, ΔG_T is the reversible work associated with an isothermal flow process, or

$$W_{rev} = -\Delta G_T \qquad (9\text{-}12)$$

Since the Gibbs free energy is a state function, it is more convenient to evaluate than work, a path function. We have shown that, neglecting surface and chemical effects,

$$dG = V\,dP - S\,dT \tag{4-20}$$

or, at constant temperature

$$dG_T = V\,dP_T$$

For an ideal gas,

$$v = \frac{RT}{P}$$

so that Eq. (4-20) can be integrated to give the result

$$\Delta g_T = RT \ln \frac{P_2}{P_1} \tag{9-13}$$

This was such a "nice" equation that people wanted to continue to use it, even though in many instances the ideal gas law did not apply to real gases. In 1901, G. N. Lewis defined "fugacity" f so that for a real gas,

$$\Delta g_T = \int_{P_1}^{P_2} v\,dP = RT \ln \frac{f_2}{f_1} \tag{9-14}$$

If we compare Eqs. (9-13) and (9-14) we observe that for an ideal gas,

$$f^* = P$$

where f^* is the fugacity of an ideal gas. Since real gases act more and more like ideal gases as pressure approaches zero, it is convenient to complete the definition of fugacity by making it numerically equal to pressure as pressure approaches zero.

$$\lim_{P \to 0} \frac{f}{P} = 1.0 \tag{9-15}$$

Equations (9-14) and (9-15), then, define fugacity.

In order to determine a numerical value of fugacity for a substance at a given pressure and temperature, we must integrate Eq. (9-14) from the limit as pressure approaches zero to the pressure of interest. Unfortunately, the specific volume of a gas approaches infinity as pressure approaches zero and the integral in Eq. (9-14) is divergent at the lower limit. However, we find that if we define a residual volume

$$\alpha_R = \frac{RT}{P} - v \tag{8-1}$$

then, experimental values of α_R approach a finite value as pressure approaches zero.

Now, by definition

$$dg_T = RT d\ln f = v\, dP_T$$

and since

$$v = \frac{RT}{P} - \alpha_R$$

we find

$$RT d\ln f = \left(\frac{RT}{P} - \alpha_R\right) dP$$

Integrating from $P_0 = 0$ to P,

$$RT \ln \frac{f}{f^*} = RT \ln \frac{P}{P_0} - \int_{P_0}^{P} \alpha_R\, dP$$

Then, as

$$f^* = P_0$$

we find

$$RT \ln \frac{f}{P} = -\int_{0}^{P} \alpha_R\, dP \tag{9-16}$$

G. Tunnel suggested that instead of Eq. (9-14), Eq. (9-16) be used in the definition of fugacity.

The ratio f/P occurs frequently and is given the name fugacity coefficient ν. Then,

$$\nu = \frac{f}{P}$$

or

$$f = \nu P \qquad (9\text{-}17)$$

This equation suggests that a fugacity may be thought of as a pressure, corrected for non-idealities in the system.

As an example of the utility of Eq. (9-16), consider a gas whose behavior can be represented by the equation of state

$$Pv = zRT \qquad (8\text{-}9)$$

In this case,

$$\alpha_R = \frac{RT}{P}(1-z)$$

and

$$\ln \nu = \int_0^P (z-1)\frac{dP}{P}$$

In reduced form,

$$\ln \nu = \int_0^{P_r} (z-1)\frac{dP_r}{P_r}$$

Then, since

$$z = z^{(0)} + \omega z^{(1)} \qquad (8\text{-}11)$$

we find

$$\ln \nu = \int_0^{P_r} (z^{(0)}-1)\frac{dP_r}{P_r} + \omega \int_0^{P_r} z^{(1)} \frac{dP_r}{P_r}$$

or

$$\log \nu = \log \nu^{(0)} + \omega \log \nu^{(1)} \qquad (9\text{-}18)$$

350 Chap. 9 *Evaluation of Functions*

The results of the evaluation of Eq. (9-18) using Eq. (8-11) are presented in Figs. A-4A and A-4B in the Appendix[27].

EXAMPLE PROBLEM 9-5. Find the minimum work required to compress one lb-mole of ethane from 1 atm and 60 °F to 100 atm and 60 °F in a flow system.

Taking the compressor as the system and basing all the calculations on one pound mole of ethane flowing, the mass balance reduces to

$$m_{in} = m_{out}$$

$$= 1 \text{ lb-mole}$$

The combined mass and energy balances reduce to

$$h_{in} - h_{out} + q - w = 0$$

The combined mass and entropy balances reduce to

$$s_{in} - s_{out} + \frac{q}{T} = 0$$

since lw is zero for minimum work. Combining the mass, energy, and entropy balances, we find

$$-w = (h_{out} - h_{in}) - T(s_{out} - s_{in})$$

or

$$-w = g_{out} - g_{in}$$

Further,

$$g_{out} - g_{in} = RT \ln \frac{v_{out} P_{out}}{v_{in} P_{in}}$$

Now, at the compressor inlet, $(P_r)_{in} = 0.0207$ and $(T_r)_{in} = 0.946$. At these conditions, ethane acts essentially as an ideal gas and

$$v_{in} = 1.0$$

However, at the outlet, $(P_r)_{out} = 2.07$ and $(T_r)_{out} = 0.945$. Using Figs. A-4A and A-4B, we find from Eq. (9-18)

$$\log v_{out} = -0.49 + (0.0980)(-0.13) = -0.503$$

where ω for ethane is 0.0980. Then

$$\nu_{out} = 0.314$$

Thus,

$$-w = (1.986)(520)\ln \frac{(0.314)(100)}{(1.0)(1.0)}$$

$$w = -3560 \text{ BTU/lb-mole of ethane flowing}$$

The negative sign indicates work will be required to compress the ethane.
Now, the equation

$$PV = nzRT \tag{8-9}$$

is a convenient equation to use in evaluating Eq. (9-14), since it can be solved directly for v as a function of P. On the other hand, it is inconvenient to integrate Eq. (9-14) when the relationship between P, v, and T is implicit in v. However, we may rewrite Eq. (9-14) in the differential form

$$nRT d\ln f = V\left(\frac{\partial P}{\partial V}\right)_{T,n} dV$$

by multiplying and dividing $VdP_{T,n}$ by $dV_{T,n}$. In this form, the equation is not integrable in the limit as P approaches zero. We can modify it to an integrable form by subtracting $nRT d\ln P$ from both sides,

$$nRT d\ln \frac{f}{P} = V\left(\frac{\partial P}{\partial V}\right)_{T,n} dV - nRT d\ln P$$

The two terms on the right hand side of this equation can be combined if we recognize that

$$\frac{d(PV)}{PV} = \frac{PdV}{PV} + \frac{VdP}{PV}$$

or

$$d\ln P = d\ln (PV) - d\ln V$$

Thus,

$$nRT d\ln \frac{f}{P} = \left[V\left(\frac{\partial P}{\partial V}\right)_{T,n} + \frac{nRT}{V}\right]dV - nRT d\ln (PV)$$

Then, integrating from zero pressure to P (i.e. from $V = \infty$ to $V = V$)

$$nRT\ln\frac{f}{P} = -\int_V^\infty \left[V\left(\frac{\partial P}{\partial V}\right)_{T,n} + \frac{nRT}{V}\right]dV - nRT\ln\frac{PV}{nRT}$$

since

$$\lim_{P\to 0}\frac{f}{P} = 1.0 \qquad (9\text{-}15)$$

and

$$\lim_{P\to 0}(PV) = nRT$$

Now,

$$\left(\frac{\partial P}{\partial V}\right)_{T,n} = -\left(\frac{\partial P}{\partial n}\right)_{T,V}\left(\frac{\partial n}{\partial V}\right)_{P,T}$$

Then, as

$$PV = nzRT \qquad (8\text{-}9)$$

we have, at constant P and T,

$$P\,dV = zRT\,dn$$

or

$$\left(\frac{\partial n}{\partial V}\right)_{P,T} = \frac{P}{zRT} = \frac{n}{V}$$

since z is a constant at constant pressure and temperature. Then,

$$\left(\frac{\partial P}{\partial V}\right)_{T,n} = -\left(\frac{\partial P}{\partial n}\right)_{T,V}\left(\frac{n}{V}\right)$$

or

$$V\left(\frac{\partial P}{\partial V}\right)_{T,n} = -n\left(\frac{\partial P}{\partial n}\right)_{T,V}$$

Substituting this result into the last integrand,

$$nRT\ln\frac{f}{P} = n\int_V^\infty \left[\left(\frac{\partial P}{\partial n}\right)_{T,V} - \frac{RT}{V}\right]dV - nRT\ln\frac{PV}{nRT}$$

or

$$RT\ln \nu = \int_V^\infty \left[\left(\frac{\partial P}{\partial n}\right)_{T,V} - \frac{RT}{V}\right]dV - RT\ln z \qquad (9\text{-}19)$$

Equation (9-19) may be used to find the fugacity coefficient of a pure component at pressure P from an equation of state implicit in v.

EFFECT OF PRESSURE AND TEMPERATURE ON FUGACITY

From the differential form of Eq. (9-14), we have at constant temperature,

$$dg_T = RTd\ln f = v\,dP_T$$

so that the effect of pressure on fugacity is given by the relationship

$$\left(\frac{d\ln f}{dP}\right)_T = \frac{v}{RT} \qquad (9\text{-}20)$$

To determine the effect of temperature on fugacity, we start with Eq. (9-14) in the form

$$\Delta g_T = RT\ln\frac{f}{f^\circ}$$

or

$$g_T - g_T^\circ = RT\ln f - RT\ln f^\circ$$

where the superscript $^\circ$ indicates a particular reference state called a standard state. Standard states frequently used are:

(1) Gases: Pure component i in the ideal gas state at system temperature and one atmosphere.
(2) Liquids: Pure component i in the liquid phase at system temperature and either its own vapor pressure, one atmosphere, or system pressure.

(3) Solids: Pure component i in the solid phase at system temperature and either its own vapor pressure, one atmosphere or system pressure.

It would be nice if only one standard state were used. Unfortunately, different standard states have been found to be most convenient for different types of problems. This is a source of some confusion.

Differentiating with respect to T at constant P, we find that since g is a function of T,

$$\left(\frac{\partial g}{\partial T}\right)_P - \left(\frac{\partial g^o}{\partial T}\right)_P = R \ln \frac{f}{f^o} + RT\left(\frac{\partial \ln f}{\partial T}\right)_P - \left(\frac{\partial \ln f^o}{\partial T}\right)_P$$

By definition,

$$g = h - Ts \tag{4-19}$$

so that

$$s = \frac{h}{T} - \frac{g}{T}$$

Further,

$$dg = v\, dP - s\, dT \tag{4-20}$$

so that

$$\left(\frac{\partial g}{\partial T}\right)_P = -s$$

Then,

$$\left(\frac{\partial g}{\partial T}\right)_P = \frac{g}{T} - \frac{h}{T}$$

Similarly,

$$\left(\frac{\partial g^o}{\partial T}\right)_P = \frac{g^o}{T} - \frac{h^o}{T}$$

From Eq. (9-14),

$$R \ln \frac{f}{f^o} = \frac{\Delta g}{T}$$

Further, $(\partial \ln f^\circ/\partial T)_P = 0$, since f° is not a function of temperature but is defined to be equal to a reference pressure at system temperature.

Then, combining these results, we find

$$\left(\frac{g}{T}-\frac{h}{T}\right) - \left(\frac{g^\circ}{T}-\frac{h^\circ}{T}\right) = \left(\frac{g}{T}-\frac{g^\circ}{T}\right) + RT\left(\frac{\partial \ln f}{\partial T}\right)_P$$

Thus, the effect of temperature on fugacity is given by the relationship

$$\left(\frac{\partial \ln f}{\partial T}\right)_P = \frac{h^\circ - h}{RT^2} \qquad (9\text{-}21)$$

where $h^\circ - h$ is the difference between the enthalpy of the substance as a gas in its standard state and the enthalpy of the substance in the given state, both at the same temperature. If the standard state is an ideal gas at one atmosphere,

$$h^\circ = h^*$$

and

$$\left(\frac{\partial \ln f}{\partial T}\right)_P = \frac{h^* - h}{RT^2} \qquad (9\text{-}22)$$

In this case, Figs. A-2A and A-2B in the Appendix can be used to evaluate the isobaric change of fugacity with temperature.

PROBLEMS

9-1. The following compressibility data was obtained for a gas at 600 K:

v(c/gm mole)	z
50	1.20
40	1.25
30	1.35
20	1.50

These data can be represented by the virial equation of state truncated after the second virial. Evaluate the second virial coefficient using the method of least squares and then find the value of the partial derivative $(\partial P/\partial v)_T$ in (atm·gm mole)/(cm³) at a specific volume of 30 cc/gm mole.

9-2. Using only *PVT* data from Table A-2, find the change in enthalpy of steam at 700 °F and 100 lb$_f$/in.^{2} as it is compressed isentropically to 500 lb$_f$/in.^{2}

9-3. A cylinder, fitted with a piston, contains 10.0 ft^3 of saturated water vapor at 400 °F. The steam then expands isothermally and reversibly to a pressure of 50 lb$_f$/in.^{2}, and in doing so does work against the piston. Determine the work in BTU and joules during this process by integrating $P\,dV$ from the steam tables.

9-4. For a gas obeying van der Waals' equation, show

(a) $(h_2 - h_1)_T = (P_2 v_2 - P_1 v_1) + a\left(\dfrac{1}{v_1} - \dfrac{1}{v_2}\right)$

(b) $(s_2 - s_1)_T = R \ln\left(\dfrac{v_2 - b}{v_1 - b}\right)$

(c) $\left(\dfrac{\partial c_V}{\partial v}\right)_T = 0$

(d) $c_P - c_V = \dfrac{\dfrac{R^2 T}{(v-b)^2}}{\dfrac{RT}{(v-b)} - \dfrac{2a}{v^3}}$

9-5. For a gas obeying the equation of state $Pv = zRT$, show

(a) $\Delta s_T = -RT \displaystyle\int_{P_1}^{P_2} \left(\dfrac{\partial z}{\partial T}\right)_P d\ln P - R \int_{P_1}^{P_2} (z-1)\,d\ln P - R \ln \dfrac{P_2}{P_1}$

$\quad = \dfrac{\Delta h}{T} - R \ln \dfrac{f_2}{f_1}$

(b) $c_P - c_V = R \dfrac{\left[z + T\left(\dfrac{\partial z}{\partial T}\right)_P\right]^2}{\left[z - P\left(\dfrac{\partial z}{\partial P}\right)_T\right]}$

9-6. Develop a relationship for the isothermal change in internal energy of a gas that follows the Redlich-Kwong equation of state.

9-7. The data in the following table was obtained for a gas whose behavior is represented by the equation $Pv = zRT$. Using *only* the data in the table, show that at

$$T_r = 1.5$$

and

$$P_r = 0.6$$

we have

$$\frac{h^* - h}{RT_c} = 0.24$$

where

h^* is the enthalpy of an ideal gas at system temperature and zero pressure, h is the enthalpy of the real gas at system temperature and pressure, and T_c is the critical temperature of the gas.

Table of Compressibility Factors

T_r/P_r	0	0.1	0.2	0.3	0.4	0.5	0.6
1.2	1.000	0.983	0.965	0.946	0.924	0.905	0.905
1.3	1.000	0.987	0.974	0.961	0.945	0.931	0.917
1.4	1.000	0.990	0.982	0.972	0.959	0.949	0.937
1.5	1.000	0.991	0.986	0.980	0.970	0.963	0.953
1.6	1.000	0.992	0.988	0.985	0.978	0.973	0.965
1.7	1.000	0.992	0.989	0.989	0.984	0.980	0.974

9-8. Methane is flowing through a well insulated, horizontal pipe, 6 in. inside diameter and 200 ft long. At the inlet of the pipe, the pressure is 2000 $lb_f/in.^2$, the temperature is $-50\,°F$, and the velocity of the methane is 55 ft/sec. At the outlet of the pipe, the pressure is 1350 $lb_f/in.^2$

Assuming that the phase behavior of methane is adequately described by the equation $Pv = zRT$ and assuming $c_P^* = 7.25$ BTU/lb mole °R, calculate the temperature and the velocity of the methane at the pipe outlet.

9-9. Saturated liquid propane at 100 °F is throttled through a well insulated globe valve to 125 $lb_f/in.^2$ If $c_P^* = 17.5$ BTU/lb mole °R, calculate the quality of the propane at the outlet of the valve using the generalized charts for simple fluids only.

9-10. An uninsulated cylinder with a volume of 5 ft^3 contains carbon dioxide at a temperature of 75 °F and a quality of 90 percent. The cylinder valve is not closed properly and carbon dioxide vapor leaks out until the pressure in the cylinder drops to 100 lb$_f$/in.^{2} The process takes place slowly, so that the temperature of the cylinder remains essentially constant. Using the generalized charts, calculate the moles of carbon dioxide which escaped from the cylinder.

9-11. According to the steam tables, the enthalpy of steam at the following points is:

$h(BTU/lb_m)$	$T(°F)$	$P(psia)$
1284.1	500	50
1406.9	750	50
1533.4	1000	50

From these data, determine the constants in the heat capacity equation, $c_p = a + bT$. Using this equation and the equation of state,

$$Pv/RT = 1 - 0.3\, P_r/T_r^4$$

calculate the change of enthalpy for an isentropic process going from 800 psia and 1000 °F to 350 psia. Compare this enthalpy change with that given by the steam tables.

9-12. Propylene (C$_3$H$_6$) is being passed from a storage cylinder through a long pipe to a chemical reaction vessel, as shown in Fig. 9-1. At the inlet of the pipe, the pressure is 195 psia and the temperature is 110 °F. The velocity at this point is known to be 55 ft/sec. At the discharge end of the pipe, the pressure is 30 psia. The pipe is well insulated. The phase behavior of propylene may be represented by

$$Pv/RT = 1 - 0.3\, P_r/T_r^4$$

For propylene:

$MW = 42$

$T_c = 365.05$ K

$P_c = 45.4$ atm

$c_p^* = 4.6 + 0.02T$ (BTU/lb mole °R), where T is in °R

What are the conditions and velocity of the propylene at the discharge end of the pipe?

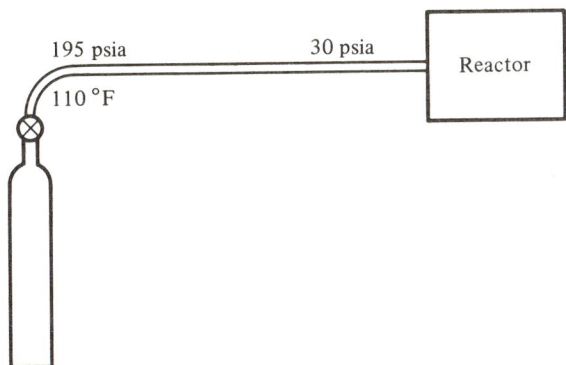

Fig. 9-4.

9-13. Estimate the fugacity of gaseous hexane at 150 $lb_f/in.^2$ and 400 °F using the following PVT data at 400 °F:

Pressure ($lb_f/in.^2$)	Specific Volume (ft^3/lb_m)
25	4.1500
50	2.0100
100	0.9400
150	0.5810
200	0.3980
300	0.0373
700	0.0352
1000	0.0336

Then, from the PVT data given, compute the fugacity of liquid hexane at 400 °F and 1000 $lb_f/in.^2$ The vapor pressure of hexane at 400 °F is 277.5 $lb_f/in.^2$

9-14. For a gas obeying van der Waals' equation, show

$$\ln(f/P) = \frac{b}{v-b} - \ln\left(\frac{v-b}{v}\right) - \frac{2a}{RTv} - \ln z$$

9-15. One hundred pounds per hour of propane is to be compressed isothermally at 285 °F from 1000 $lb_f/in.^2$ to 2000 $lb_f/in.^2$ in a centrifugal compressor. Using the generalized correlations, calculate the minimum work required for this compression process.

10

Mixtures

OBJECTIVES

After studying this chapter, the student should be able to

(1) calculate values of pressure, temperature and specific volume of a system consisting of a mixture, given any two of these three variables, using either an analytical or graphical form of an equation of state;

(2) apply equations of state in either analytical or graphical form to problems involving a mixture of ideal gases or a mixture which forms an ideal solution or a mixture which acts as a pseudo-single component to problems involving the application of the principle of conservation of mass, the principle of conservation of energy, as well as the second law of thermodynamics; and

(3) estimate partial molal properties and excess functions from physical data on non-ideal mixtures.

INTRODUCTION

Although there are some difficulties in describing the thermodynamic behavior of pure components, especially when a single equation is used to represent PVT behavior over a wide range of densities, it is possible to solve these problems with satisfactory precision. Frequently, however, the thermodynamic problem that we wish to investigate involves a mixture of components. Air, for example, is not a pure substance, but rather it is a mixture of 78.09 mol % N_2, 20.95 mol % O_2, 0.93 mol % argon, 0.03 mol % CO_2, with traces of neon, helium, krypton, hydrogen, xenon, ozone, and radon. We generally assume that air is 79 mol % N_2 and 21 mol % O_2. When we consider the problem of mixtures, we add entirely new dimensions to our problem. In this chapter, we intend to examine what effect, if any, one component in a mixture might have on another component. We will begin by considering the simplest mixtures and will then proceed to more and more complex mixtures.

As a first assumption, one might examine the possibility that the properties of a mixture are the weighted average of the properties of the individual components in the mixture. In fact, this assumption is the basis for calculating the thermodynamic properties of a multiphase single component system. The problem involving mixtures, then, would resolve itself into one of solving for the individual pure components and adding the solutions.

The assumption that the properties of a mixture are the weighted sum of the properties of the individual components in the mixture implies that all molecules act alike in their effects upon one another. The simplest mixture of different components that meets this requirement is a mixture of ideal gases. At conditions of low pressure and low density under which a gas mixture obeys the ideal gas equation of state, the molecules making up the gas are, on the average, separated by large distances. Consequently, we can assume that there is no interaction between either like pairs or unlike pairs of molecules. As the density of the gas mixture increases, the average distance between pairs of molecules decreases, and the interaction effects become important. Mixtures in which the interactions between like and unlike pairs of molecules are the same are called ideal mixtures. Mixtures in which the interactions between like and unlike pairs of molecules differ are called non-ideal mixtures.

IDEAL GAS MIXTURES

Dalton's and Amagat's Laws

There are two ways of looking at a mixture of ideal gases. One way as shown in Fig. 10-1(a), is to consider the gases at a given volume and temperature and at different pressures. Then,

$$p_1 V = n_1 RT$$

and

$$p_2 V = n_2 RT$$

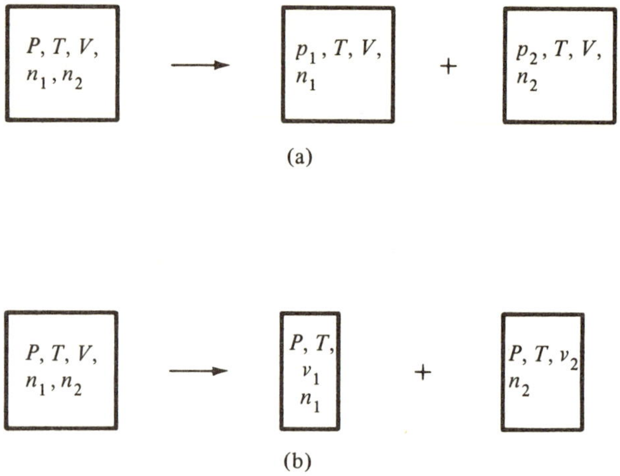

Fig. 10-1. Schematic representation of ideal gas mixtures.

Adding,

$$(p_1 + p_2)V = PV = (n_1 + n_2)RT$$

or

$$P = p_1 + p_2 = \Sigma p_i \tag{10-1}$$

where p_i is the fraction of the total pressure contributed by component i and is called the partial pressure of component i. Equation (10-1) says that total pressure of a mixture of ideal gases equals the sum of the partial pressures of the individual gases, each occupying a volume V at temperature T. This is known as Dalton's law. Now,

$$\frac{p_1}{P} = \frac{n_1}{n_1 + n_2} = y_1$$

where y_1 is the mole fraction of component i in the vapor phase. Then,

$$p_1 = y_1 P \tag{10-2}$$

Still another way of looking at a mixture of ideal gases is to assume all gases in a mixture are at the same pressure and temperature but at different volumes, as shown in Fig. 10-1(b). Then,

$$Pv_1 = n_1 RT$$

and

$$Pv_2 = n_2 RT$$

Adding,

$$P(v_1 + v_2) = PV = (n_1 + n_2)RT$$

or

$$V = v_1 + v_2 = \Sigma v_i \qquad (10\text{-}3)$$

where v_i is the portion of the total volume occupied by component i and is called the partial volume of component i. Equation (10-3) states that the total volume of a mixture of gases is equal to the sum of the partial volumes of the individual gases, each existing at pressure P and temperature T. This is known as Amagat's law or Leduc's law. Then,

$$\frac{v_1}{V} = \frac{n_1}{n_1 + n_2}$$

or

$$v_1 = y_1 V \qquad (10\text{-}4)$$

Thermodynamic Properties

A fundamental concept underlying both Dalton's law and Amagat's law is that there is no energy of attraction acting between molecules of different species. We may extend this concept to include the idea that the internal energy of ideal gas 1 is not affected by the presence of ideal gas 2, and

$$u_M = \frac{n_1 u_1 + n_2 u_2 + \ldots}{n_1 + n_2 + \ldots}$$

or

$$u_M = y_1 u_1 + y_2 u_2 + \ldots \qquad (10\text{-}5)$$

Thus, the specific molar internal energy of a mixture of ideal gases is the molar average of the specific molar internal energies of the individual species in the mixture.

By analogy with Eq. (5-3), the specific internal energy of a mixture is also given by the relationship

$$u_M = \int_{T^o}^{T} -c_{V_M} \, dT + u_M^o$$

Breaking this equation down into the contributions of the individual species, we find

$$y_1 u_1 + y_2 u_2 + \ldots = \int_{T^o}^{T} y_1 c_{V_1} \, dT + y_1 u_1^o + \int_{T^o}^{T} y_2 c_{V_2} \, dT$$

$$+ y_2 u_2^o + \ldots$$

and if we define

$$u_M^o = y_1 u_1^o + y_2 u_2^o + \ldots$$

it follows that

$$c_{V_M} = y_1 c_{V_1} + y_2 c_{V_2} + \ldots \qquad (10\text{-}6)$$

Specific heats at constant pressure are related similarly. If we apply the identity,

$$\Sigma y_i = 1.0$$

and the ideal gas relationship,

$$c_P - c_V = R$$

we find on adding the gas constant R to both sides of Eq. (10-6) that

$$\left(c_{V_M} + R\right) = y_1\left(c_{V_1} + R\right) + y_2\left(c_{V_2} + R\right) + \ldots$$

or

$$c_{P_M} = y_1 c_{P_1} + y_2 c_{P_2} + \ldots \qquad (10\text{-}7)$$

Finally, since

$$h_M = u_M + Pv_M$$

and

$$v_M = y_1 v_1 + y_2 v_2 + \ldots$$

we find that for an ideal gas at constant temperature and pressure

$$h_M = y_1 u_1 + y_2 u_2 + \ldots + P(y_1 v_1 + y_2 v_2 + \ldots)$$

or

$$h_M = y_1 h_1 + y_2 h_2 + \ldots \tag{10-8}$$

Even though Eqs. (10-5), (10-6), (10-7) and (10-8) were derived on a mole basis since we are concerned about interactions between molecules, these four equations can be shown to apply directly on a mass basis by multiplying both sides of these equations by the appropriate molecular weights.

Unfortunately, we can not extend the concept that the internal energy of a mixture of ideal gases is a molar average internal energy to entropies. The process in which two different ideal gases are mixed is an irreversible process which produces entropy ΔS_{mix} so that

$$s_M = y_1 s_1 + y_2 s_2 + \Delta s_{mix}$$

Consider a closed system in which n_1 moles of ideal gas 1 in a volume V_1 are allowed to mix isothermally and isobarically with n_2 moles of ideal gas 2 in volume V_2, such that at the end of the process, we have $n_1 + n_2$ moles of ideal gas 1 and 2 mixed in a volume of V, where

$$V = V_1 + V_2$$

This process can be carried out in three steps:

(1) Expand gas 1 in V_1 reversibly to V at constant temperature. From the energy balance equation,

$$\Delta U_{rev} = Q_{rev} - W_{rev} = 0$$

since the change in the internal energy of an ideal gas is a function of temperature only. Now, the entropy balance reduces to

$$Q_{rev} = T\Delta S_1$$

Since LW is zero in a reversible process. For a reversible expansion,

$$W_{rev} = \int P\, dV$$

so that for an ideal gas the energy balance becomes

$$Q_{rev} = W_{rev}$$

$$= n_1 RT \ln \frac{V}{V_1}$$

Combining this result with the entropy balance, we find

$$\Delta S_1 = \frac{Q_{rev}}{T} = n_1 R \ln \frac{V}{V_1}$$

Now, from Eq. (10-4),

$$\frac{V}{V_1} = \frac{n_1 + n_2}{n_1} = \frac{1}{y_1}$$

we obtain the result

$$\Delta S_1 = -n_1 R \ln y_1$$

(2) Expand gas 2 in V_2 reversibly to V at constant temperature. By analysis similar to that used for step 1, we find

$$\Delta S_2 = -n_2 R \ln y_2$$

Gas 1 and gas 2 are now in two separate boxes, each of volume V.

(3) Mix the gases reversibly and isothermally by superimposing the boxes. For this step, we find that

$$\Delta U_{rev} = Q_{rev} - W_{rev} = 0$$

Ideal Gas Mixtures

since the change in the internal energy of an ideal gas is a function of temperature only. Further, no work is required as there is no change in volume for either gas 1 or gas 2, so

$$W_{rev} = 0$$

Putting this result into the energy balance, we find

$$Q_{rev} = 0$$

and

$$\Delta S_3 = 0$$

Combining the change in entropy for the three steps,

$$\Delta S_{mix} = -n_1 R \ln y_1 - n_2 R \ln y_2$$

or for one mole of the mixture,

$$\Delta s_{mix} = -R(y_1 \ln y_1 + y_2 \ln y_2)$$

Since the value of Δs_{mix} is independent of path, this equation gives us the entropy change when two ideal gases are mixed in any manner at a specific pressure and temperature. This relationship can also be derived for ideal gases by using Dalton's law of partial pressures.

Then, the entropy of a mixture of ideal gases is given by the relationship

$$s_M = y_1 s_1 + y_2 s_2 + \ldots - R(y_1 \ln y_1 + y_2 \ln y_2 + \ldots) \tag{10-9}$$

Unfortunately, Eq. (10-9) can not be used directly on a mass basis since in general, the weight fraction of a component is not equal to the volume fraction of that component. Now, since

$$g_M = h_M - Ts_M$$

we find that

$$g_M = (y_1 h_1 + y_2 h_2 + \ldots) - T(y_1 s_1 + y_2 s_2 + \ldots)$$
$$+ RT(y_1 \ln y_1 + y_2 \ln y_2 + \ldots)$$
$$= y_1 g_1 + y_2 g_2 + \ldots + RT(y_1 \ln y_1 + y_2 \ln y_2 + \ldots) \tag{10-10}$$

Similarly,

$$a_M = u_M - Ts_M$$

$$= y_1 a_1 + y_2 a_2 + \ldots + RT(y_1 \ln y_1 + y_2 \ln y_2 + \ldots) \quad (10\text{-}11)$$

IDEAL SOLUTIONS

Thermodynamic Properties

As a natural extension of the concepts involved in calculating thermodynamic properties of ideal gas mixtures, attempts have been made to calculate thermodynamic properties of real gas mixtures using both Dalton's law of additive partial pressures and Amagat's law of additive volumes. Although both approaches lead to the same result for mixtures of ideal gases, they lead to different results for mixtures of real gases. In general, however, Amagat's law leads to a better representation of the properties of gas mixtures than Dalton's law.

The properties of mixtures which follow Amagat's law can be calculated from the properties of the individual components using equations similar to those used to calculate the properties of ideal gas mixtures. Such mixtures are called ideal solutions. In fact, a mixture of ideal gases is a special case of an ideal solution.

In order to apply the ideal solution theory to a mixture, we must first determine the thermodynamic properties which the individual components would have if they existed by themselves at the pressure and temperature of the system. Then, using y_i to represent the mole fraction of component i in such a mixture, even though both liquids and real gases may form an ideal solution, we find that

$$v_M = y_1 v_1 + y_2 v_2 + \ldots \quad (10\text{-}12)$$

$$u_M = y_1 u_1 + y_2 u_2 + \ldots \quad (10\text{-}13)$$

$$h_M = y_1 h_1 + y_2 h_2 + \ldots \quad (10\text{-}14)$$

$$s_M = y_1 s_1 + y_2 s_2 + \ldots - R(y_1 \ln y_1 + y_2 \ln y_2 + \ldots) \quad (10\text{-}15)$$

$$g_M = y_1 g_1 + y_2 g_2 + \ldots + RT(y_1 \ln y_1 + y_2 \ln y_2 + \ldots) \quad (10\text{-}16)$$

and

$$a_M = y_1 a_1 + y_2 a_2 + \ldots + RT(y_1 \ln y_1 + y_2 \ln y_2 + \ldots) \quad (10\text{-}17)$$

If, for example, the volume occupied by an individual component in an ideal solution is given by the relationship

$$v_i = \frac{z_i RT}{P}$$

then

$$v_M = \frac{RT}{P}(y_1 z_1 + y_2 z_2 + \ldots) \qquad (10\text{-}18)$$

Whereas the assumption underlying the estimation of properties of ideal gas mixtures is that the individual molecules are completely independent, in which case $z_1 = z_2 = \ldots = 1.0$ in Eq. (10-18), the assumption underlying the estimation of properties of ideal solutions is that there is an energy of interaction between molecules which is the same for like and unlike pairs. While this latter assumption is useful in many calculations, a more general assumption would take the difference of interaction between both similar and dissimilar molecules into account.

EXAMPLE PROBLEM 10-1. Calculate the rate at which heat must be added to the humidifier discussed in Example Prob. 2-1.

Taking the humidifier as the system and basing our calculations on one hour of operation, the generalized mass balance equation, Eq. (2-1), reduces to

$$m_{\text{air, in}} + m_{\text{water, in}} - m_{\text{air, out}} = 0$$

and the generalized energy balance equation, Eq. (2-25), reduces to

$$H_{\text{air, in}} + H_{\text{water, in}} - H_{\text{air, out}} + Q = 0$$

From Example Prob. 2-1, we find that

$$m_{\text{air, in}} = 5 \text{ lb}_m \text{ water} + 995 \text{ lb}_m \text{ bone dry air}$$

$$m_{\text{water, in}} = 5 \text{ lb}_m \text{ water}$$

and

$$m_{\text{air, out}} = 10 \text{ lb}_m \text{ water} + 995 \text{ lb}_m \text{ bone dry air}$$

Next, we will need to evaluate the enthalpy terms in the energy balance equation. From the steam tables, we find that at 70 °F

$$h_f = 38.05 \text{ BTU/lb}_m$$

As air-water vapor form an ideal solution, we can use Eq. (10-14) on a mass basis,

$$h_M = y_{\text{bone dry air}} h_{\text{bone dry air}} + y_{\text{water vapor}} h_{\text{water vapor}} \tag{10-14}$$

to calculate $h_{\text{air, in}}$ and $h_{\text{air, out}}$. From the steam tables, at 70 °F we find

$$h_{\text{water vapor}} = 1092.1 \text{ BTU/lb}_m$$

Assuming bone dry air acts as an ideal gas, we find

$$h_{\text{bone dry air}} = c_P (T_2 - T_1) \tag{5-5}$$

$$= \frac{7/2 \, R \, (530 - 460)}{28.96}$$

$$= 16.8 \text{ BTU/lb}_m$$

since $c_P = 7/2 \, R$ for a diatomic ideal gas. Here we have assumed

$$h_{\text{bone dry air}} = 0 \text{ at } 0 \, °F$$

Then, applying Eq. (10-14), we find

$$h_{\text{air, in}} = \frac{5}{1000} \times 1092.1 + \frac{995}{1000} \times 16.8 = 22.2 \text{ BTU/lb}_m$$

and

$$h_{\text{air, out}} = \frac{10}{1005} \times 1092.1 + \frac{995}{1005} \times 16.8 = 27.5 \text{ BTU/lb}_m$$

Substituting these results into the energy balance equation

$$Q = H_{\text{air, out}} - H_{\text{air, in}} - H_{\text{water, in}}$$

$$= 1005 \times 27.5 - 1000 \times 22.2 - 5 \times 38.05$$

or

$$Q = 5250 \text{ BTU/hr}$$

Empirical Mixing Rules

Another approach which is used to correlate the *PVT* behavior of a mixture is to assume that the mixture can be regarded as a pseudo single component. Thus, if the proper equation of state coefficients can be found, an equation of state developed for single components might be used for mixtures. As might be expected, however, the coefficients of equations of state for mixtures are functions of composition.

The simplest and most widely used rule for estimating the compressibility factor of a mixture for the equation of state

$$Pv = zRT \tag{8-9}$$

is due to Webster Kay. Kay suggests that the properties of a mixture be found by using pseudo-critical temperatures and pseudo-critical pressures to calculate the reduced pressure and temperature of the system, where

$$T_{c_M} = \sum_i y_i T_{c_i} \tag{10-19}$$

$$P_{c_M} = \sum_i y_i P_{c_i} \tag{10-20}$$

and

$$\omega_M = \sum_i y_i \omega_i \tag{10-21}$$

Van der Waals proposed that the mixture coefficients in his equation of state, Eq. (8-3), be calculated as follows:

$$a_M = \left(\sum y_1 a_i^{1/2}\right)^2 \tag{10-22}$$

or for a binary mixture,

$$a_M = y_1^2 a_1 + 2y_1 y_2 a_{12} + y_2^2 a_2$$

where

$$a_{12} = \left(a_1 a_2\right)^{1/2}$$

and

$$b_M = \sum_i y_i b_i \tag{10-23}$$

or for a binary mixture,

$$b_M = y_1 b_1 + y_2 b_2$$

Redlich and Kwong suggested that the mixture coefficients in their equation of state, Eq. (8-22), were similar to van der Waals' constants, so that

$$a_M = \left(\Sigma y_i a_i^{1/2}\right)^2 \qquad (10\text{-}24)$$

and

$$b_M = \Sigma y_i b_i \qquad (10\text{-}25)$$

Prausnitz[26] has proposed that the Redlich-Kwong interaction parameter a_{12} be calculated using an equation such as (8-23) with critical properties of the binary mixture rather than as a geometric mean. Benedict, Webb and Rubin[28] recommend that the mixture coefficients in their equation of state, Eq. (8-25), be calculated as follows:

$$B_{o,M} = \Sigma y_i B_{oi} \qquad (10\text{-}26)$$

$$A_{o,M} = \left(\Sigma_i y_i A_{oi}^{1/2}\right)^2 \qquad (10\text{-}27)$$

$$C_{o,M} = \left(\Sigma_i y_i C_{oi}^{1/2}\right)^2 \qquad (10\text{-}28)$$

$$b_M = \left(\Sigma_i y_i b_i^{1/3}\right)^3 \qquad (10\text{-}29)$$

$$a_M = \left(\Sigma_i y_i a_i^{1/3}\right)^3 \qquad (10\text{-}30)$$

$$c_M = \left(\Sigma_i y_i c_i^{1/3}\right)^3 \qquad (10\text{-}31)$$

$$\alpha_M = \left(\Sigma_i y_i \alpha_i^{1/3}\right)^3 \qquad (10\text{-}32)$$

and

$$\gamma_M = \left(\Sigma_i y_i \gamma_i^{1/2}\right)^2 \qquad (10\text{-}33)$$

Reid and Sherwood[18] present other mixing rules for the Benedict-Webb Rubin equation. The mixture coefficients for the virial equation of state, Eq. (8-26), are given by the relationships[26]

$$B_M(T) = \sum_{ij} y_i y_j B_{ij}(T) \qquad (10\text{-}34)$$

or for a binary mixture

$$B_M(T) = y_1^2 B_{11} + 2y_1 y_2 B_{12} + y_2^2 B_{22}$$

and

$$C_M(T) = \sum_{ijk} y_i y_j y_k C_{ijk}(T) \qquad (10\text{-}35)$$

or for a binary mixture

$$C_M(T) = y_1^3 C_{111} + 3y_1^2 y_2 C_{112} + 3y_1 y_2^2 C_{122} + y_2^3 C_{222}$$

Reid and Sherwood[18] discuss methods of estimating the two body interaction virial coefficients B_{12} and the three body interaction virial coefficients C_{112} and C_{122}.

Now it may be of interest to compare the results obtained in calculating the pressure exerted by a mixture of gases using various mixture combination techniques.

EXAMPLE PROBLEM 10-2. Calculate the pressure required to compress 1.0 lb mole of a mixture consisting of 75 mole percent hydrogen and 25 mole percent nitrogen into a volume of 1.5 ft³ at 122 °F using van der Waals' equation and

(a) Assuming Dalton's law of additive pressure holds;

(b) Assuming Amagat's law of additive volumes (Ideal Solution) holds; and

(c) Using van der Waals' mixture coefficients.

The following van der Waals' constants are available:

	$a[(ft^3/lb\ mole)^2\ atm]$	$b(ft^3/lb\ mole)$
Hydrogen	62.8	0.426
Nitrogen	346.0	0.618

374   Chap. 10   Mixtures

and $R = 0.7302 \dfrac{(\text{atm})(\text{ft}^3)}{(\text{lb mole})(°R)}$

(a) Using Dalton's law of additive pressure

$$P = p_{H_2} + p_{N_2}$$

we find

$$p_{N_2} = \dfrac{(0.25)(0.7302)(582)}{1.5 - (0.25)(0.618)} - \dfrac{(0.25)^2(346.0)}{1.5^2} = 69.4 \text{ atm}$$

$$p_{H_2} = \dfrac{(0.75)(0.7302)(582)}{1.5 - (0.75)(0.426)} - \dfrac{(0.75)^2(62.8)}{1.5^2} = 254.3 \text{ atm}$$

Thus

$$P = 69.4 + 254.3 = 323.7 \text{ atm}$$

(b) Using Amagat's law of additive volumes

$$v_{N_2} + v_{H_2} = V$$

we must solve

$$P = \dfrac{n_{H_2} RT}{v_{H_2} - n_{H_2} b_{H_2}} - \dfrac{n_{H_2}^2 a_{H_2}}{v_{N_2}^2}$$

$$P = \dfrac{n_{N_2} RT}{v_{N_2} - n_{N_2} b_{N_2}} - \dfrac{n_{N_2}^2 a_{N_2}}{v_{N_2}^2}$$

and

$$v_{N_2} + v_{H_2} = 1.5 \text{ ft}^3$$

simultaneously. If we assume

$$v_{N_2} = (0.25)(1.5) = 0.375 \text{ ft}^3$$

we find

$$P = \frac{(0.25)(0.7302)(582)}{0.375 - (0.25)(0.618)} - \frac{(0.25)^2(346.0)}{0.375^2} = 328.1 \text{ atm}$$

Further,

$$v_{H_2} = 1.5 - 0.375 = 1.125 \text{ ft}^3$$

so

$$P = \frac{(0.75)(0.7302)(582)}{1.125 - (0.75)(0.426)} - \frac{(0.75)^2(62.8)}{1.125^2} = 367.8 \text{ atm}$$

Since P can have only one value, we must adjust v_{N_2}. Assuming $v_{N_2} = 0.355 \text{ ft}^3$,

$$P = \frac{(0.25)(0.7302)(582)}{0.355 - (0.25)(0.618)} - \frac{(0.25)^2(346.0)}{0.355^2} = 358.3 \text{ atm}$$

Further,

$$v_{H_2} = 1.5 - 0.355 = 1.145 \text{ ft}^3$$

so

$$P = \frac{(0.75)(0.7302)(582)}{1.145 - (0.75)(0.426)} - \frac{(0.75)^2(62.8)}{1.145^2} = 359.2 \text{ atm}$$

This is close enough. A better answer is probably close to the average of these two final pressure results, or $P = 358.7$ atm.

(c) Using van der Waals' mixture coefficients

$$a_M = \left[y_{H_2} a_{H_2}^{1/2} + y_{N_2} a_{N_2}^{1/2} \right]^2 \tag{10-22}$$

$$= \left[0.75(62.8)^{1/2} + (0.25)(346.0)^{1/2} \right]^2$$

$$= 112.23 \text{ (ft}^3/\text{lb mole})^2 \text{ atm}$$

$$b_M = y_{H_2} b_{H_2} + y_{N_2} b_{N_2} \tag{10-23}$$

$$= [(0.75)(0.426) + (0.25)(0.618)]$$

$$= 0.474 \text{ ft}^3/\text{lb mole}$$

Substituting these values into van der Waals' equation, Eq. (8-3),

$$P = \frac{(0.7302)(582)}{1.5 - 0.474} - \frac{112.23}{1.5^2}$$

$$= 364.3 \text{ atm}$$

Experimentally, it has been found that for these conditions,

$$P = 348 \text{ atm}^{(29)}$$

The results of this problem are indicative of the difficulties encountered in evaluating the properties of mixtures under non-ideal conditions.

NON-IDEAL SOLUTIONS

Partial Molal Properties

In many cases, the properties of the mixture are entirely different from what would be predicted from taking a weighted average property of the pure components. Consider the problem of adding sugar, a solid at room temperature, to iced tea. Visual observation and taste indicate that after we have added the sugar, the sugar is a part of the liquid iced tea. We might ask ourselves if the sugar melted in the iced tea. This hardly seems possible if the sugar is a solid at room temperature. The answer to our dilemma of what happened to the sugar is that sugar and iced tea in a mixture or solution have properties that differ from those predicted from a weighted average of the pure component properties.

Although a thermodynamic property of a mixture Y_M is not an additive function of pure component properties, it is a function of the system pressure, temperature and composition,

$$Y_M(P, T, n_1, n_2, n_3, \ldots, n_N)$$

Taking the total derivative of Y_M

$$dY_M = \left(\frac{\partial Y_M}{\partial P}\right)_{T, n_1, n_2, \ldots} dP + \left(\frac{\partial Y_M}{\partial T}\right)_{P, n_1, n_2, \ldots} dT + \left(\frac{\partial Y_M}{\partial n_1}\right)_{P, T, n_2, n_3, \ldots} dn_1$$

$$+ \left(\frac{\partial Y_M}{\partial n_2}\right)_{P, T, n_1, n_3, \ldots} dn_2 + \ldots$$

or at constant P and T,

$$dY_M = \left(\frac{\partial Y_M}{\partial n_1}\right)_{P,T,n_2,n_3,\ldots} dn_1 + \left(\frac{\partial Y_M}{\partial n_2}\right)_{P,T,n_1,n_3,\ldots} dn_2 + \left(\frac{\partial Y_M}{\partial n_3}\right)_{P,T,n_1,n_2,\ldots} dn_3 + \ldots$$

In this equation, the coefficients of the differentials of n_i are called partial molal thermodynamic properties and are written as follows:

$$\left(\frac{\partial Y_M}{\partial n_i}\right)_{P,T,n_j} = \overline{Y}_i \qquad (10\text{-}36)$$

With this definition, we may write the total derivative of Y_M at constant P and T as

$$dY_M = \overline{Y}_1 \, dn_1 + \overline{Y}_2 \, dn_2 + \overline{Y}_3 \, dn_3 + \ldots \qquad (10\text{-}37)$$

Now, we may combine incremental amounts of solution, all of the same composition, to form a finite amount of this solution. Then, since the partial molal thermodynamic property of each increment is the same, we find

$$\Sigma dY_M = \overline{Y}_1 \Sigma dn_1 + \overline{Y}_2 \Sigma dn_2 + \overline{Y}_3 \Sigma dn_3 + \ldots$$

or

$$Y_M = \overline{Y}_1 n_1 + \overline{Y}_2 n_2 + \overline{Y}_3 n_3 + \ldots \qquad (10\text{-}38)$$

Since both the amounts of the individual components and the corresponding partial molal property can be varied, the total derivative of Eq. (10-38) is

$$dY_M = \overline{Y}_1 \, dn_1 + n_1 \, d\overline{Y}_1 + \overline{Y}_2 \, dn_2 + n_2 \, d\overline{Y}_2$$
$$+ \overline{Y}_3 \, dn_3 + n_3 \, d\overline{Y}_3 + \ldots \qquad (10\text{-}39)$$

In order that both Eqs. (10-37) and (10-39) hold simultaneously, it is necessary that

$$n_1 \, d\overline{Y}_1 + n_2 \, d\overline{Y}_2 + n_3 \, d\overline{Y}_3 + \ldots = 0 \qquad (10\text{-}40)$$

If we divide Eq. (10-40) by the derivative of n_1 and then hold P, T, n_2, and n_3 constant, we find

$$n_1 \left(\frac{\partial \overline{Y}_1}{\partial n_1}\right)_{P,T,n_2,n_3} + n_2 \left(\frac{\partial \overline{Y}_2}{\partial n_1}\right)_{P,T,n_2,n_3} + n_3 \left(\frac{\partial \overline{Y}_3}{\partial n_1}\right)_{P,T,n_2,n_3} + \ldots = 0 \qquad (10\text{-}41)$$

These relationships between partial molal properties are widely used in studies of mixtures.

As an example of the use of partial molal properties, consider the problem of estimating the volume of a non-ideal mixture. From Eq. (10-38),

$$V_M = n_1 \overline{V}_1 + n_2 \overline{V}_2 + n_3 \overline{V}_3 + \ldots \quad (10\text{-}42)$$

In order to evaluate this last equation, we need to determine the partial molal volumes. From Eq. (10-36),

$$\overline{V}_i = \left(\frac{\partial V_M}{\partial n_i}\right)_{P,T,n_j}$$

where V_M is a function of n_i. Since

$$V_M = n_T v_M$$

we find that

$$V_M = (n_1 + n_2 + n_3) v_M$$

Then differentiating V_M with respect to n_i, holding P, T, and the moles of all other components n_j constant, we find

$$\overline{V}_i = v_M \left(\frac{\partial n_T}{\partial n_i}\right)_{P,T,n_j} + n_T \left(\frac{\partial v_M}{\partial n_i}\right)_{P,T,n_j}$$

or

$$\overline{V}_i = v_M + n_T \left(\frac{\partial v_M}{\partial n_i}\right)_{P,T,n_j} \quad (10\text{-}44)$$

This relationship applies to all partial molal quantities.

For the special case of a binary mixture, these results can be represented graphically. On Fig. 10-2, the specific volume of a binary non-ideal mixture as a function of composition at constant pressure and temperature is represented by the solid line drawn as an arc between v_1 and v_2. The straight dashed line between the same two points represents the specific volume of an ideal solution since

$$v_M^I = x_1 v_1 + x_2 v_2$$

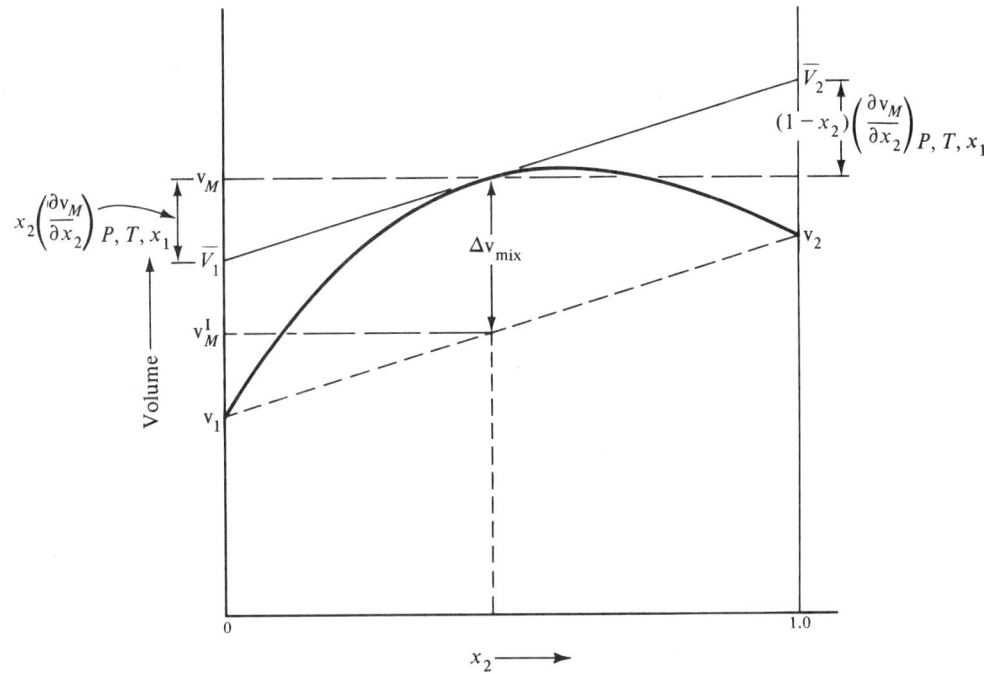

Fig. 10-2. *Volumes of a mixture at constant temperature and pressure.*

where the superscript I indicates ideal solution. As shown in Fig. 10-2, the difference between the actual specific volume of a mixture of fixed composition and the specific volume that the same mixture would have at the same pressure and temperature if it formed an ideal solution, is

$$\Delta v_{mix} = v_M - v_M^I \tag{10-45}$$

Now we can use Eq. (10-44) to locate the position of points representing $\overline{V}_1$ and $\overline{V}_2$ on Fig. 10-2. For a binary mixture,

$$x_2 = \frac{n_2}{n_1 + n_2} = \frac{n_2}{n_T}$$

Differentiating at constant n_1,

$$dx_2 = \frac{n_T - n_2}{n_T^2} dn_2$$

or

$$\frac{n_T}{dn_2} = \frac{1 - x_2}{dx_2}$$

Thus, from Eq. (10-44),

$$\overline{V}_2 = v_M + (1-x_2)\left(\frac{\partial v_M}{\partial x_2}\right)_{P,T,x_1} \tag{10-46}$$

This result has been used to locate the point representing the partial molal volume of component 2 in the binary mixture $\overline{V}_2$ on Fig. 10-2. Similarly, we can show

$$\overline{V}_1 = v_M + (1-x_1)\left(\frac{\partial v_M}{\partial x_1}\right)_{P,T,x_2}$$

But

$$x_1 + x_2 = 1.0$$

and

$$dx_1 = -dx_2$$

so, in terms of x_2, we find

$$\overline{V}_1 = v_M - x_2\left(\frac{\partial v_M}{\partial x_2}\right)_{P,T,x_1} \tag{10-47}$$

This result has been used to locate a point representing the partial molal volume of component 1 in a binary mixture $\overline{V}_1$ on Fig. 10-2. Then, the specific volume of the mixture is represented by a point on the straight line drawn between $\overline{V}_1$ and $\overline{V}_2$ on Fig. 10-2 such that

$$v_M = x_1 \overline{V}_1 + x_2 \overline{V}_2 \tag{10-48}$$

At constant pressure and temperature $\overline{V}_i$ can vary with the composition of the mixture even though the specific volume of the pure component v_i does not vary.

Excess Functions

As we have seen, thermodynamic properties of a mixture are functions not only of system pressure and temperature but also of the mixture composition. If the effect of changes in composition on the specific volume of the mixture were known, then the effect of composition on changes in other thermodynamic properties could be determined using relationships similar to those used to predict changes in thermodynamic properties of pure components.

Analytically, the specific volume of a mixture can be expressed in terms of pure component specific volumes by combining Eq. (10-12) and (10-45) to give

$$v_M = x_1 v_1 + x_2 v_2 + \Delta v_{mix} \tag{10-49}$$

or in terms of partial molal quantities,

$$v_M = x_1 \overline{V}_1 + x_2 \overline{V}_2 \tag{10-48}$$

If the mixture forms an ideal solution,

$$v_M^I = x_1 v_1 + x_2 v_2 \tag{10-12}$$

In this case,

$$\Delta v_{mix}^I = 0$$

so that for an ideal solution,

$$\overline{V}_i^I = \left(\frac{\partial (n_T v_M^I)}{\partial n_i}\right)_{P,T,n_j} \tag{10-50}$$

or

$$\overline{V}_i^I = v_i^I$$

Using this result, it can be shown that for an ideal solution,

$$\overline{U}_i^I = u_i^I \tag{10-51}$$

and

$$\overline{H}_i^I = h_i^I \tag{10-52}$$

On the other hand, if the mixture does not form an ideal solution, Δv_{mix} will not be zero. In fact, Δv_{mix} may be looked upon as a measure of the non-ideality of the mixture. We may study the non-ideality of mixtures by examining the deviation of the properties of the actual solution from the properties of an ideal solution. G. Scatchard and C. L. Raymond introduced convenient functions to use in doing this called the excess functions. These are defined by the relation

$$Y^E = Y_M - Y_M^I \tag{10-53}$$

where the superscript E refers to the excess thermodynamic property. Thus,

$$v^E = v_M - v_M^I = \Delta v_{mix} \tag{10-54}$$

$$h^E = h_M - h_M^I = \Delta h_{mix} \tag{10-55}$$

$$s^E = s_M - s_M^I = \Delta s_{mix} + R\Sigma x_i \ln x_i \tag{10-56}$$

$$g^E = g_M - g_M^I = \Delta g_{mix} - RT\Sigma x_i \ln x_i \tag{10-57}$$

and

$$a^E = a_M - a_M^I = \Delta a_{mix} - RT \Sigma x_i \ln x_i \tag{10-58}$$

Now, instead of working with a total mixture property such as v_M as a function of composition, we might examine v^E which is a much more sensitive function of composition. In cases where the deviation between the actual volume of the mixture and the volume of an ideal solution is symmetrical, we might write

$$v^E = Ax_1 x_2$$

In this case, it would be theoretically possible to determine the constant A from one experimental data point and then estimate v_M at any other composition. Much of the experimental information on non-ideal mixtures that is published in the literature is presented in the form of excess functions.

PROBLEMS

10-1. A tank having a volume of 10 ft^3 contains oxygen at 50 lb$_f$/in.^{2}, 80 °F. Nitrogen at a pressure of 100 lb$_f$/in.^{2}, 300 °F flows from a pipe into the tank until the pressure reaches 90 lb$_f$/in.^{2} The entire process is adiabatic. Assuming the gases act as ideal gases, determine the final temperature of the mixture and the change in entropy for this process.

10-2. One hundred pound moles per hour of a mixture consisting of 50 pound moles of propane and 50 pound moles of butane are to be compressed isothermally at 285 °F from 1000 lb$_f$/in.^{2} to 2000 lb$_f$/in.^{2} in a centrifugal compressor. If the mixture is an ideal gas, what is the minimum work required for this compression process?

10-3. A 10 ft^3 cylinder contains 70 mole percent methane and 30 mole percent ethane at 1000 lb$_f$/in.^{2}, 100 °F. A valve at the top of the cylinder is cracked open, the pressure in the cylinder drops slowly to 600 lb$_f$/in.^{2}, and then the valve is closed. If it is assumed that no heat is transferred through the insulation and the mixture forms an ideal gas, how many moles of gas have escaped from the cylinder? Assume c_P^* for methane is 8.3 BTU/lb mole °R; c_P^* for ethane is 10.2 BTU/lb mole °R.

10-4. Air ($MW = 28.96$) at 77 °F and 100 atmospheres total pressure is to be separated into pure oxygen ($MW = 32$) and pure nitrogen ($MW = 28$) at 77 °F and 10 atmospheres. If air is an ideal gas and consists of 79 mole percent nitrogen and 21 mole percent oxygen, what is the minimum work required to produce 1000 pounds per hour of oxygen in a flow process?

10-5. Solve Prob. 10-2 assuming the properties of the pure components are adequately represented by the equation $PV = nzRT$ and that the mixture forms an ideal solution.

10-6. Solve Prob. 10-4 assuming the properties of the pure components are adequately represented by the equation $PV = nzRT$ and that the mixture forms an ideal solution.

10-7. One hundred pound moles of the following mixture of gases is at a pressure of 1000 lb_f/in^2 and a temperature of 200 °F:

Component	Mole percent
Methane	40
Ethane	30
Propane	30

Using the equation $PV = nzRT$, calculate the volume occupied by this mixture.

10-8. What is the pressure required to compress 1.0 pound mole of a mixture of gases consisting of 75 mole percent hydrogen and 25 mole percent nitrogen into a volume of ten cubic feet at 122 °F?

Calculate your answer using

(a) Kay's rule and the relationship $PV = nzRT$

(b) Dalton's law and the relationship $PV = nzRT$

(c) The Redlich-Kwong equation and mixture coefficients. The following data may be used:

	Redlich-Kwong Constants	
Compound	$a[°R^{1/2}(ft^3/lb\ mole)^2\ atm]$	$b(ft^3/lb\ mole)$
Hydrogen	522.47	0.226
Nitrogen	5286.34	0.429

10-9. Estimate the volume of a mixture of 0.6 lb moles of methane and 0.4 lb moles of propane at 400 °F and 2000 lb_f/in^2 using

(a) The Redlich-Kwong equation of state with mixture coefficients.

(b) Kay's rule and the generalized charts.

10-10. Estimate the volume of a mixture containing 0.6 lb moles of methane and 0.4 lb moles of propane at 400 °F and 2000 lb_f/in^2 using the Benedict-Webb-Rubin equation and mixture coefficients.

Hint: This problem may be most easily solved using the digital computer.

10-11. Solve Prob. 10-2 assuming the properties of the pure components and the mixture are adequately represented by the equation $PV = nzRT$

10-12. Solve Prob. 10-3 assuming the properties of the pure components and the mixture are adequately represented by the equation $PV = nzRT$

10-13. Solve Prob. 10-4 assuming the properties of the pure components and the mixture are adequately represented by the equation $PV = nzRT$

10-14. The following mixture densities have been reported for the liquid system, ethanol–water at 1 atm and 20 °C:

Wt percent ethanol	Density(gms/cc)
0.0	0.998
10.0	0.982
20.0	0.969
30.0	0.954
40.0	0.935
50.0	0.914
60.0	0.891
70.0	0.868
80.0	0.843
90.0	0.818
100.0	0.789

(a) Using the chord area method of graphical differentiation, calculate the partial molal volume of ethanol and of water for a mixture containing 80.0 wt percent ethanol.

(b) Plot the excess volume as a function of composition.

11

Phase Equilibrium

OBJECTIVES

After studying this chapter, the student should be able to

(1) apply the Clausius-Clapeyron equation, Eq. (11-6), in estimating pure component vapor pressures and heats of vaporization;

(2) apply the Gibbs Phase Rule, Eq. (11-22), to phase equilibrium problems;

(3) apply the criteria of equilibrium in the form of Eqs. (11-2), (11-3), and (11-16) along with the principle of the conservation of mass to phase equilibrium problems;

(4) define fugacity and activity for components in a mixture; and

(5) define and evaluate numerically, the terms in the relationships for the fugacity of a component in a gas mixture.

Note that this chapter contains introductory material on the topics of fugacities, non-ideal mixture activity coefficients and multi-component phase equilibrium calculations which may not be of interest to the general reader. Therefore, this material is presented at the end of the chapter where it can be conveniently omitted.

THERMODYNAMIC EQUILIBRIUM

As was pointed out in Chap. 1, the concept of the equilibrium state is fundamental in thermodynamics. Tables of thermodynamic data as well as equations of state are applicable only to systems which may be considered to be in a state of thermodynamic equilibrium. In this state, the thermodynamic potentials or forces acting on the system as well as within the system are in a stable state of balance.

Experimentally, we might attempt to identify a system at equilibrium by observing whether or not its properties change over a very long period of time. If the properties of the system remain constant, and if then the system is slightly perturbed and the properties return to the same values they had before the perturbation, the system is in thermodynamic equilibrium. This test, however, is time consuming and subject to experimental errors. Fortunately, thermodynamics provides us with less time consuming and more precise criteria for stable equilibrium.

In order to develop the relationship between thermodynamics and the equilibrium state, let us consider the system shown in Fig. 11-1. This system has been

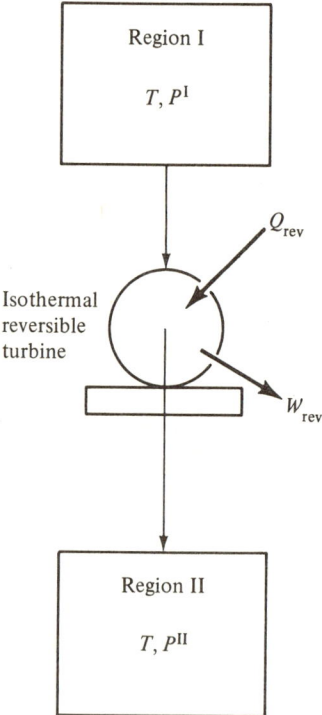

Fig. 11-1. *Thermodynamic system.*

arbitrarily divided into two elements or regions. Initially, the system is in a non-equilibrium state in which region I is at pressure P^I while region II is at pressure P^{II}. Both regions are at the same temperature T. Since there is a pressure difference within this system, we will have a net flow of mass and energy between the two regions of the system. In order to convert the flow of energy to work, let us insert an isothermal, reversible turbine between region I and region II. Then, taking the turbine as the system and basing our calculation on the time required for n moles to flow between the two systems, the work which could be attributed to the difference in pressure between region I and region II is given by the relationship

$$-W_{rev} = n(h^{II} - h^{I})_T - nT(s^{II} - s^{I})_T$$

or

$$-W_{rev} = n(g^{II} - g^{I})_T \qquad (9\text{-}12)$$

where we have neglected kinetic and potential energy effects in reducing the generalized mass, energy and entropy balance equations.

Now let us allow the system shown in Fig. 11-1 to approach equilibrium. At equilibrium, all forces within the system will be in balance and the work associated with the turbine will be zero. In this case, even though the net flow n through the turbine is zero, at any instant of time there may be a random flow of molecules either into or out of the turbine due to normal molecular motion. Therefore, in order to insure that the work associated with the turbine in the equilibrium system is zero, it is necessary to specify that

$$g_T^I = g_T^{II} \qquad (11\text{-}1)$$

Further, if the system is to be in thermal equilibrium, we require that

$$T^I = T^{II} \qquad (11\text{-}2)$$

and if the system is to be in mechanical equilibrium and if the body forces exerted by potential fields can be neglected, we require that

$$P^I = P^{II} \qquad (11\text{-}3)$$

Equations (11-1), (11-2), and (11-3) are the criteria for an equilibrium state. Although g, T, and P will be the same in all systems at equilibrium, all other intensive properties may have different values in different systems.

One broad area in which Eqs. (11-1), (11-2), and (11-3) have been successfully applied is the area of phase equilibrium. As defined in Chap. 1, a phase is any homogenous portion of matter bounded by a physical surface. Now many systems of interest contain more than one phase. The system shown in Fig. 11-1, for example, might consist of two phases arbitrarily divided so that phase I is in region I and phase II is in region II. Then, Eqs. (11-1), (11-2), and (11-3) can be applied to this heterogeneous system by assuming that the superscripts in these equations indicate different phases rather than different elements of the same phase.

In this chapter, we will wish to examine relationships between two or more phases in equilibrium within a system. In order to initiate this study, let us begin by examining phase equilibrium within a single component system.

PHASE EQUILIBRIUM IN SINGLE COMPONENT SYSTEMS

Clausius-Clapeyron Equation

From Eq. (4-20), we find that for a pure component in the vapor phase,

$$dg_g = -s_g \, dT + v_g \, dP$$

and for the pure component in the liquid phase,

$$dg_f = -s_f \, dT + v_f \, dP$$

Using the differential form of Eq. (11-1), we find that for a vapor and liquid in equilibrium,

$$(v_g - v_f) \, dP^\square = (s_g - s_f) \, dT$$

or

$$\frac{dP^\square}{dT} = \frac{s_g - s_f}{v_g - v_f} \qquad (11\text{-}4)$$

where $P^\square$ is the vapor pressure and $dP^\square/dT$ is the slope of the vapor pressure-temperature curve. Equation (11-4) is the Clapeyron equation. For a system containing both liquid and vapor, the vaporization process is reversible, and

$$h_{fg} = \Delta q_{rev}$$

Then, since

$$\Delta s = \frac{\Delta q_{rev}}{T} \qquad (3\text{-}1)$$

we find that

$$s_{fg} = \frac{h_{fg}}{T}$$

so that we may rewrite the Clapeyron equation as

$$\frac{dP^\square}{dT} = \frac{h_{fg}}{Tv_{fg}} \tag{11-5}$$

From the equation of state $Pv = zRT$, we find that

$$v_g - v_f = \frac{RT}{P^\square}(z_g - z_f)$$

or

$$v_{fg} = \frac{RTz_{fg}}{P^\square}$$

Substituting this into Eq. (11-5), we obtain the result

$$\frac{d\ln P^\square}{dT} = \frac{h_{fg}}{RT^2 z_{fg}} \tag{11-6}$$

Equation (11-6) is the Clausius-Clapeyron equation. Even though values of h_{fg} depend upon temperature, the ratio h_{fg}/z_{fg} is essentially independent of temperature. Thus, if we integrate Eq. (11-6), we obtain the result

$$\ln \frac{P_2^\square}{P_1^\square} = \frac{h_{fg}}{Rz_{fg}}\left(\frac{1}{T_1} - \frac{1}{T_2}\right) \tag{11-7}$$

This result says that if we plot the natural logarithm of vapor pressure as a function of reciprocal temperature, we should obtain a straight line of slope h_{fg}/Rz_{fg}. Such a plot is shown in Fig. 11-2.

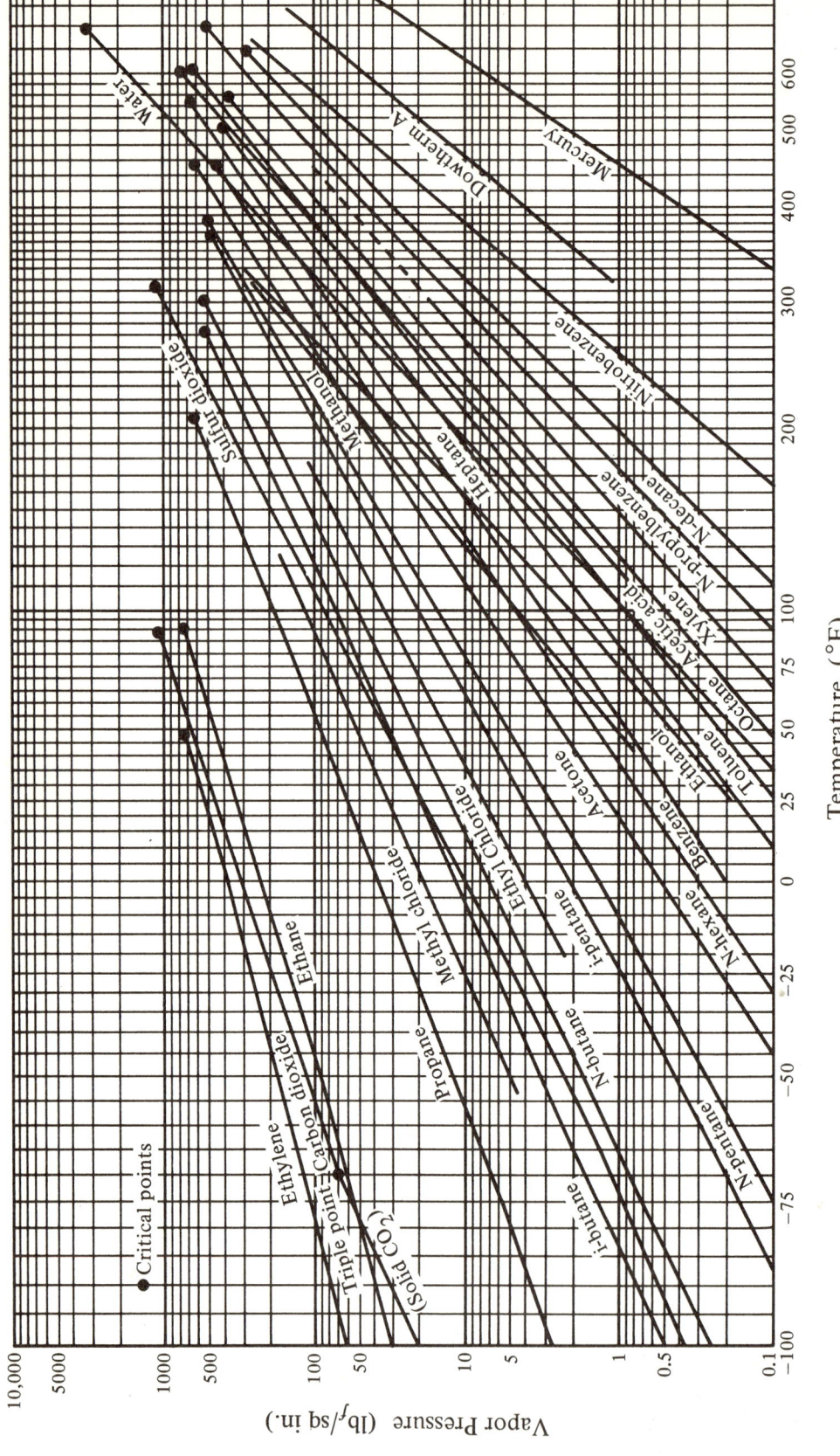

Fig. 11-2. *Vapor pressures of common liquids.*

Several equations have been developed to present vapor pressure as a function of temperature using the Clausius-Clapeyron equation as a basis. If we let $P_1^\square = 1$ atm, then $T_1 = T_b$, where T_b is the saturation temperature corresponding to a vapor pressure of one atmosphere, or the atmospheric boiling temperature, and Eq. (11-7) becomes

$$\ln P^\square = \frac{h_{fg}}{RT_b z_{fg}} - \frac{h_{fg}}{R z_{fg}}\left(\frac{1}{T}\right)$$

or

$$\ln P^\square = A - \frac{B}{T} \tag{11-8}$$

Equation (11-8) is known as the Clausius equation and is the simplest equation relating vapor pressure and temperature.

Actually, there is a very slight curvature in a plot of experimental $\ln P$ vs T^{-1} since the ratio h_{fg}/z_{fg} does vary slightly with temperature. This curvature may be taken into account if we modify the Clausius equation.

The Antoine equation,

$$\ln P^\square = A - \frac{B}{T+C} \tag{11-9}$$

is the best three-constant equation relating vapor pressure and temperature. The best four-constant equation relating vapor pressure and temperature is the Frost-Kalkwarf equation,

$$\ln P^\square = A - \frac{B}{T} - C \ln T + \frac{DP}{T^2} \tag{11-10}$$

The Martin-Shinn-Kapoor equation,

$$\ln P^\square = A - \frac{B}{T} - C \ln T + \frac{E(F-T)}{FT} \ln (F-T) \tag{11-11}$$

is one of the better five-constant equations relating vapor pressure and temperature. The particular choice of which vapor-pressure equation to use depends upon the precision of the results required.

Enthalpy of Vaporization

The parameter A of the Clausius equation is relatively constant for a wide variety of substances. Because of this relative constancy of A, F. Trouton proposed the relationship

$$\frac{h_{fg}}{T_b} = 21 \text{ (BTU/lb mole }^\circ\text{R)} \tag{11-12}$$

Implicit in Trouton's equation is the assumption that at atmospheric pressure, z_{fg} is unity.

Equation (11-12) provides us with a means of estimating enthalpies of vaporization of various compounds. Unfortunately, the ratio of h_{fg} to T_b for hydrocarbons increases with increasing molecular weight. To account for this variation, W. Kistiakowsky proposed the relationship

$$\frac{h_{fg}}{T_b} = 8.75 + R \ln \left[\frac{T_b(°R)}{1.8} \right] \left(\frac{\text{BTU}}{\text{lb mole °R}} \right) \tag{11-13}$$

This equation is a definite improvement over Trouton's rule. However, Kistiakowsky's equation has been found to be inapplicable to polar compounds. More recently, S. H. Fishtine has proposed the equation,

$$\frac{h_{fg}}{T_b} = K_F \left[8.75 + R \ln \frac{T_b(°R)}{1.8} \right] \left(\frac{\text{BTU}}{\text{lb mole °R}} \right) \tag{11-14}$$

where K_F is Fishtine's constant[30] and is presented in Table 11-1. Other methods of estimating heats of vaporization have been summarized by Reid and Sherwood[18].

For heats of vaporization at temperatures other than the atmospheric boiling temperature, K. M. Watson suggests the empirical relationship

$$\frac{(h_{fg})_b}{h_{fg}} = \left[\frac{1 - (T_b)_r}{1 - T_r} \right]^{0.38} \tag{11-15}$$

PHASE EQUILIBRIUM IN MULTI-COMPONENT SYSTEMS

The Gibbs Phase Rule

We have defined the equilibrium state of a system as a state of balance between thermodynamic forces in a system. In systems containing more than one component, we must take into account a potential which is associated with each individual species.

For a pure component system consisting of two phases, we found that at equilibrium

$$g_T^{\text{I}} = g_T^{\text{II}} \tag{11-1}$$

TABLE 11-1. Correction Factors to the Kistiakowsky Equation[30] (For Aliphatic and Alicylic Organic Compounds)

Compound type	\multicolumn{12}{c}{n, number of carbon atoms in compound, including carbon atoms of functional group}
	1	2	3	4	5	6	7	8	9	10	11	12-20
Hydrocarbons:												
n-Alkanes	0.97	1.00	1.00	1.00	1.00	1.00	1.00	1.00	1.00	1.00	1.00	1.00
Alkane isomers				0.99	0.99	0.99	0.99	0.99	0.99	0.99	0.99	0.99
Mono- and diolefins and isomers		1.01	1.01	1.01	1.01	1.01	1.01	1.01	1.01	1.01	1.01	1.00
Cyclic saturated hydrocarbons			1.00	1.00	1.00	1.00	1.00	1.00	1.00	1.00	1.00	1.00
Alkyl derivatives of cyclic saturated hydrocarbons				0.99	0.99	0.99	0.99	0.99	0.99	0.99	0.99	0.99
Halides (saturated or unsaturated):												
Monochlorides	1.05	1.04	1.03	1.03	1.03	1.03	1.03	1.03	1.02	1.02	1.02	1.01
Monobromides	1.04	1.03	1.03	1.03	1.03	1.03	1.02	1.02	1.02	1.01	1.01	1.01
Monoiodides	1.03	1.02	1.02	1.02	1.02	1.02	1.01	1.01	1.01	1.01	1.01	1.01
Polyhalides (not entirely halogenated)	1.05	1.05	1.05	1.04	1.04	1.04	1.03	1.03	1.03	1.02	1.02	1.01
Mixed Halides (completely halogenated)	1.01	1.01	1.01	1.01	1.01	1.01	1.01	1.01	1.01	1.01	1.01	1.01
Perfluorocarbons	1.00	1.00	1.00	1.00	1.00	1.00	1.00	1.00	1.00	1.00	1.00	1.00
Compounds containing the keto group:												
Esters		1.14	1.09	1.08	1.07	1.06	1.05	1.04	1.04	1.03	1.02	1.01
Ketones			1.08	1.07	1.06	1.06	1.05	1.04	1.04	1.03	1.02	1.01
Aldehydes		1.09	1.08	1.08	1.07	1.06	1.05	1.04	1.04	1.03	1.02	1.01
Nitrogen compounds:												
Primary amines	1.16	1.13	1.12	1.11	1.10	1.10	1.09	1.09	1.08	1.07	1.06	1.05†
Secondary amines		1.09	1.08	1.08	1.07	1.07	1.06	1.05	1.05	1.04	1.04	1.03†
Tertiary amines			1.01	1.01	1.01	1.01	1.01	1.01	1.01	1.01	1.01	1.01
Nitriles		1.05	1.07	1.06	1.06	1.05	1.05	1.04	1.04	1.03	1.02	1.01
Nitro compounds	1.07	1.07	1.07	1.06	1.06	1.05	1.05	1.04	1.04	1.03	1.02	1.01
Sulfur compounds:												
Mercaptans	1.05	1.03	1.02	1.01	1.01	1.01	1.01	1.01	1.01	1.01	1.01	1.01
Sulfides		1.03	1.02	1.01	1.01	1.01	1.01	1.01	1.01	1.01	1.01	1.01
Alcohols:												
Alcohols (single-OH group)	1.22	1.31	1.31	1.31	1.31	1.30	1.29	1.28	1.27	1.26	1.25	1.24†
Diols (glycols or condensed glycols)		1.33	1.33	1.33	1.33	1.33	1.33	1.33				
Triols (glycerol, etc.)			1.38	1.38	1.38							
Cyclohexanol, cyclohexyl methyl alcohol, etc.						1.20	1.20	1.21	1.24	1.26		
Miscellaneous compounds:												
Ethers (aliphatic only)		1.03	1.03	1.02	1.02	1.02	1.01	1.01	1.01	1.01	1.01	1.01
Oxides (cyclic ethers)		1.08	1.07	1.06	1.05	1.05	1.04	1.03	1.02	1.01	1.01	1.01

Notes:
1. Alicyclic compounds are carbocyclic or heterocyclic compounds having aliphatic properties.
2. Consider any phenyl group as a single carbon atom.
3. K_F factors are the same for all aliphatic isomers of a given compound. For example, $K_F = 1.31$ for n-butyl alcohol, i-butyl alcohol, t-butyl alcohol, and s-butyl alcohol.
4. In organometallic compounds, consider any metallic atom as a carbon atom.

†For $n = 12$ only; no prediction is made for K_F where $n > 12$.

K_F for non-substituted simple aromatic compounds is near unity.

Reprinted from *I & EC*, 55, 6, 47(1963). Copyright 1963 by the American Chemical Society. Reprinted by permission of the copyright owner.

By analogy, for a multi-component system consisting of two phases, we find that

$$\overline{G}_i^{\mathrm{I}} = \overline{G}_T^{\mathrm{II}}$$

where

$$\overline{G}_i = \left(\frac{\partial G}{\partial n_i}\right)_{P,T,n_j}$$

The subscript n_j indicates the moles of all components except component i. The partial molal Gibbs free energy of component i in a mixture $\overline{G}_i$ is called the chemical potential and is given the special symbol μ_i. Then, the criteria of equilibrium between two phases in a multi-component system is that

$$\mu_i^{\mathrm{I}} = \mu_i^{\mathrm{II}} \tag{11-16}$$

as well as

$$T^{\mathrm{I}} = T^{\mathrm{II}} \tag{11-2}$$

and

$$P^{\mathrm{I}} = P^{\mathrm{II}} \tag{11-3}$$

These equations can be extended to multiphase systems and used to tell us something about the phase variables in equilibrium systems. Phase variables include not only temperature and pressure but also phase concentrations in terms of mole fractions x_i where

$$x_i = \frac{n_i}{\Sigma n_i}$$

Consider a system of N components at equilibrium in π phases being acted upon only by the outside forces, pressure and temperature. Since

$$\Sigma x_i = 1.0$$

for each phase, and since we note from Eqs. (11-2) and (11-3) that the pressure and temperature will be the same in all phases, a complete definition of the system requires knowledge of the following variables:

for phase I: $x_1^I, x_2^I \ldots, x_{N-1}^I, P, T$

for phase II: $x_1^{II}, x_2^{II} \ldots, x_{N-1}^{II}, P, T$

. . .
. . .
. . .

for phase π: $x_1^\pi, x_2^\pi \ldots, x_{N-1}^\pi, P, T$

From this, we see that the total number of phase variables is

$$\pi(N-1) + 2$$

There are interrelationships between the variables, however. For π phases in equilibrium, we can write for each component,

$$\mu_1^I = \mu_1^{II} = \mu_1^{III} = \ldots = \mu_1^\pi$$
$$\mu_2^I = \mu_2^{II} = \mu_2^{III} = \ldots = \mu_2^\pi$$

. . . .
. . . .
. . . .

$$\mu_N^I = \mu_N^{II} = \mu_N^{III} = \ldots = \mu_N^\pi$$

Thus, for any one component, there occurs a set of $(\pi - 1)$ independent equations—taking the relations in pairs. For N components, we have

$$N(\pi - 1)$$

equations.

Now, from algebra, we recall that the number of independent variables F equals the total number of variables $\pi(N-1) + 2$, less number of independent relating equations $N(\pi - 1)$. That is,

$$F = \pi(N-1) + 2 - N(\pi - 1)$$

or

$$F + \pi = N + 2 \tag{11-17}$$

Equation (11-17) is known as the Gibbs Phase Rule. In this equation, F represents the number of independent phase variables or additional relationships between these variables that must be specified in addition to the restrictions imposed by the criteria of equilibrium in order to uniquely define the state of a system. An equilibrium system for which all the independent variables have been specified has zero degrees of freedom and can be represented as a point on a PT diagram. An example would be the triple point of water where three phases coexist in equilibrium. Systems with one unspecified independent variable such as saturated water where two phases are in equilibrium, have one degree of freedom and can be represented as a line while systems with two unspecified independent variables have two degrees of freedom and can be represented as a surface on a PT diagram.

However, even if all the independent variables in an equilibrium system are specified, the problem of numerically evaluating the remaining unspecified dependent variables remains. All we know at this point is that the dependent variables have some unique, but unknown values which must satisfy the equilibrium criteria relationships,

$$T^\mathrm{I} = T^\mathrm{II} = \ldots = T^\pi \tag{11-2}$$

$$P^\mathrm{I} = P^\mathrm{II} = \ldots = P^\pi \tag{11-3}$$

$$\mu_i^\mathrm{I} = \mu_i^\mathrm{II} = \ldots = \mu_i^\pi \tag{11-16}$$

In order to utilize these equilibrium criteria relationships in determining numerical values of the dependent variables, we need to have an analytical relationship between chemical potential and the phase variables; pressure, temperature and phase composition.

Partial Pressure as a Criterion of Equilibrium

Raoult's and Dalton's Law. One of the first attempts to develop analytical relationships between phase variables was based on the assumption that if the total pressure in each phase must be equal, then the partial pressure of a specific substance in all phases must be equal. That is,

$$p_i^\mathrm{I} = p_i^\mathrm{II} = \ldots = p_i^\pi \tag{11-18}$$

As we will see later, this is equivalent to the relationship

$$\mu_i^\mathrm{I} = \mu_i^\mathrm{II} = \ldots = \mu^\pi \tag{11-16}$$

for many cases of interest.

If a vapor system containing more than one component is at a pressure and temperature at which the ideal gas law applies, then Dalton's law of partial pressure holds and

$$p_{g,i} = P y_i \qquad (10\text{-}2)$$

where y_i is the mole fraction of component i in the vapor phase and P is the total pressure of the vapor phase. Raoult suggested that in a liquid or solid mixture, the partial pressure of component i is the partial pressure of that component in the vapor above the liquid or solid and is equal to the vapor pressure of component i at the system temperature times the fraction of the total liquid or solid surface area occupied by component i. That is

$$p_{f,i} = P_i^{\square} \times \frac{A_i}{A_{\text{total}}}$$

where A_i is the surface area occupied by component i and A_{total} is the total surface area.

It is difficult to measure the surface area occupied by component i. But if the components in solution are uniformly distributed, we can closely approximate this area by assuming that

$$\frac{A_i}{A_{\text{total}}} = \frac{v_i}{V_{\text{total}}}$$

where v_i is the partial volume occupied by component i and V_{total} is the total volume of the solution. Then the partial pressure of component i above the liquid or solid is

$$p_{f,i} = P_i^{\square} \times \frac{v_i}{V_{\text{total}}}$$

Further, if the solution is ideal and thus follows Amagat's law, we find from Eq. (10-4) that

$$\frac{v_i}{V_{\text{total}}} = x_i$$

so that

$$p_{f,i} = P_i^{\square} x_i \qquad (11\text{-}19)$$

where x_i is the mole fraction of component i in the liquid or solid phase. Equation (11-19) is known as Raoult's law.

Combining Eqs. (10-2) and (11-19) we find that a working relationship between phase variables for systems consisting of ideal solutions under conditions at which the ideal gas law applies to the vapor is

$$Py_i = P_i^{\square} x_i \qquad (11\text{-}20)$$

Equation (11-20) is sometimes called Raoult's and Dalton's law and is widely used to evaluate systems of vapor and liquid in equilibrium. It relates vapor and liquid phase compositions to the system pressure and the vapor pressure of each of the N components in a system.

In addition, the criteria of mechanical and thermal equilibrium, Eqs. (11-2) and (11-3), require that

$$P_f = P_g$$

$$T_f = T_g$$

If we also apply the relationships

$$\Sigma x_i = 1.0$$

to the liquid phase, and

$$\Sigma y_i = 1.0$$

to the gas phase for the system of interest, then the Gibbs Phase Rule, Eq. (11-17), tells us that the total number of additional relationships required to completely define all of the independent phase variables in the two phase equilibrium system is equal to the number of components in the system N. These additional relationships might be obtained, for example, by specifying the system pressure and temperature and $N - 2$ mole fractions in the liquid phase. In many cases, component material balances are used to provide the required additional relationships.

For the special case of a two phase binary mixture, specification of the system temperature and pressure defines all independent variables. Then the vapor and liquid phase compositions can be calculated using the relationships

$$x_1 + x_2 = 1.0$$

$$y_1 + y_2 = 1.0$$

as well as Eq. (11-20),

$$Py_1 = P_1^\square x_1$$

$$Py_2 = P_2^\square x_2$$

for each component. These four equations have been successfully applied to the design of humidifiers and driers operating at ambient temperature and atmospheric pressure by assuming that air and water form a binary mixture.

Humidification. A major application of phase equilibrium concepts is the analysis of humidification processes. These processes may be carried out to control the humidity of a space or to cool process water by contacting it with low humidity air. Dehumidification is practiced most commonly as one function in an air conditioning system.

A contribution of thermodynamics to the analysis of humidification processes is to provide a means of calculating the amount of water-vapor contained in air under a specified set of conditions. We might begin our analysis of this problem by determining the amount of water contained in air saturated with water-vapor at a given pressure and temperature. Saturation of a noncondensible gas such as air with a condensible vapor such as water implies that the water in the vapor phase is in equilibrium with condensed water in the liquid phase.

If we apply Eq. (11-20) to the air-water system at equilibrium, we find

$$y_W P = x_W P_W^\square$$

where subscript W indicates water. However, essentially no air is dissolved in the water, so

$$x_W = 1.0$$

and the mole fraction of water-vapor in the air-water mixture at equilibrium is

$$y_W = \frac{P_W^\square}{P}$$

Then, since

$$y_W + y_A = 1.0$$

we find

$$y_A = 1 - \frac{P_W^\square}{P}$$

where the subscript A indicates bone dry air. From the foregoing, it follows that at equilibrium, the moles of water in a mole of bone dry air is

$$\left(\frac{y_W}{y_A}\right)_{sat} = \frac{P_W^{\square}}{P - P_W^{\square}} = \left(\frac{n_W}{n_A}\right)_{sat} \tag{11-21}$$

In this equation, we have used the subscript sat to emphasize the fact that these are equilibrium relationships obtained at saturation conditions.

For the air-water system, a special terminology has been built up. For example, the ratio expressed by Eq. (11-21) is called the molar humidity at saturation. The mass of water contained in a unit mass of bone dry air is given by the relationship

$$\left(\frac{m_W}{m_A}\right)_{sat} = \left(\frac{y_W}{y_A}\right)_{sat} \times \frac{MW_W}{MW_A}$$

But

$$\frac{MW_W}{MW_A} = \frac{18.0}{28.96} = 0.622$$

Thus

$$\left(\frac{m_W}{m_A}\right)_{sat} = 0.622 \frac{P_W^{\square}}{P - P_W^{\square}} \tag{11-22}$$

In the air-water system, this ratio is called the humidity ratio at saturation or the specific humidity at saturation.

For the case in which the noncondensable component, air, is not saturated with the condensable phase, water, we can determine the relative amount of water in air by applying Dalton's law to the mixture. Since

$$p_W = P y_W \tag{10-2}$$

we find

$$y_W = \frac{p_W}{P}$$

Then, inasmuch as

$$y_W + y_A = 1.0$$

$$y_A = 1 - \frac{p_W}{P}$$

and

$$\frac{y_W}{y_A} = \frac{p_W}{P-p_W} = \frac{n_W}{n_A} \tag{11-23}$$

For the air-water system, this ratio is called the molar humidity. For this system, the humidity ratio or the specific humidity m_W/m_A then is

$$\frac{m_W}{m_A} = 0.622 \frac{p_W}{P-p_W} \tag{11-24}$$

Another term widely used in studies of air-water systems is relative humidity. Relative humidity is defined as the ratio of the mole fraction of water vapor in the air-water mixture at a given temperature and pressure to the mole fraction of water vapor in an air-water mixture that is saturated at the same temperature and pressure.

$$\text{relative humidity} = \frac{y_W}{y_{W_{sat}}} = \frac{p_W}{P} \times \frac{P}{P_W^\square}$$

or

$$\text{relative humidity} = \frac{p_W}{P_W^\square} \tag{11-25}$$

Relative humidity of air-water mixtures is plotted as a function of dry bulb temperature and the humidity ratio on a psychrometric chart, Fig. 11-3. The dry bulb temperature is the temperature of the surrounding air measured in a normal manner.

Also plotted on Fig. 11-3 is the wet-bulb temperature as a function of dry bulb temperature and the humidity ratio. The wet bulb temperature is the temperature indicated by a thermometer that is in equilibrium with a wet wick saturated with water that is evaporating into the surrounding air. This temperature can be related to humidity by means of an analysis of the mass transfer and heat transfer rate processes taking place between the saturated wick and the surrounding atmosphere. The use of psychrometric charts greatly facilitates the solution of humidification problems.

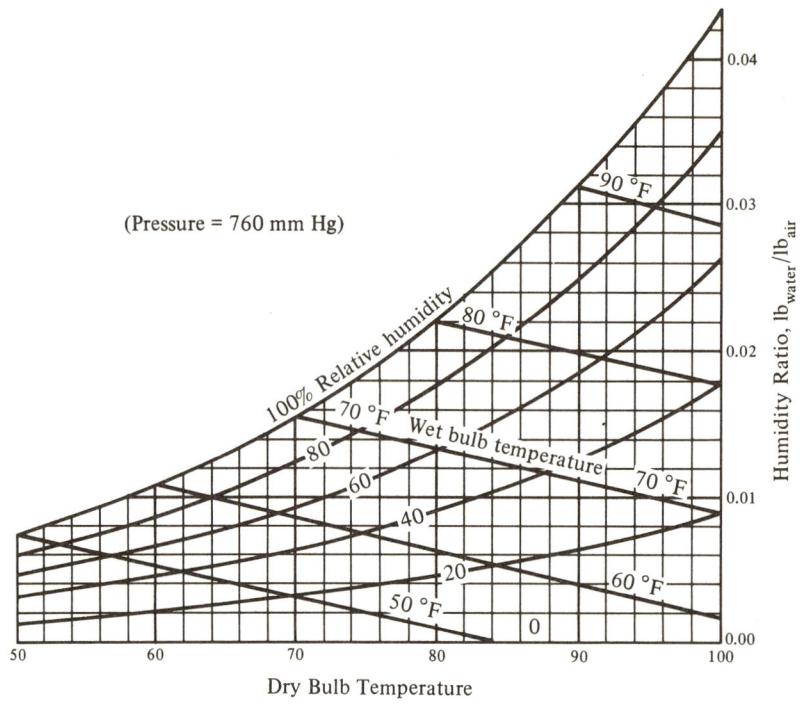

Fig. 11-3 *Psychrometric chart.*

EXAMPLE PROBLEM 11-1. The humidifier discussed in Example Prob. 2-1 is to be used to add moisture to air at 70 °F, 14.7 lb$_f$/in.² If the air entering the humidifier has a relative humidity of 20 percent and the air leaving the humidifier is saturated, how many pounds of water must be added to each pound of bone dry air entering the humidifier?

If we base our calculation on one pound of bone dry air flowing, the generalized mass balance, Eq. (2-1), reduces to

$$m_{\text{water in air, in}} + m_{\text{water added}} - m_{\text{water in air, out}} = 0$$

$$m_{\text{air, in}} - m_{\text{air, out}} = 0$$

The specific humidity of the saturated air leaving the humidifier is

$$\left(\frac{m_W}{m_A}\right)_{\text{sat}} = 0.622 \frac{P_W^\square}{P - P_W^\square} \quad (11\text{-}22)$$

From the steam tables, we find that at 70 °F,

$$P_W^\square = 0.3629 \text{ lb}_f/\text{in.}^2$$

so that

$$\frac{m_W}{m_A} = 0.622 \times \frac{0.3629}{14.7 - 0.3629}$$

$$= 0.0157 \; lb_{m_W}/lb_{m_A}$$

For each pound of bone dry air that leaves the humidifier, 0.0157 lb_m of water are carried out. Note that this result could have been obtained directly from the 100 percent relative humidity line in Fig. 11-3.

The specific humidity of the air entering the humidifier is

$$\frac{m_W}{m_A} = 0.622 \frac{p_W}{P - p_W} \qquad (11\text{-}24)$$

To find the partial pressure of water in the air entering the humidifier, we use the relationship

$$\text{relative humidity} = \frac{p_W}{p_W^{\square}} \qquad (11\text{-}25)$$

so that

$$p_W = 0.2 \times 0.3629$$

$$= 0.0726 \; lb_f/in.^2$$

Then, the specific humidity of the air entering the humidifier is

$$\frac{m_W}{m_A} = 0.622 \times \frac{0.0726}{14.7 - 0.0726}$$

$$= 0.0031 \; lb_{m_W}/lb_{m_A}$$

For each pound of air that enters the humidifier, 0.0031 lb_m of water enters the humidifier. Note that this result could have been obtained directly from Fig. 11-3.

From these results, the amount of water that must be added to the entering air then is

$$m_W = 0.0157 - 0.0031$$

$$= 0.0126 \; lb_{m_W}/lb_{m_A}$$

Bubble Points and Dew Points. Raoult's and Dalton's law, Eq. (11-20), can be used to calculate the temperature and pressure of a vapor mixture at which any further decrease in temperature at constant pressure or increase in pressure at constant temperature will result in the formation of a drop of liquid. This point is called the dew point. The dew point of a single component system is shown as point a on Fig. 11-4(a). The dew point of a multi-component mixture is shown as point a on Fig. 11-4(b).

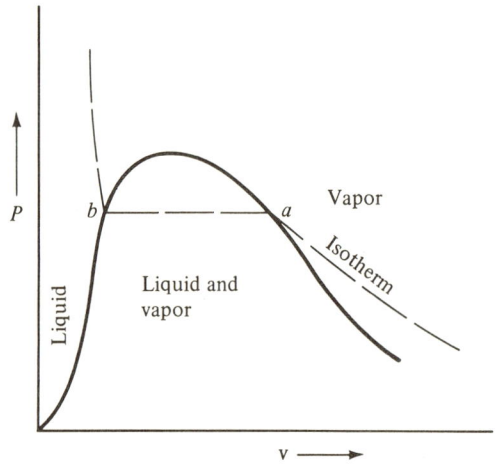

(a) Single-component system

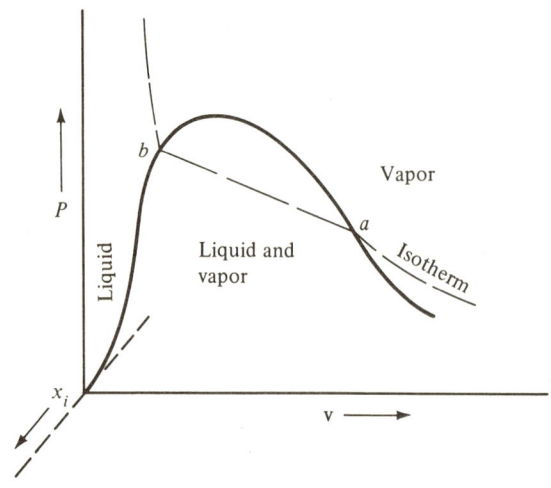

(b) Multi-component system of fixed composition

Fig. 11-4. *Isotherms on pressure-volume diagrams.*

As this is another problem involving two phases, vapor and liquid, the Gibbs Phase Rule once again indicates that the number of independent variables, including the dew point temperature and pressure, is equal to the number of components in the system N. In this case, although we may specify either the dew point pressure or temperature, we can not specify both as they are mutually dependent variables. A convenient source of the additional independent relationships required to solve for the dew point temperature or pressure is the component material balances,

$$n_i = n_{g,i} + n_{f,i}$$

or

$$n_M z_i = n_g y_i + n_f x_i \qquad (11\text{-}26)$$

where z_i is the mole fraction of component i in the original vapor mixture before the first drop of liquid has condensed and y_i is the mole fraction of component i in the vapor which is in equilibrium with liquid. Now, the composition of this drop of liquid depends upon the nature of the liquid and vapor phases as well as the system temperature and pressure. Assuming that Raoult's and Dalton's law, Eq. (11-20), adequately describes the relationship between the vapor and liquid phase compositions, we find that

$$x_i = \frac{Py_i}{P_i^\square} \qquad (11\text{-}27)$$

With this relationship, we can estimate the liquid composition at a given system temperature and pressure in terms of the equilibrium vapor composition.

Fortunately, at a dew point, only an infinitesimal amount of vapor condenses to form a minute liquid droplet. Since the amount of vapor which condenses is so small, Eq. (11-26) leads us to conclude that the composition of the vapor in equilibrium with the liquid droplet is essentially the same as the composition of the original vapor mixture so that

$$y_i = z_i$$

Thus, Eq. (11-27) may be used to determine the liquid droplet composition in terms of the composition of the original vapor mixture. Further, since the mole fractions of the components in the liquid must sum to unity,

$$\Sigma x_i = 1.0$$

we find that the criterion for a dew point pressure or temperature is

$$P\Sigma \frac{z_i}{P_i^\square} = 1.0 \tag{11-28}$$

It is also frequently necessary to calculate the pressure and temperature of a liquid mixture at which any further increase in temperature or decrease in pressure will result in the formation of a small bubble of vapor. This point is called the bubble point and is shown as point b on Figs. 11-4(a) and 11-4(b). Centrifugal pumps, for example, must operate at pressures above the bubble point of the liquid if cavitation damage is to be avoided.

A bubble point temperature or pressure can be solved for in a manner analogous to that used to solve for the dew point temperature or pressure. Using Eq. (11-20) to define the composition of this vapor bubble in terms of the liquid composition, we see that

$$y_i = \frac{P_i^\square x_i}{P} \tag{11-29}$$

From this relationship, we can estimate the composition of the bubble of vapor at a given pressure and temperature in terms of the equilibrium liquid composition.

Now at a bubble point, Eq. (11-26) leads us to conclude that since the amount of vapor formed is so small, the equilibrium liquid composition is the same as the composition of the original liquid mixture and

$$x_i = z_i$$

since only an infinitesimal amount of liquid has vaporized. Thus, Eq. (11-29) may be used to determine the vapor bubble composition in terms of the composition of the original liquid mixture. Further, since the mole fractions of the components in the vapor bubble must sum to unity,

$$\Sigma y_i = 1.0$$

the criteria for a bubble point pressure or temperature in this case is

$$\frac{1}{P} \Sigma P_i^\square z_i = 1.0 \tag{11-30}$$

EXAMPLE PROBLEM 11-2. Calculate the dew point temperature of a mixture of 30 mole percent propane, 40 mole percent n-butane, and 30 mole percent n-pentane at 14.7 $lb_f/in.^2$ for a system in which Raoult's and Dalton's law, Eq. (11-20), applies.

This problem requires that we find the temperature which satisfies the relationship

$$P\Sigma \frac{z_i}{P_i^\square} = 1.0 \qquad (11\text{-}28)$$

where

$$z_{C_3} = 0.3$$

$$z_{C_4} = 0.4$$

$$z_{C_5} = 0.3$$

By trial and error, assume $T = 55\,°F$. Then, using vapor pressures from Fig. 11-2,

$$P\Sigma \frac{z_i}{P_i^\square} = \frac{(0.3)(14.7)}{105} + \frac{(0.4)(14.7)}{24} + \frac{(0.3)(14.7)}{6.3}$$

$$= 0.99$$

This is slightly low and we must decrease the component vapor pressures. This can be done by assuming a lower temperature. Thus, assuming $T = 54\,°F$, we find

$$P\sum \frac{z_i}{P_i^\square} = \frac{(0.3)(14.7)}{100} + \frac{(0.4)(14.7)}{23} + \frac{(0.3)(14.7)}{6.0}$$

$$= 1.03$$

Thus, by interpolation, the desired dew point temperature is

$$T = 54.8\,°F$$

Of course, dew point and bubble point calculations are not limited to systems for which Raoult's and Dalton's law apply. In fact, at system pressures much greater than atmospheric, the use of Raoult's and Dalton's law could lead to serious errors.

Fugacities as a Criterion of Equilibrium

Fugacities and the Chemical Potential. As we have shown, the thermodynamic criteria for phase equilibrium is given by the relationship

$$T^{\mathrm{I}} = T^{\mathrm{II}} = \ldots = T^{\pi} \tag{11-2}$$

$$P^{\mathrm{I}} = P^{\mathrm{II}} = \ldots = P^{\pi} \tag{11-3}$$

and

$$\mu^{\mathrm{I}} = \mu^{\mathrm{II}} = \ldots = \mu_i^{\pi} \tag{11-16}$$

In developing Raoult's and Dalton's law, Eq. (11-20), we assumed that if the liquid or solid phase formed an ideal solution and that if the system temperature and pressure were such that the ideal gas law applied to the vapor phase, then we could use partial pressures as a measure of chemical potential and substitute

$$p_i^{\mathrm{I}} = p_i^{\mathrm{II}} = \ldots = p_i^{\pi} \tag{11-18}$$

for Eq. (11-16).

In order to develop a relationship between phase variables and chemical potentials for systems under conditions which do not satisfy the restrictions placed on Raoult's and Dalton's law, we must examine the concept of chemical potential more closely. The magnitude of a chemical potential of a component in a mixture, relative to the chemical potential of that same component in a standard state, can be estimated in a manner similar to that used to estimate the Gibbs free energy of a pure component relative to the Gibbs free energy of the pure component in a standard state. Standard states generally used were discussed in Chap. 9.

Recall that for a pure component i we have from Eq. (9-14)

$$g_i - g_i^{\circ} = \int_{P^{\circ}}^{P} \mathrm{v}_i \, dP = RT \ln \frac{f_i}{f_i^{\circ}} \tag{9-14}$$

By analogy, for a component i in a mixture, we have

$$\mu_i - \mu_i^{\circ} = \int_{P^{\circ}}^{P} \overline{V}_i \, dP = RT \ln \frac{\overline{f}_i}{f^{\circ}} \tag{11-31}$$

where the integration is carried out at constant temperature and composition. The superscript bar indicates the fugacities are fugacities of components in a mixture.

The ratio of the fugacity of a component in a given state and its fugacity in some other state, chosen for convenience as a standard state, is called the activity a_i. That is,

$$a_i = \frac{\bar{f}_i}{f_i^o} \tag{11-32}$$

Although the concept of activity may be associated with a pure component in a given state, it is generally associated with components in a mixture.

In order to complete the definition of the fugacity of component i in a mixture, we recall that for a pure component,

$$\lim_{P \to 0} \frac{f}{P} = 1.0 \tag{9-15}$$

since real gases act as though they were ideal at low pressures. As the fugacity of a pure component is equal to the system pressure under conditions in which the pure components act as an ideal gas, it seems reasonable to assume that the fugacity of a component in a mixture is equal to the partial pressure of that component under conditions in which the mixture acts as an ideal gas mixture. That is,

$$\lim_{P \to 0} \frac{\bar{f}_i}{p_i} = 1.0 \tag{11-33}$$

where

$$p_i = P y_i \tag{10-2}$$

Now, from Eq. (11-31) we find that for component i in phase I,

$$\mu_i^I = RT \ln \bar{f}_i^I - RT \ln \bar{f}_i^{o\,I} + \mu_i^{o\,I}$$

and for component i in phase II,

$$\mu_i^{II} = RT \ln \bar{f}_i^{II} - RT \ln \bar{f}_i^{o\,II} + \mu_i^{o\,II}$$

However, at equilibrium

$$\mu_i^I = \mu_i^{II} \tag{11-16}$$

so

$$RT \ln \bar{f}_i^I - RT \ln \bar{f}_i^{o\,I} + \mu_i^{o\,I} = RT \ln \bar{f}_i^{II} - RT \ln \bar{f}_i^{o\,II} + \mu_i^{o\,II}$$

where μ_i^o and $\bar{f}_i^o$ are properties of a mixture of ideal gases at system pressure and temperature. Since these properties are functions of composition, their values will be different in the two phases. However, at ideal gas conditions, we find from Eq. (11-33) that

$$\bar{f}_i^{o\,\mathrm{I}} = Py_i^\mathrm{I}$$

and

$$\bar{f}_i^{o\,\mathrm{II}} = Py_i^\mathrm{II}$$

so that

$$RT\ln\bar{f}_i^\mathrm{I} - RT\ln y_i^\mathrm{I} + \mu_i^{o\,\mathrm{I}} = RT\ln\bar{f}_i^\mathrm{II} - RT\ln y_i^\mathrm{II} + \mu_i^{o\,\mathrm{II}}$$

This last result can be reduced still further if we recognize that μ_i^o is a partial molal quantity of a component in a mixture of ideal gases. For such a mixture the total specific Gibbs free energy is, from Eq. (10-38),

$$g_M^* = \Sigma y_i \mu_i^o$$

where the superscript asterisk indicates a property of an ideal gas. Further, from Eq. (10-10), the total specific Gibbs free energy of a mixture of ideal gases is

$$g_M^* = \Sigma y_i (g_i^* + RT\ln y_i)$$

Equating coefficients of y_i, we find

$$\mu_i^o = g_i^* + RT\ln y_i$$

so that

$$RT\ln\bar{f}_i^\mathrm{I} - RT\ln y_i^\mathrm{I} + g_i^* + RT\ln y_i^\mathrm{I} = RT\ln\bar{f}_i^\mathrm{II} - RT\ln y_i^\mathrm{II} + g_i^* + RT\ln y_i^\mathrm{II}$$

This last equation reduces to

$$RT\ln\bar{f}_i^\mathrm{I} = RT\ln\bar{f}_i^\mathrm{II}$$

or

$$\bar{f}_i^\mathrm{I} = \bar{f}_i^\mathrm{II} \tag{11-34}$$

In general, for π phases in equilibrium,

$$\bar{f}_i^\mathrm{I} = \bar{f}_i^\mathrm{II} = \bar{f}_i^\mathrm{III} \ldots = \bar{f}_i^\pi \tag{11-35}$$

Since fugacities are more convenient to use than chemical potentials, we will use them in future calculations.

Fugacities and the Phase Variables. We might begin our examination of the relationships between fugacities and the phase variables by considering the problem of evaluating the fugacity of a component in a gas mixture. At low pressure, real gases tend to behave as ideal gases. Under these conditions, we see from Eq. (11-33) that the fugacity of a component in an ideal gas mixture is equal to the partial pressure of that component in the mixture. That is

$$\bar{f}_{g,i} = P y_i$$

At higher pressures, the gas mixture will not act as a mixture of ideal gases and we must include a correction term ν_i, called a fugacity coefficient. Then the fugacity of any component in a gas mixture is given by the relationship

$$\bar{f}_{g,i} = \nu_i P y_i \tag{11-36}$$

Fugacity coefficients of components in a mixture are calculated in a manner similar to that used to calculate fugacity coefficients of pure components. As the pressure of the system increases, this term becomes more and more important.

Obviously, the assumption of ideal gas behavior does not apply to a component in a liquid or solid phase. However, since fugacities are similar to pressure, we might assume by analogy with Eq. (11-19) that in a liquid phase which forms an ideal solution,

$$\bar{f}_{f,i} = x_i f_{f,i} \tag{11-37}$$

where $f_{f,i}$ is the fugacity of a pure liquid component at the system temperature and pressure. If the liquid mixture is not ideal, we must add a correction term γ called the liquid phase activity coefficient, to find the fugacity of a liquid component in the mixture. In this case,

$$\bar{f}_{f,i} = \gamma_i x_i f_{f,i}^o \tag{11-38}$$

where

$$\gamma_i = \frac{\bar{f}_{f,i}}{x_i f_{f,i}^o}$$

$$= \frac{a_i}{x_i}$$

and a_i is the activity of component i in a mixture relative to a standard state of the pure component at system pressure and temperature.

Pure component fugacities in the vapor phase can be estimated from Eq. (9-17) where values of the pure component fugacity coefficient v_i calculated from the equation $Pv = zRT$, are plotted as functions of reduced pressure and temperature in Figs. A-4A and A-4B in the Appendix. In cases where only approximate answers are satisfactory, it is often assumed that generalized compressibility factor data represents liquid phase behavior satisfactorily, and Figs. A-4A and A-4B may also be used to estimate the fugacity of a pure component in a liquid phase.

Unfortunately, the assumption that the equation $Pv = zRT$ represents liquid phase behavior is generally valid only as a gross approximation. A somewhat better method of estimating the fugacity of a pure component in the liquid phase is to recognize that when a liquid containing a single component is in equilibrium with its own vapor at the system temperature and the vapor pressure of the component, we have from Eq. (11-34),

$$f_{f,i}^{\Box} = f_{g,i}^{\Box}$$

$$= v_i^{\Box} P_i^{\Box}$$

where $v_i^{\Box}$ is evaluated at $P_r^{\Box}$ and T_r. As $P_r^{\Box}$ approaches zero, the vapor in equilibrium with the pure component liquid approaches ideal gas behavior and $v_i^{\Box}$ approaches unity.

If the pressure acting on the mixture of components is different from the vapor pressure of the pure component of interest, we must correct the pure component fugacity for this difference. We can account for this difference by writing

$$f_{f,i}^{o} = \left(\frac{f_{f,i}^{o}}{f_{f,i}^{\Box}}\right) f_{f,i}^{\Box}$$

But, from Eq. (9-14),

$$\ln \frac{f_{f,i}^{o}}{f_{f,i}^{\Box}} = \frac{1}{RT} \int_{P^{\Box}}^{P} v_{f,i} \, dP$$

so

$$f_{f,i}^{o} = v_i^{\Box} P_i^{\Box} \exp\left(\frac{1}{RT}\right) \int_{P^{\Box}}^{P} v_{f,i} \, dP \qquad (11\text{-}39)$$

where the term

$$\exp\left(\frac{1}{RT}\right) \int_{P^\square}^{P} v_{f,i} dP$$

is called the Poynting correction. If we assume that $v_{f,i}$ is essentially independent of pressure, we find on combining Eqs. (11-38) and (11-39), that

$$\bar{f}_{f,i} = \gamma_i x_i v_i^\square P_i^\square \exp\left[\frac{v_{f,i}(P - P_i^\square)}{RT}\right] \qquad (11\text{-}40)$$

Although developed for a component in a liquid phase, this equation can also be applied to a component in a solid phase.

In evaluating which of the quantities in Eq. (11-40) are important, we must consider the nature of the liquid mixture as well as effects of the system temperature and pressure.

The magnitude of the activity coefficient γ_i is a function of the nature of the liquid mixture. For ideal solutions, γ_i is unity. The magnitude of the pure component fugacity coefficient $v_i^\square$ and the component vapor pressure $P_i^\square$ are functions of the system temperature. If the temperature of the system is well below the critical temperature of the component so that $P_i^\square$ is low, say in the order of one atmosphere, the fugacity coefficient $v_i^\square$ will approach unity.

Finally, if the difference between the system pressure P and the vapor pressure of the pure component $P_i^\square$ is small, the Poynting correction is unity. The Poynting correction becomes increasingly important as $v_{f,i}$ or as the difference between P and $P_i^\square$ increases and as the system temperature decreases. An illustration of the use of Eq. (11-40) is presented in the following example problem.

EXAMPLE PROBLEM 11-3. What is the fugacity of hexane in an ideal liquid mixture at 300 °F and 500 psia if the mole fraction of hexane in the liquid is 0.30? Use the following data and the generalized charts:

$$P_c = 436.5 \text{ psia}$$

$$T_c = 453.6\,°F$$

$$\omega = 0.301$$

At 300 °F,

$$P_{C_6}^\square = 105 \text{ psia}$$

$$v_{f,C_6} = 0.025 \text{ ft}^3/\text{lb}_m$$

First we will use the relationship

$$\bar{f}_{f,i} = \gamma_i v_i^\square P_i^\square x_i \exp\left[\frac{v_{f,i}(P - P_i^\square)}{RT}\right] \tag{11-40}$$

to solve this problem. Now, as the liquid solution is ideal,

$$\gamma_i = 1.0$$

Next, in order to evaluate v_i, we note that

$$P_r^\square = \frac{105}{436.5} = 0.24$$

$$T_r = \frac{760}{913.3} = 0.83$$

Then, using Figs. A-4A and A-4B, we find at $P_r^\square = 0.24$, $T_r = 0.83$, $\omega = 0.301$,

$$\log v_{C_6}^\square = -0.07 + 0.301 \times -0.36 = -0.178$$

$$v_{C_6}^\square = 0.66$$

Substituting into Eq. (11-40),

$$\bar{f}_{f,C_6} = (1.0)(0.66)(105)(0.3) \exp\left[\frac{(0.025)(86)(560 - 105)}{(10.73)(760)}\right]$$

or

$$\bar{f}_{f,C_6} = 23.4 \text{ lb}_f/\text{in.}^2$$

We may also solve this problem by assuming Figs. A-4A and A-4B are applicable to liquids. Reading these figures directly for v_{f,C_6} at

$$P_r = \frac{500}{436.5} = 1.15$$

at $T_r = 0.83$ gives

$$\log v_{f,C_6} = -0.58 + 0.301 \times -0.4 = -0.700$$

$$v_{f,C_6} = 0.20$$

or that

$$f_{f,C_6} = (0.20)(500) = 100 \text{ lb}_f/\text{in}^2$$

Then, from Eq. (11-37), we find

$$\bar{f}_{f,C_6} = (0.3)(100)$$

$$\bar{f}_{f,C_6} = 30 \text{ lb}_f/\text{in}^2$$

This result is not as good as our previous result since the assumption that the equation $Pv = zRT$ represents phase behavior of a liquid phase is only gross approximation.

Phase Equilibrium in Simple Systems. As we have shown, the working criteria for vapor-liquid phase equilibrium are

$$T_f = T_g \tag{11-2}$$

$$P_f = P_g \tag{11-3}$$

and

$$\bar{f}_{g,i} = \bar{f}_{f,i} \tag{11-34}$$

where

$$\bar{f}_{g,i} = v_i P y_i \tag{11-36}$$

and

$$\bar{f}_{f,i} = \gamma_i x_i v_i^\square P_i^\square \exp \frac{v_{f,i}(P - P_i^\square)}{RT} \qquad (11\text{-}40)$$

In order to demonstrate how to apply these relationships, let us first consider a vapor-liquid system at low total pressure in which the vapor acts as an ideal gas. The temperature of this system is well below the critical temperature of all components and the liquid forms an ideal solution.

Since the pressure on the system is low, the vapor phase acts as a mixture of ideal gases and the fugacity coefficient v_i is unity. Equation (11-36) thus reduces to

$$\bar{f}_{g,i} = Py_i$$

Further, as the liquid is an ideal solution, the activity coefficient γ_i is unity. Since the system temperature is well below the critical temperature of the components in the system, $P_i^\square$ is low and $v_i^\square$ may be assumed to be unity. If we also assume that the difference between the system pressure and the vapor pressure of each component is small, the Poynting correction is also unity.

Under these conditions Eq. (11-40) reduces to

$$\bar{f}_{f,i} = P_i^\square x_i$$

Combining these results with Eq. (11-34), we find that for this particular case

$$Py_i = P_i^\square x_i \qquad (11\text{-}20)$$

Thus, we see that Eq. (11-20), Raoult's and Dalton's law, is a special case of the generalized phase equilibrium relationships. For problems in which the restrictions employed in this example do not apply, the more generalized forms of the phase equilibrium relationships must be used.

Fugacity Coefficients. As the pressure exerted on a system increases, the assumption that the vapor phase fugacity coefficient v_i in Eq. (11-36) is unity becomes less valid. However, many gas mixtures can be assumed to form ideal solutions up to pressures of 150 to 200 $lb_f/in.^2$ Therefore we might next turn our attention to the problems of estimating fugacity coefficients in ideal solutions.
If from the defining equation for the fugacity of pure component i

$$g_i - g_i^\circ = \int_{P^\circ}^{P} v_i \, dP = RT \ln \frac{f_i}{f_i^\circ} \qquad (9\text{-}14)$$

we subtract the defining equation for the fugacity of component i in a mixture

$$\mu_i - \mu_i^o = \int_{P^o}^{P} \overline{V}_i \, dP = RT \ln \frac{\overline{f}_i}{\overline{f}_i^o} \qquad (11\text{-}31)$$

we find that

$$\ln \frac{\overline{f}_i}{\overline{f}_i^o} - \ln \frac{f_1}{f_i^o} = \frac{1}{RT} \int_{P^o}^{P} (\overline{V}_i - v_i) \, dP$$

But we define an ideal solution by the relationship

$$\overline{V}_i = v_i \qquad (10\text{-}50)$$

so

$$\frac{1}{RT} \int_{P^o}^{P} (\overline{V}_i - v_i) \, dP = 0$$

and

$$\frac{\overline{f}_i}{\overline{f}_i^o} = \frac{f_i}{f_i^o}$$

From the definition of fugacity for a pure component, we observe that at pressures approaching zero,

$$f_i^o = P^o$$

Further, at low pressure,

$$\overline{f}_i^o = y_i P^o$$

Combining these results,

$$\frac{\overline{f}_{g,i}}{y_i P^o} = \frac{f_{g,i}}{P^o}$$

or

$$\overline{f}_{g,i} = y_i f_{g,i} \qquad (11\text{-}41)$$

Equation (11-41) is known as the Lewis-Randall approximation. Since

$$f_{g,i} = v_i P \qquad (9\text{-}17)$$

the Lewis-Randall approximation may also be written as

$$\overline{f}_{g,i} = v_i P y_i \qquad (11\text{-}36)$$

where the fugacity coefficient v_i is evaluated for the pure vapor component at system pressure and temperature.

The Lewis-Randall approximation is not limited to the vapor phase. It applies equally well to any phase which forms an ideal solution. For example, if the liquid phase forms an ideal solution, we find that analogous to Eq. (11-41),

$$\overline{f}_{f,i} = x_i f_{f,i} \qquad (11\text{-}37)$$

Equation (11-37) was used in the development of the general relationship for the fugacity of a component in a liquid mixture, Eq. (11-40), and in solving Example Prob. 11-3.

One soon encounters serious limitations in applying the concept of an ideal solution to the evaluation of fugacities, especially for mixtures of components which have greatly differing boiling points. For example, a significant amount of methane may dissolve in a hydrocarbon liquid mixture at temperatures well above the critical temperature of methane. In order to estimate the amount of methane that will dissolve by using the Lewis-Randall approximation, one needs to estimate the fugacity of pure liquid methane. But pure methane does not exist as a liquid above its critical point.

In this case, we may use another technique to estimate the fugacity of the methane in the liquid. Experimentally, we find that for dilute concentrations of a component in a liquid mixture at temperatures above the critical temperature of that component that

$$\overline{f}_{f,i} = H x_i \qquad (11\text{-}42)$$

The parameter H is called the Henry's law coefficient and experimental values are available. For methane dissolved in n-pentane at a low pressure and 100 °F, for example,

$$H = 19.2 \text{ atm}$$

If the concentration of the component i becomes too large, it may be necessary to correct Henry's law coefficient by means of an activity coefficient.

When Henry's law coefficients are not available, it is sometimes possible to estimate pseudo-liquid pure component fugacities by extrapolating the vapor pressure curve such as shown in Fig. 11-2 as a straight line beyond the critical point. Then, a pseudo-fugacity coefficient to be used in Eq. (11-40) can be read directly from Fig. A-4 at the pseudo-reduced vapor pressure of the component. Under these conditions, the pseudo-liquid specific volume to be used in applying the Poynting correction is also fictitious. If it can be assumed that pressure does not affect liquid volume, then the specific volume of the saturated liquid at system pressure might be used, although any other reasonable value might also be used. It should be noted, however, that there is no assurance these techniques will lead to answers that agree with experiment.

Another serious limitation in employing the ideal solution concept becomes apparent when it is necessary to evaluate the pure component vapor phase fugacity at a total system pressure greater than the vapor pressure of the component. Under these conditions, the pure component does not exist as a vapor.

The problem of estimating fugacities of pseudo-liquid and pseudo-vapor components has been circumvented to a certain extent by the introduction of pseudo-fugacity coefficient charts. These charts were prepared by extrapolating liquid phase fugacity coefficients into the vapor phase and vapor phase fugacity coefficients into the liquid phase in a manner so as to give reasonable agreement between predicted and experimental results. Empirically using the assumption of ideal solutions, plots of pseudo-fugacity coefficients for liquid and vapor components can be combined by defining a new variable

$$K_i = \frac{y_i}{x_i} \qquad (11\text{-}43)$$

where K_i is called a phase distribution coefficient. Since the equilibrium phases are vapor and liquid, K_i is also called a vapor-liquid equilibrium coefficient. On combining Eqs. (11-34), (11-36), and (11-40) we find that we can estimate K_i from the relationship

$$K_i = \left(\frac{\nu_i^\square}{\nu_i}\right)\left(\frac{P_i^\square}{P}\right) \cdot \exp\frac{v_{f,i}(P_i^\square - P)}{RT}$$

A plot in which pseudo-fugacity coefficients have been combined using vapor-liquid equilibrium coefficients is presented in nomographic form in Fig. A-5 of the Appendix. This plot combines theoretical vapor-liquid equilibrium coefficients obtained by integrating an equation similar to Eq. (9-19) using the Benedict-Webb-Rubin equation and experimental vapor-liquid equilibrium data[31]. As an example of the use of combined vapor-liquid pseudo-fugacity charts, consider the following problem.

EXAMPLE PROBLEM 11-4. Recalculate the dew point temperature of Example Prob. 11-2 using Fig. A-5.

Since

$$K_i = \frac{y_i}{x_i} = \frac{P_i^\square}{P}$$

for a system which obeys Raoult's and Dalton's law, Eq. (11-28) can be written in the more general form

$$\sum \frac{z_i}{K_i} = 1.0$$

In this problem,

$$z_{C_3} = 0.3$$

$$z_{C_4} = 0.4$$

$$z_{C_5} = 0.3$$

By trial and error, assuming $T = 52\,°F$, we find from Fig. A-5,

$$K_{C_3} = 6.3$$

$$K_{C_4} = 1.7$$

$$K_{C_5} = 0.42$$

Then

$$\sum \frac{z_i}{K_i} = \frac{0.3}{6.3} + \frac{0.4}{1.7} + \frac{0.3}{0.42}$$

$$= 0.997$$

Thus, the desired dew point temperature is

$$T = 52\,°F$$

Many of the problems encountered in using the concept of ideal solutions to evaluate fugacities of components in a mixture can be avoided if it can be assumed that an equation of state with mixture coefficients adequately represents the phase behavior of the mixture. In this case we define a residual volume of a component in a mixture as

$$\overline{\alpha}_{R,i} = \frac{RT}{P} - \overline{V}_i \tag{11-44}$$

Then, by analogy with Eq. (9-16), we find that

$$RT \ln \nu_i = -\int_0^P \overline{\alpha}_{R,i} \, dP \tag{11-45}$$

Now, in order to integrate Eq. (11-45) at fixed composition, we must know the partial molar volume as a function of pressure. Unfortunately, this is not generally the case, and we must modify the equation. This can be done as follows: Consider μ_i as a function of volume and temperature. Taking the total derivative of μ_i, then holding temperature constant, we find

$$d\mu_i = \left(\frac{\partial \mu_i}{\partial V}\right)_T dV = RT d\ln \overline{f}_i$$

Now, using the definition of the Helmholtz work function,

$$A = U - TS \tag{4-21}$$

we may write the property relationship in the form

$$dA = S \, dT - P \, dV + \mu_i \, dn_i + \mu_j \, dn_j \tag{11-46}$$

Taking a Maxwell of Eq. (11-46) gives us the relationship

$$\left(\frac{\partial \mu_i}{\partial V}\right)_{T,n_i,n_j} = -\left(\frac{\partial P}{\partial n_i}\right)_{T,V,n_j}$$

Thus,

$$RT d\ln \overline{f}_i = -\left(\frac{\partial P}{\partial n_i}\right)_{T,V,n_j} dV$$

This last equation is not integrable in the limit as P approaches zero. We can modify it to an integrable form by noting that at constant mole fraction,

$$RTd\ln(Py_i) = RTd\ln P + RTd\ln y_i$$

$$= RTd\ln P$$

Subtracting from both sides of the last equation,

$$RTd\ln \frac{\bar{f}_i}{y_i P} = -\left(\frac{\partial P}{\partial n_i}\right)_{T,V,n_j} dV - RTd\ln P$$

The last terms on the right hand side can be combined if we recognize that

$$\frac{d(PV)}{PV} = \frac{P\,dV}{PV} + \frac{V\,dP}{PV}$$

or

$$d\ln P = d\ln(PV) - d\ln V$$

Thus

$$RTd\ln \frac{\bar{f}_i}{y_i P} = \left[\frac{RT}{V} - \left(\frac{\partial P}{\partial n_i}\right)_{T,V,n_j}\right] dV - RTd\ln(PV)$$

Integrating from $P = 0$ to P, and noting that at $P = 0$, $V = \infty$ and $PV = nRT$, we find

$$RT\ln \nu_i = \int_V^\infty \left[\left(\frac{\partial P}{\partial n_i}\right)_{T,V,n_j} - \frac{RT}{V}\right] dV - RT\ln \frac{Pv_M}{RT} \qquad (11\text{-}47)$$

This equation is useful when we know pressure as a function of volume. For example, the Redlich-Kwong equation of state is of this form. The integration of Eq. (11-47) using the Redlich-Kwong equation of state gives us

$$\ln \nu_i = (z-1)\frac{B_i}{B_M} - \ln(z - B_M P) - \frac{A_M^2}{B_M}\left(\frac{2A_i}{A_M} - \frac{B_i}{B_M}\right)\ln\left(1 + \frac{B_M P}{z}\right)$$

$$(11\text{-}48)$$

where

$$z = \frac{Pv_M}{RT} \tag{8-9}$$

At moderate pressure, the integrated form of the Redlich-Kwong equation is

$$\ln \nu_i = B_i - A_i^2 + (A_i - A_M)^2 P \tag{11-49}$$

Integration of Eq. (11-47) using the Benedict-Webb-Rubin equation of state gives the following result:

$$RT\ln \nu_i = \left[(B_{oM} + B_{oi})RT - 2(A_{oM}A_{oi})^{1/2} - 2(C_{oM}C_{oi})^{1/2} T^{-2}\right] v^{-1}$$

$$+ 3/2\left[RT(b_M^2 b_i)^{1/3} - (a_M^2 a_i)^{1/3}\right] v^{-2}$$

$$+ 3/5\left[a_M (\alpha_M^2 \alpha_i)^{1/3} + \alpha_M (a_M^2 a_i)^{1/3}\right] v^{-5}$$

$$+ \left[3v^{-2} (c_M^2 c_i)^{1/3} T^{-2}\right]\left\{\left[1 - \exp(-\gamma_M v^{-2})\right]/\gamma_M v^{-2}\right.$$

$$\left. - \frac{1}{2}\exp(-\gamma_M v^{-2})\right\} - (2v^{-2} c_M T^{-2})\left(\frac{\gamma_i}{\gamma_M}\right)\left\{\left[1 - \exp(-\gamma_M v^{-2})\right]/\gamma_M v^{-2}\right.$$

$$\left. - \exp(-\gamma_M v^{-2})(1 + \gamma_M v^{-2}/2)\right\} \tag{11-50}$$

Equation (11-50) has been applied successfully to mixtures of light hydrocarbons in both the vapor and liquid phases. Finally, the integration of Eq. (11-47) using the virial equation of state gives the result

$$\ln \nu_i = \frac{2}{v_M}\sum_j y_j B_{ij}(T) + \frac{3}{2v_M^2}\sum_{jk} y_j y_k C_{ijk}(T) + \ldots - \ln \frac{Pv_M}{RT} \tag{11-51}$$

A major dilemma confronts us immediately as we attempt to use any of the integrated forms of Eq. (11-47). Both the mixture coefficients in any of these integrated equations, as well as the specific volume of the vapor mixture v_M, are functions of composition. However, initially, the phase compositions are not known.

In order to approximate the required phase compositions, we might first assume that either Raoult's and Dalton's law applies or that both the liquid phase and vapor phase form ideal solutions. Combining the vapor-liquid equilibrium coefficients obtained from either

$$Py_i = x_i P_i^\square \tag{11-20}$$

or from Fig. A-5 in the Appendix with component material balances, as was done for the dew point calculations in Example Probs. 11-2 and 11-4, provides us with a means of estimating, to a first approximation, the required phase composition. Mixture coefficients, such as those presented in Eqs. (10-22) through (10-35), can then be calculated. Then, required specific volumes can be estimated using techniques developed in Example Prob. 8-3. Improved estimates of the vapor-liquid equilibrium coefficients can next be made using Eq. (11-40) and an integrated form of Eq. (11-47). The entire procedure is then repeated until the phase compositions used to estimate mixture coefficients are the same as the calculated phase compositions. This procedure is most conveniently carried out using a digital computer.

Activity Coefficients.

Activity coefficients and the phase variables. The activity coefficient γ_i in Eq. (11-40) is a parameter used to account for the deviations of real mixture behavior from ideal solution behavior. As liquids frequently form non-ideal solutions, activity coefficients are used primarily in the analysis of liquid phase mixtures.

Now, from our previous discussion of the defining equations for the activity coefficient,

$$\gamma_i = \frac{a_i}{x_i} = \frac{\overline{f}_{f,i}}{x_i f_{f,i}^o} \tag{11-38}$$

we saw that the standard state used as a basis for the activity coefficient is the pure liquid at the mixture pressure and temperature. With this choice of a standard state, values of activity coefficients depend only upon the nature and the composition of the mixture. Experimentally, we find that liquid mixtures vary over the entire range from essentially ideal solutions to extremely non-ideal solutions. In the latter extreme, two liquid components such as water and oil may not mix to any appreciable extent at all. These components form two distinct liquid phases and are said to be immiscible. Each component exerts its own vapor pressure, independent of the relative amount of the other component present. Therefore, the total pressure above two immiscible liquids is

$$P = P_1^\square + P_2^\square$$

where the vapor pressures are functions only of the system temperature.

In the following discussion, we will first examine the thermodynamic relationships between activity coefficients and the composition of liquid mixtures which form homogeneous phases. Then we will consider some of the analytical relationships that have been developed between activity coefficients and the composition of liquid mixtures which form homogeneous phases. In addition, various techniques using experimental data to evaluate the parameters in these analytical relationships will be examined.

The Gibbs-Duhem equation. Fortunately, the activity coefficients of individual components in a mixture are not independent. Since the chemical potential,

$$\mu_i = \overline{G}_i = \left(\frac{\partial G}{\partial n_i}\right)_{P,T,n_j}$$

is a partial molal quantity, we find on applying Eq. (10-40) at constant pressure and temperature that

$$n_1 \, d\mu_1 + n_2 \, d\mu_2 + n_3 \, d\mu_3 + \ldots = 0 \qquad (11\text{-}52)$$

Equation (11-52) is known as the Gibbs-Duhem equation or sometimes as the equation of Gibbs, Duhem and Margules. It shows that the changes in the partial molal Gibbs free energies, or chemical potentials, with respect to composition, are not independent. It is important to note that this equation is written for components in a single phase. For binary mixture Eq. (11-52) may be written

$$x_1 \, d\mu_1 + x_2 \, d\mu_2 = 0 \qquad (11\text{-}53)$$

Now, Eq. (11-53) can be written in several forms. Using the definition of activity as a ratio of fugacity of a component in a mixture to its fugacity in a standard state, we find from Eqs. (11-31) and (11-32)

$$\mu_i - \mu_i^\circ = RT \ln \frac{\overline{f}_i}{f_i^\circ} = RT \ln a_i$$

Then, as μ_i° and f_i° are constants whose values depend only on the standard state selected,

$$d\mu_i = RT d \ln \overline{f}_i = RT d \ln a_i$$

Thus, we may write the Gibbs-Duhem equation for a binary mixture either in the form

$$x_1 \, d\ln \bar{f}_1 + x_2 \, d\ln \bar{f}_2 = 0 \qquad (11\text{-}54)$$

or

$$x_1 \, d\ln a_1 + x_2 \, d\ln a_2 = 0 \qquad (11\text{-}55)$$

Further, since

$$\gamma_i = \frac{a_i}{x_i} \qquad (11\text{-}38)$$

we find

$$x_1 \, d\ln(\gamma_1 x_1) + x_2 \, d\ln(\gamma_2 x_2) = 0 \qquad (11.56)$$

Now,

$$\ln(\gamma_1 x_1) = \ln x_1 + \ln \gamma_1$$

Then,

$$x_1 d\ln x_1 + x_2 d\ln x_2 = dx_1 + dx_2$$

But

$$x_1 + x_2 = 1$$

and

$$dx_1 = -dx_2$$

so

$$x_1 d\ln x_1 + x_2 d\ln x_2 = 0$$

and in terms of activity coefficients, the Gibbs-Duhem equation becomes

$$x_1 d\ln \gamma_1 + x_2 d\ln \gamma_2 = 0 \qquad (11\text{-}57)$$

Equations (11-54), (11-55), and (11-57) can be modified by dividing by dx_1 at constant pressure, temperature and x_2 to give the results

$$x_1\left(\frac{\partial \ln \bar{f}_1}{\partial x_1}\right)_{P,T,x_2} + x_2\left(\frac{\partial \ln \bar{f}_2}{\partial x_1}\right)_{P,T,x_2} = 0 \qquad (11\text{-}58)$$

$$x_1\left(\frac{\partial \ln a_1}{\partial x_1}\right)_{P,T,x_2} + x_2\left(\frac{\partial \ln a_2}{\partial x_1}\right)_{P,T,x_2} = 0 \qquad (11\text{-}59)$$

$$x_1\left(\frac{\partial \ln \gamma_1}{\partial x_1}\right)_{P,T,x_2} + x_2\left(\frac{\partial \ln \gamma_2}{\partial x_1}\right)_{P,T,x_2} = 0 \qquad (11\text{-}60)$$

An application of these latter forms of the Gibbs-Duhem equation is shown in the next example problem.

EXAMPLE PROBLEM 11-5. The following activity coefficient data have been reported for the sodium-lead system at 1126 K[32].

x_{Na}	γ_{Na}	γ_{Pb}
0.1	0.047	0.982
0.2	0.070	0.915
0.3	0.113	0.767
0.4	0.182	0.592
0.5	0.370	0.337
0.6	0.504	0.225
0.7	0.690	0.127
0.8	0.852	0.065
0.9	1.000	0.027

Evaluate the thermodynamic consistency of these results using the Gibbs-Duhem equation at x_{Na} equal to 0.25, 0.5, and 0.75.

Let us write Eq. (11-60) in the form

$$x_{Na}\left(\frac{\partial \ln \gamma_{Na}}{\partial x_{Na}}\right)_{P,T,x_{Pb}} + (1 - x_{Na})\left(\frac{\partial \ln \gamma_{Pb}}{\partial x_{Na}}\right)_{P,T,x_{Pb}} = D$$

If the data are consistent, then the deviation D must be equal to zero. The chord-area method will be employed to evaluate the necessary partial derivatives. First, we must find the average change in $\ln \gamma_i$ corresponding to a change in x_{Na} as shown in the following table.

x_{Na}	Δx_{Na}	$\ln \gamma_{Na}$	$\Delta \ln \gamma_{Na}$	$\dfrac{\Delta \ln \gamma_{Na}}{\Delta x_{Na}}$	$\ln \gamma_{Pb}$	$\Delta \ln \gamma_{Pb}$	$\dfrac{\Delta \ln \gamma_{Pb}}{\Delta x_{Na}}$
0.1		−3.09			−0.02		
	0.1		0.43	4.3		−0.07	−0.7
0.2		−2.66			−0.09		
	0.1		0.48	4.8		−0.17	−1.7
0.3		−2.18			−0.26		
	0.1		0.47	4.7		−0.26	−2.6
0.4		−1.71			−0.52		
	0.1		0.72	7.2		−0.57	−5.7
0.5		−0.99			−1.09		
	0.1		0.30	3.0		−0.40	−4.0
0.6		−0.69			−1.49		
	0.1		0.32	3.2		−0.57	−5.7
0.7		−0.37			−2.06		
	0.1		0.21	2.1		−0.67	−6.7
0.8		−0.16			−2.73		
	0.1		0.16	1.6		−0.88	−8.8
0.9		−0.00			−3.61		

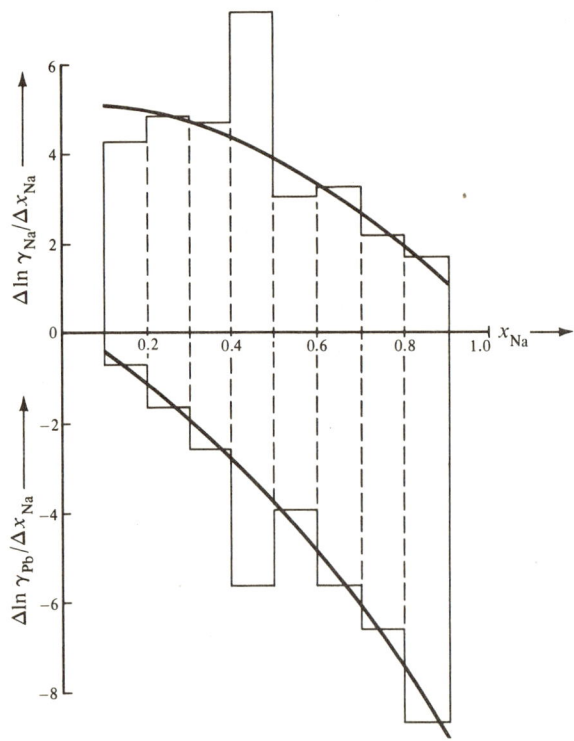

Fig. 11-5. *Graphical differentiation of activity coefficients.*

Values of the average change in $\ln \gamma_i$ with respect to changes in x_{Na} are plotted as rectangles in Fig. 11-5. In order to obtain the necessary partial derivatives, we draw a smooth curve through the rectangles such that the area under the smooth curve is equal to the area under the rectangles and equal in value to $\Delta \ln \gamma_i$. In doing this, we have ignored an apparent experimental discrepancy at x_{Na} of 0.5.

From the smooth curves and the Gibbs-Duhem equation, we obtain the following results:

x_{Na}	$\left(\dfrac{\partial \ln \gamma_{Na}}{\partial x_{Na}}\right)$	$\left(\dfrac{\partial \ln \gamma_{Pb}}{\partial x_{Na}}\right)$	D
0.25	4.8	−1.7	−0.08
0.50	3.9	−4.0	−0.05
0.75	2.2	−6.8	−0.05

Although D does not equal zero exactly, this is a reasonably consistent set of phase equilibrium data within the range of compositions checked.

Closer consideration of the Gibbs-Duhem equation shows that it must be slightly modified if we are to use it to evaluate equilibrium data for a two component system. In a binary system in which the vapor and liquid are in equilibrium, the Gibbs Phase Rule tells us that there are two degrees of freedom. Thus, if we fix pressure and temperature, the system is completely defined. This means that for a specified pressure and temperature, the compositions of the liquid and vapor phases are fixed. Yet, we would like to evaluate equilibrium data over a complete range of compositions. This can be accomplished if we obtain the data from either an isothermal or an isobaric experiment.

When applied to binary isothermal equilibrium data, the Gibbs-Duhem equation is used in the form[26]

$$x_1 \left(\frac{\partial \ln \gamma_1}{\partial x_1}\right)_{T,n_2} + x_2 \left(\frac{\partial \ln \gamma_2}{\partial x_2}\right)_{T,n_2} = \frac{v^E}{RT}\left(\frac{\partial P}{\partial x_1}\right)_{T,n_2} \qquad (11\text{-}61)$$

where the excess volume is defined by the relationship

$$v^E = v_M - v_M^I = \Delta v_{mix} \qquad (10\text{-}54)$$

and the superscript I indicates a property of an ideal solution. The term

$$\frac{v^E}{RT}\left(\frac{\partial P}{\partial x_1}\right)_{T,n_2}$$

may be considered to be a correction term which takes into account the variation of pressure with composition for isothermal data. The excess volume of liquid mixtures is usually numerically much smaller than RT. Therefore, the correction for varying pressure is usually negligible.

When applied to binary isobaric equilibrium data, The Gibbs-Duhem equation is used in the form[26]

$$x_1 \left(\frac{\partial \ln \gamma_1}{\partial x_1} \right)_{P,n_2} + x_2 \left(\frac{\partial \ln \gamma_2}{\partial x_1} \right)_{P,n_2} = \frac{h^E}{RT} \left(\frac{\partial T}{\partial x_1} \right)_{P,n_2} \qquad (11\text{-}62)$$

where h^E is the excess enthalpy. Here

$$\frac{h^E}{RT} \left(\frac{\partial T}{\partial x_1} \right)_{P,n_2}$$

is a correction term which takes into account the variation of temperature with composition for isobaric data. Generally, this correction term will be important if the boiling point spread between the pure components is more than a few degrees unless the two components are from the same homologous series and therefore have a very low excess enthalpy or enthalpy of mixing.

Analytical relationships. Now, the Gibbs-Duhem equation is the basis for correlations which can be used to predict the effect of mixture composition on activity coefficients. By examining data shown schematically in Fig. 11-6, M.

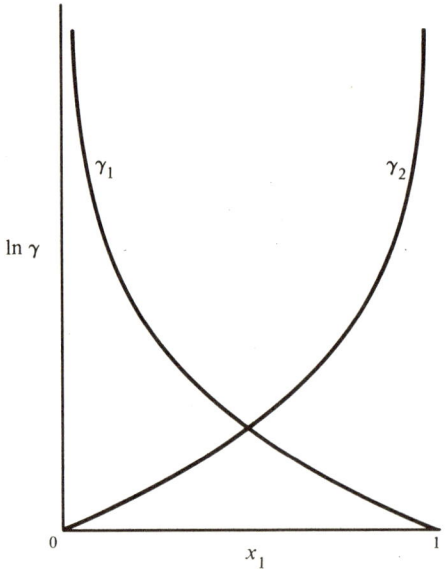

Fig. 11-6. *Activity coefficients.*

Margules concluded that activity coefficients could be represented by a polynomial of composition variables. He suggested that we set

$$\ln \gamma_1 = ax_2 + bx_2^2 + cx_2^3$$

$$\ln \gamma_2 = a'x_1 + b'x_1^2 + c'x_1^3$$

The number of constants in these equations can be reduced by employing the Gibbs-Duhem equation in the form

$$x_1\left(\frac{\partial \ln \gamma_1}{\partial x_1}\right)_{P,T,x_2} + x_2\left(\frac{\partial \ln \gamma_2}{\partial x_1}\right)_{P,T,x_2} = 0 \qquad (11\text{-}60)$$

Now, since

$$x_1 + x_2 = 0$$

we find

$$dx_1 = -dx_2$$

and

$$x_1\left(\frac{\partial \ln \gamma_1}{\partial x_1}\right)_{P,T,x_2} = -x_1\left(\frac{\partial \ln \gamma_1}{\partial x_2}\right)_{P,T,x_2}$$

$$= -ax_1 - 2bx_1x_2 - 3cx_1x_2^2$$

Further,

$$x_2\left(\frac{\partial \ln \gamma_2}{\partial x_1}\right)_{P,T,x_2} = a'x_2 + 2b'x_1x_2 + 3c'x_1^2x_2$$

Then, letting

$$x_1 = 1 - x_2$$

we have from the Gibbs-Duhem equation,

$$a'x_2 + 2b'(1-x_2)x_2 + 3c'(1-x_2)^2x_2 = a(1-x_2) + 2b(1-x_2)x_2$$

$$+ 3c(1-x_2)x_2^2$$

Equating coefficients of x_2, we obtain the result

$$a = 0$$

$$c' = -c$$

$$a' = 0$$

$$b' = \frac{2b + 3c}{2}$$

Substituting these values into the original equations, we find

$$\frac{\ln \gamma_1}{x_2^2} = b + cx_2 \tag{11-63}$$

$$\frac{\ln \gamma_2}{x_1^2} = b + c(x_2 + 0.5) \tag{11-64}$$

These equations are known as the Margules equations. For systems at conditions under which Raoult's and Dalton's law should apply to ideal liquid solutions, values of γ_i for non-ideal liquid solutions can be calculated from the relationship

$$Py_i = \gamma_i x_i P_i^{\square} \tag{11-65}$$

at each experimental point. If the experimental activity coefficients are plotted in the form of the Margules equation, a straight line should be obtained with a slope of c and an intercept of b.

Experimentally, we find that the constants in the Margules equations are functions of temperature. Unfortunately, no generalized relationship exists between these constants and temperature. Nevertheless, it has been suggested that the Margules constants be taken as inversely proportional to $T^{1/4}$. On this basis, we may write the Margules equations in the following form:

$$\frac{T^{1/4} \ln \gamma_1}{x_2^2} = b' + c'x_2 \tag{11-66}$$

and

$$\frac{T^{1/4} \ln \gamma_1}{x_1^2} = b' + c'(0.5 + x_2) \tag{11-67}$$

The constants in Eqs. (11-66) and (11-67) will have values differing from the constants in Eqs. (11-63) and (11-64) as a result of the addition of the temperature term.

Other integrated forms of the Gibbs-Duhem equation can be developed by analysis of the excess Gibbs free energy g^E where

$$g^E = g_M - g_M^I \qquad (10\text{-}57)$$

Now the Gibbs free energy of a non-ideal binary mixture is given by the relationship

$$g_M = x_1 \overline{G}_1 + x_2 \overline{G}_2$$

while the Gibbs free energy of an ideal binary mixture is given by the relationship

$$g_M^I = x_1 g_1 + x_2 g_2 + RT(x_1 \ln x_1 + x_2 \ln x_2) \qquad (10\text{-}16)$$

Then

$$g^E = x_1(\overline{G}_1 - g_1) + x_2(\overline{G}_2 - g_2) - RT(x_1 \ln x_1 + x_2 \ln x_2)$$

But

$$\overline{G}_i - g_i = \mu_i - \mu_i^\circ$$

$$= RT \ln a_i$$

$$= RT \ln \gamma_i + RT \ln x_1$$

Thus

$$g^E = RT(x_1 \ln \gamma_1 + x_2 \ln \gamma_2)$$

We now define the partial molal excess Gibbs free energy by the relationship

$$\bar{G}_i^E = \left(\frac{\partial G^E}{\partial n_i}\right)_{P,T,n_j}$$

$$= \left(\frac{\partial n_M g^E}{\partial n_i}\right)_{P,T,n_j}$$

In this case,

$$n_M = n_1 + n_2$$

so

$$n_M g_1^E = RT(n_1 \ln \gamma_1 + n_2 \ln \gamma_2)$$

Then

$$\bar{G}_1^E = RT\ln \gamma_1 + RT\left[n_1\left(\frac{\partial \ln \gamma_1}{\partial n_1}\right)_{P,T,n_2} + n_2\left(\frac{\partial \ln \gamma_2}{\partial n_1}\right)_{P,T,n_2}\right]$$

which we find on using the Gibbs-Duhem relationship reduces to

$$\bar{G}_1^E = RT\ln \gamma_1 \tag{11-68}$$

Again note that γ_1 is based on a reference state of the pure liquid at system temperature and pressure. As an example of the use of Eq. (11-68) in calculating activity coefficients, consider the following problem.

EXAMPLE PROBLEM 11-6. Calculate the activity coefficient for component 1 in a binary solution for which

$$g^E = (x_1 v_{f,1} + x_2 v_{f,2})A_{12}\Phi_1 \Phi_2$$

where

$$A_{12} = \text{molecular interaction energy term}$$

$$\Phi_i = \text{volume fraction of component } i$$

$$= \frac{x_i v_{f,i}}{\Sigma x_j v_{f,j}}$$

Substituting

$$\frac{n_1}{n_1 + n_2} = x_1$$

and

$$\frac{n_2}{n_1 + n_2} = x_2$$

into the definition of g^E, we find

$$g^E = \frac{n_1 n_2 v_{f,1} v_{f,2} A_{12}}{(n_1 + n_2)(n_1 v_{f,1} + n_2 v_{f,2})}$$

or, multiplying both sides of this equation by the total moles in the system $n_1 + n_2$,

$$G^E = \frac{n_1 n_2 v_{f,1} v_{f,2} A_{12}}{n_1 v_{f,1} + n_2 v_{f,2}}$$

and

$$RT \ln \gamma_1 = \left(\frac{\partial G^E}{\partial n_1}\right)_{P,T,n_2}$$

$$= \frac{v_{f,1}(n_2 v_{f,2})^2 A_{12}}{(n_1 v_{f,1} + n_2 v_{f,2})^2}$$

Finally, multiplying and dividing by $(n_1 + n_2)^2$, we obtain the result

$$RT \ln \gamma_1 = v_{f,1} \Phi_2^2 A_{12}$$

Equation (11-68) is the starting point for many current correlations of liquid phase activity coefficients. The approach taken is to simplify the relationship

$$g^E = h^E - Ts^E \qquad (11\text{-}69)$$

by assuming either that

$$h^E = 0$$

or by assuming that the solution is "regular" as defined by J. Hildebrand[33] and both

$$v^E = 0$$

and

$$s^E = 0$$

Solutions for which h^E is zero are called "athermal" solutions. Athermal behavior is approximated by mixtures of polymer molecules in solvents of similar chemical characteristics. Many nonpolar solutions seem to approach "regular" solution behavior since orienting and chemical effects which increase the non-ideality of a mixture are absent.

As an example of this approach, van Laar developed integrated forms of the Gibbs-Duhem equation for a regular solution whose phase behavior could be represented by the van der Waals equation of state,

$$P = \frac{RT}{v - b_M} - \frac{a_M}{v^2} \qquad (8\text{-}3)$$

Details of these calculations are presented by Prausnitz[26].

Prausnitz shows that van Laar's results can be written in the form

$$\ln \gamma_1 = \frac{b_1}{RT} \left(\frac{x_2 b_2}{x_1 b_1 + x_2 b_2} \right)^2 \left(\frac{a_1^{1/2}}{b_1} - \frac{a_2^{1/2}}{b_2} \right)^2$$

$$= \frac{A}{\left(1 + \frac{A}{B} \frac{x_1}{x_2}\right)^2} \qquad (11\text{-}70)$$

and
$$\ln \gamma_2 = \frac{b_2}{RT}\left(\frac{x_1 b_1}{x_1 b_1 + x_2 b_2}\right)^2 \left(\frac{a_1^{1/2}}{b_1} - \frac{a_2^{1/2}}{b_2}\right)^2$$

$$= \frac{B}{\left(1 + \frac{B}{A}\frac{x_2}{x_1}\right)^2} \tag{11-71}$$

where

$$A = \frac{b_1}{RT}\left(\frac{a_1^{1/2}}{b_1} - \frac{a_2^{1/2}}{b_2}\right)^2$$

and

$$B = \frac{b_2}{RT}\left(\frac{a_1^{1/2}}{b_1} - \frac{a_2^{1/2}}{b_2}\right)^2$$

Equations (11-70) and (11-71) can be rearranged in the forms

$$\left(\frac{1}{\ln \gamma_1}\right)^{1/2} = \left(\frac{1}{A}\right)^{1/2} + \left(\frac{A}{B^2}\right)^{1/2}\left(\frac{x_1}{x_2}\right) \tag{11-72}$$

$$\left(\frac{1}{\ln \gamma_2}\right)^{1/2} = \left(\frac{1}{B}\right)^{1/2} + \left(\frac{B}{A^2}\right)^{1/2}\left(\frac{x_2}{x_1}\right) \tag{11-73}$$

A plot of experimental data in the form of Eq. (11-72) and (11-73) should yield two different straight lines. The values of the constants A and B obtained from the slopes and intercepts of these two different straight lines should be the same if the experimental data is consistent.

Hildebrand and G. Scatchard recognized that van Laar's theory could be greatly improved if it could be freed from the limitations of the van der Waals equation of state. For regular solutions, they developed the relationship

$$\ln \gamma_1 = \frac{v_{f,1}}{RT}\left(\frac{x_2 v_{f,2}}{x_1 v_{f,1} + x_2 v_{f,2}}\right)(\delta_1 - \delta_2)^2 \tag{11-74}$$

where δ_1 is a parameter called the solubility parameter and

$$\delta_1 = \left(\frac{u_{fg,i}}{v_{f,i}}\right)^{1/2}$$

Values of specific liquid molar volumes and solubility parameters for many components are available in the literature[18,26,33]. The solubility parameter relationship, Eq. (11-74), is quite similar to the van Laar relationship, Eq. (11-70), except that experimental values of $v_{f,i}$ are used in place of values of the van der Waals constant b_i and experimental values of δ_i are used in place of the combined van der Waals constants $a_i^{1/2}/b_i$. For a multi-component mixture, Eq. (11-74) takes the form

$$\ln \gamma_1 = \frac{v_{f,1}}{RT} (\delta_1 - \delta_M)^2 \tag{11-75}$$

where

$$\delta_M = \frac{\Sigma x_i v_{f,i} \delta_i}{\Sigma x_i v_{f,i}}$$

If the constants in the various forms of the integrated Gibbs-Duhem equations were available, we could use these analytical expressions to predict activity coefficients for any phase of known composition. Indeed, the analytical form of these equations is ideally suited for predicting activity coefficients with a digital computer. It should be pointed out however, that if the phase compositions are unknown, we are faced with the same dilemma we encountered previously in attempting to evaluate fugacity coefficients using equations of state with mixture coefficients.

If several experimental activity coefficients are available, the constants in equations such as the Margules equations can be determined by using the method of least squares. Another method of determining the necessary constants is based on the use of azeotropic data. An azeotrope may be defined as that mixture for which the composition of the vapor and the liquid in equilibrium are identical. Thus, at an azeotrope of a binary mixture, $x_1 = y_1$ and $x_2 = y_2$. Consequently, from Eq. (11-65), we find that at an azeotrope,

$$\gamma_i = \frac{P}{P_i^\square} \tag{11-76}$$

The use of azeotropic data in Eq. (11-76) is a convenient way of finding γ_1 and γ_2. These values may then be used to solve for the two constants in the Margules relationships, Eqs. (11-63) and (11-64). A knowledge of the Margules constants permits us to develop the entire vapor-liquid equilibrium curve.

Azeotropic data can also be used to evaluate the van Laar constants. The calculations of these constants can be simplified if we first rearrange the van Laar equations. Dividing Eq. (11-70) by Eq. (11-71) we find

$$\frac{\ln \gamma_1}{\ln \gamma_2} = \frac{B}{A} \left(\frac{x_2}{x_1}\right)^2$$

so that

$$B = A \frac{\ln \gamma_1}{\ln \gamma_2}\left(\frac{x_1}{x_2}\right)^2 \qquad (11\text{-}77)$$

If this result is used to eliminate B from Eq. (11-70), we find

$$A = \ln \gamma_1 \left(1 + \frac{x_2 \ln \gamma_2}{x_1 \ln \gamma_1}\right)^2 \qquad (11\text{-}78)$$

Values of A calculated using Eq. (11-78) may then be substituted into Eq. (11-77) to calculate B.

EXAMPLE PROBLEM 11-7. For the system ethanol-water at a total pressure of 760 mm and a temperature of 175.6 °F, calculate the mole fraction of ethanol in the vapor when the mole fraction of ethanol in the liquid is 0.5,

(a) neglecting the non-ideality of the liquid, and
(b) considering the non-ideality of the liquid using activity coefficients calculated from the Margules equations.

Ethanol and water form an azeotrope of 760 mm of Hg total pressure and 172.7 °F when the mole fraction of ethanol in the liquid is 0.8943.

The following vapor pressure data is available in the literature:

	Vapor Pressure in mm of Hg	
$T(°F)$	Ethanol	Water
172.7	755	330.9
175.6	803	346.4

(a) Since the vapor pressure of both components and the system pressure is low and we can neglect the non-ideality of the liquid, Raoult's and Dalton's law

$$Py_i = x_i P_i^\square \qquad (11\text{-}20)$$

applies, and we find

$$y_E = \frac{(0.5)(803)}{760} = 0.53$$

where the subscript E indicates ethanol.

(b) From Eq. (11-76), we find from azeotropic data that

$$\gamma_E = \frac{760}{755} = 1.0066$$

and

$$\gamma_W = \frac{760}{330.9}$$

$$= 2.297$$

where the subscript W indicates water.

Neglecting the effect of temperature on the Margules constants, we find using Eqs. (11-63) and (11-64) that

$$\frac{\ln 1.0066}{(0.1057)^2} = b + 0.1057c$$

and

$$\frac{\ln 2.297}{(0.8943)^2} = b + 0.6057c$$

Solving,

$$b = 0.495$$

$$c = 0.9$$

Then, substituting $x_E = 0.5$ into Eq. (11-63), we find

$$\ln \gamma_E = 0.5^2 \, [0.495 + 0.9(0.5)]$$

$$= 0.236$$

Thus,

$$\gamma_E = 1.266$$

Then, from Eq. (11-65),

$$y_E = \frac{(1.266)(803)(0.5)}{760}$$

$$= 0.67$$

Experimental data indicates that at the conditions of the problem,

$$y_E = 0.66$$

There is one major flaw in the solution to this problem as we have presented it. Using Eq. (11-20), we find for water,

$$y_W = \frac{(0.5)(346.4)}{760}$$

$$= 0.23$$

By definition

$$y_E + y_W = 1.0$$

However, if we use the values we have calculated, we find

$$y_E + y_W = 0.53 + 0.23 \neq 1.0$$

Here we have encountered an example of an overspecified problem. The Gibbs Phase Rule indicates that for a binary system in vapor-liquid equilibrium, we can specify only two of the three variables P, T, or y_E. The use of estimated activity coefficients does not alter this conclusion. Thus, we must vary either P, T, or y_E until the relationship

$$y_E + y_W = 0$$

is satisfied. An approximation which satisfies this relationship may be obtained by normalizing the calculated values of y_E and y_W. However, it is probably more realistic to vary the assumed system pressure or temperature.

Further Applications of Phase Equilibrium Relationships

In the more general phase equilibrium problem, we not only would like to calculate the composition of each phase but also the relative amounts of each phase. This problem can be solved by combining phase equilibrium relationships with component mass balances.

As an example of the approach to be taken, let us consider a system in which a vapor and a liquid are in equilibrium. If the total number of moles in the system n_M of known composition is distributed between n_g moles of vapor and n_f moles of liquid, the over-all mass balance equation reduces to

$$n_M = n_f + n_g$$

Similarly, the component material balance for component i distributed between a vapor and liquid reduces to

$$n_M z_i = n_f x_i + n_g y_i \tag{11-26}$$

where

z_i = mole fraction of component i in the total mixture,

x_i = mole fraction of component i in the equilibrium liquid, and

y_i = mole fraction of component i in the equilibrium vapor.

Now, the vapor and liquid phase compositions are related by the equation

$$K_i = \frac{y_i}{x_i} \tag{11-43}$$

Eliminating y_i from Eq. (11-26), we find

$$n_M z_i = n_f x_i + n_g K_i x_i$$

or

$$x_i = \frac{z_i}{\left(\dfrac{n_f}{n_M}\right)(1 - K_i) + K_i} \tag{11-79}$$

where we have made use of the overall mass balance, to eliminate n_g. Then, since

$$y_i = K_i x_i$$

we find

$$y_i = \frac{K_i z_i}{\left(\dfrac{n_f}{n_M}\right)(1 - K_i) + K_i} \tag{11-80}$$

Now, we define a function of n_f/n_M as

$$\Phi\left(\frac{n_f}{n_M}\right) = \Sigma x_i - \Sigma y_i$$

$$= \Sigma \frac{(1 - K_i) z_i}{\left(\dfrac{n_f}{n_M}\right)(1 - K_i) + K_i} = 0 \tag{11-81}$$

The number of moles of liquid n_f per number of total moles in the mixture n_M is found when the value of this function is zero. At a bubble point,

$$n_f = n_M$$

and Eq. (11-81) reduces to

$$\Sigma K_i z_i = 1.0 \qquad (11\text{-}82)$$

Similarly, at a dew point

$$n_f = 0$$

and Eq. (11-81) reduces to

$$\Sigma \frac{z_i}{K_i} = 1.0 \qquad (11\text{-}83)$$

Special techniques are available for solving Eq. (11-81). For example, the correct values of n_f/n_M may be found graphically by plotting $\Phi(n_f/n_M)$ from Eq. (11-81) at several values of n_f/n_M. It may also be solved by trial and error.

A somewhat more direct method of solving this equation is based on the use of a Taylors series expansion,

$$\Phi(x) = \Phi(x_0) + \frac{(x-x_0)}{1!}\frac{d\Phi(x_0)}{dx} + \frac{(x-x_0)^2}{2!}\frac{d^2\Phi(x_0)}{dx^2} + \cdots$$

Now, the problem is to find an x such that

$$\Phi(x) = 0$$

If we ignore all terms in Taylors series beyond the second, we find that at $\Phi(x) = 0$,

$$\Phi(x_0) + (x - x_0)\,\Phi'(x_0) = 0$$

or

$$x = x_0 - \frac{\Phi(x_0)}{\Phi'(x_0)} \qquad (11\text{-}84)$$

where

$$\Phi'(x_0) = \frac{d\Phi(x_0)}{dx}$$

Equation (11-84) may be used to solve an equation such as Eq. (11-81). One first assumes a value of n_f/n_M, denoted by the symbol x_0. If $\Phi(x_0)$ does not equal zero, then a better estimate of the correct value of n_f/n_M, denoted by the symbol x, can be obtained by the use of Eq. (11-84). If $\Phi(x)$ based on the new value of x does not equal zero, the procedure is repeated, this time using the last calculated value of x as x_0. This procedure must be repeated until $\Phi(x)$ is as close to zero as desired. This method of solving an equation is known as the Newton-Raphson method and can be programmed on a digital computer.

EXAMPLE PROBLEM 11-8. One pound mole of a mixture consisting of 30 mole percent propane, 40 mole percent n-butane and 30 mole percent n-pentane is allowed to reach equilibrium at 150 °F and 100 lb$_f$/in.² Calculate the moles and composition of liquid and vapor present in the equilibrium mixture using the following vapor-liquid equilibrium coefficients:

$$K_{C_3} = 2.9; \quad K_{nC_4} = 1.05; \quad K_{nC_5} = 0.41$$

Let us use Newton's method to solve this problem. In this case

$$\Phi\left(\frac{n_f}{n_M}\right) = \sum \frac{(1-K_i)z_i}{\left(\frac{n_f}{n_M}\right)(1-K_i) + K_i} \tag{11-81}$$

$$\Phi'\left(\frac{n_f}{n_M}\right) = \sum \frac{(-1)(1-K_i)^2 z_i}{\left[\left(\frac{n_f}{n_M}\right)(1-K_i) + K_i\right]^2} \tag{11-85}$$

Note that in this case, $\Phi'(n_f/n_M)$ is always negative.

Now, we can facilitate the selection of our first approximation of n_f/n_M by observing that

$$0 < \frac{n_f}{n_M} < 1.0$$

If we assume $n_f/n_M = 1.0$, we find

$$\Phi\left(\frac{n_f}{n_M}\right) = \frac{(-1.9)(0.3)}{(1.0)(-1.9) + 2.9} + \frac{(-0.05)(0.4)}{(1.0)(-0.05) + 1.05} + \frac{(0.59)(0.3)}{(1.0)(0.59) + 0.41}$$

$$= -0.413$$

This is not very close to zero. Thus, substituting into Eq. (11-85), we find

$$\Phi'\left(\frac{n_f}{n_M}\right) = -\frac{(-1.9)^2(0.3)}{[(1.0)(-1.9) + 2.9]^2} - \frac{(-0.05)^2(0.4)}{[(1.0)(-0.05) + 1.05]^2}$$

$$-\frac{(0.59)^2(0.3)}{[(1.0)(0.59) + 0.41]^2} = -1.188$$

These results may be substituted into Eq. (11-84) to provide a better estimate of n_f/n_M

$$\left(\frac{n_f}{n_M}\right)_{\text{calculated}} = 1.0 - \frac{-0.413}{-1.188} = 0.65$$

This procedure is continued as shown in the following table:

$\frac{n_f}{n_M}$ (assumed)	$\Phi\left(\frac{n_f}{n_M}\right)$	$\Phi'\left(\frac{n_f}{n_M}\right)$	$\frac{n_f}{n_M}$ (calculated)
1.00	−0.413	−1.188	0.65
0.65	−0.139	−0.557	0.40
0.40	−0.012	−0.488	0.37
0.37	−0.003		

Then, since $n_M = 1.0$ mole, we find

$$n_f = 0.37 \text{ moles}$$

$$n_g = 1.0 - 0.37 = 0.63 \text{ moles}$$

These are two of the desired answers. Finally, with the calculated value of n_f/n_M, we can use Eq. (11-79) to find the liquid phase composition and Eq. (11-80) to find the vapor phase composition. These results are summarized below:

Component	x	y
Propane	0.14	0.40
n-Butane	0.39	0.41
n-Pentane	0.47	0.19

As demonstrated in the preceding example problem, the computational procedures to be followed in solving phase equilibrium problems are well developed.

446   Chap. 11   Phase Equilibrium

The only limitation in the solution of these problems is the estimation of correct values of the fugacities of components in a mixture and the related vapor-liquid equilibrium coefficients. This in turn depends upon our ability to characterize the phase behavior of mixtures.

PROBLEMS

11-1. Use the Gibbs Phase Rule to show whether we would expect isothermal condensation or nonisothermal condensation of the following mixtures flowing through a heat exchanger. The pressure drop of the mixtures through the heat exchanger may be assumed to be negligible:

   (a) Butane—pentane

   (b) Steam—air—water

   (c) Pentane—steam—water

11-2. Estimate the critical pressure and the heat of vaporization of isopropyl amine,

$$\begin{array}{c} NH_2 \\ H \ | \ H \\ HC-C-CH \\ H \ H \ H \end{array}$$

at 20 °C using the following vapor pressure data:

$T(°C)$	$P(lb_f/in.^2)$
0	3.7
10	5.9
20	9.0
30	13.0
40	19.0

The critical temperature of isopropyl amine is 203 °C and it has a molecular weight of 59.11 gm per gm-mole.

11-3. The following data is available from the literature:

Compound	Molecular Weight	Boiling Point, °R	h_{fg} (BTU/lb mole)
Butane	58	490.7	9,614.2
Benzene	78	635.6	13,217.2
Ethyl alcohol	46	632.3	16,912.7
Freon-12	121	438.3	9,064.0

Calculate the heat of vaporization for one pound mole of each of the above compounds using:

(a) Trouton's rule
(b) Kistiakowsky's equation
(c) Fishtine's equation

11-4. Show that

$$\left(\frac{\partial G}{\partial n_i}\right)_{P,T,n_j} = \left(\frac{\partial U}{\partial n_i}\right)_{S,V,n_j} = \mu_i$$

11-5. When limited quantities of cooling water for a condenser are available, a cooling tower (such as shown in Fig. 11-7) is often used. Consider the case where 20,000

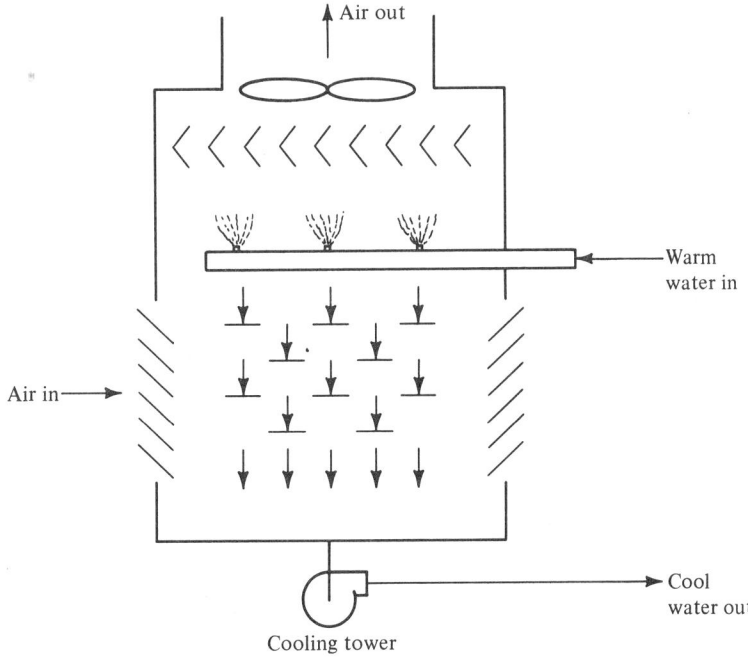

Fig. 11-7. *Cooling tower.*

lb_m/hr of water at 105 °F enters the top of the cooling tower, and the cool water leaves the bottom at 65 °F. The air-water vapor mixture enters the bottom of the cooling tower at 14.7 $lb_f/in.^2$, and has a temperature of 72 °F and a relative humidity of 50 percent. The air-water vapor mixture leaving the tower has a pressure of 14.3 $lb_f/in.^2$, a temperature of 90 °F, and a relative humidity of 80 percent.

Determine the lb_m of dry air per minute that must be used and the fraction of the incoming water that evaporates. Assume the process to be adiabatic. The enthalpy of air may be calculated relative to a reference temperature of 0 °F by assuming c_p = 0.24 BTU/lb_m °F, and the enthalpy of liquid water and water vapor may be obtained from steam tables.

11-6. Assuming the vapor acts as an ideal gas and that the activity coefficient of the liquid is unity, calculate, for the mixture presented below:

(a) The bubble point temperature at a system pressure of 14.7 $lb_f/in.^2$
(b) The dew point pressure at a system temperature of 70 °F.
(c) The bubble point pressure at a system temperature of 70 °F.

Mixture

Component	Mole Fraction
Propane	0.3
n-Butane	0.4
n-Pentane	0.3

11-7. A technician has determined that the bubble point of a mixture known to contain propane, n-butane, and n-pentane is 21 °F and 1 atm. In analyzing the individual components, the analytical equipment fails after indicating only the amount of n-butane in the original sample. This amount was found to be 67.3 mole percent. What are the mole percentages of the propane and normal pentane in this mixture? Assume that Raoult's and Dalton's law applies in this case.

11-8. Repeat Prob. 11-6 using Fig. A-5.

11-9. It is proposed to recover methyl chloride from a stream of air at 140 °F and 400 $lb_f/in.^2$ by compressing the mixture reversibly and isothermally to 1000 $lb_f/in.^2$. If the air entering and leaving the compression process is saturated with methyl chloride and if $v^\square_{methyl\ chloride}$ is 0.851, what fraction of the methyl chloride fed to the process is recovered as liquid? You may assume that methyl chloride and air form ideal solutions and that air is insoluble in the liquid methyl chloride.

11-10. A mixture consisting of 120 lb-moles of ethane and 80 lb-moles of propane is to be compressed isothermally and reversibly from 10 atm and 70 °F to the dew point of the mixture in a cylinder fitted with a piston.

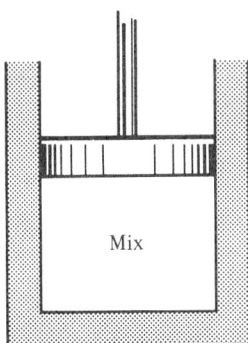

Fig. 11-8

(a) Assuming that Raoult's and Dalton's law adequately represents the equilibrium phase behavior of these components, calculate the final pressure of the mixture.

(b) Assuming the gaseous mixture follows the relationship

$$PV = nzRT$$

calculate the work required to compress the mixture in part (a) of this problem. How much heat was transferred during this process?

11-11. Calculate the vapor-liquid equilibrium constants for ethane-pentane at 160 °F and

(a) 100 psia
(b) 1000 psia

using

(a) Raoult's and Dalton's law;
(b) The Lewis-Randall rule and fugacities where the specific volume of liquid ethane can be assumed to be 0.04 ft^3/lb$_m$; of liquid n-pentane, 0.03 ft^3/lb$_m$;
(c) Fig. A-5.

11-12. For a binary mixture of gases obeying van der Waals' equation of state, show

$$\ln \frac{\bar{f}_1}{y_1 P} = \frac{n_M b_1}{V - n_M b_M} - \frac{2 n_M a_1^{1/2} a_M^{1/2}}{RTV} + \ln \frac{V}{(V - n_M b_M)} - \ln z$$

where

$$n_M = n_1 + n_2$$

Hint: Show

$$RT\ln \frac{\bar{f}_1}{y_1 P} = \int_V^\infty \left(\frac{RT[V + n_2(b_1 - b_2)]}{(V - n_M b_M)^2} - \frac{2n_M a_1^{1/2} a_M^{1/2}}{V^2} - \frac{RT}{V} \right) dV - RT\ln z$$

11-13. Using the equation of state for a mixture of 1 and 2

$$\frac{PV}{n_M RT} = 1 + \frac{n_M B_M(T)}{V}$$

where

$$n_M = n_1 + n_2$$

$$B_M(T) = y_1^2 B_{11} + 2y_1 y_2 B_{12} + y_2^2 B_{22}$$

and

$$y_1 = \frac{n_1}{n_M}, \quad y_2 = \frac{n_2}{n_M}$$

show that

$$\ln \frac{\bar{f}_1}{y_1 P} = \frac{2}{v} (y_1 B_{11} + y_2 B_{12}) - \ln z$$

11-14. Using the Benedict-Webb-Rubin equation of state, show that

$$RT\ln \nu = 2(B_0 RT - A_0 - C_0/T^2)/v + 3(bRT - a)/2v^2 + 6a\alpha/5v^5$$

$$+ \frac{c}{v^2 T^2} \left\{ \left[1 - \exp\left(-\frac{\gamma}{v^2}\right) \right] \left(\frac{v^2}{\gamma}\right) + \left(\frac{1}{2} + \frac{\gamma}{v^2}\right) \exp\left(-\frac{\gamma}{v^2}\right) \right\} - RT\ln z$$

for a single component system.
Hint: Integrals involving exponential functions in this problem may be more readily solved by making the substitution

$$x = \frac{n_M^2 \gamma}{V^2}$$

11-15. Using the result obtained in Prob. 11-14 and assuming the Lewis-Randall rule holds, estimate the vapor liquid equilibrium coefficient for propane at 70 °F and 30 lb$_f$/in.^{2} The volume of liquid propane may be assumed to be constant and equal to 1.41 ft^3/lb mole. Compare your results with results obtained from Fig. A-5.

11-16. D. L. Stinson of the University of Wyoming Mineral Engineering Department has demonstrated a process using atmospheric freezing to separate salt from water for a cost of approximately $0.50 per 1000 gallons of brine treated (Canadian Patent No. 782,794). We would like to compare this with other water desalinization processes which use more conventional sources of energy.

A device using electrical power will be used to separate out solid sodium chloride from a 5 mole percent salt in water solution. The products will be pure sodium chloride and pure water, both at 100 °C. The device will be fed continuously with the 5 percent solution at 100 °C.

The vapor pressure of water over aqueous sodium chloride solutions has been determined experimentally at 100 °C. The results may be represented by the following equation:

Vapor pressure = 760 − (1560)(mole NaCl/mole H$_2$O)[mm Hg]. A saturated solution of sodium chloride in water at 100 °C contains 0.121 mole NaCl/mole H$_2$O.

If power is available at 30.0 mills per kWhr, what is the minimum power cost of producing 1000 gallons of pure water in the proposed equipment? Has Stinson developed a perpetual motion machine of the second kind?

11-17. The partial pressure of acetone over mixtures of acetone and water has been determined at 60 °C as follows:

Mole percent Acetone	Partial Pressure Acetone, mm Hg
0	0
3.3	190
11.7	443
31.8	588
55.4	672
73.6	711
100.0	860

Using the Gibbs-Duhem equation and assuming that the fugacity of the liquid phase does not change with total pressure and that the vapor obeys the ideal gas law, calculate the partial pressure of water over a solution containing 55.4 mole percent acetone at 60 °C.

11-18. The following data has been reported for vapor-liquid equilibrium studies on the methanol-benzene system at one atmosphere pressure [Jr. of Chem. and Eng. Data, 14, 4, 418 (Oct., 1969)].

T°C	x_1	y_1	γ_1	γ_2
62.0	0.109	0.475	4.798	1.060
59.3	0.226	0.538	2.908	1.181
59.1	0.311	0.554	2.192	1.290
58.4	0.529	0.597	1.426	1.751
58.4	0.715	0.643	1.135	2.569
58.9	0.811	0.694	1.058	3.271
60.7	0.912	0.795	1.002	4.444

NOTE: Subscript 1 indicates methanol, subscript 2 indicates benzene.

(a) Check the thermodynamic consistency of this data at 0.25, 0.50 and 0.75 mole percent methanol using the Gibbs-Duhem equation.

(b) Calculate the Margules equation constants from this data using the method of least squares.

(c) Calculate the van Laar equation constants from this data using the method of least squares.

(d) Methanol and benzene form an azeotrope at one atmosphere pressure and 58.3 °C when the mole fraction of methanol in the liquid phase is 0.610. Calculate the Margules equation constants and the van Laar equation constants using this azeotropic data and compare with results obtained in parts (b) and (c).

11-19. At 760 mm mercury, total pressure, n-heptane and 1-propanol form an azeotrope at 84.8 °C and 0.496 mole fraction n-heptane. At this temperature, the vapor pressure of n-heptane is 512.0 mm of Hg and the vapor pressure of 1-propanol is 463.8 mm of Hg. Calculate the mole fraction of n-heptane in the vapor when the mole fraction of n-heptane in the liquid is 0.207 using

(a) Margules' equations

(b) van Laar's equations

11-20. Assuming that liquid benzene and liquid hexane form a regular solution, estimate the activity of benzene in a mixture containing 2.0 moles of liquid benzene and 3.0 moles of liquid hexane at 100 °F. Use the following data:

Component	Solubility Parameter, $\delta (cal/cc)^{1/2}$	$v_f (cc/gm\ mole)$
Benzene	9.16	89.4
Hexane	7.27	131.6

11-21. One pound mole of a mixture consisting of 30 mole percent propane, 40 mole percent n-butane, and 30 mole percent n-pentane is allowed to reach equilibrium at 50 °F and 30 lb_f/in^2. Assuming the vapor acts as an ideal gas and that the activity coefficient of the liquid is unity, calculate

(a) The moles of liquid and vapor present in the equilibrium mixture.
(b) The mole fractions in the liquid and vapor phases.

12

Chemical Reactions

OBJECTIVES

After studying this chapter, the student should be able to

(1) apply the principle of conservation of mass and the principle of conservation of energy to systems which involve chemical reactions;
(2) define and numerically evaluate the reaction free energy ΔG_R^o;
(3) calculate the equilibrium composition of a reacting mixture using Eq. (12-16); and
(4) estimate the electromotive potential as well as the changes in thermodynamic properties of a simple electrochemical cell.

INTRODUCTION

In our study of the thermodynamics of heat engines in Chap. 7, we indicated that energy in the less useful form of internal energy could first be converted to heat and then into the more useful forms of mechanical and electrical energy. The conversion of internal energy into heat involves a chemical reaction.

In a chemical reaction, one or more chemical species are converted into different species generally with the simultaneous evolution or adsorption of energy in the form of heat. While the desired products of many reactions may be specific chemical entities such as plastics or synthetic fibers, the desired product from many other reactions is energy. For example, the desired product from the reaction between oxygen and a fuel such as wood, coal or natural gas is heat. Unfortunately, some reactions such as the burning of high sulfur coal produce not only desired products such as heat but also undesired products such as sulfur dioxide and nitrogen oxides which pollute our atmosphere.

Whereas our emphasis in Chap. 7 was the conversion of heat evolved in a chemical reaction to mechanical or electrical energy, our emphasis in this chapter will be on the chemical reaction itself.

MASS BALANCE

Consider a process in which methane, the primary constituent of natural gas, and oxygen are introduced into a gas burner, shown in Fig. 12-1, and react to form

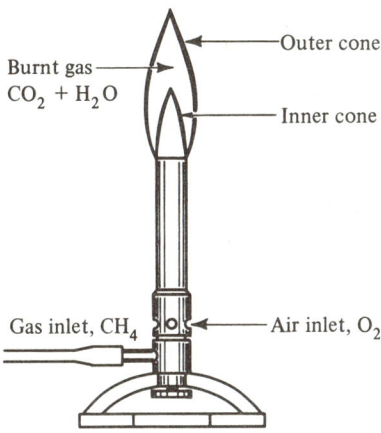

Fig. 12-1. *The Bunsen burner.*

carbon dioxide and water.

$$CH_4 + 2O_2 \rightarrow CO_2 + 2H_2O$$

A mass balance can be applied to the total feed entering the burner and the total effluent leaving the burner. If all the methane and oxygen react, the mass balance, based on a unit time, becomes

$$\left(m_{CH_4} + m_{O_2}\right)_{in} - \left(m_{CO_2} + m_{H_2O}\right)_{out} = 0$$

A mass balance applied to the total feed entering the total effluent leaving a system is called an overall mass balance.

However, if we attempt to apply the mass balance to a specific molecule such as methane, we are immediately faced with a problem. Methane entering the burner does not accumulate in the burner nor does it appear in the burner effluent since it has been converted to carbon dioxide and water. However, we can apply the mass balance to the individual atoms which make up the molecules. The mass of carbon atoms entering the burner in the methane does equal the mass of carbon atoms leaving the burner in the carbon dioxide. Similarly, the mass of hydrogen and oxygen atoms entering the burner equals the mass of hydrogen and oxygen atoms leaving the burner. This fact is the basis for "balancing" equations representing chemical reactions. Since we are dealing with reactions between molecules or atoms, it has been found to be more convenient to base these mass balance relationships on the number of atoms or moles which make up the molecules rather than on the mass of the atoms. An equation representing a chemical reaction is said to be balanced when the number of atoms of a particular element on the left hand side and the right hand side are equal.

Although, strictly speaking, the mass balance equation for individual species applies only to the atoms in a reacting system, it is still possible to apply it to molecular components if we introduce a term in the balance equation to represent the moles or mass of a particular chemical species formed or consumed in the reaction. Consider the reaction in which components A and B react to form components R and S.

$$aA + bB \rightleftharpoons rR + sS$$

The lower case letters are called molecular stoichiometric coefficients and have values as close to unity as possible such that the number of atoms or elements of a specific type are equal on both sides of the equation. Now we define the extent to which A is produced by the equation

$$\alpha = \frac{\text{moles } A \text{ in product} - \text{moles } A \text{ in reactants}}{-a} \qquad (12\text{-}1)$$

where α is called the extent of the reaction. Then, the number of moles of A that disappear due to chemical reaction is $a\alpha$. Further, the balanced chemical reaction tells us that when $a\alpha$ moles of A are consumed, i.e. when $-a\alpha$ moles of A are produced, $-b\alpha$ moles of B, $r\alpha$ moles of R and $s\alpha$ moles of S are also produced. Then, the generalized mass balance equation based on moles of component A in our example reaction becomes

$$\delta n_{A_{\text{in}}} - \delta n_{A_{\text{out}}} - a \, d\alpha = dn_{A_{\text{sys}}} \qquad (12\text{-}2)$$

HEATS OF REACTION

Energy Balance for Reacting Mixtures

Chemical reactions are generally accompanied by the evolution or absorption of heat because of differences in molecular structure of the products and reactants. For example, if products of a reaction possess greater energy as a result of their structure than the reactants, it will be necessary to supply energy to accomplish the reaction. This energy is generally added to the reacting mixture in the form of heat, called the heat of reaction.

The amount of energy which can be attributed to a chemical reaction can be calculated using the general mass and energy balance equations. In order to do this, Canjar and Manning[4] found it convenient to refer the enthalpy of all components to atoms at an arbitrary reference temperature and pressure. Then, the enthalpy of any component is the sum of the energy required to form the component from the atoms at some reference temperature and pressure h_F plus the energy required to change the state of the molecule from the reference temperature and pressure to the system temperature and pressure h. On this basis, we may write the generalized balance equation on a mole basis as follows:

$$\left[\left(h + h_F + \frac{v^2}{2} + gz\right)\delta n\right]_{in} - \left[\left(h + h_F + \frac{v^2}{2} + gz\right)\delta n\right]_{out} + \delta Q - \delta W$$

$$= d\left[\left(u + u_F + \frac{v^2}{2} + gz\right)n\right]_{sys} \qquad (12\text{-}3)$$

where the subscript F refers to the energy associated with the structure of the components. The difference between h_F of the products and reactants is commonly called the enthalpy of reaction ΔH_R or the heat of reaction.

The reason for this particular choice of terms may become apparent if we consider the reaction

$$aA \rightarrow bB$$

taking place in the gas phase. Assume that the extent of reaction is α and that both the reactant A and the product B are at the reference temperature and pressure. If this reaction takes place in a flow system and if we neglect kinetic and potential energy terms, Eq. (12-3) reduces to

$$\left(n_A h_{F_A} + n_B h_{F_B}\right)_{in} - \left(n_A h_{F_A} + n_B h_{F_B}\right)_{out} + Q = 0$$

on a unit time basis. In this case, we have eliminated the gas enthalpies of the reacting molecules h_{in} and the product molecules h_{out} since these terms represent the enthalpy required to change the state of the molecules from the reference state to the state of the system and in this example the state of the system is the reference state. Then the heat flow associated with this reaction is

$$Q = h_{F_B}\left(n_{B_{out}} - n_{B_{in}}\right) + h_{F_A}\left(n_{A_{out}} - n_{A_{in}}\right)$$

Applying Eq. (12-2), we note that over a unit time period the moles of B produced are

$$n_{B_{out}} - n_{B_{in}} = b\alpha$$

and the moles of A produced are

$$n_{A_{out}} - n_{A_{in}} = -a\alpha$$

so

$$Q = b\alpha h_{F_B} - a\alpha h_{F_A}$$

$$= \alpha\left(bh_{F_B} - ah_{F_A}\right)$$

At this point, we will find it convenient to define enthalpy of reaction by the relationship

$$\Delta H_R = \Sigma a_i h_{F_i}(\text{products}) - \Sigma a_i h_{F_i}(\text{reactants}) \tag{12-4}$$

where the a_i are stoichiometric coefficients in the balanced chemical reaction equation. It should be noted that when an enthalpy of reaction is reported for a particular reaction, it applies for the molecular stoichiometric numbers a_i as written. If each stoichiometric number is doubled, the enthalpy of the reaction is doubled. Then, in this case,

$$\Delta H_R = bh_{F_B} - ah_{F_A}$$

so that the heat flow resulting from the reaction is

$$Q = \alpha \Delta H_R$$

On the other hand, if the reaction in this example takes place reversibly in a closed system at constant pressure, Eq. (12-3) reduces to

$$Q = b\alpha u_{F_B} - a\alpha u_{F_A} + W$$

In this case,
$$W = P(V_{final} - V_{initial})$$

Since
$$V = n_A v_A + n_B v_B$$

this result can be combined with the mole balance relationships to give
$$W = P\left[\left(n_{B_{final}} - n_{B_{initial}}\right) v_B + \left(n_{A_{final}} - n_{A_{initial}}\right) v_A\right]$$

or
$$W = P(b\alpha v_B - a\alpha v_A)$$

Thus
$$Q = \alpha\left[b\left(u_{F_B} + Pv_B\right) - a\left(u_{F_A} + Pv_A\right)\right]$$
$$= \alpha\left(bh_{F_B} - ah_{F_A}\right)$$

or
$$Q = \alpha \Delta H_R$$

which is the same result we obtained for the flow system.

By convention, Q is positive when heat enters the system (endothermic) and negative when it leaves the system (exothermic). Thus, ΔH_R of an endothermic reaction is positive while ΔH_R of an exothermic reaction is negative.

Finally, for an open system at steady state the concept of enthalpy of reaction can be combined with the energy balance in the form of Eq. (12-3) to yield the result,

$$\left[\left(h + \frac{v^2}{2} + gz\right)\delta n\right]_{in} - \left[\left(h + \frac{v^2}{2} + gz\right)\delta n\right]_{out} + \delta Q - \delta W - \Delta H_R \, d\alpha = 0 \quad (12\text{-}5)$$

If enthalpy values of components found in tables or calculated using equations of state and specific heat data include the enthalpy associated with the structure of the component, it is not necessary to include an enthalpy of reaction term in the energy balance equation.

Enthalpy of Formation

It may prove instructive to examine enthalpies associated with molecular structure somewhat more closely. In order to do this, let us consider a special reaction called the formation reaction. We define a formation reaction as a reaction which forms one mole of a single compound from the elements, or molecules in the case of diatomic gases such as O_2, H_2, Cl_2 and N_2, which make it up. For example, the reaction

$$C + \tfrac{1}{2}O_2 + 2H_2 \rightarrow CH_3OH$$

is the formation reaction for methyl alcohol. However, the reaction

$$H_2O + SO_3 \rightarrow H_2SO_4$$

is not a formation reaction of sulfuric acid, since water and sulfur trioxide are not elements.

The enthalpy of formation, then, is defined by the relationship

$$\Delta h_F^\circ = h_F^\circ(\text{compound}) - \Sigma b_i h_{F_i}^\circ(\text{elements}) \tag{12-6}$$

where both the compound and its elements are in their standard states and at the same reference temperature. Standard states commonly used were discussed in Chap. 9. The symbol b_i in Eq. (12-6) is a stoichiometric coefficient and refers to the number of elements, or molecules in the case of O_2, H_2, Cl_2, and N_2, contained in the compound. The enthalpy of the elements at the reference state is taken as zero. Representative heats of formation of compounds in their standard state at one atmosphere and at a reference temperature of 298 K are presented in Table 12-1 [34-39]. The superscript $\circ$ in the table indicates that the reactants and products are in their standard states; the subscript F indicates that the reaction is a formation reaction; and the sub-subscript 298 is the reference temperature in kelvins.

TABLE 12-1.[†] Standard Heats of Formation and Combustion at 25 °C [34-39]

Notes: 1. Units are calories per gram mole of listed compound.
2. Combustion reaction products are $H_2O(l)$ and $CO_2(g)$.
3. Reference states: pure compound at 25°C, 1.0 atm.

Substance	Formula	State	$\Delta h_{F_{298}}^\circ$	$-\Delta h_{C_{298}}^\circ$
Normal paraffins:				
Methane	CH_4	g	−17,889	212,800
Ethane	C_2H_6	g	−20,236	372,820
Propane	C_3H_8	g	−24,820	530,600
n-Butane	C_4H_{10}	g	−30,150	687,640
n-Pentane	C_5H_{12}	g	−35,000	845,160

TABLE 12-1. (continued)

Substance	Formula	State	$\Delta h^o_{F_{298}}$	$-\Delta h^o_{C_{298}}$
n-Hexane	C_6H_{14}	g	−39,960	1,002,570
n-Heptane	C_7H_{16}	g	−44,890	1,160,010
Increment per C atom above C_7			−4,930	157,440
Normal monoolefins (1-alkenes):				
Ethylene	C_2H_4	g	12,496	337,150
Propylene	C_3H_6	g	4,879	491,990
1-Butene	C_4H_8	g	−30	649,380
1-Pentene	C_5H_{10}	g	−5,000	806,700
1-Hexene	C_6H_{12}	g	−9,960	964,240
Miscellaneous hydrocarbons:				
Acetylene	C_2H_2	g	54,194	310,620
Benzene	C_6H_6	g	19,820	789,080
Benzene	C_6H_6	l	11,718	780,980
1,3-Butadiene	C_4H_6	g	26,330	607,490
Cyclohexane	C_6H_{12}	g	−29,430	944,770
Cyclohexane	C_6H_{12}	l	−37,330	936,860
Ethylbenzene	C_8H_{10}	g	7,120	1,101,130
Isopropene	C_5H_8	g	18,100	761,610
Methylcyclohexane	C_7H_{14}	g	−36,990	1,099,580
Methylcyclohexane	C_7H_{14}	l	−45,450	1,091,130
Styrene	C_8H_8	g	35,220	1,060,900
Toluene	C_7H_8	g	11,950	943,580
Toluene	C_7H_8	l	2,867	934,500
Miscellaneous oxygenated organic compounds:				
Acetaldehyde	C_2H_4O	g	−39,720	
Acetic acid	$C_2H_4O_2$	l	−115,800	
Ethanol	C_2H_6O	g	−56,190	
Ethanol	C_2H_6O	l	−66,370	
Ethylene glycol	$C_2H_6O_2$	l	−108,700	
Ethylene oxide	C_2H_4O	g	−12,580	
Methanol	CH_4O	g	−47,960	
Methanol	CH_4O	l	−57,040	
Miscellaneous inorganic compounds:				
Ammonia	NH_3	g	−10,980	
Calcium carbonate	$CaCO_3$	s	−288,460	

TABLE 12-1. (continued)

Substance	Formula	State	$\Delta h^o_{F_{298}}$	$-\Delta h^o_{C_{298}}$
Calcium chloride	$CaCl_2$	s	−190,200	
Calcium hydroxide	$Ca(OH)_2$	s	−235,680	
Calcium oxide	CaO	s	−151,790	
Carbon, graphite	C	s	0	94,051
Carbon dioxide	CO_2	g	−94,051	
Carbon monoxide	CO	g	−26,417	67,636
Chlorine	Cl_2	g	0	
Hydrochloric acid	HCl	g	−22,063	
Hydrogen	H_2	g	0	68,315
Hydrogen sulfide	H_2S	g	−4,930	
Iron chloride	$FeCl_2$	s	−81,690	
Iron oxide, magnetite	Fe_3O_4	s	−267,300	
Iron oxide, hematite	Fe_2O_3	s	−197,000	
Iron sulfate	$FeSO_4$	s	−221,900	
Nitric acid	HNO_3	l	−41,610	
Nitrogen	N_2	g	0	
Nitrogen oxides	NO	g	21,570	
	NO_2	g	7,930	
	N_2O	g	19,513	
	N_2O_4	g	2,190	
Oxygen	O_2	g	0	
Sodium carbonate	Na_2CO_3	s	−270,260	
Sodium carbonate, deca-hydrate	$Na_2CO_3 \cdot 10H_2O$	s	−975,600	
Sodium chloride	$NaCl$	s	−98,260	
Sodium hydroxide	$NaCH$	s	−101,720	
Sulfur dioxide	SO_2	g	−70,940	
Sulfur trioxide	SO_3	g	−94,580	
Sulfur trioxide	SO_3	l	−105,410	
Sulfuric acid	H_2SO_4	l	−194,548	
Water	H_2O	g	−57,795	
Water	H_2O	l	−68,315	

[†]Table prepared by Dr. Jing Chao, Thermodynamics Research Center, Texas A & M University. (1974) Data used with permission of the American Petroleum Institute and the Thermodynamics Research Center.

Now, the enthalpy of a reaction is defined by the relationship

$$\Delta H_R = \Sigma a_i h_{F_i}(\text{products}) - \Sigma a_i \Delta h_{F_i}^\circ(\text{reactants}) \qquad (12\text{-}4)$$

If we introduce Eq. (12-6), we find that Eq. (12-4) can be written in the form

$$\Delta H_R^\circ = \Sigma a_i \Delta h_{F_i}^\circ(\text{products}) - \Sigma a_i \Sigma b_j \Delta h_{F_j}(\text{elements in products})$$

$$- \Sigma a_i \Delta h_{F_i}^\circ(\text{reactants}) + \Sigma a_i \Sigma b_j \Delta h_{F_j}^\circ(\text{elements in reactants})$$

or

$$\Delta H_R^\circ = \Sigma a_i \Delta h_{F_i}^\circ(\text{products}) - \Sigma a_i h_{F_i}^\circ(\text{reactants}) \qquad (12\text{-}7)$$

since the elements making up the products and reactants must be the same. This result could have been obtained more directly by making use of the fact that the enthalpies associated with the structure of the elements are taken to be zero.

EXAMPLE PROBLEM 12-1. Calculate the enthalpy of reaction for the reaction

$$CO_2(g) + H_2(g) \to CO(g) + H_2O(g)$$

where the products and reactants are in their standard states at one atmosphere and at 298 K using enthalpy of formation data.

The formation reactions of the components involved in this reaction and their enthalpies of formation from Table 12-1 are as follows:

$CO_2(g)$: $\quad C(s) + O_2(g) \to CO_2(g), \quad \Delta h_{F_{298}}^\circ = -94{,}051$ cal/gm mole

$H_2(g)$: $\quad$ Since H_2 is an "element," its heat of formation is zero.

$CO(g)$: $\quad C(s) + \frac{1}{2}O_2(g) \to CO(g), \quad \Delta h_{F_{298}}^\circ = -26{,}417$ cal/gm mole

$H_2O(g)$: $\quad H_2(g) + \frac{1}{2}O_2(g) \to H_2O(g), \quad \Delta h_{F_{298}}^\circ = -57{,}795$ cal/gm mole

Now, we may rewrite these reactions so that their sum gives us the desired reaction. This is equivalent to assuming that a reaction consists of two steps–breaking the reactants into elements and recombining the elements into the desired products. Then, we may write the formation reactions in such a manner that their sum gives us the desired reaction as follows:

	$h^o_{F_{298}}$
$CO_2(g) \to C(s) + O_2(g)$	+94,051 cal/gm mole
$C(s) + \frac{1}{2}O_2(g) \to CO(g)$	−26,417 cal/gm mole
$H_2(g) + \frac{1}{2}O_2(g) \to H_2O(g)$	−57,795 cal/gm mole

Adding,

$$CO_2(g) + H_2(g) \to CO(g) + H_2O(g), \quad \Delta H^o_{R_{298}} = 9839 \text{ cal/gm mole}$$

Note that since the formation reaction of CO_2 was reversed, the sign of the enthalpy of formation of CO_2 was changed. Now, the statement of the heat of reaction calculated in this example is not quite complete. A more complete answer would be

$$\Delta H^o_{R_{298}} = 9839 \text{ cal/gm mole of } CO_2 \text{ reacted}$$

Also, in this example, the heat of formation of water vapor at 25 °C was available. This is a hypothetical state, since water does not exist as a vapor at 25 °C and one atmosphere pressure. The difference between the enthalpy of formation of water vapor and liquid water at 25 °C is the enthalpy of vaporization of water at 25 °C.

Effect of Temperature and Pressure

Once the value of the enthalpy of reaction is available at any temperature and pressure, the enthalpy of reaction can be calculated at any other temperature and pressure. Following procedures developed in Chap. 4, it can be shown that

$$dh(P, T) = \left[v - T\left(\frac{\partial v}{\partial T}\right)_P\right] dP + c_P \, dT$$

This equation may be integrated as a line integral over any convenient path. For example, in order to calculate the effect of changing temperature on the enthalpy of an isobaric reaction, we may select a path consisting of three steps.

1. Cool reactants from the temperature of interest to the reference temperature

$$\Delta H = \int_T^{T^o} c_P(\text{reactants}) \, dT$$

2. Let the reaction proceed at the reference temperature

$$\Delta H = \Delta H_{R_{T^o}}$$

3. Heat the products from the reference temperature to the temperature of interest

$$\Delta H = \int_{T^o}^{T} c_p(\text{products})\, dT$$

Then, for the reaction $aA + bB \to rR + sS$ taking place at temperature T,

$$\Delta H_{R_T} = \int_{T}^{T^o} \left(ac_{P_A} + bc_{P_B}\right) dT + \Delta H_{R_{T^o}}$$

$$+ \int_{T^o}^{T} \left(rc_{P_R} + sc_{P_S}\right) dT$$

$$= \Delta H_{R_{T^o}} + \int_{T^o}^{T} \left(rc_{P_R} + sc_{P_S} - ac_{P_A} - bc_{P_B}\right) dT$$

This last equation is generally written as

$$\Delta H_{R_T} = \Delta H_{R_{T^o}} + \int_{T^o}^{T} \Delta c_p\, dT \tag{12-8}$$

where $\Delta c_p = rc_{P_R} + sc_{P_S} - ac_{P_A} - bc_{P_B}$

We may also find the effect of pressure on the enthalpy of an isothermal reaction. In this case, we might select a path as follows:

1. Expand reactants from pressure P to one atmosphere.
2. Let reaction proceed at one atmosphere.
3. Compress products from one atmosphere to pressure P.

Along an isothermal path the change in enthalpy is given by the relationship

$$dh = \left[v - T\left(\frac{\partial v}{\partial T}\right)_P\right]dP$$

This integral has been evaluated using the equation of state

$$Pv = zRT \tag{8-9}$$

and the results presented in Fig. A-2.

Enthalpy of Combustion

Unfortunately, only a few formation reactions are physically possible. It is very difficult to make all possible hydrocarbons starting with only carbon and hydrogen. One type of reaction that lends itself to experimental investigation is the combustion reaction. The combustion reaction is defined as the reaction of one mole of an element or compound and oxygen to form specified combustion products, both reactants and products being in their standard states and at the same reference temperature. For organic compounds made up only of carbon, hydrogen, and oxygen, the products are specified to be gaseous CO_2 and H_2O, either as a liquid or a vapor. An example of a combustion reaction is

$$CH_4(g) + 2O_2(g) \rightarrow CO_2(g) + 2H_2O(l)$$

where $CO_2(g)$ and $H_2O(l)$ are the specified products of combustion.

Since combustion is a chemical reaction, the enthalpy of combustion may be determined from enthalpy of formation data. Thus, using Eq. (12-4), we find

$$\Delta h_C^o = \Sigma a_i \Delta h_{F_i} \text{(combustion products)} - \Delta h_F^o \text{(compound)} \tag{12-9}$$

since

$$h_F^o(\text{oxygen}) = 0$$

and the stoichiometric coefficient of the compound is unity. The subscript C indicates that this is the enthalpy of a combustion reaction. Table 12-1 shows the enthalpies of combustion for some representative compounds at one atmosphere and at a reference temperature of 298 K when gaseous CO_2 and liquid H_2O are the specified reaction products.

Data from tables of enthalpy of combustion may be used to calculate the enthalpy of any desired reaction. If we combine Eqs. (12-7) and (12-9), we find that

$$\Delta H_R^o = \Sigma a_i \Delta h_{C_i}^o (\text{reactants}) - \Sigma a_i \Delta h_{C_i}^o (\text{products}) \qquad (12\text{-}10)$$

since the enthalpies of formation of the combustion products must cancel. Because the compound we are evaluating in the combustion reaction is a reactant while it is a product in the formation reaction, we find that the relationship between the enthalpy of the reaction and enthalpies of combustion appears to be the reverse of the relationship between the enthalpy of the same reaction and the enthalpies of formation. In order to maintain the definition $\Delta = $ (products − reactants) we must use the negative of the enthalpy of combustion in Eq. (12-10). Then,

$$\Delta H_R^o = \Sigma a_i \left(-\Delta h_{C_i}^o\right)(\text{products}) - \Sigma a_i \left(-\Delta h_{C_i}^o\right)(\text{reactants}) \qquad (12\text{-}11)$$

Values of the negative of the enthalpy of combustion are presented in Table 12-1. The reason we can think of the desired reaction as either a series of formation reactions or combustion reactions is that the enthalpy of reaction is independent of path.

EXAMPLE PROBLEM 12-2. Recalculate the enthalpy of reaction at 298 K for the chemical reaction in Example Prob. 12-1 using enthalpy of combustion data.

In order to calculate the enthalpy of the reaction

$$CO_2(g) + H_2(g) \to CO(g) + H_2O(g)$$

we select the path to be combustion of the reactants and the recombination of the products of combustion to form the desired products. $CO_2(g)$ and $H_2O(g)$ are already final products of combustion and thus have no enthalpies of combustion. Then, noting that the formation reaction of H_2O is the combustion reaction of H_2, the desired combustion reactions may be written as follows:

$$CO_2(g) \to CO(g) + \tfrac{1}{2}O_2(g); \quad \Delta h_{C_{298}}^o = +67{,}636 \text{ cal/mole } CO_2 \text{ unburned}$$

$$H_2(g) + \tfrac{1}{2}O_2(g) \to H_2O(g); \quad \Delta h_{C_{298}}^o = -57{,}795 \text{ cal/mole } H_2 \text{ burned}$$

where the enthalpies of combustion have been obtained from Table 12-1. Adding,

$$CO_2(g) + O_2(g) \to CO(g) + H_2O(g); \quad \Delta H_R^o = 9841 \text{ cal/mole } CO_2 \text{ reacted}$$

This may be compared to the result 9839 cal/mole CO_2 reacted obtained in Example Prob. 12-1. The insignificant difference is due to round-off errors.

Although one of the primary uses of enthalpy of combustion data is in estimating enthalpies of reaction, this data can be used directly as enthalpy of reaction data for combustion reactions where we are interested in such things as heat liberated and flame temperatures.

Frequently, the negative of the enthalpy of combustion of a compound is called the "heating value" of that compound because of its relationship to the heat liberated during a combustion reaction. Heating values determined for combustion reactions in which the water produced is in the liquid phase are called higher or gross heating values. In actual operation of boilers, the water vapor in the combustion products is not condensed to the liquid phase and its latent heat of vaporization is not available for making steam. Thus, the latent heat of vaporization of water must be subtracted from the higher or gross heating value to give the lower or net heating value.

For anthracite and bituminous coals, the higher heating value in BTU's per pound of fuel can be approximated using Dulong's formula,

$$-\Delta h_C^\circ = 14{,}544 \, C + 62{,}028 \left(H - \frac{O}{8}\right) + 4050 \, S \qquad (12\text{-}12)$$

where C, H, O and S are the carbon, hydrogen, oxygen and sulfur in the coal expressed as fractions of a pound. For lignites and sub-bituminous coals, the heating values obtained from this formula are generally too low.

Enthalpy of combustion data is also useful in predicting flame temperatures. As an example of this, let us consider the problem of calculating the maximum temperature that can be obtained when methane is burned with air.

EXAMPLE PROBLEM 12-3. Methane is to be burned using a stoichiometric amount of air in a well insulated burner.

$$CH_4(g) + 2O_2(g) \to CO_2(g) + 2H_2O(g)$$

If the methane and air are assumed to form an ideal solution and if the mixture is fed to the burner at 25 °C and 20 atm pressure, what is the maximum temperature that could be attained by the final products leaving the burner at one atm pressure?

We will take the burner as our system and base our calculations on one gram mole of methane fed. Now, if air is used as the source of oxygen, 3.76 moles of nitrogen must be added to each side of the stoichiometric equation for each mole of oxygen fed to the burner, since air is 21 mole percent oxygen. Then, the combustion equation becomes:

$$CH_4(g) + 2O_2(g) + (2)(3.76) \, N_2(g) \to CO_2(g) + 2H_2O(g) + (2)(3.76) \, N_2(g)$$

As we are using a stoichiometric amount of air (called 100 percent theoretical air), the reactants fed to the burner are 1.0 mole methane, 2.0 moles oxygen, and 7.52 moles nitrogen. Since the reaction is assumed to go to completion, the products from the burner are 1.0 mole carbon dioxide, 2.0 moles water vapor, and 7.52 moles nitrogen.

Now, the energy balance for this system reduces to

$$H_{in} - H_{out} - \Delta H_R = 0$$

In evaluating the terms of this energy balance, we must evaluate the enthalpies of the products and reactants using the reference state at which the enthalpy of reaction is calculated. Since we have enthalpy of formation and enthalpy of combustion data available in Table 12-1 at 25 °C and 1.0 atmosphere, we will use these conditions as the reference states for the enthalpies of the reactants and products. Then, for either a reactant or a product,

$$\Delta h = \int_{P^o}^{P} \left[v - T \left(\frac{\partial v}{\partial T} \right)_P \right] dP + \int_{T^o}^{T} c_P \, dT$$

where the pressure integral can be evaluated using Fig. A-2, and the temperature integral can be evaluated using specific heat data from Table A-3. First, we will estimate the enthalpy of the reactants relative to 25 °C and 1.0 atm pressure. In this case, there is no temperature effect on enthalpy, but there is a pressure effect. At 25 °C and 20 atm, we find using critical constants from Table A-4,

$$P_{r,O_2} = \frac{20}{49.8} = 0.402 \qquad T_{r,O_2} = \frac{298}{154.6} = 1.93 \qquad \omega_{O_2} = 0.0230$$

$$P_{r,N_2} = \frac{20}{34.0} = 0.588 \qquad T_{r,N_2} = \frac{298}{126.2} = 2.36 \qquad \omega_{N_2} = 0.0373$$

$$P_{r,CH_4} = \frac{20}{45.4} = 0.441 \qquad T_{r,CH_4} = \frac{298}{190.5} = 1.56 \qquad \omega_{CH_4} = 0.0109$$

Then, from Figs. A-2A and A-2B,

$$\left(\frac{h^* - h}{RT_c} \right)_{O_2} = 0.13 + (0.0230)(-0.03) = 0.129$$

$$\left(\frac{h^* - h}{RT_c} \right)_{N_2} = 0.14 + (0.0373)(-0.009) = 0.140$$

$$\left(\frac{h^* - h}{RT_c} \right)_{CH_4} = 0.18 + (0.0109)(0) = 0.180$$

Thus,

$$H_{in} = -(1.0)(0.180)(190.5)(1.986) - (2.0)(0.129)(154.6)(1.986)$$

$$-(7.52)(0.140)(126.2)(1.986)$$

$$= -411 \text{ cal}$$

Now, the nitrogen does not play any role in the enthalpy of reaction term. From Table 12-1, we find

$$\begin{array}{cc} & \Delta h^o_{C_{298}} \\ & (cal/gm\ mole) \\ CH_4(g) + 2O_2(g) \to CO_2(g) + 2H_2O(l) & -212{,}800 \\ 2H_2O(l) \to 2H_2O(g) & +2(10{,}520) \end{array}$$

Adding,

$$CH_4(g) + 2O_2(g) \to CO_2(g) + 2H_2O(g), \quad \Delta H_R = -191{,}760. \text{ Then,}$$

$$H_{out} - H_{in} = -\Delta H_R = +191{,}760 \text{ cal}$$

Finally, for the products being heated isobarically,

$$H_{out} = \int_{298}^{T} nc_P\, dT$$

Using specific heat data from Table A-3, we find

$$H_{out} = \int_{298}^{T} \Big[1.0(6.214 + 10.396 \times 10^{-3}\, T - 3.545 \times 10^{-6} T^2)$$

$$+ 2.0(7.256 + 2.298 \times 10^{-3} T + 0.283 \times 10^{-6} T^2)$$

$$+ (7.52)(6.524 + 1.250 \times 10^{-3} T - 0.001 \times 10^{-6} T^2) \Big] dT$$

$$= \int_{298}^{T} (69.786 + 24.392 \times 10^{-3} T - 2.987 \times 10^{-6} T^2)\, dT$$

Integrating,

$$H_{out} = 69.786T + 12.196 \times 10^{-3}T^2 - 0.996 \times 10^{-6}T^3 - 21853 \text{ cal}$$

Now, we may substitute into the reduced energy balance

$$-\Delta H_R = H_{out} - H_{in}$$

to obtain

$$191{,}760 = (69.786T + 12.196 \times 10^{-3}T^2 - 0.996 \times 10^{-6}T^3 - 21853)$$

$$- (-411)$$

or

$$69.786T + 12.196 \times 10^{-3}T^2 - 0.996 \times 10^{-6}T^3 - 213202 = 0$$

We must solve this last equation to obtain the required temperature.

This can be conveniently done by using Newton's method of solving equations. This technique, developed in Chap. 11, is a method by which an estimate of the correct temperature denoted by the symbol x_o, can be improved by repeated application of the equation

$$x = x_o - \frac{\Phi(x_o)}{\Phi'(x_o)} \tag{11-84}$$

where x is an improved estimate of the maximum final temperature. In this case, let

$$\Phi(x) = 69.786T + 12.196 \times 10^{-3}T^2 - 0.996 \times 10^{-6}T^3 - 213{,}202$$

Then,

$$\frac{d\Phi(x)}{dx} = 69.786 + 24.392 \times 10^{-3}T - 2.987 \times 10^{-6}T^2$$

Assume as a first estimate that $T = x_o = 2000$ K. Then,

$$\Phi(x_o) = (69.786)(2000) + 12.196 \times 10^{-3}(2000)^2 - 0.996 \times 10^{-6}$$

$$\times (2000)^3 - 213{,}202$$

$$= -32{,}814$$

and

$$\Phi'(x_o) = 69.786 + 24.392 \times 10^{-3}(2000) - 2.987 \times 10^{-6}(2000)^2$$

$$= 106.6$$

Then, an improved estimate of T would be

$$T = x = 2000 - \frac{-32{,}814}{106.6} = 2308 \text{ K}$$

Now, we must check this result by seeing if $\Phi(x)$ is zero:

$$\Phi(x) = (69.786)(2308) + 12.196 \times 10^{-3}(2308)^2 - 0.996 \times 10^{-6}$$

$$\times (2308)^3 - 213{,}202$$

$$= 585$$

Since $\Phi(x)$ does not equal zero, we must further improve our estimate of T. Assume $T = x_o = 2308$ K. Then,

$$\Phi'(x) = 69.786 + 24.392 \times 10^{-3}(2308) - 2.987 \times 10^{-6}(2308)^2$$

$$= 110$$

So, an improved estimate of T would be

$$T = x = 2308 - \frac{585}{110} = 2303 \text{ K}$$

This answer is within the accuracy of the data used, so

$$T_{out} = 2303 \text{ K}$$

Now, the maximum temperature we have just calculated in Example Prob. 12-3 when methane is burned with air is sometimes called the uncorrected theoretical flame temperature for perfect combustion. For several reasons, methane does not develop a flame temperature of 2303 K. First, no combustion process can be carried out without some loss of heat due to temperature gradients and radiation from hot gases. Second, at high temperatures, carbon dioxide and water dissociate according to the reactions

$$CO_2(g) \rightleftharpoons CO(g) + \tfrac{1}{2}O_2(g)$$

$$H_2O(g) \rightleftharpoons H_2(g) + \tfrac{1}{2}O_2(g)$$

This, in effect, is equivalent to incomplete combustion, and lowers the actual temperature attained when methane is burned.

In order to determine the extent to which carbon dioxide and water dissociate at a given temperature and pressure we will have to consider the topic of chemical equilibrium.

CHEMICAL EQUILIBRIUM

Reaction Free Energy

As shown in Chap. 11, the criterion for equilibrium in an isothermal flow process based on the general mass and energy and entropy balance is

$$\Delta G_T = 0 \qquad (11\text{-}1)$$

This equation is also the basis for estimating the extent to which a chemical reaction will proceed before reaching chemical equilibrium.

Consider the following reaction taking place in a homogeneous phase

$$aA + bB \rightleftharpoons rR + sS$$

For this reaction,

$$dG = V\,dP - S\,dT + \mu_A\,dn_A + \mu_B\,dn_B + \mu_R\,dn_R + \mu_S\,dn_S$$

or at constant pressure and temperature,

$$dG = \mu_A\,dn_A + \mu_B\,dn_B + \mu_R\,dn_R + \mu_S\,dn_S$$

Now, when dn_A moles of A react, the stoichiometric equation tells us that $(b/a)dn_A$ moles of B react. Then,

$$(b/a)dn_A = dn_B$$

or

$$\frac{dn_A}{a} = \frac{dn_B}{b}$$

Similarly,

$$-\frac{dn_A}{a} = \frac{dn_R}{r}$$

and

$$-\frac{dn_A}{a} = \frac{dn_S}{s}$$

Thus, we see that

$$-\frac{dn_A}{a} = -\frac{dn_B}{b} = \frac{dn_R}{r} = \frac{dn_S}{s} = d\alpha$$

where $d\alpha$ is called the differential extent of reaction. Then, we may write

$$dn_A = -a\, d\alpha$$

$$dn_B = -b\, d\alpha$$

$$dn_R = r\, d\alpha$$

$$dn_S = s\, d\alpha$$

so that

$$dG = (r\mu_R + s\mu_S - a\mu_A - b\mu_B)\, d\alpha$$

or

$$\left(\frac{\partial G}{\partial \alpha}\right)_{P,T} = (r\mu_R + s\mu_S) - (a\mu_A + b\mu_B)$$

For convenience, we define

$$\left(\frac{\partial G}{\partial \alpha}\right)_{P,T} = \Delta G_R = \text{``reaction potential'' or ``reaction free energy''} \quad (12\text{-}13)$$

From Eq. (11-31), at constant temperature we find that at the pressure of the system,

$$\mu_i - \mu_i^o = RT \ln \overline{f_i} - RT \ln \overline{f_i^o}$$

or

$$\mu_i = RT \ln a_i + \mu_i^o$$

where μ_i^o is the chemical potential of pure component i in its standard state and a_i is its activity. Then, our equation for the reaction free energy can be written

$$\Delta G_R = rRT\ln a_R + sRT\ln a_S - aRT\ln a_A - bRT\ln a_B$$

$$+ r\mu_R^o + s\mu_S^o - a\mu_A^o - b\mu_B^o$$

$$= RT\ln \frac{a_R^r a_S^s}{a_A^a a_B^b} + r\mu_R^o + s\mu_S^o - a\mu_A^o - b\mu_B^o$$

This last equation may be written

$$\Delta G_R = RT\ln J_A + \Delta G_R^o \tag{12-14}$$

where

$$\Delta G_R^o = r\mu_R^o + s\mu_S^o - a\mu_A^o - b\mu_B^o$$

$$= \left(\frac{\partial G^o}{\partial \alpha}\right)_{P,T} \tag{12-15}$$

and

$$J_A = \text{activity quotient} = \frac{a_R^r a_S^s}{a_A^a a_B^b}$$

At equilibrium, $\Delta G_R = 0$ and Eq. (12-14) reduces to

$$\Delta G_R^o = -RT\ln K_A \tag{12-16}$$

where

$$\Delta G_R^o = \text{reaction free energy in the standard state}$$

and represents the change in the Gibbs free energy for a reaction in which each of the reactants and products is in its standard state and

$$K_A = \text{activity quotient at equilibrium} = \frac{a_R^r a_S^s}{a_A^a a_B^b}$$

at equilibrium. Values of K_A can be used to determine the extent to which a reaction, such as the dissociation of H_2O to H_2 and O_2, take place if allowed to proceed to equilibrium.

Equilibrium Conversion

Simple Reactions. For the gas-phase reaction $aA \rightarrow bB$ at chemical equilibrium,

$$K_A = \frac{a_B^b}{a_A^a}$$

Now, activity has been defined as

$$a_i = \frac{\bar{f}_i}{f_i^o} \tag{11-32}$$

If we define the standard state as the pure component in the ideal gas state at system temperature and one atmosphere pressure,

$$\bar{f}_A^o = f_A^o = 1 \text{ atm}$$

$$\bar{f}_B^o = f_B^o = 1 \text{ atm}$$

Then our chemical equilibrium relationship becomes

$$K_A = \frac{\bar{f}_B^b}{\bar{f}_A^a}$$

where the units of these fugacities are atmospheres. Further, if the gas mixture is an ideal solution, the Lewis-Randall rule applies, and

$$\bar{f}_i = y_i f_i \tag{11-41}$$

Values of the fugacity of the pure component may be found using Figs. A-4A and A-4B and the relationship

$$f_i = \nu_i P \tag{9-17}$$

Substituting these values into our equation for chemical equilibrium, we find

$$K_A = \frac{y_B^b}{y_A^a} \frac{\nu_B^b}{\nu_A^a} P^{b-a}$$

This equation may be written

$$K_A = K_y K_v P^{b-a} \tag{12-17}$$

where

$$K_y = \frac{y_B^b}{y_A^a}$$

$$K_v = \frac{v_B^b}{v_A^a}$$

and the pressure is given in atmospheres. Since the fugacity of a component in the ideal gas vapor phase is given by Dalton's law,

$$\bar{f}_{g,i} = y_i P$$

some authors write the chemical equilibrium relationship as

$$K_A = K_P K_v \tag{12-18}$$

where

$$K_P = K_y \cdot P^{b-a}$$

Another form of the chemical equilibrium relationship is frequently used for reactions taking place in the liquid phase. In this case, a convenient standard state is pure liquid at system temperature and system pressure. In this case, we may use the relationship

$$a_i = \gamma_i x_i \tag{11-38}$$

where γ_i is an activity coefficient based on mole fraction. Then

$$K_A = K_x K_\gamma \tag{12-19}$$

where

$$K_x = \frac{x_B^b}{x_A^a}$$

$$K_\gamma = \frac{\gamma_B^b}{\gamma_A^a}$$

For an ideal solution, $\gamma_i = 1.0$ and $K_A = K_x$. Now, if we have a value of K_A available, the only unknowns in Eqs. (12-17), (12-18), and (12-19) are the mole fractions. For a two component system,

$$x_A + x_B = 1.0$$

is the only other equation we would need to solve for the chemical equilibrium composition of a reacting mix. For a system containing more than two components, however, we must make use of the extent of the reaction α. Thus, in the reaction

$$A + B \rightarrow R + S$$

if we start with A_o moles of A and B_o moles of B, at equilibrium we will have

$$A_o - \alpha \text{ moles of } A$$

$$B_o - \alpha \text{ moles of } B$$

$$\alpha \text{ moles of } R$$

$$\alpha \text{ moles of } S$$

where α is the number of moles of A that have reacted. The total moles of reactant and product is found to be

$$(A_o - \alpha) + (B_o - \alpha) + \alpha + \alpha = A_o + B_o$$

so that equilibrium

$$x_A = \frac{A_o - \alpha}{A_o + B_o}$$

$$x_B = \frac{B_o - \alpha}{A_o + B_o}$$

$$x_R = \frac{\alpha}{A_o + B_o}$$

$$x_S = \frac{\alpha}{A_o + B_o}$$

Then,

$$K_y = \frac{\alpha^2}{(A_o - \alpha)(B_o - \alpha)}$$

and we have reduced the number of variables in K_y from four to one. When this equation is substituted into Eq. (12-17), or (12-18), we can solve for α. The required mole fractions can then be calculated from the stoichiometric relationships.

EXAMPLE PROBLEM 12-4. For the reaction

$$H_2O(g) \rightleftharpoons H_2(g) + \tfrac{1}{2}O_2(g)$$

$$\Delta G_R^o = 29 \text{ k cal/gm mole } H_2O$$

at 2300 K and one atmosphere pressure. What is the extent to which $H_2O(g)$ has dissociated?

We will use one gram mole of $H_2O(g)$ as the basis for our calculations. Then, if α is the number of moles of water dissociated, we find that

$$\text{moles } H_2O(g) = 1 - \alpha$$

$$\text{moles } H_2(g) = \alpha$$

$$\text{moles } O_2(g) = 0.5\alpha$$

and

$$\text{the total moles} = 1 + 0.5\alpha$$

Thus,

$$K_A = \frac{y_{H_2} \cdot y_{O_2}^{1/2}}{y_{H_2O}} \cdot \frac{\nu_{H_2} \cdot \nu_{O_2}^{1/2}}{\nu_{H_2O}}$$

and if we assume that all the fugacity coefficients are equal to unity,

$$K_A = \frac{\left(\dfrac{\alpha}{1 + 0.5\alpha}\right)\left(\dfrac{0.5\alpha}{1 + 0.5\alpha}\right)^{0.5}}{\left(\dfrac{1-\alpha}{1 + 0.5\alpha}\right)} = \left(\dfrac{\alpha}{1-\alpha}\right)\left(\dfrac{0.5\alpha}{1 + 0.5\alpha}\right)^{0.5}$$

Now, we must calculate K_A. Since

$$\Delta G_R^\circ = -RT \ln K_A \tag{12-16}$$

we have

$$\ln K_A = -\frac{29{,}000}{(1.986)(2300)} = -6.35$$

Then,

$$\log K_A = -\frac{6.35}{2.303} = -2.76$$

and

$$K_A = 0.00175$$

so that

$$\left(\frac{\alpha}{1-\alpha}\right)\left(\frac{0.5\alpha}{1+0.5\alpha}\right)^{0.5} = 0.00175$$

Solving for α gives $\alpha = 0.018$. Thus, we see that at 2300 K, 1.8 mole percent of the water vapor will dissociate.

Complex Reactions. We might now turn our attention from the calculation of equilibrium compositions for reactions which occur in a single step to the calculation of equilibrium composition for complex reactions which occur in a series of steps. Consider the reaction

$$A \rightleftharpoons B$$

with chemical equilibrium coefficient

$$K_{A_1} = \frac{a_B}{a_A}$$

which really proceeds by means of two reactions

$$A \rightleftharpoons R$$

with chemical equilibrium coefficient

$$K_{A_2} = \frac{a_R}{a_A}$$

and

$$R \rightleftharpoons B$$

with chemical equilibrium coefficient

$$K_{A_3} = \frac{a_B}{a_R}$$

Consider first the case in which we can not detect the presence of R in the product. In order to solve for the concentration of B at equilibrium, we must first estimate the activity of R. This can be done by noting that

$$a_R = K_{A_2} a_A$$

Then, we find

$$K_{A_3} = \frac{a_B}{K_{A_2} a_A}$$

or

$$K_{A_2} \cdot K_{A_3} = \frac{a_B}{a_A}$$

But

$$\frac{a_B}{a_A} = K_{A_1}$$

so
$$K_{A_2} \cdot K_{A_3} = K_{A_1}$$

This result indicates that the equilibrium composition of B does not depend upon the path by which A is converted to B. This is not surprising if we recognize that the chemical equilibrium coefficient is a single valued function of the reaction free energy ΔG_R^o which is a point function.

On the other hand, both B and R may be present as reaction products. In this case, we must use two of the three chemical equilibrium coefficients to solve for the two unknown equilibrium concentrations. It is a general rule that the number of independent chemical reactions which must be considered in a system of reacting components is the least number which includes every reactant and every product present in measurable amounts and accounts for the formation of each product from the original reactants. Thus, the reaction

$$A \rightleftharpoons B$$

and

$$A \rightleftharpoons R$$

are two independent reactions which account for the presence of both B and R in the reaction products.

As there are two chemical reactions involved in this example, we let

α = extent to which A reacts in the first reaction

β = extent to which A reacts in the second reaction

If the initial amount of A is A_o moles, then, at equilibrium we will have

$$A_o - \alpha - \beta \text{ moles of } A$$

$$\alpha \text{ moles of } B$$

$$\beta \text{ moles of } R$$

The total number of moles at equilibrium will be

$$(A_o - \alpha - \beta) + \alpha + \beta = A_o$$

Then,

$$x_A = \frac{n_A}{n_A + n_B + n_R} = \frac{A_o - \alpha - \beta}{A_o}$$

$$x_B = \frac{n_B}{n_A + n_B + n_R} = \frac{\alpha}{A_o}$$

$$x_R = \frac{n_R}{n_A + n_B + n_R} = \frac{\beta}{A_o}$$

If the reacting mixture is an ideal solution, we then have two equations:

$$K_{A_1} = \frac{a_B}{a_A} = \left(\frac{\alpha}{A_o}\right)\left(\frac{A_o}{A_o - \alpha - \beta}\right)\left(\frac{\nu_B}{\nu_A}\right)\left(\frac{P}{P}\right)$$

$$K_{A_2} = \frac{a_R}{a_A} = \left(\frac{\beta}{A_o}\right)\left(\frac{A_o}{A_o - \alpha - \beta}\right)\left(\frac{\nu_R}{\nu_A}\right)\left(\frac{P}{P}\right)$$

which we can solve for the two unknowns α and β.

One technique used in solving this type of problem involves a comparison of the relative magnitudes of the chemical equilibrium coefficients. If, for example, $(K_A)_1$ is several orders of magnitude larger than $(K_A)_2$, it is reasonable to assume that β is essentially zero. This reduces the problem to one equation with one unknown. On the other hand, if $(K_A)_1$ and $(K_A)_2$ are approximately the same order of magnitude, one can divide the two equations to obtain the result

$$\beta = \frac{v_B}{v_R} \cdot \frac{K_{A_2}}{K_{A_1}} \cdot \alpha$$

This result can then be substituted into either of the original equations to provide one equation with one unknown which can be solved using either graphical or analytical techniques.

Unfortunately, the task of solving a more complex set of non-linear equations may not be so straight forward. For example, in order to calculate the high temperature equilibrium composition resulting from the combustion of a hydrocarbon in air, it might be necessary to consider 20 or more chemical reactions because of the formation of ions in the product. Computer programs, involving iterative techniques, have been developed to calculate equilibrium compositions for many complex reactions.

Of the various procedures proposed to calculate equilibrium compositions for multiple reactions, a relatively simple method proposed by Meissner[40] appears to be one of the more useful. This method involves a stepwise procedure in which the initial reactants are allowed to reach equilibrium in the first reaction while all other reactions are assumed not to occur. The equilibrium products from the first reaction are then assumed to be the reactants for the second reaction. The second reaction is then allowed to reach equilibrium while all other reactions are assumed not to occur. This procedure is repeated until all the reactions in sequence have been allowed to take place. The equilibrium products from the last reaction are then recycled as reactants to the first reaction and the entire procedure repeated until convergence is obtained on products from the last reaction.

In addition to chemical equilibrium problems involving more than one reaction, we may also encounter problems in which reactions take place in more than one phase. In this case, we must consider not only chemical equilibrium but also phase equilibrium. Then, since

$$\bar{f}_i^{\mathrm{I}} = \bar{f}_i^{\mathrm{II}} = \bar{f}_i^{\mathrm{III}} = \ldots = \bar{f}_i^{\pi} \tag{11-35}$$

we find that if we select the same standard state for component i in all phases such that

$$(f_i^o)^{\mathrm{I}} = (f_i^o)^{\mathrm{II}} = (f_i^o)^{\mathrm{III}} = \ldots = (f_i^o)^{\pi}$$

then
$$a_i^{\mathrm{I}} = a_i^{\mathrm{II}} = a_i^{\mathrm{III}} = \ldots = a_i^{\pi} \qquad (12\text{-}20)$$

With this choice of standard states, we see that the value of the component activities used in chemical equilibrium calculations is independent of phase.

As an example of a reaction which takes place in two phases, consider the combustion of graphite,

$$\mathrm{C(s)} + \mathrm{O_2(g)} \rightarrow \mathrm{CO_2(g)}$$

where

$$K_A = \frac{a_{\mathrm{CO_2}}}{a_{\mathrm{C}} \cdot a_{\mathrm{O_2}}}$$

In this case, since the gas phase does not contain any solid carbon it is most convenient to calculate the activities of the $\mathrm{O_2(g)}$ and the $\mathrm{CO_2(g)}$ in the gas phase and the activity of the C(s) in the solid phase. Then, the standard state for the gases is the pure component as an ideal gas at system temperature and one atmosphere pressure, and

$$\bar{f}^{\,\circ}_{\mathrm{CO_2}} = \bar{f}^{\,\circ}_{\mathrm{O_2}} = 1.0 \text{ atm}$$

Further, the standard state of the solid is the pure component as a solid at system temperature and pressure, so that

$$a_{\mathrm{C}} = 1.0$$

since the solid phase is pure C(s) at system temperature and pressure and

$$K_A = \frac{\bar{f}_{\mathrm{CO_2}}}{\bar{f}_{\mathrm{O_2}}}$$

In this example, we found it most convenient to use activities of components in different phases. Alternately, we might consider a multiphase chemical reaction to take place in a single phase and calculate the activities of all components in terms of the composition of that particular phase. Compositions of other phases are then related to the composition of the first phase through phase relationships developed in Chap. 11 by using the same standard state for any one component in all phases. Finally, we should point out that if a reaction takes place in a multiphase system, the number of degrees of freedom in the system at equilibrium predicted by the Gibbs Phase Rule must be reduced by the number of independent chemical reactions we can write for the system.

Calculation of Reaction Free Energy

Just as the enthalpy of a reaction has been calculated using enthalpies of formation, the Gibbs free energy of a reaction can also be calculated using Gibbs free energies of the formation reaction. For the reaction

$$aA + bB \rightleftharpoons rR + sS$$

we have defined the reaction free energy in the standard state by the relationship

$$\Delta G_R^o = r\mu_R^o + s\mu_S^o - a\mu_A^o - b\mu_B^o \qquad (12\text{-}15)$$

Now μ_i^o is the chemical potential associated with the molecular structure of pure component i in its standard state. Although we can not assign an absolute value to this function, we can obtain its relative value by measuring the reaction free energy of a formation reaction. Applying Eq. (12-15) to the formation reaction of component i, we find that

$$\Delta g_{F_i}^o = \mu_i^o \text{ (compound)} - \Sigma b_i \mu_i^o \text{ (elements)}$$

or

$$\mu_i^o \text{ (compound)} = \Delta g_{F_i}^o + \Sigma b_i \mu_i^o \text{ (elements)}$$

Then, Eq. (12-15) may be rewritten in terms of $\Delta g_{F_i}^o$ as

$$\Delta G_R^o = \Sigma a_i \Delta g_{F_i}^o \text{ (products)} + \Sigma a_i \Sigma b_j \mu_j^o \text{(elements in products)}$$

$$- \Sigma a_i \Delta g_{F_i}^o \text{ (reactants)} - \Sigma a_i \Sigma b_j \mu_j^o \text{ (elements in reactants)}$$

or

$$\Delta G_R^o = \Sigma a_i \Delta g_{F_i}^o \text{ (products)} - \Sigma a_i \Delta g_{F_i}^o \text{ (reactants)} \qquad (12\text{-}21)$$

since the Gibbs free energy of formation of the elements must cancel.

Values of $(\Delta g_F^o)_{298}$ for compounds referred to their elements are presented in Table 12-2[34-37]. Here, superscript ^{o} indicates that the compound and elements are in their standard states; subscript F indicates the reaction is a formation reaction; and sub-subscript 298 indicates a reference temperature of 298 K.

TABLE 12-2.[†] Standard Free Energies of Formation at 25 °C [34-37]

Notes: 1. Units are calories per gram mole of listed compound.
2. Reference states: pure compound at 25 °C, 1.0 atm.

Substance	Formula	State	$\Delta g^o_{F_{298}}$
Normal paraffins:			
Methane	CH_4	g	−12,140
Ethane	C_2H_6	g	−7,860
Propane	C_3H_8	g	−5,614
n-Butane	C_4H_{10}	g	−4,100
n-Pentane	C_5H_{12}	g	−2,000
n-Hexane	C_6H_{14}	g	−70
n-Heptane	C_7H_{16}	g	1,920
Increment per C atom above C_7			2,009
Normal monoolefins (1-alkenes):			
Ethylene	C_2H_4	g	16,282
Propylene	C_3H_6	g	14,990
1-Butene	C_4H_8	g	17,090
1-Pentene	C_5H_{10}	g	18,960
1-Hexene	C_6H_{12}	g	20,940
Miscellaneous hydrocarbons:			
Acetylene	C_2H_2	g	50,000
Benzene	C_6H_6	l	29,756
Benzene	C_6H_6	g	30,989
1,3-Butadiene	C_4H_6	g	36,010
Cyclohexane	C_6H_{12}	g	7,590
Cyclohexane	C_6H_{12}	l	6,370
Ethylbenzene	C_8H_{10}	g	31,208
Isoprene	C_5H_8	g	34,870
Methylcyclohexane	C_7H_{14}	g	6,520
Methylcyclohexane	C_7H_{14}	l	4,860
Styrene	C_8H_8	g	51,100
Toluene	C_7H_8	g	29,228
Toluene	C_7H_8	l	27,282
Miscellaneous oxygenated organic compounds:			
Acetaldehyde	C_2H_4O	g	−30,810
Acetic acid	$C_2H_4O_2$	l	−93,200

TABLE 12-2. (continued)

Substance	Formula	State	$\Delta g^o_{F_{298}}$
Ethanol	C_2H_6O	g	−40,290
Ethanol	C_2H_6O	l	−41,800
Ethylene glycol	$C_2H_6O_2$	l	−77,250
Ethylene oxide	C_2H_4O	g	−3,120
Methanol	CH_4O	g	−38,720
Methanol	CH_4O	l	−39,760
Miscellaneous inorganic compounds:			
Ammonia	NH_3	g	−3,930
Calcium carbonate	$CaCO_3$	s	−269,800
Calcium chloride	$CaCl_2$	s	−178,800
Calcium hydroxide	$Ca(OH)_2$	s	−214,760
Calcium oxide	CaO	s	−144,370
Carbon, graphite	C	s	0
Carbon dioxide	CO_2	g	−94,254
Carbon monoxide	CO	g	−32,780
Chlorine	Cl_2	g	0
Hydrochloric acid	HCl	g	−22,778
Hydrogen	H_2	g	0
Hydrogen sulfide	H_2S	g	−8,020
Iron chloride	$FeCl_2$	s	−72,260
Iron oxide, magnetite	Fe_3O_4	s	−242,700
Iron oxide, hematite	Fe_2O_3	s	−177,400
Iron sulfate	$FeSO_4$	s	−196,200
Nitric acid	HNO_3	l	−19,310
Nitrogen	N_2	g	0
Nitrogen oxides	NO	g	20,690
	NO_2	g	12,242
	N_2O	g	24,933
	N_2O_4	g	23,350
Oxygen	O_2	g	0
Sodium carbonate	Na_2CO_3	s	−250,496
Sodium carbonate, deca-hydrate	$Na_2CO_3 \cdot 10H_2O$	s	—
Sodium chloride	$NaCl$	s	−91,788
Sodium hydroxide	$NaOH$	s	−90,867

TABLE 12-2. (continued)

Substance	Formula	State	$\Delta g^\circ_{F_{298}}$
Sulfur dioxide	SO_2	g	−71,748
Sulfur trioxide	SO_3	g	−88,690
Sulfur trioxide	SO_3	l	−88,040
Sulfuric acid	H_2SO_4	l	−164,938
Water	H_2O	g	−54,634
Water	H_2O	l	−56,687

[†]Table prepared by Dr. Jing Chao, Thermodynamics Research Center, Texas A & M University. (1974) Data used with permission of the American Petroleum Institute and the Thermodynamics Research Center.

EXAMPLE PROBLEM 12-5. Calculate the Gibbs free energy of reaction for the reaction considered in Example Prob.(12-1),

$$CO_2(g) + H_2(g) \rightleftharpoons CO(g) + H_2O(g)$$

at 298 K and 1.0 atm using Gibbs' free energy of formation data.

The formation reactions of the components involved in this reaction and the corresponding Gibbs free energy of formations from Table 12-2 are

	$\Delta g^\circ_{F_{298}}$ (cal/mole)
$CO_2(g)$: $C(s) + O_2(g) \rightarrow CO_2(g)$	−94,254
$H_2(g)$: Since H_2 is an element, $\Delta g^\circ_{F_{298}} = 0$	0
$CO(g)$: $C(s) + \frac{1}{2}O_2(g) \rightarrow CO(g)$	−32,780
$H_2O(g)$: $H_2(g) + \frac{1}{2}O_2(g) \rightarrow H_2O(g)$	−54,634

These equations may be written so that their sum gives the desired reaction as follows:

	$\Delta g^\circ_{F_{298}}$ (cal/mole)
$CO_2(g) \rightarrow C(s) + O_2(g)$	+94,254
$C(s) + \frac{1}{2}O_2(g) \rightarrow CO(g)$	−32,780
$H_2(g) + \frac{1}{2}O_2(g) \rightarrow H_2O(g)$	−54,634

Adding,

$$CO_2(g) + H_2(g) \to CO(g) + H_2O(g), \quad \Delta G^o_{R_{298}} = +6840 \text{ cal/mole of } CO_2 \text{ reacted}$$

In this example, the Gibbs free energy of formation of $H_2O(g)$ was available. Again, note that this is a hypothetical state, since H_2O does not exist as a gas at 25 °C and one atmosphere.

Still another method of calculating the Gibbs free energy of an isothermal reaction is based on the relationship

$$\Delta g^o_F = \Delta h^o_F - T\Delta s^o_F \tag{12-22}$$

Enthalpies of formation can be found in Table 12-1, and

$$\Delta s^o_F = s^o_F \text{ (compound)} - \Sigma b_i s^o_{F_i} \text{ (elements)} \tag{12-23}$$

The entropy values of elements at 298 K are

C(graphite)	1.372 cal/gm mole–K
$H_2(g)$	31.207 cal/gm mole–K
$O_2(g)$	49.005 cal/gm mole–K

A more extensive list of entropy of formation values is presented in Table 12-3[34,35,36,37,38].

By combining enthalpies and entropies of formation, we can calculate Δg^o_F for each reactant and for each product. These can be combined to find ΔG^o_R, since

$$\Delta G^o_R = \Sigma a_i \Delta g^o_{F_i} \text{ (products)} - \Sigma a_i \Delta g^o_{F_i} \text{ (reactants)} \tag{12-21}$$

EXAMPLE PROBLEM 12-6. Recalculate the Gibbs free energy of reaction calculated in Example Prob. 12-5 for the reaction

$$CO_2(g) + H_2(g) \to CO(g) + H_2O(g)$$

at reference conditions using enthalpy and entropy of formation data.

TABLE 12-3.[†] Standard Entropies of Formation at 25 °C[(34-38)]

Note: Entropy units e.u. are calories per gram mole, K.

Substance	Formula	State	$s^o_{F_{298}}$
Normal paraffins:			
Methane	CH_4	g	44.50
Ethane	C_2H_6	g	54.85
Propane	C_3H_8	g	64.51
n-Butane	C_4H_{10}	g	74.12
n-Pentane	C_5H_{12}	g	83.40
n-Hexane	C_6H_{14}	g	92.83
n-Heptane	C_7H_{16}	g	102.24
Increment per C atom above C_7			9.31
Normal monoolefins (1-alkenes):			
Ethylene	C_2H_4	g	52.45
Propylene	C_3H_6	g	63.80
1-Butene	C_4H_8	g	73.04
1-Pentene	C_5H_{10}	g	82.65
1-Hexene	C_6H_{12}	g	91.93
Miscellaneous hydrocarbons:			
Acetylene	C_2H_2	g	47.997
Benzene	C_6H_6	g	64.34
Benzene	C_6H_6	l	41.30
1,3-Butadiene	C_4H_6	g	66.62
Cyclohexane	C_6H_{12}	g	71.28
Cyclohexane	C_6H_{12}	l	48.85
Ethylbenzene	C_8H_{10}	g	86.15
Isoprene	C_5H_8	g	75.44
Methylcyclohexane	C_7H_{14}	g	82.06
Methylcyclohexane	C_7H_{14}	l	59.26
Styrene	C_8H_8	g	82.48
Toluene	C_7H_8	g	76.42
Toluene	C_7H_8	l	52.48
Miscellaneous oxygenated organic compounds:			
Acetaldehyde	C_2H_4O	g	59.8
Acetic acid	$C_2H_4O_2$	l	38.2
Ethanol	C_2H_6O	g	67.54
Ethanol	C_2H_6O	l	38.4

TABLE 12-3. (continued)

Substance	Formula	State	$s^o_{F_{298}}$
Ethylene glycol	$C_2H_6O_2$	l	39.9
Ethylene oxide	C_2H_4O	g	57.94
Methanol	CH_4O	g	57.29
Methanol	CH_4O	l	30.3
Miscellaneous inorganic compounds:			
Ammonia	NH_3	g	46.05
Calcium carbonate	$CaCO_3$	s	22.2
Calcium chloride	$CaCl_2$	s	25.0
Calcium hydroxide	$Ca(OH)_2$	s	19.93
Calcium oxide	CaO	s	9.50
Carbon, graphite	C	s	1.372
Carbon dioxide	CO_2	g	51.070
Carbon monoxide	CO	g	47.217
Chlorine	Cl_2	g	53.290
Hydrochloric acid	HCl	g	44.643
Hydrogen	H_2	g	31.207
Hydrogen sulfide	H_2S	g	49.16
Iron chloride	$FeCl_2$	s	28.19
Iron oxide, magnetite	Fe_3O_4	s	35.0
Iron oxide, hematite	Fe_2O_3	s	20.89
Iron sulfate	$FeSO_4$	s	25.7
Nitric acid	HNO_3	l	37.19
Nitrogen	N_2	g	45.770
Nitrogen oxides	NO	g	50.335
	NO_2	g	57.421
	N_2O	g	52.084
	N_2O_4	g	72.810
Oxygen	O_2	g	49.005
Sodium carbonate	Na_2CO_3	s	33.173
Sodium carbonate, deca-hydrate	$Na_2CO_3 \cdot 10H_2O$	s	—
Sodium chloride	$NaCl$	s	17.236
Sodium hydroxide	$NaOH$	s	15.400
Sulfur dioxide	SO_2	g	59.30

TABLE 12-3. (*continued*)

Substance	Formula	State	$s^o_{F_{298}}$
Sulfur trioxide	SO_3	g	61.34[†]
Sulfur trioxide	SO_3	l	22.85
Sulfuric acid	H_2SO_4	l	37.501
Water	H_2O	g	45.106
Water	H_2O	l	16.718

[†]Table prepared by Dr. Jing Chao, Thermodynamics Research Center, Texas A & M University. (1974) Data used with permission of the American Petroleum Institute and the Thermodynamics Research Center.

From Table 12-3, we find

for $CO_2(g)$, $\Delta s^o_{F_{298}} = 51.07 - 1.37 - 49.01 = 0.69$ cal/gm mole–K

for $H_2(g)$, $\Delta s^o_{F_{298}} = 31.21 - 31.21 = 0.00$ cal/gm mole–K

for $CO(g)$, $\Delta s^o_{F_{298}} = 47.22 - 1.37 - \frac{1}{2}(49.01) = 21.34$ cal/gm mole–K

for $H_2O(g)$, $\Delta s^o_{F_{298}} = 45.11 - 31.21 - \frac{1}{2}(49.01) = -10.61$ cal/gm mole–K

Combining these results with enthalpies of formation from Table 12-1,

for $CO_2(g)$, $\Delta g^o_{F_{298}} = -94,051 - 298(0.69) = -94,256$ cal/gm mole

for $H_2(g)$, $\Delta g^o_{F_{298}} = 0 - 298(0) = 0$ cal/gm mole

for $CO(g)$, $\Delta g^o_{F_{298}} = -26,417 - 298(21.34) = -32,776$ cal/gm mole

for $H_2O(g)$, $\Delta g^o_{F_{298}} = -57,795 - 298(-10.61) = -54,633$ cal/gm mole

These results may be compared with the data shown in Table 12-2. Then, from Eq. (12-21),

$$\Delta G^o_{R_{298}} = -32,776 - 54,633 + 94,256$$

$$= +6847 \text{ cal/gm mole } CO_2 \text{ reacted}$$

The insignificant difference in this result and the result obtained in Example Prob. 12-5 is due to round-off error.

It is not necessary to calculate Δg_F^o for each reactant and for each product in order to calculate ΔG_R^o for an isothermal reaction. From the relationship

$$\Delta G_R^o = \Sigma a_i \Delta g_{F_i}^o \text{ (products)} - \Sigma a_j \Delta g_{F_j}^o \text{ (reactants)} \qquad (12\text{-}21)$$

we find on expanding that

$$\Delta G_R^o = \Sigma a_i \Delta h_{F_i}^o \text{ (products)} - T\Sigma a_i \Delta s_{F_i}^o \text{ (products)}$$

$$- \Sigma a_i \Delta h_{F_i}^o \text{ (reactants)} + T\Sigma a_i \Delta s_{F_i}^o \text{ (reactants)}$$

or

$$\Delta G_R^o = \Delta H_R^o - T\Delta S_R^o \qquad (12\text{-}24)$$

Now ΔH_R^o can be obtained from either

$$\Delta H_R^o = \Sigma a_i \Delta h_{F_i}^o \text{ (products)} - \Sigma a_i \Delta h_{F_i}^o \text{ (reactants)} \qquad (12\text{-}7)$$

or

$$\Delta H_R^o = \Sigma a_i(-\Delta h_{C_i}^o)\text{(products)} - \Sigma a_i(-\Delta h_{C_i}^o)\text{(reactants)} \qquad (12\text{-}11)$$

by using data from Table 12-1. Further, we may define ΔS_R^o by the relationship

$$\Delta S_R^o = \Sigma a_i s_{F_i}^o \text{ (products)} - \Sigma a_i s_{F_i}^o \text{ (reactants)} \qquad (12\text{-}25)$$

since the entropies of the elements in the products and reactants cancel. This simplifies the steps required to estimate ΔG_R^o.

EXAMPLE PROBLEM 12-7. Repeat Example Prob. 12-6 using Eq. (12-25).

We have shown in Example Prob. 12-1 that for the reaction

$$CO_2(g) + H_2(g) \rightarrow CO(g) + H_2O(g)$$

$$\Delta H_{R_{298}}^o = 9839 \text{ cal/gm mole}$$

Now, from Eq. (12-25),

$$\Delta S_R^o = 47.22 + 45.11 - 51.07 - 31.21 = 10.05 \text{ cal/gm mole–K}$$

Thus,

$$\Delta G_R^o = \Delta H_R^o - T\Delta S_R^o \qquad (12\text{-}24)$$

$$= 9839.0 - (298)(10.05)$$

$$= 6844 \text{ cal/gm mole } CO_2 \text{ reacted}$$

This compares to 6840 and 6847 cal/gm mole calculated in the preceding examples. The insignificant difference is due to round-off errors.

Third Law of Thermodynamics

An inspection of Table 12-3 shows that values of s_F^o are tabulated and not Δs_F^o. This can be done since entropy, as distinct from energy functions, has an absolute value.

In 1902, T. W. Richards investigated the change of the Gibbs free energy and the change in the enthalpy of a reaction. Figure 12-2 is a schematic plot of his data.

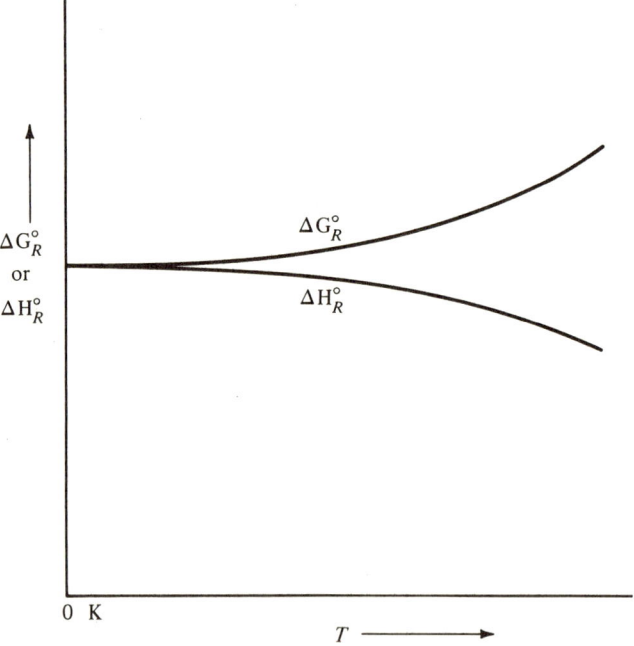

Fig. 12-2. *Limiting approach of ΔG_R^o and ΔH_R^o as the temperature approaches absolute zero.*

W. Nernst, using some of his own data, but mostly Richards' data, noticed that since

$$\Delta G_R^o = \Delta H_R^o - T\Delta S_R^o \qquad (12\text{-}24)$$

we could write

$$T\Delta S_R^o = \Delta H_R^o - \Delta G_R^o$$

Thus, Richards' experiment shows that as the reaction temperature approaches 0 K,

$$\Delta H_R^o - \Delta G_R^o \to 0$$

or

$$T\Delta S_R^o \to 0$$

This means that ΔS_R^o does not approach infinity as the temperature of the reaction approach 0 K. Now, Planck noticed that not only did the curves representing free energy and enthalpy of the reaction approach one another as temperature approached 0 K, but the curves approached each other with zero slopes. Thus, if we examine the derivative of Eq. (12-24) with respect to temperature as temperature approaches zero, we find

$$\lim_{T \to 0} \left[T\left(\frac{\partial \Delta S_R^o}{\partial T}\right)_P + \Delta S_R^o \right] = \lim_{T \to 0} \left[\left(\frac{\partial (\Delta H_R^o)}{\partial T}\right)_P - \left(\frac{\partial \Delta G_R^o}{\partial T}\right)_P \right]$$

but as temperature approaches zero, we see from Fig. 12-2 that

$$\left(\frac{\partial \Delta H_R^o}{\partial T}\right)_P = 0$$

$$\left(\frac{\partial \Delta G_R^o}{\partial T}\right)_P = 0$$

Thus, $\Delta S_R^o = 0$ at 0 K, or

$$\Sigma a_i s_{F_i}^o \text{ (products)} - \Sigma a_i s_{F_i}^o \text{ (reactants)} = 0 \text{ at 0 K} \tag{12-26}$$

This result is called the Nernst heat theorem.

Upon studying the Nernst heat theorem, Planck felt that it wasn't logical for the entropy of the products to have some finite value just balanced by the entropy of the reactants. He decided it would be more consistent to assume that both the entropy of the products and reactants were zero at 0 K.

Planck's ideas were accepted only skeptically for about 20 years. But as time went on, more and more data became available, and it was found that almost all reactions followed Planck's hypothesis. There were discrepancies, however.

The discrepancies led to the generalization that if a substance has a crystalline form at 0 K, its entropy is zero. But if it does not become crystalline at 0 K (such as sulfur which becomes non-crystalline or glassy), its entropy is not zero. We say that compounds that are glassy at 0 K have a definite amount of disorder. Also, mixtures will have an entropy of mixing at 0 K. This led to the introduction of the Third Law of thermodynamics,

The entropy of a perfect crystal at absolute zero is equal to zero.

The Third Law tells us that at any pressure or temperature the entropy of a component, unlike the energy of the component, has an absolute value which may be calculated using equations of state, specific heat data, and phase transition enthalpy data.

Effect of Temperature on Chemical Equilibrium

Frequently, we find it desirable to calculate ΔG_R^o at some temperature other than the reference temperature. A convenient procedure is to note that at constant temperature,

$$\Delta G_R^o = \Delta H_R^o - T\Delta S_R^o \qquad (12\text{-}24)$$

We have shown that $(\Delta H_R^o)_T$ can be calculated for any temperature T by picking a path in which the reactants are first cooled from T to T_0; the reaction then proceeds at T_0, and finally the products are heated from T_0 to T.

$$\Delta H_{R_T}^o = \int_T^{T_0} \Sigma a_i c_{P_i} \text{ (reactants) } dT + \Delta H_{R_{T_0}}^o + \int_{T_0}^T \Sigma a_i c_{P_i} \text{ (products)}$$

$$= \Delta H_{R_{T_0}}^o + \int_{T_0}^T \Delta c_P \, dT \qquad (12\text{-}8)$$

Similarly, along the same path,

$$\Delta S_{R_T}^o = \int_T^{T_0} \frac{\Sigma a_i c_{P_i} \text{ (reactants)}}{T} dT + \Delta S_{R_{T_0}}^o + \int_{T_0}^T \frac{\Sigma a_i c_{P_i} \text{ (products)}}{T} dT$$

or

$$\Delta S_{R_T}^o = \Delta S_{R_{T_0}}^o + \int_{T_0}^T \frac{\Delta c_P}{T} dT \qquad (12\text{-}27)$$

Then, at temperature T

$$\Delta G_{R_T}^o = \Delta H_{R_T}^o - T\Delta S_{R_T}^o \qquad (12\text{-}24)$$

Instead of looking at the effect of temperature on ΔG_R^o, we might examine the effect of temperature on the chemical equilibrium coefficient K_A. We showed that for a system at constant temperature and pressure

$$\Delta G_{R_T}^o = -RT \ln K_A \qquad (12\text{-}16)$$

and

$$\ln K_A = -\frac{\Delta G_R^o}{RT}$$

Now, let's permit temperature to vary, but hold pressure constant. Then, as both K_A and ΔG_R^o are functions of temperature,

$$\left(\frac{\partial \ln K_A}{\partial T}\right)_P = \frac{\Delta G_R^o}{RT^2} - \frac{1}{RT}\left(\frac{\partial \Delta G_R^o}{\partial T}\right)_P$$

Furthermore, as

$$d(\Delta G_R^o) = (\Delta V_R^o)dP - (\Delta S_R^o)dT$$

we find that at constant pressure,

$$\left(\frac{\partial \Delta G_R^o}{\partial T}\right)_P = -\Delta S_R^o$$

Further,

$$\Delta G_R^o = \Delta H_R^o - T\Delta S_R^o$$

So,

$$-\Delta S_R^o = \frac{\Delta G_R^o - \Delta H_R^o}{T} = \left(\frac{\partial \Delta G_R^o}{\partial T}\right)_P$$

From these results, we find

$$\left(\frac{\partial \ln K_A}{\partial T}\right)_P = \frac{\Delta G_R^o}{RT^2} - \frac{1}{RT}\left(\frac{\Delta G_R^o - \Delta H_R^o}{T}\right)$$

or

$$\left(\frac{\partial \ln K_A}{\partial T}\right)_P = \frac{\Delta H_R^o}{RT^2} \qquad (12\text{-}28)$$

or

$$\int_{T_1}^{T_2} d\ln K_A = \int_{T_1}^{T_2} \frac{\Delta H_R^o}{RT^2} dT \qquad (12\text{-}29)$$

In order to integrate Eq. (12-29), we must find ΔH_R^o as a function of temperature, $\Delta H_R^o(T)$. This can be done by integrating Eq. (12-8) between the limits T_0 and a variable T. For example, if Δc_P is a constant, we have

$$\Delta H_R^o(T) = \Delta H_{R_{T_0}}^o + \Delta c_P(T - T_0)$$

or

$$\Delta H_R^o(T) = (\Delta c_P)T + C_1$$

Using this function of $\Delta H_R^o(T)$ in Eq. (12-29) yields the relationship

$$\int_{T_1}^{T_2} d\ln K_A = \int_{T_1}^{T_2} \frac{\Delta c_P}{RT} dT + \int_{T_1}^{T_2} \frac{C_1}{RT^2} dT$$

so that

$$\ln \frac{K_A(T_2)}{K_A(T_1)} = \frac{\Delta c_P}{R} \ln \frac{T_2}{T_1} - \frac{C_1}{R}\left(\frac{1}{T_2} - \frac{1}{T_1}\right) \qquad (12\text{-}30)$$

However, if $c_P = a + bT + cT^2$, and if we define $\Delta c_P = \Delta a + \Delta bT + \Delta cT^2$, we find by integrating Eq. (12-8) that

$$\Delta H_R^o(T) = \Delta H_{R_0}^o + \Delta a(T - T_0) + \tfrac{1}{2}\Delta b(T^2 - T_0^2) + 1/3\,\Delta c(T^3 - T_0^3)$$

If $T_0 = 0$ K, then

$$\Delta H_R^o(T) = \Delta aT + \tfrac{1}{2}\Delta bT^2 + 1/3\,\Delta cT^3 + \Delta H_{R_0}^o.$$

Thus, integrating Eq. (12-29), we find that

$$\ln \frac{K_A(T_2)}{K_A(T_1)} = \frac{\Delta a}{R} \ln \frac{T_2}{T_1} + \frac{\Delta b}{2R}(T_2 - T_1) + \frac{\Delta c}{6R}(T_2^2 - T_1^2)$$

$$- \frac{\Delta H_{R_0}^o}{R}\left(\frac{1}{T_2} - \frac{1}{T_1}\right)$$

or, combining the constant terms evaluated at T_1, we find that at $T_2 = T$,

$$\ln K_A = -\frac{\Delta H^o_{R_0}}{RT} + \frac{\Delta a}{R} \ln T + \frac{\Delta b}{2R} T + \frac{\Delta c}{6R} T^2 + C_2 \qquad (12\text{-}31)$$

If $(\Delta H^o_R)_0$ at 0 K is known, the integration constant C_2 can be found from a single value of K_A.

Estimation Methods for Reaction Free Energy Data

Frequently, thermodynamic data is not readily available for compounds that may be of possible interest. There are several methods available for estimating the desired properties.

Most of these methods are based on the observation that there is a regularity in the changes of thermodynamic properties. For example, in Table 12-1, we see that to find Δh^o_F for normal paraffins larger than n-heptane, for every carbon atom in the molecule over seven, we add -4930 cal/gm mole to the $(\Delta h^o_F)_{298}$ of heptane. This is an extrapolation method. Another commonly used method for estimating thermodynamic properties is the method of interpolation. If values of the desired thermodynamic properties are available for compounds that are reasonably similar to the compound of interest, an educated guess may be made of the value of the property for the compound of interest. A common method of making an estimate of hydrocarbon properties is to plot the values of the desired property versus the number of carbon atoms in the hydrocarbons. Then, the value of the property of the hydrocarbon of interest may be interpolated directly from the plot.

Another method of estimating Δh^o_F, s^o_F and c_P is the group contribution method of Andersen, Beyer and Watson[41]. In this method, the enthalpy and entropy of formation of any compound in the ideal gas state at 298 K as well as the coefficients a, b, and c in the equation

$$c_P \left(\frac{\text{cal}}{\text{gm-mole } ^\circ K}\right) = a + bT + cT^2$$

are estimated by assuming that the compound can be built up from the base groups listed in Table 12-4, by appropriate substitutions. For example, all paraffin hydrocarbons may be considered to be derived from methane.

In order to build the desired chemical species from a base group, hydrogen atoms on the base group are first replaced by methyl ($-CH_3$) radicals. The first substitution of a hydrogen atom by a $-CH_3$ radical is called a primary substitution. The incremental contributions of this primary substitution to the ideal gas state thermodynamic properties of the base group are presented in Table 12-5.

TABLE 12-4. Base Group Properties[41]

Group	$\Delta h^o_{F_{298}}(g)$ kcal/g-mole	$s^o_{F_{298}}(g)$ cal/(g-mole)(K)	a	b(10³)	c(10⁶)
				Ideal Gas at T(K)	
Methane	−17.9	44.5	3.42	17.85	−4.16
Cyclopentane	−21.4	70.7	2.62	82.67	−24.72
Benzene	18.1	64.4	0.23	77.83	−27.16
Naphthalene	35.4	80.7	3.15	109.40	−34.79
Methylamine	−7.1	57.7	4.02	30.72	−8.70
Dimethylamine	−7.8	65.2	3.92	48.31	−14.09
Trimethylamine	−10.9	—	3.93	65.85	−19.48
Dimethyl ether	−46.0	63.7	6.42	39.64	−11.45
Formamide	−49.5	—	6.51	25.18	−7.47

Reprinted by permission from *National Petroleum News*, 36, 27, R-476 (July 5, 1944).

TABLE 12-5. Contribution of Primary CH_3 Substitution Groups Replacing Hydrogen[41]

Base Group	$\Delta(\Delta h^o_{F_{298}})(g)$ kcal/g-mole	$\Delta s^o_{F_{298}}(g)$ cal/(g-mole)(K)	Δa	$\Delta b(10^3)$	$\Delta c(10^6)$
				Ideal Gas at T(K)	
1. Methane	−2.2	10.4	−2.04	24.00	−9.67
2. Cyclopentane					
(a) Enlargement of ring	−9.3	0.7	−1.04	19.30	−5.79
(b) First substitution	−5.2	11.5	−0.07	18.57	−5.77
(c) Second substitution:					
ortho	−12.2	—			
meta	−8.4	—	−0.24	16.56	−5.05
para	−7.1	—			
(d) Third substitution	−7.0	—	—	—	—
3. Benzene and naphthalene					
(a) First substitution	−4.5	12.0	0.36	17.65	−5.88
(b) Second substitution:					
ortho	−6.3	8.1	5.20	6.02	1.18
meta	−6.5	9.2	1.72	14.18	−3.76
para	−8.0	7.8	1.28	14.57	−3.98
(c) Third substitution (sym)	—	8.0	0.57	16.51	−5.19
4. Methylamine	−5.7	—			
5. Dimethylamine	−6.3	—	−0.10	17.52	−5.35
6. Trimethylamine	−4.1	—			
7. Formamide					
Substitution on C atom	−9.0	—	6.11	−1.75	4.75

Reprinted by permission from *National Petroleum News*, 36, 27, R-476 (July 5, 1944).

Further replacements of any hydrogen atoms by —CH_3 radicals on the primary substituted base group are called secondary methyl substitutions. The contribution of secondary methyl substitutions to the ideal gas state thermodynamic properties, summarized in Table 12-6, depends both on A, the type of carbon atom upon which the substitution is made, and B, the highest type of carbon atom adjacent to the substituted carbon atom.

TABLE 12-6. Contribution of Secondary CH_3 Substitutions Replacing Hydrogen[41]

A	B	$\Delta(\Delta h^o_{F_{298}})(g)$ kcal/g-mole	$\Delta s^o_{F_{298}}(g)$ cal/(g-mole)(K)	Δa	$\Delta b(10^3)$	$\Delta c(10^6)$
					Ideal Gas at $T(K)$	
1	1	−4.5	9.8	−0.97	22.86	−8.75
1	2	−5.2	9.2	1.11	18.47	−6.85
1	3	−5.5	9.5	1.00	19.88	−8.03
1	4	−5.0	11.0	1.39	17.12	−5.88
1	5	−6.1	10.0	0.10	17.18	−5.20
2	1	−6.6	5.8	1.89	17.60	−6.21
2	2	−6.8	7.0	1.52	19.95	−8.57
2	3	−6.8	6.3	1.01	19.69	−7.83
2	4	−5.1	6.0	2.52	16.11	−5.88
2	5	−5.8	2.7	0.01	17.42	−5.33
3	1	−8.1	2.7	−0.96	27.47	−12.38
3	2	−8.0	4.8	−1.19	28.77	−12.71
3	3	−6.9	5.8	−3.27	30.96	−14.06
3	4	−5.7	1.7	−0.14	24.57	−10.27
3	5	−9.2	1.3	0.42	16.20	−4.68
1	—O—in ester or ether	−7.0	14.4	−0.01	17.58	−5.33
Substitution of H of OH group to form ester		+9.5	16.7	0.44	16.63	−4.95

Reprinted by permission from *National Petroleum News*, 36, 27, R-476 (July 5, 1944).

Carbon atom types are defined by Andersen, Beyer and Watson on the basis of the number of hydrogen atoms attached to the carbon atom. Thus,

$$
\begin{array}{ll}
\text{Type 1} & -CH_3 \\
2 & -CH_2- \\
3 & -CH\text{\textless} \\
4 & -C\text{\textless} \\
5 & \text{C atom in benzene or napthalene ring}
\end{array}
$$

Two special secondary substitutions are listed in Table 12-6; one is used in forming a methyl ester from carboxylic acid and the other in forming an ethyl ester or ether from a methyl ester or ether.

Table 12-7 lists the incremental contributions resulting from the replacement of single carbon-carbon bonds with multiple carbon-carbon bonds. Table 12-8 lists the incremental contributions resulting from the replacement of one or two methyl radicals by the groups listed. Although only one methyl group would be replaced by an —OH radical, two methyl groups would be replaced by a =O radical.

TABLE 12-7. Multiple-Bond Contributions Replacing Single Bonds[41]

Type of Bond A	B		$\Delta(\Delta h^o_{F_{298}})(g)$ kcal/g-mole	$\Delta s^o_{F_{298}}(g)$ cal/(g-mole)(K)	Δa	$\Delta b(10^3)$	$\Delta c(10^6)$
						Ideal Gas at $T(K)$	
1	=	1	32.8	−2.1	1.33	−12.69	+4.77
1	=	2	30.0	0.8	1.56	−14.87	+5.57
1	=	3	28.2	2.2	0.63	−23.65	+13.10
2	=	2	28.0	−0.9	0.40	−18.87	+9.89
2	=	2 cis	28.4	−0.6	0.40	−18.87	+9.89
2	=	2 trans	27.5	−1.2	0.40	−18.87	+9.89
2	=	3	26.7	1.6	0.63	−23.65	+13.10
3	=	3	25.5	—	−4.63	−17.84	+11.88
Additional correction for each pair of conjugated double bonds			−3.8	−10.4	Approximately zero		
1	≡	1	74.4	−6.8	5.58	−31.19	+11.19
1	≡	2	69.1	−7.8	6.42	−36.41	+14.53
2	≡	2	65.1	−6.3	4.66	−36.10	+15.28
Correction for double bond adjacent to aromatic ring			−5.1	−4.3	Approximately zero		

Reprinted by permission from *National Petroleum News*, 36, 27, R-476 (July 5, 1944).

The contributions to heats of formation resulting from substitution of chlorine for methyl groups vary with the number of substitutions made on a single carbon atom. Corresponding variations were not found for the contributions to entropy or heat capacity or for the substitution of the other halogens. As noted in Table 12-8 a correction must be applied to the entropies calculated for the halogenated methanes. In general the calculated results tend to be uncertain for single carbon-atom compounds.

TABLE 12-8. Substitution Group Contributions Replacing CH_3 Group[41]

Group	$\Delta(\Delta h^o_{F_{298}})(g)$ kcal/g-mole	$\Delta s^o_{F_{298}}(g)$ cal/(g-mole)(K)	Δa	$\Delta b(10^3)$	$\Delta c(10^6)$
				Ideal Gas at $T(K)$	
—OH (aliphatic, meta, para)	−32.7	2.6	3.17	−14.86	5.59
—OH ortho	−47.7	—	—	—	—
—NO_2	1.2	2.0	6.3	−19.53	10.36
—CN	39.0	4.0	3.64	−13.92	4.53
—Cl	0‡ for first Cl on a carbon; 4.5 for each additional	0	2.19	−18.85	6.26
—Br	10.0	3.0‡	2.81	−19.41	6.33
—F	−35.0	−1.0‡	2.24	−23.61	11.79
—I	24.8	5.0‡	2.73	−17.37	4.09
=aldehyde	−12.9	−12.3	3.61	−55.72	22.72
=ketone	−13.2	−2.4	5.02	−66.08	30.21
—COOH	−87.0	15.4	8.50	−15.07	7.94
—SH	15.8	5.2	4.07	−24.96	12.37
—C_6H_5	32.3	21.7	−0.79	53.63	−19.21
—NH_2	12.3	−4.8	1.26	−7.32	2.23

‡Add 1.0 to the calculated entropy contributions of halides for methyl derivatives; for example, methyl chloride = 44.5(base) + 10.4(primary CH_3) − 0.0(Cl substitution) + 1.0.

Reprinted by permission from *National Petroleum News*, 36, 27, R-476 (July 5, 1944).

Andersen, Beyer and Watson recommend the following procedure when estimating the thermodynamic properties of a complex compound:

(1) Select the base group and determine its properties from Table 12-4. Where a choice of base group is possible, select the group having the largest entropy. Proceed to build up the desired compound with as few substitutions as possible.

(2) Add the contributions given in Tables 12-5 and 12-6 which result from all —CH_3 substitutions replacing hydrogen required to establish the carbon skeleton of the compound. In this operation, the longest straight chain should be built up first and then the longest side chain. Where the same compound may be arrived at by alternate substitutions, an average result is used.

(3) Add the contributions from Table 12-6 which result from additional CH_3 substitutions replacing hydrogen in the positions occupied in the compound by other groups which are listed in Table 12-8.

(4) Add the contributions given in Table 12-7 which result from multiple bonds.

(5) Add the incremental contributions given in Table 12-8 which result from replacement of CH_3 groups by substitution groups.

This method and several other methods of estimating thermodynamic properties are summarized and compared by Reid and Sherwood[18].

EXAMPLE PROBLEM 12-8. Using the method of Andersen, Beyers, and Watson, calculate

$$\Delta h^o_{F_{298}}$$

for acetaldehyde. The structural formula for acetaldehyde is

```
      H  H
      HC — C = 0
      H
```

This problem will be solved by building on methane as a base group. From Table 12-4, we have for $(\Delta h^o_F)_{298}$ of our base group is equal to -17.9 k cal/gm mole. From Table 12-5, we find that for

```
  H       H⁺         H   H
  HCH  +  CH   →     HC — CH  +  H⁺
  H       H          H   H
```

the change in enthalpy is -2.2 k cal/gm mole. From Table 12-6, since

```
  H   H    H⁺         H   H   H
  HC — CH  + CH   →   HC — C — CH  +  H⁺
  H   H    H          H   H   H
```

is a 1-1 addition, we find the change in enthalpy is -4.5 k cal/gm mole. Further,

```
                              H
                              HCH
  H   H   H    H⁺        H    |    H
  HC — C — CH  + CH  →   HC — C — CH  +  H⁺
  H   H   H    H         H    H    H
```

is a 2-1 addition, so the change in enthalpy is -6.6 k cal/gm mole. Now, using Table 12-8, we find that the change in enthalpy accompanying the replacement of the last two methyl groups with =O is -12.9 k cal/gm mole. Adding, we find $(\Delta h^o_F)_{298}$ for acetaldehyde is -44.1 k cal/gm mole. From Table 12-1, the value of $(\Delta h^o_F)_{298}$ for acetaldehyde is -39.72 k cal/gm mole. The calculated value

ELECTROCHEMICAL CELLS

Electrochemical cells consist of a positive and negative electrode separated by an electrolyte. There are two types of electrochemical cells. The galvanic cell, or voltaic cell, is used to obtain electrical work from a chemical reaction by drawing current from the electrodes. This type of cell is used in automobiles, flashlights, etc. Naturally occurring galvanic cells are the primary cause of corrosion.

A fuel cell is a special type of galvanic cell. It also consists of a positive and a negative electrode separated by an electrolyte. However, the reactants are not stored in the cell but are fed continuously, and the products of the reaction are continuously withdrawn. Thus, the fuel cell operates as an open system converting energy from a chemical reaction directly to electrical energy.

The second type of electrochemical cell is called an electrolytic cell. In this second type of cell, electrical work is used to promote a chemical reaction. For example, in some electrolytic cells, chlorine is produced while in other electrolytic cells, electroplating takes place.

A diagram of a galvanic cell is shown in Fig. 12-3. At the anode, a reaction

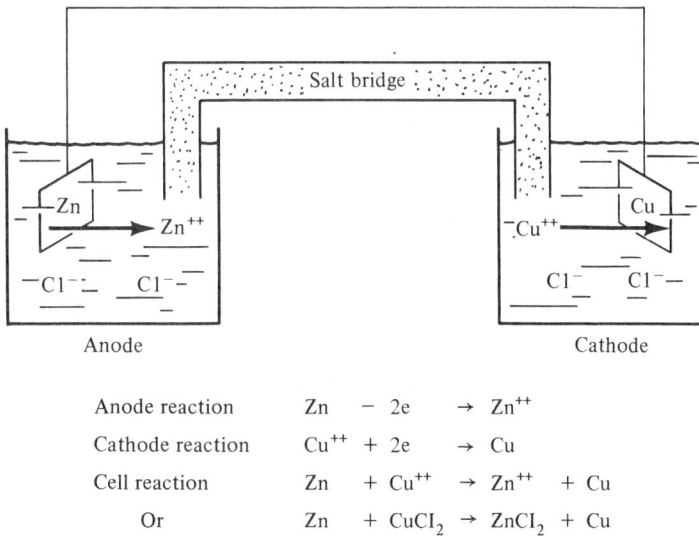

Anode reaction	Zn	$- 2e$	$\rightarrow$	Zn^{++}
Cathode reaction	Cu^{++}	$+ 2e$	$\rightarrow$	Cu
Cell reaction	Zn	$+ Cu^{++}$	$\rightarrow$	$Zn^{++} + Cu$
Or	Zn	$+ CuCl_2$	$\rightarrow$	$ZnCl_2 + Cu$

Fig. 12-3. *Galvanic cell.*

called the oxidation reaction takes place. Oxidation is the removal of electrons. An example of the oxidation reaction is

$$Zn - 2e \rightarrow Zn^{++}$$

where two electrons have been removed from the zinc atom Zn to form one zinc cation Zn^{++}. The number of positive charges on the zinc cation is called the valence of zinc z_{Zn}.

At the cathode, a reaction called the reduction reaction takes place. Reduction is the addition of electrons. An example of a reduction reaction is

$$Cu^{++} + 2e \rightarrow Cu$$

Now, the cell reaction can be found by adding the oxidation and reduction reactions,

$$Zn + Cu^{++} \rightarrow Zn^{++} + Cu$$

or for convenience adding the chlorine anion Cl^-

$$Zn + CuCl_2 \rightarrow ZnCl_2 + Cu$$

In this case, the number of electrons removed from one zinc atom exactly balances the number of electrons added to the copper cation. In general, if we take any portion of a galvanic cell as a system, there will be no accumulation of electrons within the system at steady state. This means that the number of electrons produced in the oxidation reaction must always be identical to the number of electrons consumed in the associated reduction reaction. That is, if the reduction reaction taking place in conjunction with the oxidation of zinc were

$$Ag^+ + 1e \rightarrow Ag$$

then, in order to equalize the number of electrons transferred in the oxidation and reduction reactions, we must either multiply the reduction reaction by two to obtain the cell reaction

$$Zn + 2Ag^+ \rightarrow Zn^{++} + 2Ag$$

or multiply the oxidation reaction by ½ to obtain the cell reaction

$$\tfrac{1}{2}Zn + Ag^+ \rightarrow \tfrac{1}{2}Zn^{++} + Ag$$

In the first cell reaction, two electrons are transferred in the reaction, as written, while in the second cell reaction, one electron is transferred in the reaction, as written.

The change in the Gibbs free energy of the cell shown in Fig. 12-3 is obtained by combining the definition of G with the property relationship. The result is

$$dG = V\,dP - S\,dT + \mu_{Zn}dn_{Zn} + \mu_{CuCl_2}dn_{CuCl_2}$$

$$+ \mu_{ZnCl_2}dn_{ZnCl_2} + \mu_{Cu}dn_{Cu} + \epsilon\,dq$$

At constant pressure, temperature and electric charge, we find

$$dG_{P,T,q} = \mu_{Zn}dn_{Zn} + \mu_{Cu}dn_{CuCl_2} + \mu_{ZnCl_2}dn_{ZnCl_2} + \mu_{Cu}dn_{Cu} = \Delta G_R\,d\alpha$$

Further, the electrical work term consists of the product of the potential across the cell,

$$\epsilon = E_{cell}$$

and dq the quantity of charge transferred during a change of dn_{Zn} moles of zinc. Michael Faraday showed that there are 96,487 coulombs of electrical charge associated with one gram mole of positive charge. This quantity is called the Faraday constant F. If we now introduce a quantity n_{e_i} called the number of equivalents of component i and defined as

$$n_{e_i} = n_i z_i \qquad (12\text{-}32)$$

where z_i is the valence of compound i, we see that one gram equivalent of ions is equivalent to one gram mole of univalent positive charges. Further, since the valence of the silver ion z_{Ag} is unity, one gram mole of silver is the same as one gram equivalent of silver. Thus, there are 96,487 coulombs of charge associated with one gram mole of silver ions. On the other hand, since the valence of zinc z_{Zn} is two, one gram mole of zinc is the same as two gram equivalents of zinc. There are 192,974 coulombs of charge associated with one gram mole of zinc ions.

Now, when $d\alpha$ moles of zinc are oxidized, $z_{Zn}\,d\alpha$ gram equivalents of zinc ions are produced. The amount of charge transferred during this differential reaction is

$$dq = zF\,d\alpha$$

so that

$$\epsilon\,dq = zFE_{cell}\,d\alpha$$

Note that z is numerically equal to the number of electrons or equivalent charges transferred per mole of zinc oxidized in the balanced cell reaction.

Thus, the change in the Gibbs free energy of the cell shown in Fig. 12-3 resulting from the reaction of $d\alpha$ moles of zinc at constant pressure and temperature is

$$dG_{P,T} = (\Delta G_R + zFE_{cell})\, d\alpha_{P,T}$$

If the system is at equilibrium,

$$dG_{P,T} = 0$$

and

$$\Delta G_R = -zFE_{cell} \tag{12-33}$$

In this case, ΔG_R would be the increase in free energy when z units of positive electricity are carried through the cell from the anode to the cathode, and $-zFE_{cell}$ is the reversible work that must be performed by an external agency in driving the same z units of positive electricity from the cathode to the anode. Consequently, Eq. (12-33) applies when there is no net flow of current in the cell.

Now, if all the components in the electrochemical cell are in their standard states, Eq. (12-33) becomes

$$\Delta G_R^\circ = -zFE_{cell}^\circ \tag{12-34}$$

Then, since

$$\Delta G_R = \Delta G_R^\circ + RT \ln J_A \tag{12-14}$$

if we substitute the electrochemical equivalents of ΔG_R into Eq. (12-14), we obtain the relationship

$$E_{cell} = E_{cell}^\circ - \frac{RT}{zF} \ln J_A$$

or

$$E_{cell} = E_{cell}^\circ - \frac{2.303\, RT}{zF} \log J_A \tag{12-35}$$

This equation relates the reversible potential of an electrochemical cell to the activities of the reactants and products.

In Eq. (12-35), E_{cell}° is the sum of the oxidation potential at the anode and the reduction potential at the cathode,

$$E_{cell}^\circ = E_{\text{oxid. at anode}}^\circ + E_{\text{red. at cathode}}^\circ \tag{12-36}$$

Values of standard oxidation potentials for a series of elements in their standard states have been tabulated by W. M. Latimer [42]. The reduction potential of these elements is the negative of their oxidation potentials.

We evaluate the effect of component concentrations on the standard cell potential by observing first that at 25 °C,

$$\frac{2.303\,RT}{F} = 0.0592 \text{ volts}$$

The activity quotient term in Eq. (12-35) can then be calculated by noting that in a solid phase,

$$a_{solid} = 1.0$$

To calculate the activity of an electrolyte in the liquid phase, we use the relationship

$$a_{solute} = m\gamma$$

where

m = molality of the solute

$$= \frac{\text{gram moles of solute}}{1000 \text{ grams of solvent, usually water}}$$

γ = activity coefficient based on solute molality and equal to unity at infinite dilution

For a solute that decomposes into x cations and y anions, the activity of the solute is

$$a = (a_{\pm})^{x+y}$$

where

$$a_{\pm} = \gamma_{\pm} m_{\pm}$$

Values of $\gamma_{\pm}$ have been tabulated by Latimer [42] and

$$m_{\pm} = m(x^x y^y)^{1/x+y}$$

Finally, galvanic cells can be made to approach a reversible process. Thus, it is a convenient device to use for measuring thermodynamic properties. For example, at constant temperature and pressure,

$$\Delta G_R^o = -zFE_{cell}^o \tag{12-34}$$

and at constant pressure,

$$\Delta S_R^o = -\left(\frac{\partial \Delta G_R^o}{\partial T}\right)_P$$

Combining these results, we find that

$$\Delta S_R^o = +zF\left(\frac{\partial E_{cell}^o}{\partial T}\right)_P \tag{12-37}$$

Further, from the isothermal relationship

$$\Delta H_R^o = \Delta G_R^o + T\Delta S_R^o \tag{12-24}$$

we find

$$\Delta H_R^o = -zF\left[E_{cell}^o - T\left(\frac{\partial E_{cell}^o}{\partial T}\right)_P\right] \tag{12-38}$$

EXAMPLE PROBLEM 12-9. What is the potential of the cell shown in Fig. 12-3 at 25 °C if the molality of the zinc chloride solution is 1.0 and the molality of the cupric chloride solution is 0.1? The following data is available in the literature[42].

For the anode reaction

$$Zn - 2e \rightarrow Zn^{++}, E_{oxid}^o = +0.763 \text{ volts}$$

and the cathode reaction

$$Cu^{++} + 2e \rightarrow Cu, E_{red}^o = +0.337 \text{ volts}$$

Further,

$\gamma_\pm = 0.33$ for a one molal solution of zinc chloride

$\gamma_\pm = 0.52$ for a one tenth molal solution of cupric chloride

The cell reaction in this problem is the sum of the anode and cathode reactions. That is,

$$Zn + Cu^{++} \rightarrow Cu + Zn^{++}$$

or

$$Zn + CuCl_2 \rightarrow Cu + ZnCl_2$$

Now, we must evaluate the Nernst equation,

$$E_{cell} = E^o_{cell} - \frac{2.303\,RT}{zF} \log \frac{a_{ZnCl_2} a_{Cu}}{a_{CuCl_2} a_{Zn}}$$

For this reaction, the number of gram equivalents transferred when one mole of zinc reacts z is 2.0. Further, we see from Eq. (12-36)

$$E^o_{cell} = (+0.763) + (+0.337)$$

$$= 1.100 \text{ V}$$

Now, we must evaluate the activity coefficient. By definition,

$$a_{Zn} = a_{Cu} = 1.0$$

For a one molal solution of zinc chloride,

$$\gamma_\pm = 0.33$$

Then, zinc chloride breaks up into one cation and two anions,

$$ZnCl_2 \rightarrow Zn^{++} + 2Cl^-$$

Thus, $x = 1$ and $y = 2$, so that

$$m_\pm = [(1^1 \cdot 2^2)^{1/1+2}](1.0)$$

$$= 1.587$$

Then,

$$a_{ZnCl_2} = [(0.33)(1.587)]^{(1+2)}$$

$$= 0.1436$$

Similarly,

$$a_{CuCl_2} = 0.000562$$

Substituting these values into the Nernst equation,

$$E_{cell} = 1.100 - \frac{0.0592}{2} \log \frac{0.1436}{0.000562}$$

$$= 1.029 \text{ volts}$$

PROBLEMS

12-1. One step in the manufacture of CCl_4 involves the reaction

$$CS_2(l) + 3Cl_2(g) \rightarrow CCl_4(l) + S_2Cl_2(l)$$

which takes place in a water-cooled reactor at 25 °C. The enthalpies associated with the structure of the components at this temperature are CS_2 (l),21(k cal/gm mole); Cl_2 (g),0(k cal/gm mole); CCl_4 (l),–33.3(k cal/gm mole); S_2Cl_2 (l),–14.4 (k cal/gm mole). How many kg of cooling water at 10 °C must pass through the coils in the reactor for each kg of Cl_2 reacting to keep the temperature at 25 °C?

12-2. Calculate the heat of formation of gaseous benzene, C_6H_6, at 25 °C and one atmosphere using heat of combustion data.

12-3. Calculate the heat of combustion of gaseous n-octane, C_8H_{18}, at 25 °C and one atmosphere using heat of formation data.

12-4. Calculate the enthalpy of the reaction

$$H_2(g) + \tfrac{1}{2}O_2(g) \rightarrow H_2O(g)$$

at 100 atmospheres and 500 K.

12-5. Gaseous propane, C_3H_8, at 25 °C and one atmosphere pressure is mixed with air at 125 °C and burned. If 300 percent theoretical air is used, and if the water formed is in the vapor phase, what is the maximum theoretical temperature of the combustion products?

12-6. It is proposed to study the reaction between methane and oxygen in a well insulated flow reactor

$$CH_4(g) + 2O_2(g) \rightarrow CO_2(g) + H_2O(g)$$

During one specific test, 1.0 gram mole of methane and 0.2 gram moles of oxygen, both at 25 °C and 1 atm, are fed to the reactor in one hour. If the oxygen is completely reacted, what is the maximum theoretical temperature of the product gases leaving the reactor at 1 atmosphere pressure?

12-7. It is proposed to obtain energy for a rocket by burning hydrogen with a stoichiometric amount of oxygen

$$H_2(g) + \tfrac{1}{2}O_2(g) \rightarrow H_2O(g)$$

In order to compare this fuel with other fuels, we must calculate the maximum energy that will be released from a combustion chamber per pound of fuel.

If the reactants, $H_2(g)$ and $O_2(g)$, flow into the combustion chamber at 77 °F and one atmosphere pressure, what is the heat rejected from the chamber, per pound of hydrogen, if the water vapor flows from the chamber at 900 °F and 100 atm pressure? Neglect velocity and potential energy changes as well as dissociation of the water vapor.

12-8. For the reaction

$$H_2(g) + CO_2(g) \rightarrow H_2O(g) + CO(g)$$

$$\Delta G_R^o = 3{,}200 \text{ cal/gm mole}$$

at 700 K and 1.0 atmosphere total pressure:

(a) If one gram mole of $H_2(g)$ and 2 gram moles of $CO_2(g)$ are allowed to come to chemical equilibrium, what would be the mole fraction of CO_2 in the final mixture?

(b) What would be the value of ΔG_R^o for this reaction at 700 K if the pressure of the reaction is increased to 100 atmospheres total pressure?

12-9. It is proposed to produce ammonia NH_3 at 500 °C and 100 atm pressure, according to the following reaction:

$$3H_2 + N_2 \rightarrow 2NH_3$$

Starting with one gram mole of nitrogen and three gram moles of hydrogen, calculate the equilibrium composition of the mixture.

12-10. In the study of the conversion of coal to liquid hydrocarbons, we must consider the hydrogenation of the atomatic ring compounds in the coal. The reaction in which benzene C_6H_6 is hydrogenated to cyclohexane C_6H_{12}

$$C_6H_6(g) + 3H_2(g) \rightarrow C_6H_{12}(g)$$

is representative of the reactions involved.

Calculate the moles of cyclohexane that would be formed at equilibrium from the hydrogenation of one mole of benzene with three moles of hydrogen at 1000 K and 10 atmospheres. The fugacity coefficient of hydrogen at these conditions is 1.01. Unfortunately, the hydrogen available is only 75 mole percent pure, so that for every 3 moles of hydrogen introduced, we also introduce 1 mole of an inert. Therefore, you must take the inerts into account in your calculation.

12-11. Calculate the equilibrium composition for the simultaneous reactions

$$2CO(g) + 2H_2(g) \rightarrow CH_4(g) + CO_2(g) \qquad (1)$$

$$CO(g) + H_2(g) \rightarrow C(s) + H_2O(g) \qquad (2)$$

at 1000 K and 500 atm when stoichiometric amounts of CO and H_2 are reacted. At these conditions

$$\Delta G_R^o = 5884 \text{ cal/gm mole for reaction (1)}$$

$$\Delta G_R^o = 1906 \text{ cal/gm mole for reaction (2)}$$

$$\nu_{CH_4} = 1.2$$

$$\nu_{CO_2} = 1.1$$

$$\nu_{CO} = 1.2$$

$$\nu_{H_2} = 1.1$$

12-12. In the oxidation of FeS(s) by O_2, the following additional solid phases might be anticipated: $FeSO_4$, $Fe_2(SO_4)_3$, Fe_2O_3, FeO, Fe_3O_4. While only experiment can decide what associations of phases actually occur, estimate the maximum number of phases which could exist in equilibrium at a given temperature and pressure.

12-13. Using the method of Andersen, Beyers, and Watson, calculate

$$\Delta h_{F_{298}}^o, s_{F_{298}}^o, \text{ and } \Delta g_{F_{298}}^o \text{ for}$$

(a) 2-2 dimethyl propane:

$$\begin{array}{c} \text{H} \\ \text{HCH} \\ \text{H} \quad | \quad \text{H} \\ \text{HC} - \text{C} - \text{CH} \\ \text{H} \quad | \quad \text{H} \\ \text{HCH} \\ \text{H} \end{array}$$

(b) isopropyl alcohol:

$$\begin{array}{c} \text{H} \\ \text{O} \\ \text{H} \quad | \quad \text{H} \\ \text{HC} - \text{C} - \text{CH} \\ \text{H} \quad \text{H} \quad \text{H} \end{array}$$

(c) 1-3 butadiene:

$$\begin{array}{c} \text{H} \quad \text{H} \quad \text{H} \quad \text{H} \\ \text{HC} = \text{C} - \text{C} = \text{CH} \end{array}$$

12-14. A plant produces methanol from carbon monoxide and hydrogen. In the process, stoichiometric proportions of CO and H_2 are compressed isothermally and reversibly at 25 °C from 1 atm to 310 atm. The compressor discharges the gas continuously to a well insulated catalytic reactor where the following reaction takes place:

$$CO(g) + 2H_2(g) \rightarrow CH_3OH(g)$$

The methanol leaving the reactor is in equilibrium with the CO and H_2 in the effluent stream. There is negligible pressure drop across the reactor.

Assuming the gases act as ideal gases, calculate the work done by the compressor in calories per gram mole of carbon monoxide fed, the percent conversion of carbon monoxide to methanol, and the temperature of the reactor effluent stream.

12-15. Calculate the potential of the cell shown in Fig. 12-3 at 40 °C if the anode is aluminum immersed in 0.5 molal aluminum nitrate, and the cathode is copper immersed in 0.5 molal cupric chloride. The required potentials and activity coefficients can be found in the literature.[42]

Appendix

TABLE A-1. Conversion Factors

Some useful constants and conversion factors are:

Length	1.0 meter	= 100 cm
		= 10^{+10} Angstroms
	1.0 in.	= 2.54 cm
	1.0 ft	= 12 inches = 30.48 cm
Mass	1.0 kilogram	= 1000 grams
	1.0 pound mass	= 0.45359237 kilograms
	1.0 slug	= 32.1740 lb_m
Force	1.0 newton	= 10^5 dynes
		= $10^5 \dfrac{(gm)(cm)}{sec^2}$
	1.0 pound force	= 32.1740 poundals
		= 4.44822 newtons
Pressure	1.0 standard atmosphere	= 14.69595 lb_f/in^2
		= 101,325 newtons/meter^2
		= 29.92126 in. of Hg at 0 °C
		= 760 mm of Hg at 0 °C
		= 760 torr
		= 1.013250 bars
Temperature	T(°F) + 459.67	= T(°R)
	T(°C) + 273.15	= T(K)
	T(°R)	= 1.8 T(K)
Volume	1.0 liter	= 1000 cm^3
	1.0 ft^3	= 7.480520 gallons
Density	1.0 gram/cm^3	= 62.42795 lb_m/ft^3
		= 8.345403 lb_m/gal
Energy	1.0 joule	= 1.0 newton-meter
		= 1.0 volt-coulomb
		= 1.0 watt-second
		= 10^7 ergs or 10^7 dyne-cm
	1.0 calorie	= 4.1868 joules
	1.0 BTU	= 251.9958 calories
		= 778.1693 ft-lb_f
		= 1055.056 joules
	1.0 ft-lb_f	= 1.355818 joules
Entropy	1.0 entropy unit (e.u.)	= $1.0 \dfrac{\text{calorie}}{(\text{gm mole})(K)}$
		= $1.0 \dfrac{\text{BTU}}{(\text{lb mole})(°R)}$
		= $4.1868 \dfrac{\text{joules}}{(\text{gm mole})(K)}$

TABLE A-1. (*continued*)

Power	1.0 horsepower	= 550 ft-lb$_f$/sec
		= 42.406 BTU/min
		= 2544.33 BTU/hr
		= 745.6999 watts
	1.0 kilowatt	= 737.56 ft-lb$_f$/sec
		= 56.869 BTU/min
		= 3412.142 BTU/hr

Gas Constant, R: R

$$= 1.98584 \frac{\text{calories}}{(\text{gm mole})(K)}$$

$$= 1.98584 \frac{\text{BTU}}{(\text{lb-mole})(°R)}$$

$$= 10.7314 \frac{(\text{lb}_f/\text{in}^2)(\text{ft}^3)}{(\text{lb-mole})(°R)}$$

$$= 0.730227 \frac{(\text{atm})(\text{ft}^3)}{(\text{lb-mole})(°R)}$$

$$= 82.0560 \frac{(\text{atm})(\text{cm}^3)}{(\text{gm mole})(K)}$$

$$= 8.31433 \frac{\text{joules}}{(\text{gm mole})(K)}$$

Conversion Factor between Force and Mass Units, g_c: g_c

$$= 1.0 \frac{(\text{kg})(\text{meter})}{(\text{newton})(\text{sec}^2)}$$

$$= 32.174 \frac{(\text{lb}_m)(\text{ft})}{(\text{lb}_f)(\text{sec}^2)}$$

Other Constants

Boltzmann constant, k_B = $1.38054 \times 10^{-16} \frac{\text{erg}}{\text{K-molecule}}$

Faraday's constant, F = 96,487 coulombs/gram equivalent

Avogadro Number, N_A = 6.02252×10^{23} molecules/gm mole

Note: Calories in this table are international steam table calories, where 1.0 international steam table calorie = 1.000669 thermal calories.

For an extensive listing of conversion factors based on the SI system of units, see "Metric Practice Guide" ASTM E 380-70 published by the American Society for Testing and Materials, 1916 Race Street, Philadelphia, PA, 19103.

TABLE A-2. Saturated Steam: Temperature Table†

Temp Fahr t	Abs Press. Lb per Sq In. p	Specific Volume			Enthalpy			Entropy			Temp Fahr t
		Sat. Liquid v_f	Evap v_{fg}	Sat. Vapor v_g	Sat. Liquid h_f	Evap h_{fg}	Sat. Vapor h_g	Sat. Liquid s_f	Evap s_{fg}	Sat. Vapor s_g	
32.0*	0.08859	0.016022	3304.7	3304.7	−0.0179	1075.5	1075.5	0.0000	2.1873	2.1873	32.0*
34.0	0.09600	0.016021	3061.9	3061.9	1.996	1074.4	1076.4	0.0041	2.1762	2.1802	34.0
36.0	0.10395	0.016020	2839.0	2839.0	4.008	1073.2	1077.2	0.0081	2.1651	2.1732	36.0
38.0	0.11249	0.016019	2634.1	2634.2	6.018	1072.1	1078.1	0.0122	2.1541	2.1663	38.0
40.0	0.12163	0.016019	2445.8	2445.8	8.027	1071.0	1079.0	0.0162	2.1432	2.1594	40.0
42.0	0.13143	0.016019	2272.4	2272.4	10.035	1069.8	1079.9	0.0202	2.1325	2.1527	42.0
44.0	0.14192	0.016019	2112.8	2112.8	12.041	1068.7	1080.7	0.0242	2.1217	2.1459	44.0
46.0	0.15314	0.016020	1965.7	1965.7	14.047	1067.6	1081.6	0.0282	2.1111	2.1393	46.0
48.0	0.16514	0.016021	1830.0	1830.0	16.051	1066.4	1082.5	0.0321	2.1006	2.1327	48.0
50.0	0.17796	0.016023	1704.8	1704.8	18.054	1065.3	1083.4	0.0361	2.0901	2.1262	50.0
52.0	0.19165	0.016024	1589.2	1589.2	20.057	1064.2	1084.2	0.0400	2.0798	2.1197	52.0
54.0	0.20625	0.016026	1482.4	1482.4	22.058	1063.1	1085.1	0.0439	2.0695	2.1134	54.0
56.0	0.22183	0.016028	1383.6	1383.6	24.059	1061.9	1086.0	0.0478	2.0593	2.1070	56.0
58.0	0.23843	0.016031	1292.2	1292.2	26.060	1060.8	1086.9	0.0516	2.0491	2.1008	58.0
60.0	0.25611	0.016033	1207.6	1207.6	28.060	1059.7	1087.7	0.0555	2.0391	2.0946	60.0
62.0	0.27494	0.016036	1129.2	1129.2	30.059	1058.5	1088.6	0.0593	2.0291	2.0885	62.0
64.0	0.29497	0.016039	1056.5	1056.5	32.058	1057.4	1089.5	0.0632	2.0192	2.0824	64.0
66.0	0.31626	0.016043	989.0	989.1	34.056	1056.3	1090.4	0.0670	2.0094	2.0764	66.0
68.0	0.33889	0.016046	926.5	926.5	36.054	1055.2	1091.2	0.0708	1.9996	2.0704	68.0
70.0	0.36292	0.016050	868.3	868.4	38.052	1054.0	1092.1	0.0745	1.9900	2.0645	70.0
72.0	0.38844	0.016054	814.3	814.3	40.049	1052.9	1093.0	0.0783	1.9804	2.0587	72.0
74.0	0.41550	0.016058	764.1	764.1	42.046	1051.8	1093.8	0.0821	1.9708	2.0529	74.0
76.0	0.44420	0.016063	717.4	717.4	44.043	1050.7	1094.7	0.0858	1.9614	2.0472	76.0
78.0	0.47461	0.016067	673.8	673.9	46.040	1049.5	1095.6	0.0895	1.9520	2.0415	78.0
80.0	0.50683	0.016072	633.3	633.3	48.037	1048.4	1096.4	0.0932	1.9426	2.0359	80.0
82.0	0.54093	0.016077	595.5	595.5	50.033	1047.3	1097.3	0.0969	1.9334	2.0303	82.0
84.0	0.57702	0.016082	560.3	560.3	52.029	1046.1	1098.2	0.1006	1.9242	2.0248	84.0
86.0	0.61518	0.016087	227.5	527.5	54.026	1045.0	1099.0	0.1043	1.9151	2.0193	86.0
88.0	0.65551	0.016093	496.8	496.8	56.022	1043.9	1099.9	0.1079	1.9060	2.0139	88.0
90.0	0.69813	0.016099	468.1	468.1	58.018	1042.7	1100.8	0.1115	1.8970	2.0086	90.0
92.0	0.74313	0.016105	441.3	441.3	60.014	1041.6	1101.6	0.1152	1.8881	2.0033	92.0
94.0	0.79062	0.016111	416.3	416.3	62.010	1040.5	1102.5	0.1188	1.8792	1.9980	94.0
96.0	0.84072	0.016117	392.8	392.9	64.006	1039.3	1103.3	0.1224	1.8704	1.9928	96.0
98.0	0.89356	0.016123	370.9	370.9	66.003	1038.2	1104.2	0.1260	1.8617	1.9876	98.0
100.0	0.94924	0.016130	350.4	350.4	67.999	1037.1	1105.1	0.1295	1.8530	1.9825	100.0
102.0	1.00789	0.016137	331.1	331.1	69.995	1035.9	1105.9	0.1331	1.8444	1.9775	102.0
104.0	1.06965	0.016144	313.1	313.1	71.992	1034.8	1106.8	0.1366	1.8358	1.9725	104.0
106.0	1.1347	0.016151	296.16	296.18	73.99	1033.6	1107.6	0.1402	1.8273	1.9675	106.0
108.0	1.2030	0.016158	280.28	280.30	75.98	1032.5	1108.5	0.1437	1.8188	1.9626	108.0
110.0	1.2750	0.016165	265.37	265.39	77.98	1031.4	1109.3	0.1472	1.8105	1.9577	110.0
112.0	1.3505	0.016173	251.37	251.38	79.98	1030.2	1110.2	0.1507	1.8021	1.9528	112.0
114.0	1.4299	0.016180	238.21	238.22	81.97	1029.1	1111.0	0.1542	1.7938	1.9480	114.0
116.0	1.5133	0.016188	225.84	225.85	83.97	1027.9	1111.9	0.1577	1.7856	1.9433	116.0
118.0	1.6009	0.016196	214.20	214.21	85.97	1026.8	1112.7	0.1611	1.7774	1.9386	118.0
120.0	1.6927	0.016204	203.25	203.26	87.97	1025.6	1113.6	0.1646	1.7693	1.9339	120.0
122.0	1.7891	0.016213	192.94	192.95	89.96	1024.5	1114.4	0.1680	1.7613	1.9293	122.0
124.0	1.8901	0.016221	183.23	183.24	91.96	1023.3	1115.3	0.1715	1.7533	1.9247	124.0
126.0	1.9959	0.016229	174.08	174.09	93.96	1022.2	1116.1	0.1749	1.7453	1.9202	126.0
128.0	2.1068	0.016238	165.45	165.47	95.96	1021.0	1117.0	0.1783	1.7374	1.9157	128.0
130.0	2.2230	0.016247	157.32	157.33	97.96	1019.8	1117.8	0.1817	1.7295	1.9112	130.0
132.0	2.3445	0.016256	149.64	149.66	99.95	1018.7	1118.6	0.1851	1.7217	1.9068	132.0
134.0	2.4717	0.016265	142.40	142.41	101.95	1017.5	1119.5	0.1884	1.7140	1.9024	134.0
136.0	2.6047	0.016274	135.55	135.57	103.95	1016.4	1120.3	0.1918	1.7063	1.8980	136.0
138.0	2.7438	0.016284	129.09	129.11	105.95	1015.2	1121.1	0.1951	1.6986	1.8937	138.0
140.0	2.8892	0.016293	122.98	123.00	107.95	1014.0	1122.0	0.1985	1.6910	1.8895	140.0
142.0	3.0411	0.016303	117.21	117.22	109.95	1012.9	1122.8	0.2018	1.6834	1.8852	142.0
144.0	3.1997	0.016312	111.74	111.76	111.95	1011.7	1123.6	0.2051	1.6759	1.8810	144.0
146.0	3.3653	0.016322	106.58	106.59	113.95	1010.5	1124.5	0.2084	1.6684	1.8769	146.0
148.0	3.5381	0.016332	101.68	101.70	115.95	1009.3	1125.3	0.2117	1.6610	1.8727	148.0
150.0	3.7184	0.016343	97.05	97.07	117.95	1008.2	1126.1	0.2150	1.6536	1.8686	150.0
152.0	3.9065	0.016353	92.66	92.68	119.95	1007.0	1126.9	0.2183	1.6463	1.8646	152.0
154.0	4.1025	0.016363	88.50	88.52	121.95	1005.8	1127.7	0.2216	1.6390	1.8606	154.0
156.0	4.3068	0.016374	84.56	84.57	123.95	1004.6	1128.6	0.2248	1.6318	1.8566	156.0
158.0	4.5197	0.016384	80.82	80.83	125.96	1003.4	1129.4	0.2281	1.6245	1.8526	158.0
160.0	4.7414	0.016395	77.27	77.29	127.96	1002.2	1130.2	0.2313	1.6174	1.8487	160.0
162.0	4.9722	0.016406	73.90	73.92	129.96	1001.0	1131.0	0.2345	1.6103	1.8448	162.0
164.0	5.2124	0.016417	70.70	70.72	131.96	999.8	1131.8	0.2377	1.6032	1.8409	164.0
166.0	5.4623	0.016428	67.67	67.68	133.97	998.8	1132.6	0.2409	1.5961	1.8371	166.0
168.0	5.7223	0.016440	64.78	64.80	135.97	997.4	1133.4	0.2441	1.5892	1.8333	168.0
170.0	5.9926	0.016451	62.04	62.06	137.97	996.2	1134.2	0.2473	1.5822	1.8295	170.0
172.0	6.2736	0.016463	59.43	59.45	139.98	995.0	1135.0	0.2505	1.5753	1.8258	172.0
174.0	6.5656	0.016474	56.95	56.97	141.98	993.8	1135.8	0.2537	1.5684	1.8221	174.0
176.0	6.8690	0.016486	54.59	54.61	143.99	992.6	1136.6	0.2568	1.5616	1.8184	176.0
178.0	7.1840	0.016498	52.35	52.36	145.99	991.4	1137.4	0.2600	1.5548	1.8147	178.0

*The states shown are mestable

TABLE A-2. Saturated Steam: Temperature Table (*continued*)

Temp Fahr t	Abs Press. Lb per Sq In. p	Specific Volume			Enthalpy			Entropy			Temp Fahr t
		Sat. Liquid v_f	Evap v_{fg}	Sat. Vapor v_g	Sat. Liquid h_f	Evap h_{fg}	Sat. Vapor h_g	Sat. Liquid s_f	Evap s_{fg}	Sat. Vapor s_g	
180.0	7.5110	0.016510	50.21	50.22	148.00	990.2	1138.2	0.2631	1.5480	1.8111	180.0
182.0	7.850	0.016522	48.172	48.189	150.01	989.0	1139.0	0.2662	1.5413	1.8075	182.0
184.0	8.203	0.016534	46.232	46.249	152.01	987.8	1139.8	0.2694	1.5346	1.8040	184.0
186.0	8.568	0.016547	44.383	44.400	154.02	986.5	1140.5	0.2725	1.5279	1.8004	186.0
188.0	8.947	0.016559	42.621	42.638	156.03	985.3	1141.3	0.2756	1.5213	1.7969	188.0
190.0	9.340	0.016572	40.941	40.957	158.04	984.1	1142.1	0.2787	1.5148	1.7934	190.0
192.0	9.747	0.016585	39.337	39.354	160.05	982.8	1142.9	0.2818	1.5082	1.7900	192.0
194.0	10.168	0.016598	37.808	37.824	162.05	981.6	1143.7	0.2848	1.5017	1.7865	194.0
196.0	10.605	0.016611	36.348	36.364	164.06	980.4	1144.4	0.2879	1.4952	1.7831	196.0
198.0	11.058	0.016624	34.954	34.970	166.08	979.1	1145.2	0.2910	1.4888	1.7798	198.0
200.0	11.526	0.016637	33.622	33.639	168.09	977.9	1146.0	0.2940	1.4824	1.7764	200.0
204.0	12.512	0.016664	31.135	31.151	172.11	975.4	1147.5	0.3001	1.4697	1.7698	204.0
208.0	13.568	0.016691	28.862	28.878	176.14	972.8	1149.0	0.3061	1.4571	1.7632	208.0
212.0	14.696	0.016719	26.782	26.799	180.17	970.3	1150.5	0.3121	1.4447	1.7568	212.0
216.0	15.901	0.016747	24.878	24.894	184.20	967.8	1152.0	0.3181	1.4323	1.7505	216.0
220.0	17.186	0.016775	23.131	23.148	188.23	965.2	1153.4	0.3241	1.4201	1.7442	220.0
224.0	18.556	0.016805	21.529	21.545	192.27	962.6	1154.9	0.3300	1.4081	1.7380	224.0
228.0	20.015	0.016834	20.056	20.073	196.31	960.0	1156.3	0.3359	1.3961	1.7320	228.0
232.0	21.567	0.016864	18.701	18.718	200.35	957.4	1157.8	0.3417	1.3842	1.7260	232.0
236.0	23.216	0.016895	17.454	17.471	204.40	954.8	1159.2	0.3476	1.3725	1.7201	236.0
240.0	24.968	0.016926	16.304	16.321	208.45	952.1	1160.6	0.3533	1.3609	1.7142	240.0
244.0	26.826	0.016958	15.243	15.260	212.50	949.5	1162.0	0.3591	1.3494	1.7085	244.0
248.0	28.796	0.016990	14.264	14.281	216.56	946.8	1163.4	0.3649	1.3379	1.7028	248.0
252.0	30.883	0.017022	13.358	13.375	220.62	944.1	1164.7	0.3706	1.3266	1.6972	252.0
256.0	33.091	0.017055	12.520	12.538	224.69	941.4	1166.1	0.3763	1.3154	1.6917	256.0
260.0	35.427	0.017089	11.745	11.762	228.76	938.6	1167.4	0.3819	1.3043	1.6862	260.0
264.0	37.894	0.017123	11.025	11.042	232.83	935.9	1168.7	0.3876	1.2933	1.6808	264.0
268.0	40.500	0.017157	10.358	10.375	236.91	933.1	1170.0	0.3932	1.2823	1.6755	268.0
272.0	43.249	0.017193	9.738	9.755	240.99	930.3	1171.3	0.3987	1.2715	1.6702	272.0
276.0	46.147	0.017228	9.162	9.180	245.08	927.5	1172.5	0.4043	1.2607	1.6650	276.0
280.0	49.200	0.017264	8.627	8.644	249.17	924.6	1173.8	0.4098	1.2501	1.6599	280.0
284.0	52.414	0.01730	8.1280	8.1453	253.3	921.7	1175.0	0.4154	1.2395	1.6548	284.0
288.0	55.795	0.01734	7.6634	7.6807	257.4	918.8	1176.2	0.4208	1.2290	1.6498	288.0
292.0	59.350	0.01738	7.2301	7.2475	261.5	915.9	1177.4	0.4263	1.2186	1.6449	292.0
296.0	63.084	0.01741	6.8259	6.8433	265.6	913.0	1178.6	0.4317	1.2082	1.6400	296.0
300.0	67.005	0.01745	6.4483	6.4658	269.7	910.0	1179.7	0.4372	1.1979	1.6351	300.0
304.0	71.119	0.01749	6.0955	6.1130	273.8	907.0	1180.9	0.4426	1.1877	1.6303	304.0
308.0	75.433	0.01753	5.7655	5.7830	278.0	904.0	1182.0	0.4479	1.1776	1.6256	308.0
312.0	79.953	0.01757	5.4566	5.4742	282.1	901.0	1183.1	0.4533	1.1676	1.6209	312.0
316.0	84.688	0.01761	5.1673	5.1849	286.3	897.9	1184.1	0.4586	1.1576	1.6162	316.0
320.0	89.643	0.01766	4.8961	4.9138	290.4	894.8	1185.2	0.4640	1.1477	1.6116	320.0
324.0	94.826	0.01770	4.6418	4.6595	294.6	891.6	1186.2	0.4692	1.1378	1.6071	324.0
328.0	100.245	0.01774	4.4030	4.4208	298.7	888.5	1187.2	0.4745	1.1280	1.6025	328.0
332.0	105.907	0.01779	4.1788	4.1966	302.9	885.3	1188.2	0.4798	1.1183	1.5981	332.0
336.0	111.820	0.01783	3.9681	3.9859	307.1	882.1	1189.1	0.4850	1.1086	1.5936	336.0
340.0	117.992	0.01787	3.7699	3.7878	311.3	878.8	1190.1	0.4902	1.0990	1.5892	340.0
344.0	124.430	0.01792	3.5834	3.6013	315.5	875.5	1191.0	0.4954	1.0894	1.5849	344.0
348.0	131.142	0.01797	3.4078	3.4258	319.7	872.2	1191.1	0.5006	1.0799	1.5806	348.0
352.0	138.138	0.01801	3.2423	3.2603	323.9	868.9	1192.7	0.5058	1.0705	1.5763	352.0
356.0	145.424	0.01806	3.0863	3.1044	328.1	865.5	1193.6	0.5110	1.0611	1.5721	356.0
360.0	153.010	0.01811	2.9392	2.9573	332.3	862.1	1194.4	0.5161	1.0517	1.5678	360.0
364.0	160.903	0.01816	2.8002	2.8184	336.5	858.6	1195.2	0.5212	1.0424	1.5637	364.0
368.0	169.113	0.01821	2.6691	2.6873	340.8	855.1	1195.9	0.5263	1.0332	1.5595	368.0
372.0	177.648	0.01826	2.5451	2.5633	345.0	851.6	1196.7	0.5314	1.0240	1.5554	372.0
376.0	186.517	0.01831	2.4279	2.4462	349.3	848.1	1197.4	0.5365	1.0148	1.5513	376.0
380.0	195.729	0.01836	2.3170	2.3353	353.6	844.5	1198.0	0.5416	1.0057	1.5473	380.0
384.0	205.294	0.01842	2.2120	2.2304	357.9	840.8	1198.7	0.5466	0.9966	1.5432	384.0
388.0	215.220	0.01847	2.1126	2.1311	362.2	837.2	1199.3	0.5516	0.9876	1.5392	388.0
392.0	225.516	0.01853	2.0184	2.0369	366.5	833.4	1199.9	0.5567	0.9786	1.5352	392.0
396.0	236.193	0.01858	1.9291	1.9477	370.8	829.7	1200.4	0.5617	0.9696	1.5313	396.0
400.0	247.259	0.01864	1.8444	1.8630	375.1	825.9	1201.0	0.5667	0.9607	1.5274	400.0
404.0	258.725	0.01870	1.7640	1.7827	379.4	822.0	1201.5	0.5717	0.9518	1.5234	404.0
408.0	270.600	0.01875	1.6877	1.7064	383.8	818.2	1201.9	0.5766	0.9429	1.5195	408.0
412.0	282.894	0.01881	1.6152	1.6340	388.1	814.2	1202.4	0.5816	0.9341	1.5157	412.0
416.0	295.617	0.01887	1.5463	1.5651	392.5	810.2	1202.8	0.5866	0.9253	1.5118	416.0
420.0	308.780	0.01894	1.4808	1.4997	396.9	806.2	1203.1	0.5915	0.9165	1.5080	420.0
424.0	322.391	0.01900	1.4184	1.4374	401.3	802.2	1203.5	0.5964	0.9077	1.5042	424.0
428.0	336.463	0.01906	1.3591	1.3782	405.7	798.0	1203.7	0.6014	0.8990	1.5004	428.0
432.0	351.00	0.01913	1.30266	1.32179	410.1	793.9	1204.0	0.6063	0.8903	1.4966	432.0
436.0	366.03	0.01919	1.24887	1.26806	414.6	789.7	1204.2	0.6112	0.8816	1.4928	436.0
440.0	381.54	0.01926	1.19761	1.21687	419.0	785.4	1204.4	0.6161	0.8729	1.4890	440.0
444.0	397.56	0.01933	1.14874	1.16806	423.5	781.1	1204.6	0.6210	0.8643	1.4853	444.0
448.0	414.09	0.01940	1.10212	1.12152	428.0	776.7	1204.7	0.6259	0.8557	1.4815	448.0
452.0	431.14	0.01947	1.05764	1.07711	432.5	772.3	1204.8	0.6308	0.8471	1.4778	452.0
456.0	448.73	0.01954	1.01518	1.03472	437.0	767.8	1204.8	0.6356	0.8385	1.4741	456.0

TABLE A-2. Saturated Steam: Temperature Table (*continued*)

Temp Fahr t	Abs Press. Lb per Sq In. p	Specific Volume			Enthalpy			Entropy			Temp Fahr t
		Sat. Liquid v_f	Evap v_{fg}	Sat. Vapor v_g	Sat. Liquid h_f	Evap h_{fg}	Sat. Vapor h_g	Sat. Liquid s_f	Evap s_{fg}	Sat. Vapor s_g	
460.0	466.87	0.01961	0.97463	0.99424	441.5	763.2	1204.8	0.6405	0.8299	1.4704	460.0
464.0	485.56	0.01969	0.93588	0.95557	446.1	758.6	1204.7	0.6454	0.8213	1.4667	464.0
468.0	504.83	0.01976	0.89885	0.91862	450.7	754.0	1204.6	0.6502	0.8127	1.4629	468.0
472.0	524.67	0.01984	0.86345	0.88329	455.2	749.3	1204.5	0.6551	0.8042	1.4592	472.0
476.0	545.11	0.01992	0.82958	0.84950	459.9	744.5	1204.3	0.6599	0.7956	1.4555	476.0
480.0	566.15	0.02000	0.79716	0.81717	464.5	739.6	1204.1	0.6648	0.7871	1.4518	480.0
484.0	587.81	0.02009	0.76613	0.78622	469.1	734.7	1203.8	0.6696	0.7785	1.4481	484.0
488.0	610.10	0.02017	0.73641	0.75658	473.8	729.7	1203.5	0.6745	0.7700	1.4444	488.0
492.0	633.03	0.02026	0.70794	0.72820	478.5	724.6	1203.1	0.6793	0.7614	1.4407	492.0
496.0	656.61	0.02034	0.68065	0.70100	483.2	719.5	1202.7	0.6842	0.7528	1.4370	496.0
500.0	680.86	0.02043	0.65448	0.67492	487.9	714.3	1202.2	0.6890	0.7443	1.4333	500.0
504.0	705.78	0.02053	0.62938	0.64991	492.7	709.0	1201.7	0.6939	0.7357	1.4296	504.0
508.0	731.40	0.02062	0.60530	0.62592	497.5	703.7	1201.1	0.6987	0.7271	1.4258	508.0
512.0	757.72	0.02072	0.58218	0.60289	502.3	698.2	1200.5	0.7036	0.7185	1.4221	512.0
516.0	784.76	0.02081	0.55997	0.58079	507.1	692.7	1199.8	0.7085	0.7099	1.4183	516.0
520.0	812.53	0.02091	0.53864	0.55956	512.0	687.0	1199.0	0.7133	0.7013	1.4146	520.0
524.0	841.04	0.02102	0.51814	0.53916	516.9	681.3	1198.2	0.7182	0.6926	1.4108	524.0
528.0	870.31	0.02112	0.49843	0.51955	521.8	675.5	1197.3	0.7231	0.6839	1.4070	528.0
532.0	900.34	0.02123	0.47947	0.50070	526.8	669.6	1196.4	0.7280	0.6752	1.4032	532.0
536.0	931.17	0.02134	0.46123	0.48257	531.7	663.6	1195.4	0.7329	0.6665	1.3993	536.0
540.0	962.79	0.02146	0.44367	0.46513	536.8	657.5	1194.3	0.7378	0.6577	1.3954	540.0
544.0	995.22	0.02157	0.42677	0.44834	541.8	651.3	1193.1	0.7427	0.6489	1.3915	544.0
548.0	1028.49	0.02169	0.41048	0.43217	546.9	645.0	1191.9	0.7476	0.6400	1.3876	548.0
552.0	1062.59	0.02182	0.39479	0.41660	552.0	638.5	1190.6	0.7525	0.6311	1.3837	552.0
556.0	1097.55	0.02194	0.37966	0.40160	557.2	632.0	1189.2	0.7575	0.6222	1.3797	556.0
560.0	1133.38	0.02207	0.36507	0.38714	562.4	625.3	1187.7	0.7625	0.6132	1.3757	560.0
564.0	1170.10	0.02221	0.35099	0.37320	567.6	618.5	1186.1	0.7674	0.6041	1.3716	564.0
568.0	1207.72	0.02235	0.33741	0.35975	572.9	611.5	1184.5	0.7725	0.5950	1.3675	568.0
572.0	1246.26	0.02249	0.32429	0.34678	578.3	604.5	1182.7	0.7775	0.5859	1.3634	572.0
576.0	1285.74	0.02264	0.31162	0.33426	583.7	597.2	1180.9	0.7825	0.5766	1.3592	576.0
580.0	1326.17	0.02279	0.29937	0.32216	589.1	589.9	1179.0	0.7876	0.5673	1.3550	580.0
584.0	1367.7	0.02295	0.28753	0.31048	594.6	582.4	1176.9	0.7927	0.5580	1.3507	584.0
588.0	1410.0	0.02311	0.27608	0.29919	600.1	574.7	1174.8	0.7978	0.5485	1.3464	588.0
592.0	1453.3	0.02328	0.26499	0.28827	605.7	566.8	1172.6	0.8030	0.5390	1.3420	592.0
596.0	1497.8	0.02345	0.25425	0.27770	611.4	558.8	1170.2	0.8082	0.5293	1.3375	596.0
600.0	1543.2	0.02364	0.24384	0.26747	617.1	550.6	1167.7	0.8134	0.5196	1.3330	600.0
604.0	1589.7	0.02382	0.23374	0.25757	622.9	542.2	1165.1	0.8187	0.5097	1.3284	604.0
608.0	1637.3	0.02402	0.22394	0.24796	628.8	533.6	1162.4	0.8240	0.4997	1.3238	608.0
612.0	1686.1	0.02422	0.21442	0.23865	634.8	524.7	1159.5	0.8294	0.4896	1.3190	612.0
616.6	1735.9	0.02444	0.20516	0.22960	640.8	515.6	1156.4	0.8348	0.4794	1.3141	616.0
620.0	1786.9	0.02466	0.19615	0.22081	646.9	506.3	1153.2	0.8403	0.4689	1.3092	620.0
624.0	1839.0	0.02489	0.18737	0.21226	653.1	496.6	1149.8	0.8458	0.4583	1.3041	624.0
628.0	1892.4	0.02514	0.17880	0.20394	659.5	486.7	1146.1	0.8514	0.4474	1.2988	628.0
632.0	1947.0	0.02539	0.17044	0.19583	665.9	476.4	1142.2	0.8571	0.4364	1.2934	632.0
636.0	2002.8	0.02566	0.16226	0.18792	672.4	465.7	1138.1	0.8628	0.4251	1.2879	636.0
640.0	2059.9	0.02595	0.15427	0.18021	679.1	454.6	1133.7	0.8686	0.4134	1.2821	640.0
644.0	2118.3	0.02625	0.14644	0.17269	685.9	443.1	1129.0	0.8746	0.4015	1.2761	644.0
648.0	2178.1	0.02657	0.13876	0.16534	692.9	431.1	1124.0	0.8806	0.3893	1.2699	648.0
652.0	2239.2	0.02691	0.13124	0.15816	700.0	418.7	1118.7	0.8868	0.3767	1.2634	652.0
656.0	2301.7	0.02728	0.12387	0.15115	707.4	405.7	1113.1	0.8931	0.3637	1.2567	656.0
660.0	2365.7	0.02768	0.11663	0.14431	714.9	392.1	1107.0	0.8995	0.3502	1.2498	660.0
664.0	2431.1	0.02811	0.10947	0.13757	722.9	377.7	1100.6	0.9064	0.3361	1.2425	664.0
668.0	2498.1	0.02858	0.10229	0.13087	731.5	362.1	1093.5	0.9137	0.3210	1.2347	668.0
672.0	2566.6	0.02911	0.09514	0.12424	740.2	345.7	1085.9	0.9212	0.3054	1.2266	672.0
676.0	2636.8	0.02970	0.08799	0.11769	749.2	328.5	1077.6	0.9287	0.2892	1.2179	676.0
680.0	2708.6	0.03037	0.08080	0.11117	758.5	310.1	1068.5	0.9365	0.2720	1.2086	680.0
684.0	2782.1	0.03114	0.07349	0.10463	768.2	290.2	1058.4	0.9447	0.2537	1.1984	684.0
688.0	2857.4	0.03204	0.06595	0.09799	778.8	268.2	1047.0	0.9535	0.2337	1.1872	688.0
692.0	2934.5	0.03313	0.05797	0.09110	790.5	243.1	1033.6	0.9634	0.2110	1.1744	692.0
696.0	3013.4	0.03455	0.04916	0.08371	804.4	212.8	1017.2	0.9749	0.1841	1.1591	696.0
700.0	3094.3	0.03662	0.03857	0.07519	822.4	172.7	995.2	0.9901	0.1490	1.1390	700.0
702.0	3135.5	0.03824	0.03173	0.06997	835.0	144.7	979.7	1.0006	0.1246	1.1252	702.0
704.0	3177.2	0.04108	0.02192	0.06300	854.2	102.0	956.2	1.0169	0.0876	1.1046	704.0
705.0	3198.3	0.04427	0.01304	0.05730	873.0	61.4	934.4	1.0329	0.0527	1.0856	705.0
705.47*	3208.2	0.05078	0.00000	0.05078	906.0	0.0	906.0	1.0612	0.0000	1.0612	705.47*

*Critical temperature

†Steam Table Values reprinted from *1967 ASME Steam Tables* by Combustion Engineering, Inc. Reproduced with permission of Combustion Engineering, Inc. and ASME.

TABLE A-2. Saturated Steam: Pressure Table

Abs Press. Lb/Sq In. p	Temp Fahr t	Sat. Liquid v_f	Specific Volume Evap v_{fg}	Sat. Vapor v_g	Sat. Liquid h_f	Enthalpy Evap h_{fg}	Sat. Vapor h_g	Sat. Liquid s_f	Entropy Evap s_{fg}	Sat. Vapor s_g	Abs Press. Lb/Sq In. p
0.08865	32.018	0.016022	3302.4	3302.4	0.0003	1075.5	1075.5	0.0000	2.1872	2.1872	0.08865
0.25	59.323	0.016032	1235.5	1235.5	27.382	1060.1	1087.4	0.0542	2.0425	2.0967	0.25
0.50	79.586	0.016071	641.5	641.5	47.623	1048.6	1096.3	0.0925	1.9446	2.0370	0.50
1.0	101.74	0.016136	333.59	333.60	69.73	1036.1	1105.8	0.1326	1.8455	1.9781	1.0
5.0	162.24	0.016407	73.515	73.532	130.20	1000.9	1131.1	0.2349	1.6094	1.8443	5.0
10.0	193.21	0.016592	38.404	38.420	161.26	982.1	1143.3	0.2836	1.5043	1.7879	10.0
14.696	212.00	0.016719	26.782	26.799	180.17	970.3	1150.5	0.3121	1.4447	1.7568	14.696
15.0	213.03	0.016726	26.274	26.290	181.21	969.7	1150.9	0.3137	1.4415	1.7552	15.0
20.0	227.96	0.016834	20.070	20.087	196.27	960.1	1156.3	0.3358	1.3962	1.7320	20.0
30.0	250.34	0.017009	13.7266	13.7436	218.9	945.2	1164.1	0.3682	1.3313	1.6995	30.0
40.0	267.25	0.017151	10.4794	10.4965	236.1	933.6	1169.8	0.3921	1.2844	1.6765	40.0
50.0	281.02	0.017274	8.4967	8.5140	250.2	923.9	1174.1	0.4112	1.2474	1.6586	50.0
60.0	292.71	0.017383	7.1562	7.1736	262.2	915.4	1177.6	0.4273	1.2167	1.6440	60.0
70.0	302.93	0.017482	6.1875	6.2050	272.7	907.8	1180.6	0.4411	1.1905	1.6316	70.0
80.0	312.04	0.017573	5.4536	5.4711	282.1	900.9	1183.1	0.4534	1.1675	1.6208	80.0
90.0	320.28	0.017659	4.8779	4.8953	290.7	894.6	1185.3	0.4643	1.1470	1.6113	90.0
100.0	327.82	0.017740	4.4133	4.4310	298.5	888.6	1187.2	0.4743	1.1284	1.6027	100.0
110.0	334.79	0.01782	4.0306	4.0484	305.8	883.1	1188.9	0.4834	1.1115	1.5950	110.0
120.0	341.27	0.01789	3.7097	3.7275	312.6	877.8	1190.4	0.4919	1.0960	1.5879	120.0
130.0	347.33	0.01796	3.4364	3.4544	319.0	872.8	1191.7	0.4998	1.0815	1.5813	130.0
140.0	353.04	0.01803	3.2010	3.2190	325.0	868.0	1193.0	0.5071	1.0681	1.5752	140.0
150.0	358.43	0.01809	2.9958	3.0139	330.6	863.4	1194.1	0.5141	1.0554	1.5695	150.0
160.0	363.55	0.01815	2.8155	2.8336	336.1	859.0	1195.1	0.5206	1.0435	1.5641	160.0
170.0	368.42	0.01821	2.6556	2.6738	341.2	854.8	1196.0	0.5269	1.0322	1.5591	170.0
180.0	373.08	0.01827	2.5129	2.5312	346.2	850.7	1196.9	0.5328	1.0215	1.5543	180.0
190.0	377.53	0.01833	2.3847	2.4030	350.9	846.7	1197.6	0.5384	1.0113	1.5498	190.0
200.0	381.80	0.01839	2.2689	2.2873	355.5	842.8	1198.3	0.5438	1.0016	1.5454	200.0
210.0	385.91	0.01844	2.16373	2.18217	359.9	839.1	1199.0	0.5490	0.9923	1.5413	210.0
220.0	389.88	0.01850	2.06779	2.08629	364.2	835.4	1199.6	0.5540	0.9834	1.5374	220.0
230.0	393.70	0.01855	1.97991	1.99846	368.3	831.8	1200.1	0.5588	0.9748	1.5336	230.0
240.0	397.39	0.01860	1.89909	1.91769	372.3	828.4	1200.6	0.5634	0.9665	1.5299	240.0
250.0	400.97	0.01865	1.82452	1.84317	376.1	825.0	1201.1	0.5679	0.9585	1.5264	250.0
260.0	404.44	0.01870	1.75548	1.77418	379.9	821.6	1201.5	0.5722	0.9508	1.5230	260.0
270.0	407.80	0.01875	1.69137	1.71013	383.6	818.3	1201.9	0.5764	0.9433	1.5197	270.0
280.0	411.07	0.01880	1.63169	1.65049	387.1	815.1	1202.3	0.5805	0.9361	1.5166	280.0
290.0	414.25	0.01885	1.57597	1.59482	390.6	812.0	1202.6	0.5844	0.9291	1.5135	290.0
300.0	417.35	0.01889	1.52384	1.54274	394.0	808.9	1202.9	0.5882	0.9223	1.5105	300.0
350.0	431.73	0.01912	1.30642	1.32554	409.8	794.2	1204.0	0.6059	0.8909	1.4968	350.0
400.0	444.60	0.01934	1.14162	1.16095	424.2	780.4	1204.6	0.6217	0.8630	1.4847	400.0
450.0	456.28	0.01954	1.01224	1.03179	437.3	767.5	1204.8	0.6360	0.8378	1.4738	450.0
500.0	467.01	0.01975	0.90787	0.92762	449.5	755.1	1204.7	0.6490	0.8148	1.4639	500.0
550.0	476.94	0.01994	0.82183	0.84177	460.9	743.3	1204.3	0.6611	0.7936	1.4547	550.0
600.0	486.20	0.02013	0.74962	0.76975	471.7	732.0	1203.7	0.6723	0.7738	1.4461	600.0
650.0	494.89	0.02032	0.68811	0.70843	481.9	720.9	1202.8	0.6828	0.7552	1.4381	650.0
700.0	503.08	0.02050	0.63505	0.65556	491.6	710.2	1201.8	0.6928	0.7377	1.4304	700.0
750.0	510.84	0.02069	0.58880	0.60949	500.9	699.8	1200.7	0.7022	0.7210	1.4232	750.0
800.0	518.21	0.02087	0.54809	0.56896	509.8	689.6	1199.4	0.7111	0.7051	1.4163	800.0
850.0	525.24	0.02105	0.51197	0.53302	518.4	679.5	1198.0	0.7197	0.6899	1.4096	850.0
900.0	531.95	0.02123	0.47968	0.50091	526.7	669.7	1196.4	0.7279	0.6753	1.4032	900.0
950.0	538.39	0.02141	0.45064	0.47205	534.7	660.0	1194.7	0.7358	0.6612	1.3970	950.0
1000.0	544.58	0.02159	0.42436	0.44596	542.6	650.4	1192.9	0.7434	0.6476	1.3910	1000.0
1050.0	550.53	0.02177	0.40047	0.42224	550.1	640.9	1191.0	0.7507	0.6344	1.3851	1050.0
1100.0	556.28	0.02195	0.37863	0.40058	557.5	631.5	1189.1	0.7578	0.6216	1.3794	1100.0
1150.0	561.82	0.02214	0.35859	0.38073	564.8	622.2	1187.0	0.7647	0.6091	1.3738	1150.0
1200.0	567.19	0.02232	0.34013	0.36245	571.9	613.0	1184.8	0.7714	0.5969	1.3683	1200.0
1250.0	572.38	0.02250	0.32306	0.34556	578.8	603.8	1182.6	0.7780	0.5850	1.3630	1250.0
1300.0	577.42	0.02269	0.30722	0.32991	585.6	594.6	1180.2	0.7843	0.5733	1.3577	1300.0
1350.0	582.32	0.02288	0.29250	0.31537	592.3	585.4	1177.8	0.7906	0.5620	1.3525	1350.0
1400.0	587.07	0.02307	0.27871	0.30178	598.8	576.5	1175.3	0.7966	0.5507	1.3474	1400.0
1450.0	591.70	0.02327	0.26584	0.28911	605.3	567.4	1172.8	0.8026	0.5397	1.3423	1450.0
1500.0	596.20	0.02346	0.25372	0.27719	611.7	558.4	1170.1	0.8085	0.5288	1.3373	1500.0
1550.0	600.59	0.02366	0.24235	0.26601	618.0	549.4	1167.4	0.8142	0.5182	1.3324	1550.0
1600.0	604.87	0.02387	0.23159	0.25545	624.2	540.3	1164.5	0.8199	0.5076	1.3274	1600.0
1650.0	609.05	0.02407	0.22143	0.24551	630.4	531.3	1161.6	0.8254	0.4971	1.3225	1650.0
1700.0	613.13	0.02428	0.21178	0.23607	636.5	522.2	1158.6	0.8309	0.4867	1.3176	1700.0
1750.0	617.12	0.02450	0.20263	0.22713	642.5	513.1	1155.6	0.8363	0.4765	1.3128	1750.0
1800.0	621.02	0.02472	0.19390	0.21861	648.5	503.8	1152.3	0.8417	0.4662	1.3079	1800.0
1850.0	624.83	0.02495	0.18558	0.21052	654.5	494.6	1149.0	0.8470	0.4561	1.3030	1850.0
1900.0	628.56	0.02517	0.17761	0.20278	660.4	485.2	1145.6	0.8522	0.4459	1.2981	1900.0
1950.0	632.22	0.02541	0.16999	0.19540	666.3	475.8	1142.0	0.8574	0.4358	1.2931	1950.0
2000.0	635.80	0.02565	0.16266	0.18831	672.1	466.2	1138.3	0.8625	0.4256	1.2881	2000.0
2100.0	642.76	0.02615	0.14885	0.17501	683.8	446.7	1130.5	0.8727	0.4053	1.2780	2100.0
2200.0	649.45	0.02669	0.13603	0.16272	695.5	426.7	1122.2	0.8828	0.3848	1.2676	2200.0
2300.0	655.89	0.02727	0.12406	0.15133	707.2	406.0	1113.2	0.8929	0.3640	1.2569	2300.0
2400.0	662.11	0.02790	0.11287	0.14076	719.0	384.8	1103.7	0.9031	0.3430	1.2460	2400.0
2500.0	668.11	0.02859	0.10209	0.13068	731.7	361.6	1093.3	0.9139	0.3206	1.2345	2500.0
2600.0	673.91	0.02938	0.09172	0.12110	744.5	337.6	1082.0	0.9247	0.2977	1.2225	2600.0
2700.0	679.53	0.03029	0.08165	0.11194	757.3	312.3	1069.7	0.9356	0.2741	1.2097	2700.0
2800.0	684.96	0.03134	0.07171	0.10305	770.7	285.1	1055.8	0.9468	0.2491	1.1958	2800.0
2900.0	690.22	0.03262	0.06158	0.09420	785.1	254.7	1039.8	0.9588	0.2215	1.1803	2900.0
3000.0	695.33	0.03428	0.05073	0.08500	801.8	218.4	1020.3	0.9728	0.1891	1.1619	3000.0
3100.0	700.28	0.03681	0.03771	0.07452	824.0	169.3	993.3	0.9914	0.1460	1.1373	3100.0
3200.0	705.08	0.04472	0.01191	0.05663	875.5	56.1	931.6	1.0351	0.0482	1.0832	3200.0
3208.2*	705.47	0.05078	0.00000	0.05078	906.0	0.0	906.0	1.0612	0.0000	1.0612	3208.2*

*Critical pressure

TABLE A-2. Superheated Steam

Abs Press. Lb/Sq In. (Sat. Temp)		Sat. Water	Sat. Steam	Temperature — Degrees Fahrenheit													
				200	250	300	350	400	450	500	600	700	800	900	1000	1100	1200
1 (101.74)	Sh v h s	0.01614 69.73 0.1326	333.6 1105.8 1.9781	98.26 392.5 1150.2 2.0509	148.26 422.4 1172.9 2.0841	198.26 452.3 1195.7 2.1152	248.26 482.1 1218.7 2.1445	298.26 511.9 1241.8 2.1722	348.26 541.7 1265.1 2.1985	398.26 571.5 1288.6 2.2237	498.26 631.1 1336.1 2.2708	598.26 690.7 1384.5 2.3144	698.26 750.3 1433.7 2.3551	798.26 809.8 1483.8 2.3934	898.26 869.4 1534.9 2.4296	998.26 929.0 1586.8 2.4640	1098.26 988.6 1639.7 2.4969
5 (162.24)	Sh v h s	0.01641 130.20 0.2349	73.53 1131.1 1.8443	37.76 78.14 1148.6 1.8716	87.76 84.21 1171.7 1.9054	137.76 90.24 1194.8 1.9369	187.76 96.25 1218.0 1.9664	237.76 102.24 1241.3 1.9943	287.76 108.23 1264.7 2.0208	337.76 114.21 1288.2 2.0460	437.76 126.15 1335.9 2.0932	537.76 138.08 1384.3 2.1369	637.76 150.01 1433.6 2.1776	737.76 161.94 1483.7 2.2159	837.76 173.86 1534.5 2.2521	937.76 185.78 1586.7 2.2866	1037.76 197.70 1639.6 2.3194
10 (193.21)	Sh v h s	0.01659 161.26 0.2836	38.42 1143.3 1.7879	6.79 38.84 1146.6 1.7928	56.79 41.93 1170.2 1.8273	106.79 44.98 1193.7 1.8593	156.79 48.02 1217.1 1.8892	206.79 51.03 1240.6 1.9173	256.79 54.04 1264.1 1.9439	306.79 57.04 1287.8 1.9692	406.79 63.03 1335.5 2.0166	506.79 69.00 1384.0 2.0603	606.79 74.98 1433.4 2.1011	706.79 80.94 1483.5 2.1394	806.79 86.91 1534.5 2.1757	906.79 92.87 1586.6 2.2101	1006.79 98.84 1639.5 2.2430
14.696 (212.00)	Sh v h s	0.0167 180.17 .3121	26.799 1150.5 1.7568		38.00 28.42 1168.8 1.7833	88.00 30.52 1192.6 1.8158	138.00 32.60 1216.3 1.8459	188.00 34.67 1239.9 1.8743	238.00 36.72 1263.6 1.9010	288.00 38.77 1287.4 1.9265	388.00 42.86 1335.2 1.9739	488.00 46.93 1383.8 2.0177	588.00 51.00 1433.2 2.0585	688.00 55.06 1483.4 2.0969	788.00 59.13 1534.5 2.1332	888.00 63.19 1586.6 2.1676	988.00 67.25 1639.4 2.2005
15 (213.03)	Sh v h s	0.01673 181.21 0.3137	26.290 1150.9 1.7552		36.97 27.837 1168.7 1.7809	86.97 29.899 1192.5 1.8134	136.97 31.939 1216.2 1.8437	186.97 33.963 1239.9 1.8720	236.97 35.977 1263.6 1.8988	286.97 37.985 1287.4 1.9242	386.97 41.986 1335.2 1.9717	486.97 45.978 1383.8 2.0155	586.97 49.964 1433.2 2.0563	686.97 53.946 1483.4 2.0946	786.97 57.926 1534.5 2.1309	886.97 61.905 1586.5 2.1653	986.97 65.882 1639.4 2.1982
20 (227.96)	Sh v h s	0.01683 196.27 0.3358	20.087 1156.3 1.7320		22.04 20.788 1167.1 1.7475	72.04 22.356 1191.4 1.7805	122.04 23.900 1215.4 1.8111	172.04 25.428 1239.2 1.8397	222.04 26.946 1263.0 1.8666	272.04 28.457 1286.9 1.8921	372.04 31.466 1334.9 1.9397	472.04 34.465 1383.5 1.9836	572.04 37.458 1432.9 2.0244	672.04 40.447 1483.2 2.0628	772.04 43.435 1534.3 2.0991	872.04 46.420 1586.3 2.1336	972.04 49.405 1639.3 2.1665
25 (240.07)	Sh v h s	0.01693 208.52 0.3535	16.301 1160.6 1.7141		9.93 16.558 1165.6 1.7212	59.93 17.829 1190.2 1.7547	109.93 19.076 1214.5 1.7856	159.93 20.307 1238.5 1.8145	209.93 21.527 1262.5 1.8415	259.93 22.740 1286.4 1.8672	359.93 25.153 1334.6 1.9149	459.93 27.557 1383.3 1.9588	559.93 29.954 1432.7 1.9997	659.93 32.348 1483.0 2.0381	759.93 34.740 1534.2 2.0744	859.93 37.130 1586.2 2.1089	959.93 39.518 1639.2 2.1418
30 (250.34)	Sh v h s	0.01701 218.93 0.3682	13.744 1164.1 1.6995			49.66 14.810 1189.0 1.7334	99.66 15.859 1213.6 1.7647	149.66 16.892 1237.8 1.7937	199.66 17.914 1261.9 1.8210	249.66 18.929 1286.0 1.8467	349.66 20.945 1334.2 1.8946	449.66 22.951 1383.0 1.9386	549.66 24.952 1432.5 1.9795	649.66 26.949 1482.8 2.0179	749.66 28.943 1534.0 2.0543	849.66 30.936 1586.1 2.0888	949.66 32.927 1639.0 2.1217
35 (259.29)	Sh v h s	0.01708 228.03 0.3809	11.896 1167.1 1.6872			40.71 12.654 1187.8 1.7152	90.71 13.562 1212.7 1.7468	140.71 14.453 1237.1 1.7761	190.71 15.334 1261.3 1.8035	240.71 16.207 1285.5 1.8294	340.71 17.939 1333.9 1.8774	440.71 19.662 1382.8 1.9214	540.71 21.379 1432.3 1.9624	640.71 23.092 1482.7 2.0009	740.71 24.803 1533.9 2.0372	840.71 26.512 1586.0 2.0717	940.71 28.220 1638.9 2.1046
40 (267.25)	Sh v h s	0.01715 236.14 0.3921	10.497 1169.8 1.6765			32.75 11.036 1186.6 1.6992	82.75 11.838 1211.7 1.7312	132.75 12.624 1236.4 1.7608	182.75 13.398 1260.8 1.7883	232.75 14.165 1285.0 1.8143	332.75 15.685 1333.6 1.8624	432.75 17.195 1382.5 1.9065	532.75 18.699 1432.1 1.9476	632.75 20.199 1482.5 1.9860	732.75 21.697 1533.7 2.0224	832.75 23.194 1585.8 2.0569	932.75 24.689 1638.8 2.0899
45 (274.44)	Sh v h s	0.01721 243.49 0.4021	9.399 1172.1 1.6671			25.56 9.777 1185.4 1.6849	75.56 10.497 1210.8 1.7173	125.56 11.201 1235.7 1.7471	175.56 11.892 1260.2 1.7748	225.56 12.577 1284.6 1.8010	325.56 13.932 1333.3 1.8492	425.56 15.276 1382.3 1.8934	525.56 16.614 1431.9 1.9345	625.56 17.950 1482.3 1.9730	725.56 19.282 1533.6 2.0093	825.56 20.613 1585.7 2.0439	925.56 21.943 1638.7 2.0768
50 (281.02)	Sh v h s	0.1727 250.21 0.4112	8.514 1174.1 1.6586			18.98 8.769 1184.1 1.6720	68.98 9.424 1209.9 1.7048	118.98 10.062 1234.9 1.7349	168.98 10.688 1259.6 1.7628	218.98 11.306 1284.1 1.7890	318.98 12.529 1332.9 1.8374	418.98 13.741 1382.0 1.8816	518.98 14.947 1431.7 1.9227	618.98 16.150 1482.2 1.9613	718.98 17.350 1533.4 1.9977	818.98 18.549 1585.6 2.0322	918.98 19.746 1638.6 2.0652
55 (287.07)	Sh v h s	0.01733 256.43 0.4196				12.93 7.945 1182.9 1.6601	62.93 8.546 1208.9 1.6933	112.93 9.130 1234.2 1.7237	162.93 9.702 1259.1 1.7518	212.93 10.267 1283.6 1.7781	312.93 11.381 1332.6 1.8266	412.93 12.485 1381.8 1.8710	512.93 13.581 1431.5 1.9121	612.93 14.677 1482.0 1.9507	712.93 15.769 1533.3 1.987	812.93 16.859 1585.5 2.022	912.93 17.948 1638.5 2.055
60 (292.71)	Sh v h s	0.1738 262.21 0.4273	7.174 1177.6 1.6440			7.29 7.257 1181.6 1.6492	57.29 7.815 1208.0 1.6934	107.29 8.354 1233.5 1.7134	157.29 8.881 1258.5 1.7417	207.29 9.400 1283.2 1.7681	307.29 10.425 1332.3 1.8168	407.29 11.438 1381.5 1.8612	507.29 12.446 1431.3 1.9024	607.29 13.450 1481.8 1.9410	707.29 14.452 1533.2 1.9774	807.29 15.452 1585.3 2.0120	907.29 16.450 1638.4 2.0450
65 (297.98)	Sh v h s	0.01743 267.63 0.4344	6.653 1179.1 1.6375			2.02 6.675 1180.3 1.6390	52.02 7.195 1207.0 1.6731	102.02 7.697 1232.7 1.7040	152.02 8.186 1257.9 1.7324	202.02 8.667 1282.7 1.7590	302.02 9.615 1331.9 1.8077	402.02 10.552 1381.3 1.8522	502.02 11.484 1431.1 1.8935	602.02 12.412 1481.6 1.9321	702.02 13.337 1533.0 1.9685	802.02 14.261 1585.2 2.0031	902.02 15.183 1638.3 2.0361
70 (302.93)	Sh v h s	0.01748 272.74 0.4411	6.205 1180.6 1.6316				47.07 6.664 1206.0 1.6640	97.07 7.133 1232.0 1.6951	147.07 7.590 1257.3 1.7237	197.07 8.039 1282.2 1.7504	297.07 8.922 1331.6 1.7993	397.07 9.793 1381.0 1.8439	497.07 10.659 1430.9 1.8852	597.07 11.522 1481.5 1.9238	697.07 12.382 1532.9 1.9603	797.07 13.240 1585.1 1.9949	897.07 14.097 1638.2 2.0279
75 (307.61)	Sh v h s	0.01753 277.56 0.4474	5.814 1181.9 1.6260				42.39 6.204 1205.0 1.6554	92.39 6.645 1231.2 1.6868	142.39 7.074 1256.7 1.7156	192.39 7.494 1281.7 1.7424	292.39 8.320 1331.3 1.7915	392.39 9.135 1380.7 1.8361	492.39 9.945 1430.7 1.8774	592.39 10.750 1481.3 1.9161	692.39 11.553 1532.7 1.9526	792.39 12.355 1585.0 1.9872	892.39 13.155 1638.1 2.0202

Sh = superheat, F
v = specific volume, cu ft per lb
h = enthalpy, Btu per lb
s = entropy, Btu per F per lb

TABLE A-2. Superheated Steam (*continued*)

Abs Press. Lb/Sq In. (Sat. Temp)		Sat. Water	Sat. Steam	Temperature — Degrees Fahrenheit												
				350	400	450	500	550	600	700	800	900	1000	1100	1200	1300	1400
80 (312.04)	Sh v h s	0.01757 282.15 0.4534	5.471 1183.1 1.6208	37.96 5.801 1204.0 1.6473	87.96 6.218 1230.5 1.6790	137.96 6.622 1256.1 1.7080	187.96 7.018 1281.3 1.7349	237.96 7.408 1306.2 1.7602	287.96 7.794 1330.9 1.7842	387.96 8.560 1380.5 1.8289	487.96 9.319 1430.5 1.8702	587.96 10.075 1481.1 1.9089	687.96 10.829 1532.6 1.9454	787.96 11.581 1584.9 1.9800	887.96 12.331 1638.0 2.0131	987.96 13.081 1692.0 2.0446	1087.96 13.829 1746.8 2.0750
85 (316.26)	Sh v h s	0.01762 286.52 0.4590	5.167 1184.2 1.6159	33.74 5.445 1203.0 1.6396	83.74 5.840 1229.7 1.6716	133.74 6.223 1255.5 1.7008	183.74 6.597 1280.8 1.7279	233.74 6.966 1305.8 1.7532	283.74 7.330 1330.6 1.7772	383.74 8.052 1380.2 1.8220	483.74 8.768 1430.3 1.8634	583.74 9.480 1481.0 1.9021	683.74 10.190 1532.4 1.9386	783.74 10.898 1584.7 1.9733	883.74 11.604 1637.9 2.0063	983.74 12.310 1691.9 2.0379	1083.74 13.014 1746.8 2.0682
90 (320.28)	Sh v h s	0.01766 290.69 0.4643	4.895 1185.3 1.6113	29.72 5.128 1202.0 1.6323	79.72 5.505 1228.9 1.6646	129.72 5.869 1254.9 1.6940	179.72 6.223 1280.3 1.7212	229.72 6.572 1305.4 1.7467	279.72 6.917 1330.2 1.7707	379.72 7.600 1380.0 1.8156	479.72 8.277 1430.1 1.8570	579.72 8.950 1480.8 1.8957	679.72 9.621 1532.3 1.9323	779.72 10.290 1584.6 1.9669	879.72 10.958 1637.8 2.0000	979.72 11.625 1691.8 2.0316	1079.72 12.290 1746.7 2.0619
95 (324.13)	Sh v h s	0.01770 294.70 0.4694	4.651 1186.2 1.6069	25.87 4.845 1200.9 1.6253	75.87 5.205 1228.1 1.6580	125.87 5.551 1254.3 1.6876	175.87 5.889 1279.8 1.7149	225.87 6.221 1305.0 1.7404	275.87 6.548 1329.9 1.7645	375.87 7.196 1379.7 1.8094	475.87 7.838 1429.9 1.8509	575.87 8.477 1480.6 1.8897	675.87 9.113 1532.1 1.9262	775.87 9.747 1584.5 1.9609	875.87 10.380 1637.7 1.9940	975.87 11.012 1691.7 2.0256	1075.87 11.643 1746.6 2.0559
100 (327.82)	Sh v h s	0.01774 298.54 0.4743	4.431 1187.2 1.6027	22.18 4.590 1199.9 1.6187	72.18 4.935 1227.4 1.6516	122.18 5.266 1253.7 1.6814	172.18 5.588 1279.3 1.7088	222.18 5.904 1304.6 1.7344	272.18 6.216 1329.6 1.7586	372.18 6.833 1379.5 1.8036	472.18 7.443 1429.7 1.8451	572.18 8.050 1480.4 1.8839	672.18 8.655 1532.0 1.9205	772.18 9.258 1584.4 1.9552	872.18 9.860 1637.6 1.9883	972.18 10.460 1691.6 2.0199	1072.18 11.060 1746.5 2.0502
105 (331.37)	Sh v h s	0.01778 302.24 0.4790	4.231 1188.0 1.5988	18.63 4.359 1198.8 1.6122	68.63 4.690 1226.6 1.6455	118.63 5.007 1253.1 1.6755	168.63 5.315 1278.8 1.7031	218.63 5.617 1304.2 1.7288	268.63 5.915 1329.2 1.7530	368.63 6.504 1379.2 1.7981	468.63 7.086 1429.4 1.8396	568.63 7.665 1480.3 1.8785	668.63 8.241 1531.8 1.9151	768.63 8.816 1584.2 1.9498	868.63 9.389 1637.5 1.9828	968.63 9.961 1691.5 2.0145	1068.63 10.532 1746.4 2.0448
110 (334.79)	Sh v h s	0.01782 305.80 0.4834	4.048 1188.9 1.5950	15.21 4.149 1197.7 1.6061	65.21 4.468 1225.8 1.6396	115.21 4.772 1252.5 1.6698	165.21 5.068 1278.3 1.6975	215.21 5.357 1303.8 1.7233	265.21 5.642 1328.9 1.7476	365.21 6.205 1379.0 1.7928	465.21 6.761 1429.2 1.8344	565.21 7.314 1480.1 1.8732	665.21 7.865 1531.7 1.9099	765.21 8.413 1584.1 1.9446	865.21 8.961 1637.4 1.9777	965.21 9.507 1691.4 2.0093	1065.21 10.053 1746.4 2.0397
115 (338.08)	Sh v h s	0.01785 309.25 0.4877	3.881 1189.6 1.5913	11.92 3.957 1196.7 1.6001	61.92 4.265 1225.0 1.6340	111.92 4.558 1251.8 1.6644	161.92 4.841 1277.9 1.6922	211.92 5.119 1303.3 1.7181	261.92 5.392 1328.5 1.7425	361.92 5.932 1378.7 1.7877	461.92 6.465 1429.0 1.8294	561.92 6.994 1479.9 1.8682	661.92 7.521 1531.6 1.9049	761.92 8.046 1584.0 1.9396	861.92 8.570 1637.2 1.9727	961.92 9.093 1691.4 2.0044	1061.92 9.615 1746.3 2.0347
120 (341.27)	Sh v h s	0.01789 312.58 0.4919	3.7275 1190.4 1.5879	8.73 3.7815 1195.6 1.5943	58.73 4.0786 1224.1 1.6286	108.73 4.3610 1251.2 1.6592	158.73 4.6341 1277.4 1.6872	208.73 4.9009 1302.9 1.7132	258.73 5.1637 1328.2 1.7376	358.73 5.6813 1378.4 1.7829	458.73 6.1928 1428.8 1.8246	558.73 6.7006 1479.8 1.8635	658.73 7.2060 1531.4 1.9001	758.73 7.7096 1583.9 1.9349	858.73 8.2119 1637.1 1.9680	958.73 8.7130 1691.3 1.9996	1058.73 9.2134 1746.2 2.0300
130 (347.33)	Sh v h s	0.01796 318.95 0.4998	3.4544 1191.7 1.5813	2.67 3.4699 1193.4 1.5833	52.67 3.7489 1222.5 1.6182	102.67 4.0129 1249.9 1.6493	152.67 4.2672 1276.4 1.6775	202.67 4.5151 1302.1 1.7037	252.67 4.7589 1327.5 1.7283	352.67 5.2384 1377.9 1.7737	452.67 5.7118 1428.4 1.8155	552.67 6.1814 1479.4 1.8545	652.67 6.6486 1531.1 1.8911	752.67 7.1140 1583.6 1.9259	852.67 7.5781 1636.9 1.9591	952.67 8.0411 1691.1 1.9907	1052.67 8.5033 1746.1 2.0211
140 (353.04)	Sh v h s	0.01803 324.96 0.5071	3.2190 1193.0 1.5752		46.96 3.4661 1220.8 1.6085	96.96 3.7143 1248.7 1.6400	146.96 3.9526 1275.3 1.6686	196.96 4.1844 1301.3 1.6949	246.96 4.4119 1326.8 1.7196	346.96 4.8588 1377.4 1.7652	446.96 5.2995 1428.0 1.8071	546.96 5.7364 1479.1 1.8461	646.96 6.1709 1530.8 1.8828	746.96 6.6036 1583.4 1.9176	846.96 7.0349 1636.7 1.9508	946.96 7.4652 1690.9 1.9825	1046.96 7.8946 1745.9 2.0129
150 (358.43)	Sh v h s	0.01809 330.65 0.5141	3.0139 1194.1 1.5695		41.57 3.2208 1219.1 1.5993	91.57 3.4555 1247.4 1.6313	141.57 3.6799 1274.3 1.6602	191.57 3.8974 1300.5 1.6867	241.57 4.1112 1326.2 1.7115	341.57 4.5298 1376.9 1.7573	441.57 4.9421 1427.6 1.7992	541.57 5.3507 1478.7 1.8383	641.57 5.7568 1530.5 1.8751	741.57 6.1612 1583.1 1.9099	841.57 6.5642 1636.5 1.9431	941.57 6.9661 1690.7 1.9748	1041.57 7.3671 1745.7 2.0052
160 (363.55)	Sh v h s	0.01815 336.07 0.5206	2.8336 1195.1 1.5641		36.45 3.0060 1217.4 1.5906	86.45 3.2288 1246.0 1.6231	136.45 3.4413 1273.3 1.6522	186.45 3.6469 1299.6 1.6790	236.45 3.8480 1325.4 1.7039	336.45 4.2220 1376.4 1.7499	436.45 4.6295 1427.2 1.7919	536.45 5.0132 1478.4 1.8310	636.45 5.3945 1530.3 1.8678	736.45 5.7741 1582.9 1.9027	836.45 6.1522 1636.3 1.9359	936.45 6.5293 1690.5 1.9676	1036.45 6.9055 1745.6 1.9980
170 (368.42)	Sh v h s	0.01821 341.24 0.5269	2.6738 1196.0 1.5591		31.58 2.8162 1215.6 1.5823	81.58 3.0288 1244.7 1.6152	131.58 3.2306 1272.2 1.6447	181.58 3.4255 1298.8 1.6717	231.58 3.6158 1324.7 1.6968	331.58 3.9879 1375.8 1.7428	431.58 4.3536 1426.8 1.7850	531.58 4.7155 1478.0 1.8241	631.58 5.0749 1530.0 1.8610	731.58 5.4325 1582.6 1.8959	831.58 5.7888 1636.1 1.9291	931.58 6.1440 1690.4 1.9608	1031.58 6.4983 1745.4 1.9913
180 (373.08)	Sh v h s	0.01827 346.19 0.5328	2.5312 1196.9 1.5543		26.92 2.6474 1213.8 1.5743	76.92 2.8508 1243.4 1.6078	126.92 3.0433 1271.2 1.6376	176.92 3.2286 1297.9 1.6647	226.92 3.4093 1324.0 1.6900	326.92 3.7621 1375.3 1.7362	426.92 4.1084 1426.3 1.7784	526.92 4.4508 1477.7 1.8176	626.92 4.7907 1529.7 1.8545	726.92 5.1289 1582.4 1.8894	826.92 5.4657 1635.9 1.9227	926.92 5.8014 1690.2 1.9545	1026.92 6.1363 1745.3 1.9849
190 (377.53)	Sh v h s	0.01833 350.94 0.5384	2.4030 1197.6 1.5498		22.47 2.4961 1212.0 1.5667	72.47 2.6915 1242.0 1.6006	122.47 2.8756 1270.1 1.6308	172.47 3.0525 1297.1 1.6581	222.47 3.2246 1323.2 1.6835	322.47 3.5601 1374.8 1.7299	422.47 3.8889 1425.9 1.7722	522.47 4.2140 1477.4 1.8115	622.47 4.5365 1529.4 1.8484	722.47 4.8572 1582.1 1.8834	822.47 5.1766 1635.7 1.9166	922.47 5.4949 1690.0 1.9484	1022.47 5.8124 1745.1 1.9789
200 (381.80)	Sh v h s	0.01839 355.51 0.5438	2.2873 1198.3 1.5454		18.20 2.3598 1210.1 1.5593	68.20 2.5480 1240.6 1.5938	118.20 2.7247 1269.0 1.6242	168.20 2.8939 1296.2 1.6518	218.20 3.0583 1322.6 1.6773	318.20 3.3783 1374.3 1.7239	418.20 3.6915 1425.5 1.7663	518.20 4.0008 1477.0 1.8057	618.20 4.3077 1529.1 1.8426	718.20 4.6128 1581.9 1.8776	818.20 4.9165 1635.4 1.9109	918.20 5.2191 1689.8 1.9427	1018.20 5.5209 1745.0 1.9732

Sh = superheat, F
v = specific volume, cu ft per lb
h = enthalpy, Btu per lb
s = entropy, Btu per F per lb

TABLE A-2. Superheated Steam (*continued*)

Abs Press. Lb/Sq In. (Sat. Temp)		Sat. Water	Sat. Steam	\multicolumn{13}{c}{Temperature — Degrees Fahrenheit}
				400	450	500	550	600	700	800	900	1000	1100	1200	1300	1400	1500
210 (385.91)	Sh v h s	0.01844 359.91 0.5490	2.1822 1199.0 1.5413	14.09 2.2364 1208.02 1.5522	64.09 2.4181 1239.2 1.5872	114.09 2.5880 1268.0 1.6180	164.09 2.7504 1295.3 1.6458	214.09 2.9078 1321.9 1.6715	314.09 3.2137 1373.7 1.7182	414.09 3.5128 1425.1 1.7607	514.09 3.8080 1476.7 1.8001	614.09 4.1007 1528.8 1.8371	714.09 4.3915 1581.6 1.8721	814.09 4.6811 1635.2 1.9054	914.09 4.9695 1689.6 1.9372	1014.09 5.2571 1744.8 1.9677	1114.09 5.5440 1800.8 1.9970
220 (389.88)	Sh v h s	0.01850 364.17 0.5540	2.0863 1199.6 1.5374	10.12 2.1240 1206.3 1.5453	60.12 2.2999 1237.8 1.5808	110.12 2.4638 1266.9 1.6120	160.12 2.6199 1294.5 1.6400	210.12 2.7710 1321.2 1.6658	310.12 3.0642 1373.2 1.7128	410.12 3.3504 1424.7 1.7553	510.12 3.6327 1476.3 1.7948	610.12 3.9125 1528.5 1.8318	710.12 4.1905 1581.4 1.8668	810.12 4.4671 1635.0 1.9002	910.12 4.7426 1689.4 1.9320	1010.12 5.0173 1744.7 1.9625	1110.12 5.2913 1800.6 1.9919
230 (393.70)	Sh v h s	0.01855 368.28 0.5588	1.9985 1200.1 1.5336	6.30 2.0212 1204.4 1.5385	56.30 2.1919 1236.3 1.5747	106.30 2.3503 1265.7 1.6062	156.30 2.5008 1293.6 1.6344	206.30 2.6461 1320.4 1.6604	306.30 2.9276 1372.7 1.7075	406.30 3.2020 1424.2 1.7502	506.30 3.4726 1476.0 1.7897	606.30 3.7406 1528.2 1.8268	706.30 4.0068 1581.1 1.8618	806.30 4.2717 1634.8 1.8952	906.30 4.5355 1689.3 1.9270	1006.30 4.7984 1744.5 1.9576	1106.30 5.0606 1800.5 1.9869
240 (397.39)	Sh v h s	0.01860 372.27 0.5634	1.9177 1200.6 1.5299	2.61 1.9268 1202.4 1.5320	52.61 2.0928 1234.9 1.5687	102.61 2.2462 1264.6 1.6006	152.61 2.3915 1292.7 1.6291	202.61 2.5316 1319.7 1.6552	302.61 2.8024 1372.1 1.7025	402.61 3.0661 1423.8 1.7452	502.61 3.3259 1475.6 1.7848	602.61 3.5831 1527.9 1.8219	702.61 3.8385 1580.9 1.8570	802.61 4.0926 1634.6 1.8904	902.61 4.3456 1689.1 1.9223	1002.61 4.5977 1744.3 1.9528	1102.61 4.8492 1800.4 1.9822
250 (400.97)	Sh v h s	0.01865 376.14 0.5679	1.8432 1201.1 1.5264		49.03 2.0016 1233.4 1.5629	99.03 2.1504 1263.5 1.5951	149.03 2.2909 1291.8 1.6239	199.03 2.4262 1319.0 1.6502	299.03 2.6872 1371.6 1.6976	399.03 2.9410 1423.4 1.7405	499.03 3.1909 1475.3 1.7801	599.03 3.4382 1527.6 1.8173	699.03 3.6837 1580.6 1.8524	799.03 3.9278 1634.4 1.8858	899.03 4.1709 1688.9 1.9177	999.03 4.4131 1744.2 1.9482	1099.03 4.6546 1800.2 1.9776
260 (404.44)	Sh v h s	0.01870 379.90 0.5722	1.7742 1201.5 1.5230		45.56 1.9173 1231.9 1.5573	95.56 2.0619 1262.4 1.5899	145.56 2.1981 1290.9 1.6189	195.56 2.3289 1318.2 1.6453	295.56 2.5808 1371.1 1.6930	395.56 2.8256 1423.0 1.7359	495.56 3.0663 1474.9 1.7756	595.56 3.3044 1527.3 1.8128	695.56 3.5408 1580.4 1.8480	795.56 3.7758 1634.2 1.8814	895.56 4.0097 1688.7 1.9133	995.56 4.2427 1744.0 1.9439	1095.56 4.4750 1800.1 1.9732
270 (407.80)	Sh v h s	0.01875 383.56 0.5764	1.7101 1201.9 1.5197		42.20 1.8391 1230.4 1.5518	92.20 1.9799 1261.2 1.5848	142.20 2.1121 1290.0 1.6140	192.20 2.2388 1317.5 1.6406	292.20 2.4824 1370.5 1.6885	392.20 2.7186 1422.6 1.7315	492.20 2.9509 1474.6 1.7713	592.20 3.1806 1527.1 1.8085	692.20 3.4084 1580.1 1.8437	792.20 3.6349 1634.0 1.8771	892.20 3.8603 1688.5 1.9090	992.20 4.0849 1743.9 1.9396	1092.20 4.3087 1800.0 1.9690
280 (411.07)	Sh v h s	0.01880 387.12 0.5805	1.6505 1202.3 1.5166		38.93 1.7665 1228.8 1.5464	88.93 1.9037 1260.0 1.5798	138.93 2.0322 1289.1 1.6093	188.93 2.1551 1316.8 1.6361	288.93 2.3909 1370.0 1.6841	388.93 2.6194 1422.1 1.7273	488.93 2.8437 1474.2 1.7671	588.93 3.0655 1526.8 1.8043	688.93 3.2855 1579.9 1.8395	788.93 3.5042 1633.8 1.8730	888.93 3.7217 1688.4 1.9050	988.93 3.9384 1743.7 1.9356	1088.93 4.1543 1799.8 1.9649
290 (414.25)	Sh v h s	0.01885 390.60 0.5844	1.5948 1202.6 1.5135		35.75 1.6988 1227.3 1.5412	85.75 1.8327 1258.9 1.5750	135.75 1.9578 1288.1 1.6048	185.75 2.0772 1316.0 1.6317	285.75 2.3058 1369.5 1.6799	385.75 2.5269 1421.7 1.7232	485.75 2.7440 1473.9 1.7630	585.75 2.9585 1526.5 1.8003	685.75 3.1711 1579.6 1.8356	785.75 3.3824 1633.5 1.8690	885.75 3.5926 1688.2 1.9010	985.75 3.8019 1743.6 1.9316	1085.75 4.0106 1799.7 1.9610
300 (417.35)	Sh v h s	0.01889 393.99 0.5882	1.5427 1202.9 1.5105		32.65 1.6356 1225.7 1.5361	82.65 1.7665 1257.7 1.5703	132.65 1.8883 1287.2 1.6003	182.65 2.0044 1315.2 1.6274	282.65 2.2263 1368.9 1.6758	382.65 2.4407 1421.3 1.7192	482.65 2.6509 1473.6 1.7591	582.65 2.8585 1526.2 1.7964	682.65 3.0643 1579.4 1.8317	782.65 3.2688 1633.3 1.8652	882.65 3.4721 1688.0 1.8972	982.65 3.6746 1743.4 1.9278	1082.65 3.8764 1799.6 1.9572
310 (420.36)	Sh v h s	0.01894 397.30 0.5920	1.4939 1203.2 1.5076		29.64 1.5763 1224.1 1.5311	79.64 1.7044 1256.5 1.5657	129.64 1.8233 1286.2 1.5960	179.64 1.9363 1314.5 1.6233	279.64 2.1520 1368.4 1.6719	379.64 2.3600 1420.9 1.7153	479.64 2.5638 1473.2 1.7553	579.64 2.7650 1525.9 1.7927	679.64 2.9644 1579.2 1.8280	779.64 3.1625 1633.1 1.8615	879.64 3.3594 1687.8 1.8935	979.64 3.5555 1743.3 1.9241	1079.64 3.7509 1799.4 1.9536
320 (423.31)	Sh v h s	0.01899 400.53 0.5956	1.4480 1203.4 1.5048		26.69 1.5207 1222.5 1.5261	76.69 1.6462 1255.2 1.5612	126.69 1.7623 1285.3 1.5918	176.69 1.8725 1313.7 1.6192	276.69 2.0823 1367.8 1.6680	376.69 2.2843 1420.5 1.7116	476.69 2.4821 1472.9 1.7516	576.69 2.6774 1525.6 1.7890	676.69 2.8708 1578.9 1.8243	776.69 3.0628 1632.9 1.8579	876.69 3.2538 1687.6 1.8899	976.69 3.4438 1743.1 1.9206	1076.69 3.6332 1799.3 1.9500
330 (426.18)	Sh v h s	0.01903 403.70 0.5991	1.4048 1203.6 1.5021		23.82 1.4684 1220.9 1.5213	73.82 1.5915 1254.0 1.5568	123.82 1.7050 1284.4 1.5876	173.82 1.8125 1313.0 1.6153	273.82 2.0168 1367.3 1.6643	373.82 2.2132 1420.0 1.7079	473.82 2.4054 1472.5 1.7480	573.82 2.5950 1525.3 1.7855	673.82 2.7828 1578.7 1.8208	773.82 2.9692 1632.7 1.8544	873.82 3.1545 1687.5 1.8864	973.82 3.3389 1742.9 1.9171	1073.82 3.5227 1799.2 1.9466
340 (428.99)	Sh v h s	0.01908 406.80 0.6026	1.3640 1203.8 1.4994		21.01 1.4191 1219.2 1.5165	71.01 1.5399 1252.8 1.5525	121.01 1.6511 1283.4 1.5836	171.01 1.7561 1312.2 1.6114	271.01 1.9552 1366.7 1.6606	371.01 2.1463 1419.6 1.7044	471.01 2.3333 1472.2 1.7445	571.01 2.5175 1525.0 1.7820	671.01 2.7000 1578.4 1.8174	771.01 2.8811 1632.5 1.8510	871.01 3.0611 1687.3 1.8831	971.01 3.2402 1742.8 1.9138	1071.01 3.4186 1799.0 1.9432
350 (431.73)	Sh v h s	0.01912 409.83 0.6059	1.3255 1204.0 1.4968		18.27 1.3725 1217.5 1.5119	68.27 1.4913 1251.5 1.5483	118.27 1.6002 1282.4 1.5797	168.27 1.7028 1311.4 1.6077	268.27 1.8970 1366.2 1.6571	368.27 2.0832 1419.2 1.7009	468.27 2.2652 1471.8 1.7411	568.27 2.4445 1524.7 1.7787	668.27 2.6219 1578.2 1.8141	768.27 2.7980 1632.3 1.8477	868.27 2.9730 1687.1 1.8798	968.27 3.1471 1742.6 1.9105	1068.27 3.3205 1798.9 1.9400
360 (434.41)	Sh v h s	0.01917 412.81 0.6092	1.2891 1204.1 1.4943		15.59 1.3285 1215.8 1.5073	65.59 1.4454 1250.3 1.5441	115.59 1.5521 1281.5 1.5758	165.59 1.6525 1310.6 1.6040	265.59 1.8421 1365.6 1.6536	365.59 2.0237 1418.7 1.6976	465.59 2.2009 1471.5 1.7379	565.59 2.3755 1524.4 1.7754	665.59 2.5482 1577.9 1.8109	765.59 2.7196 1632.1 1.8445	865.59 2.8898 1686.9 1.8766	965.59 3.0592 1742.5 1.9073	1065.59 3.2279 1798.8 1.9368
380 (439.61)	Sh v h s	0.01925 418.59 0.6156	1.2218 1204.4 1.4894		10.39 1.2472 1212.4 1.4982	60.39 1.3606 1247.7 1.5360	110.39 1.4635 1279.5 1.5683	160.39 1.5598 1309.0 1.5969	260.39 1.7410 1364.5 1.6470	360.39 1.9139 1417.9 1.6911	460.39 2.0825 1470.8 1.7315	560.39 2.2484 1523.8 1.7692	660.39 2.4124 1577.4 1.8047	760.39 2.5750 1631.6 1.8384	860.39 2.7366 1686.5 1.8705	960.39 2.8973 1742.2 1.9012	1060.39 3.0572 1798.5 1.9307

Sh = superheat, F
v = specific volume, cu ft per lb
h = enthalpy, Btu per lb
s = entropy, Btu per F per lb

TABLE A-2. Superheated Steam (*continued*)

Abs Press. Lb/Sq In. (Sat. Temp)		Sat. Water	Sat. Steam	Temperature — Degrees Fahrenheit												
				450	500	550	600	650	700	800	900	1000	1100	1200	1300	1400	1500
400 (444.60)	Sh v h s	0.01934 424.17 0.6217	1.1610 1204.6 1.4847	5.40 1.1738 1208.8 1.4894	55.40 1.2841 1245.1 1.5282	105.40 1.3836 1277.5 1.5611	155.40 1.4763 1307.4 1.5901	205.40 1.5646 1335.9 1.6163	255.40 1.6499 1363.4 1.6406	355.40 1.8151 1417.0 1.6850	455.40 1.9759 1470.1 1.7255	555.40 2.1339 1523.1 1.7632	655.40 2.2901 1576.9 1.7988	755.40 2.4450 1631.2 1.8325	855.40 2.5987 1686.2 1.8647	955.40 2.7515 1741.9 1.8955	1055.40 2.9037 1798.2 1.9250
420 (449.40)	Sh v h s	0.01942 429.56 0.6276	1.1057 1204.7 1.4802	.60 1.1071 1205.2 1.4808	50.60 1.2148 1242.4 1.5206	100.60 1.3113 1275.4 1.5542	150.60 1.4007 1305.8 1.5835	200.60 1.4856 1334.5 1.6100	250.60 1.5676 1362.3 1.6345	350.60 1.7258 1416.2 1.6791	450.60 1.8795 1469.4 1.7197	550.60 2.0304 1522.7 1.7575	650.60 2.1795 1576.4 1.7932	750.60 2.3273 1630.8 1.8269	850.60 2.4739 1685.8 1.8591	950.60 2.6196 1741.6 1.8899	1050.60 2.7647 1798.0 1.9195
440 (454.03)	Sh v h s	0.01950 434.77 0.6332	1.0554 1204.8 1.4759		45.97 1.1517 1239.7 1.5132	95.97 1.2454 1273.4 1.5474	145.97 1.3319 1304.2 1.5772	195.97 1.4138 1333.2 1.6040	245.97 1.4926 1361.1 1.6286	345.97 1.6445 1415.3 1.6734	445.97 1.7918 1468.7 1.7142	545.97 1.9363 1522.1 1.7521	645.97 2.0790 1575.9 1.7878	745.97 2.2203 1630.4 1.8216	845.97 2.3605 1685.5 1.8538	945.97 2.4998 1741.2 1.8847	1045.97 2.6384 1797.7 1.9143
460 (458.50)	Sh v h s	0.01959 439.83 0.6387	1.0092 1204.8 1.4718		41.50 1.0939 1236.9 1.5060	91.50 1.1852 1271.3 1.5409	141.50 1.2691 1302.5 1.5711	191.50 1.3482 1331.8 1.5982	241.50 1.4242 1360.0 1.6230	341.50 1.5703 1414.4 1.6680	441.50 1.7117 1468.0 1.7089	541.50 1.8504 1521.5 1.7469	641.50 1.9872 1575.4 1.7826	741.50 2.1226 1629.9 1.8165	841.50 2.2569 1685.1 1.8488	941.50 2.3903 1740.9 1.8797	1041.50 2.5230 1797.4 1.9093
480 (462.82)	Sh v h s	0.01967 444.75 0.6439	0.9668 1204.8 1.4677		37.18 1.0409 1234.1 1.4990	87.18 1.1300 1269.1 1.5346	137.18 1.2115 1300.8 1.5652	187.18 1.2881 1330.5 1.5925	237.18 1.3615 1358.8 1.6176	337.18 1.5023 1413.6 1.6628	437.18 1.6384 1467.3 1.7038	537.18 1.7716 1520.9 1.7419	637.18 1.9030 1574.9 1.7777	737.18 2.0330 1629.5 1.8116	837.18 2.1619 1684.7 1.8439	937.18 2.2900 1740.6 1.8748	1037.18 2.4173 1797.2 1.9045
500 (467.01)	Sh v h s	0.01975 449.52 0.6490	0.9276 1204.7 1.4639		32.99 0.9919 1231.2 1.4921	82.99 1.0791 1267.0 1.5284	132.99 1.1584 1299.1 1.5595	182.99 1.2327 1329.1 1.5871	232.99 1.3037 1357.7 1.6123	332.99 1.4397 1412.7 1.6578	432.99 1.5708 1466.6 1.6990	532.99 1.6992 1520.3 1.7371	632.99 1.8256 1574.4 1.7730	732.99 1.9507 1629.1 1.8069	832.99 2.0746 1684.4 1.8393	932.99 2.1977 1740.3 1.8702	1032.99 2.3200 1796.9 1.8998
520 (471.07)	Sh v h s	0.01982 454.18 0.6540	0.8914 1204.5 1.4601		28.93 0.9466 1228.3 1.4853	78.93 1.0321 1264.8 1.5223	128.93 1.1094 1297.4 1.5539	178.93 1.1816 1327.7 1.5818	228.93 1.2504 1356.5 1.6072	328.93 1.3819 1411.8 1.6530	428.93 1.5085 1465.9 1.6943	528.93 1.6323 1519.7 1.7325	628.93 1.7542 1573.9 1.7684	728.93 1.8746 1628.7 1.8024	828.93 1.9940 1684.0 1.8348	928.93 2.1125 1740.0 1.8657	1028.93 2.2302 1796.7 1.8954
540 (475.01)	Sh v h s	0.01990 458.71 0.6587	0.8577 1204.4 1.4565		24.99 0.9045 1225.3 1.4786	74.99 0.9884 1262.5 1.5164	124.99 1.0640 1295.7 1.5485	174.99 1.1342 1326.3 1.5767	224.99 1.2010 1355.3 1.6023	324.99 1.3284 1410.9 1.6483	424.99 1.4508 1465.1 1.6897	524.99 1.5704 1519.1 1.7280	624.99 1.6880 1573.4 1.7640	724.99 1.8042 1628.2 1.7981	824.99 1.9193 1683.6 1.8305	924.99 2.0336 1739.7 1.8615	1024.99 2.1471 1796.4 1.8911
560 (478.84)	Sh v h s	0.01998 463.14 0.6634	0.8264 1204.2 1.4529		21.16 0.8653 1222.2 1.4720	71.16 0.9479 1260.3 1.5106	121.16 1.0217 1293.9 1.5431	171.16 1.0902 1324.9 1.5717	221.16 1.1552 1354.2 1.5975	321.16 1.2787 1410.0 1.6438	421.16 1.3972 1464.4 1.6853	521.16 1.5129 1518.6 1.7237	621.16 1.6266 1572.9 1.7598	721.16 1.7388 1627.8 1.7939	821.16 1.8500 1683.3 1.8263	921.16 1.9603 1739.4 1.8573	1021.16 2.0699 1796.1 1.8870
580 (482.57)	Sh v h s	0.02006 467.47 0.6679	0.7971 1203.9 1.4495		17.43 0.8287 1219.1 1.4654	67.43 0.9100 1258.0 1.5049	117.43 0.9824 1292.1 1.5380	167.43 1.0492 1323.4 1.5668	217.43 1.1125 1353.0 1.5929	317.43 1.2324 1409.2 1.6394	417.43 1.3473 1463.7 1.6811	517.43 1.4593 1518.0 1.7196	617.43 1.5693 1572.4 1.7556	717.43 1.6780 1627.4 1.7898	817.43 1.7855 1682.9 1.8223	917.43 1.8921 1739.1 1.8533	1017.43 1.9980 1795.9 1.8831
600 (486.20)	Sh v h s	0.02013 471.70 0.6723	0.7697 1203.7 1.4461		13.80 0.7944 1215.9 1.4590	63.80 0.8746 1255.6 1.4993	113.80 0.9456 1290.3 1.5329	163.80 1.0109 1322.0 1.5621	213.80 1.0726 1351.8 1.5884	313.80 1.1892 1408.3 1.6351	413.80 1.3008 1463.0 1.6769	513.80 1.4093 1517.4 1.7155	613.80 1.5160 1571.9 1.7517	713.80 1.6211 1627.0 1.7859	813.80 1.7252 1682.6 1.8184	913.80 1.8284 1738.8 1.8494	1013.80 1.9309 1795.6 1.8792
650 (494.89)	Sh v h s	0.02032 481.89 0.6828	0.7084 1202.8 1.4381		5.11 0.7173 1207.6 1.4430	55.11 0.7954 1249.6 1.4858	105.11 0.8634 1285.7 1.5207	155.11 0.9254 1318.3 1.5507	205.11 0.9835 1348.7 1.5775	305.11 1.0929 1406.0 1.6249	405.11 1.1969 1461.2 1.6671	505.11 1.2979 1515.9 1.7059	605.11 1.3969 1570.7 1.7422	705.11 1.4944 1625.9 1.7765	805.11 1.5909 1681.6 1.8092	905.11 1.6864 1738.0 1.8403	1005.11 1.7813 1794.9 1.8701
700 (503.08)	Sh v h s	0.02050 491.60 0.6928	0.6556 1201.8 1.4304			46.92 0.7271 1243.4 1.4726	96.92 0.7928 1281.0 1.5090	146.92 0.8520 1314.6 1.5399	196.92 0.9072 1345.6 1.5673	296.92 1.0102 1403.7 1.6154	396.92 1.1078 1459.4 1.6580	496.92 1.2023 1514.4 1.6970	596.92 1.2948 1569.4 1.7335	696.92 1.3858 1624.8 1.7679	796.92 1.4757 1680.7 1.8006	896.92 1.5647 1737.2 1.8318	996.92 1.6530 1794.3 1.8617
750 (510.84)	Sh v h s	0.02069 500.89 0.7022	0.6095 1200.7 1.4232			39.16 0.6676 1236.9 1.4598	89.16 0.7313 1276.1 1.4977	139.16 0.7882 1310.7 1.5296	189.16 0.8409 1342.5 1.5577	289.16 0.9386 1401.5 1.6065	389.16 1.0306 1457.6 1.6494	489.16 1.1195 1512.9 1.6886	589.16 1.2063 1568.2 1.7252	689.16 1.2916 1623.8 1.7598	789.16 1.3759 1679.8 1.7926	889.16 1.4592 1736.4 1.8239	989.16 1.5419 1793.6 1.8538
800 (518.21)	Sh v h s	0.02087 509.81 0.7111	0.5690 1199.4 1.4163			31.79 0.6151 1230.1 1.4472	81.79 0.6774 1271.1 1.4869	131.79 0.7323 1306.8 1.5198	181.79 0.7828 1339.3 1.5484	281.79 0.8759 1399.1 1.5980	381.79 0.9631 1455.8 1.6413	481.79 1.0470 1511.4 1.6807	581.79 1.1289 1566.9 1.7175	681.79 1.2093 1622.7 1.7522	781.79 1.2885 1678.9 1.7851	881.79 1.3669 1735.7 1.8164	981.79 1.4446 1792.9 1.8464
850 (525.24)	Sh v h s	0.02105 518.40 0.7197	0.5330 1198.0 1.4096			24.76 0.5683 1223.0 1.4347	74.76 0.6296 1265.9 1.4763	124.76 0.6829 1302.8 1.5102	174.76 0.7315 1336.0 1.5396	274.76 0.8205 1396.8 1.5899	374.76 0.9034 1454.0 1.6336	474.76 0.9830 1510.0 1.6733	574.76 1.0606 1565.7 1.7102	674.76 1.1366 1621.6 1.7450	774.76 1.2115 1678.0 1.7780	874.76 1.2855 1734.9 1.8094	974.76 1.3588 1792.3 1.8395
900 (531.95)	Sh v h s	0.02123 526.70 0.7279	0.5009 1196.4 1.4032			18.05 0.5263 1215.5 1.4223	68.05 0.5869 1260.6 1.4659	118.05 0.6388 1298.6 1.5010	168.05 0.6858 1332.7 1.5311	268.05 0.7713 1394.4 1.5822	368.05 0.8504 1452.2 1.6263	468.05 0.9262 1508.5 1.6662	568.05 0.9998 1564.4 1.7033	668.05 1.0720 1620.6 1.7382	768.05 1.1430 1677.1 1.7713	868.05 1.2131 1734.1 1.8028	968.05 1.2825 1791.6 1.8329

Sh = superheat, F
v = specific volume, cu ft per lb
h = enthalpy, Btu per lb
s = entropy, Btu per F per lb

TABLE A-2. Superheated Steam (*continued*)

Abs Press. Lb/Sq In. (Sat. Temp)		Sat. Water	Sat. Steam	\multicolumn{13}{c	}{Temperature — Degrees Fahrenheit}
				550	600	650	700	750	800	850	900	1000	1100	1200	1300	1400	1500
950 (538.39)	Sh v h s	0.02141 534.74 0.7358	0.4721 1194.7 1.3970	11.61 0.4883 1207.6 1.4098	61.61 0.5485 1255.1 1.4557	111.61 0.5993 1294.4 1.4921	161.61 0.6449 1329.3 1.5228	211.61 0.6871 1361.5 1.5500	261.61 0.7272 1392.0 1.5748	311.61 0.7656 1421.5 1.5977	361.61 0.8030 1450.3 1.6193	461.61 0.8753 1507.0 1.6595	561.61 0.9455 1563.2 1.6967	661.61 1.0142 1619.5 1.7317	761.61 1.0817 1676.2 1.7649	861.61 1.1484 1733.3 1.7965	961.61 1.2143 1791.0 1.8267
1000 (544.58)	Sh v h s	0.02159 542.55 0.7434	0.4460 1192.9 1.3910	5.42 0.4535 1199.3 1.3973	55.42 0.5137 1249.3 1.4457	105.42 0.5636 1290.1 1.4833	155.42 0.6080 1325.9 1.5149	205.42 0.6489 1358.7 1.5426	255.42 0.6875 1389.6 1.5677	305.42 0.7245 1419.4 1.5908	355.42 0.8295 1448.5 1.6126	455.42 0.8966 1505.4 1.6530	555.42 0.9622 1561.9 1.6905	655.42 1.0266 1618.4 1.7256	755.42 1.0901 1675.3 1.7589	855.42 1.1529 1732.5 1.7905	955.42 1.1529 1790.3 1.8207
1050 (550.53)	Sh v h s	0.02177 550.15 0.7507	0.4222 1191.0 1.3851		49.47 0.4821 1243.4 1.4358	99.47 0.5312 1285.7 1.4748	149.47 0.5745 1322.4 1.5072	199.47 0.6142 1355.8 1.5354	249.47 0.6515 1387.2 1.5608	299.47 0.6872 1417.3 1.5842	349.47 0.7216 1446.6 1.6062	449.47 0.7881 1503.9 1.6469	549.47 0.8524 1560.7 1.6845	649.47 0.9151 1617.4 1.7197	749.47 0.9767 1674.4 1.7531	849.47 1.0373 1731.8 1.7848	949.47 1.0973 1789.6 1.8151
1100 (556.28)	Sh v h s	0.02195 557.55 0.7578	0.4006 1189.1 1.3794		43.72 0.4531 1237.3 1.4259	93.72 0.5017 1281.2 1.4664	143.72 0.5440 1318.8 1.4996	193.72 0.5826 1352.9 1.5284	243.72 0.6188 1384.7 1.5542	293.72 0.6533 1415.2 1.5779	343.72 0.6865 1444.7 1.6000	443.72 0.7505 1502.4 1.6410	543.72 0.8121 1559.4 1.6787	643.72 0.8723 1616.3 1.7141	743.72 0.9313 1673.5 1.7475	843.72 0.9894 1731.0 1.7793	943.72 1.0468 1789.0 1.8097
1150 (561.82)	Sh v h s	0.02214 564.78 0.7647	0.3807 1187.0 1.3738		39.18 0.4263 1230.9 1.4160	89.18 0.4746 1276.6 1.4582	139.18 0.5162 1315.2 1.4923	189.18 0.5538 1349.9 1.5216	239.18 0.5889 1382.2 1.5478	289.18 0.6223 1413.0 1.5717	339.18 0.6544 1442.8 1.5941	439.18 0.7161 1500.9 1.6353	539.18 0.7754 1558.1 1.6732	639.18 0.8332 1615.2 1.7087	739.18 0.8899 1672.6 1.7422	839.18 0.9456 1730.2 1.7741	939.18 1.0007 1788.3 1.8045
1200 (567.19)	Sh v h s	0.02232 571.85 0.7714	0.3624 1184.8 1.3683		32.81 0.4016 1224.2 1.4061	82.81 0.4497 1271.8 1.4501	132.81 0.4905 1311.5 1.4851	182.81 0.5273 1346.9 1.5150	232.81 0.5615 1379.7 1.5415	282.81 0.5939 1410.8 1.5658	332.81 0.6250 1440.9 1.5883	432.81 0.6845 1499.4 1.6298	532.81 0.7418 1556.9 1.6679	632.81 0.7974 1614.2 1.7035	732.81 0.8519 1671.6 1.7371	832.81 0.9055 1729.4 1.7691	932.81 0.9584 1787.6 1.7996
1300 (577.42)	Sh v h s	0.02269 585.58 0.7843	0.3299 1180.2 1.3577		22.58 0.3570 1209.9 1.3860	72.58 0.4052 1261.9 1.4340	122.58 0.4451 1303.9 1.4711	172.58 0.4804 1340.8 1.5022	222.58 0.5129 1374.6 1.5296	272.58 0.5436 1406.4 1.5544	322.58 0.5729 1437.1 1.5773	422.58 0.6287 1496.5 1.6194	522.58 0.6822 1554.3 1.6578	622.58 0.7341 1612.0 1.6937	722.58 0.7847 1669.8 1.7275	822.58 0.8345 1727.9 1.7596	922.58 0.8836 1786.3 1.7902
1400 (587.07)	Sh v h s	0.02307 598.83 0.7966	0.3018 1175.3 1.3474		12.93 0.3176 1194.1 1.3652	62.93 0.3667 1251.4 1.4181	112.93 0.4059 1296.1 1.4575	162.93 0.4400 1334.5 1.4900	212.93 0.4712 1369.3 1.5182	262.93 0.5004 1402.0 1.5436	312.93 0.5282 1433.2 1.5670	412.93 0.5809 1493.2 1.6096	512.93 0.6311 1551.8 1.6484	612.93 0.6798 1609.9 1.6845	712.93 0.7272 1668.0 1.7185	812.93 0.7737 1726.3 1.7508	912.93 0.8195 1785.0 1.7815
1500 (596.20)	Sh v h s	0.02346 611.68 0.8085	0.2772 1170.1 1.3373		3.80 0.2820 1176.3 1.3431	53.80 0.3328 1240.2 1.4022	103.80 0.3717 1287.9 1.4443	153.80 0.4049 1328.0 1.4782	203.80 0.4350 1364.0 1.5073	253.80 0.4629 1397.4 1.5333	303.80 0.4894 1429.2 1.5572	403.80 0.5394 1490.1 1.6004	503.80 0.5869 1549.2 1.6395	603.80 0.6327 1607.7 1.6759	703.80 0.6773 1666.2 1.7101	803.80 0.7210 1724.8 1.7425	903.80 0.7639 1783.7 1.7734
1600 (604.87)	Sh v h s	0.02387 624.20 0.8199	0.2555 1164.5 1.3274			45.13 0.3026 1228.3 1.3861	95.13 0.3415 1279.4 1.4312	145.13 0.3741 1321.4 1.4667	195.13 0.4032 1358.5 1.4968	245.13 0.4301 1392.8 1.5235	295.13 0.4555 1425.2 1.5478	395.13 0.5031 1486.9 1.5916	495.13 0.5482 1546.6 1.6312	595.13 0.5915 1605.6 1.6678	695.13 0.6336 1664.3 1.7022	795.13 0.6748 1723.2 1.7347	895.13 0.7153 1782.3 1.7657
1700 (613.13)	Sh v h s	0.02428 636.45 0.8309	0.2361 1158.6 1.3176			36.87 0.2754 1215.3 1.3697	86.87 0.3147 1270.5 1.4183	136.87 0.3468 1314.5 1.4555	186.87 0.3751 1352.9 1.4867	236.87 0.4011 1388.1 1.5140	286.87 0.4255 1421.2 1.5388	386.87 0.4711 1483.8 1.5833	486.87 0.5140 1544.0 1.6232	586.87 0.5552 1603.4 1.6601	686.87 0.5951 1662.5 1.6947	786.87 0.6341 1721.7 1.7274	886.87 0.6724 1781.0 1.7585
1800 (621.02)	Sh v h s	0.02472 648.49 0.8417	0.2186 1152.3 1.3079			28.98 0.2505 1201.2 1.3526	78.98 0.2906 1261.1 1.4054	128.98 0.3223 1307.4 1.4446	178.98 0.3500 1347.2 1.4768	228.98 0.3752 1383.3 1.5049	278.98 0.3988 1417.1 1.5302	378.98 0.4426 1480.6 1.5753	478.98 0.4836 1541.4 1.6156	578.98 0.5229 1601.2 1.6528	678.98 0.5609 1660.7 1.6876	778.98 0.5980 1720.1 1.7204	878.98 0.6343 1779.7 1.7516
1900 (628.56)	Sh v h s	0.02517 660.36 0.8522	0.2028 1145.6 1.2981			21.44 0.2274 1185.7 1.3346	71.44 0.2687 1251.3 1.3925	121.44 0.3004 1300.2 1.4338	171.44 0.3275 1341.4 1.4672	221.44 0.3521 1378.4 1.4960	271.44 0.3749 1412.9 1.5219	371.44 0.4171 1477.4 1.5677	471.44 0.4565 1538.8 1.6084	571.44 0.4940 1599.1 1.6458	671.44 0.5303 1658.8 1.6808	771.44 0.5656 1718.6 1.7138	871.44 0.6002 1778.4 1.7451
2000 (635.80)	Sh v h s	0.02565 672.11 0.8625	0.1883 1138.3 1.2881			14.20 0.2056 1168.3 1.3154	64.20 0.2488 1240.9 1.3794	114.20 0.2805 1292.6 1.4231	164.20 0.3072 1335.4 1.4578	214.20 0.3312 1373.5 1.4874	264.20 0.3534 1408.7 1.5138	364.20 0.3942 1474.1 1.5603	464.20 0.4320 1536.2 1.6014	564.20 0.4680 1596.9 1.6391	664.20 0.5027 1657.0 1.6743	764.20 0.5365 1717.0 1.7075	864.20 0.5695 1771.1 1.7389
2100 (642.76)	Sh v h s	0.02615 683.79 0.8727	0.1750 1130.5 1.2780			7.24 0.1847 1148.5 1.2942	57.24 0.2304 1229.8 1.3661	107.24 0.2624 1284.9 1.4125	157.24 0.2888 1329.3 1.4486	207.24 0.3123 1368.4 1.4790	257.24 0.3339 1404.4 1.5060	357.24 0.3734 1470.9 1.5532	457.24 0.4099 1533.6 1.5948	557.24 0.4445 1594.7 1.6327	657.24 0.4778 1655.2 1.6681	757.24 0.5101 1715.4 1.7014	857.24 0.5418 1775.7 1.7330
2200 (649.45)	Sh v h s	0.02669 695.46 0.8828	0.1627 1122.2 1.2676			.55 0.1636 1123.9 1.2691	50.55 0.2134 1218.0 1.3523	100.55 0.2458 1276.8 1.4020	150.55 0.2720 1323.1 1.4395	200.55 0.2950 1363.3 1.4708	250.55 0.3161 1400.0 1.4984	350.55 0.3545 1467.6 1.5463	450.55 0.3897 1530.9 1.5883	550.55 0.4231 1592.5 1.6266	650.55 0.4551 1653.3 1.6622	750.55 0.4862 1713.9 1.6956	850.55 0.5165 1774.4 1.7273
2300 (655.89)	Sh v h s	0.02727 707.18 0.8929	0.1513 1113.2 1.2569				44.11 0.1975 1205.3 1.3381	94.11 0.2305 1268.4 1.3914	144.11 0.2566 1316.7 1.4305	194.11 0.2793 1358.1 1.4628	244.11 0.2999 1395.7 1.4910	344.11 0.3372 1464.2 1.5397	444.11 0.3714 1528.3 1.5821	544.11 0.4035 1590.3 1.6207	644.11 0.4344 1651.5 1.6565	744.11 0.4643 1712.3 1.6901	844.11 0.4935 1773.1 1.7219

Sh = superheat, F
v = specific volume, cu ft per lb
h = enthalpy, Btu per lb
s = entropy, Btu per F per lb

TABLE A-2. Superheated Steam (*continued*)

Abs Press. Lb/Sq In. (Sat. Temp)		Sat. Water	Sat. Steam	\multicolumn{13}{c}{Temperature – Degrees Fahrenheit}
				700	750	800	850	900	950	1000	1050	1100	1150	1200	1300	1400	1500
2400 (662.11)	Sh v h s	0.02790 718.95 0.9031	0.1408 1103.7 1.2460	37.89 0.1824 1191.6 1.3232	87.89 0.2164 1259.7 1.3808	137.89 0.2424 1310.1 1.4217	187.89 0.2648 1352.8 1.4549	237.89 0.2850 1391.2 1.4837	287.89 0.3037 1426.9 1.5095	337.89 0.3214 1460.9 1.5332	387.89 0.3382 1493.7 1.5553	437.89 0.3545 1525.6 1.5761	487.89 0.3703 1557.0 1.5959	537.89 0.3856 1588.1 1.6149	637.89 0.4155 1649.6 1.6509	737.89 0.4443 1710.8 1.6847	837.89 0.4724 1771.8 1.7167
2500 (668.11)	Sh v h s	0.02859 731.71 0.9139	0.1307 1093.3 1.2345	31.89 0.1681 1176.7 1.3076	81.89 0.2032 1250.6 1.3701	131.89 0.2293 1303.4 1.4129	181.89 0.2514 1347.4 1.4472	231.89 0.2712 1386.7 1.4766	281.89 0.2896 1423.1 1.5029	331.89 0.3068 1457.5 1.5269	381.89 0.3232 1490.7 1.5492	431.89 0.3390 1522.9 1.5703	481.89 0.3543 1554.6 1.5903	531.89 0.3692 1585.9 1.6094	631.89 0.3980 1647.8 1.6456	731.89 0.4259 1709.2 1.6796	831.89 0.4529 1770.4 1.7116
2600 (673.91)	Sh v h s	0.02938 744.47 0.9247	0.1211 1082.0 1.2225	26.09 0.1544 1160.2 1.2908	76.09 0.1909 1241.1 1.3592	126.09 0.2171 1296.5 1.4042	176.09 0.2390 1341.9 1.4395	226.09 0.2585 1382.1 1.4696	276.09 0.2765 1419.2 1.4964	326.09 0.2933 1454.1 1.5208	376.09 0.3093 1487.7 1.5434	426.09 0.3247 1520.2 1.5646	476.09 0.3395 1552.2 1.5848	526.09 0.3540 1583.7 1.6040	626.09 0.3819 1646.0 1.6405	726.09 0.4088 1707.7 1.6746	826.09 0.4350 1769.1 1.7068
2700 (679.53)	Sh v h s	0.03029 757.34 0.9356	0.1119 1069.7 1.2097	20.47 0.1411 1142.0 1.2727	70.47 0.1794 1231.1 1.3481	120.47 0.2058 1289.5 1.3954	170.47 0.2275 1336.3 1.4319	220.47 0.2468 1377.5 1.4628	270.47 0.2644 1415.2 1.4900	320.47 0.2809 1450.7 1.5148	370.47 0.2965 1484.6 1.5376	420.47 0.3114 1517.5 1.5591	470.47 0.3259 1549.8 1.5794	520.47 0.3399 1581.5 1.5988	620.47 0.3670 1644.1 1.6355	720.47 0.3931 1706.1 1.6697	820.47 0.4184 1767.8 1.7021
2800 (684.96)	Sh v h s	0.03134 770.69 0.9468	0.1030 1055.8 1.1958	15.04 0.1278 1121.2 1.2527	65.04 0.1685 1220.6 1.3368	115.04 0.1952 1282.2 1.3867	165.04 0.2168 1330.7 1.4245	215.04 0.2358 1372.8 1.4561	265.04 0.2531 1411.2 1.4838	315.04 0.2693 1447.2 1.5089	365.04 0.2845 1481.6 1.5321	415.04 0.2991 1514.8 1.5537	465.04 0.3132 1547.3 1.5742	515.04 0.3268 1579.3 1.5938	615.04 0.3532 1642.2 1.6306	715.04 0.3785 1704.5 1.6651	815.04 0.4030 1766.5 1.6975
2900 (690.22)	Sh v h s	0.03262 785.13 0.9588	0.0942 1039.8 1.1803	9.78 0.1138 1095.3 1.2283	59.78 0.1581 1209.6 1.3251	109.78 0.1853 1274.7 1.3780	159.78 0.2068 1324.9 1.4171	209.78 0.2256 1368.0 1.4494	259.78 0.2427 1407.2 1.4777	309.78 0.2585 1443.7 1.5032	359.78 0.2734 1478.5 1.5266	409.78 0.2877 1512.1 1.5485	459.78 0.3014 1544.9 1.5692	509.78 0.3147 1577.0 1.5889	609.78 0.3403 1640.4 1.6259	709.78 0.3649 1703.0 1.6605	809.78 0.3887 1765.2 1.6931
3000 (695.33)	Sh v h s	0.03428 801.84 0.9728	0.0850 1020.3 1.1619	4.67 0.0982 1060.5 1.1966	54.67 0.1483 1197.9 1.3131	104.67 0.1759 1267.0 1.3692	154.67 0.1975 1319.0 1.4097	204.67 0.2161 1363.2 1.4429	254.67 0.2329 1403.1 1.4717	304.67 0.2484 1440.2 1.4976	354.67 0.2630 1475.4 1.5213	404.67 0.2770 1509.4 1.5434	454.67 0.2904 1542.4 1.5642	504.67 0.3033 1574.8 1.5841	604.67 0.3282 1638.5 1.6214	704.67 0.3522 1701.4 1.6561	804.67 0.3753 1763.8 1.6888
3100 (700.28)	Sh v h s	0.03681 823.97 0.9914	0.0745 993.3 1.1373		49.72 0.1389 1185.4 1.3007	99.72 0.1671 1259.1 1.3604	149.72 0.1887 1313.0 1.4024	199.72 0.2071 1358.4 1.4364	249.72 0.2237 1399.0 1.4658	299.72 0.2390 1436.7 1.4920	349.72 0.2533 1472.3 1.5161	399.72 0.2670 1506.6 1.5384	449.72 0.2800 1539.9 1.5594	499.72 0.2927 1572.6 1.5794	599.72 0.3170 1636.7 1.6169	699.72 0.3403 1699.8 1.6518	799.72 0.3628 1762.5 1.6847
3200 (705.08)	Sh v h s	0.04472 875.54 1.0351	0.0566 931.6 1.0832		44.92 0.1300 1172.3 1.2877	94.92 0.1588 1250.9 1.3515	144.92 0.1804 1306.9 1.3951	194.92 0.1987 1353.4 1.4300	244.92 0.2151 1394.9 1.4600	294.92 0.2301 1433.1 1.4866	344.92 0.2442 1469.2 1.5110	394.92 0.2576 1503.8 1.5335	444.92 0.2704 1537.4 1.5547	494.92 0.2827 1570.3 1.5749	594.92 0.3065 1634.8 1.6126	694.92 0.3291 1698.3 1.6477	794.92 0.3510 1761.2 1.6806
3300	Sh v h s				0.1213 1158.2 1.2742	0.1510 1242.5 1.3425	0.1727 1300.7 1.3879	0.1908 1348.4 1.4237	0.2070 1390.7 1.4542	0.2218 1429.5 1.4813	0.2357 1466.1 1.5059	0.2488 1501.0 1.5287	0.2613 1534.9 1.5501	0.2734 1568.1 1.5704	0.2966 1632.9 1.6084	0.3187 1696.7 1.6436	0.3400 1759.9 1.6767
3400	Sh v h s				0.1129 1143.2 1.2600	0.1435 1233.7 1.3334	0.1653 1294.3 1.3807	0.1834 1343.4 1.4174	0.1994 1386.4 1.4486	0.2140 1425.9 1.4761	0.2276 1462.9 1.5010	0.2405 1498.3 1.5240	0.2528 1532.4 1.5456	0.2646 1565.8 1.5660	0.2872 1631.1 1.6042	0.3088 1695.1 1.6396	0.3296 1758.5 1.6728
3500	Sh v h s				0.1048 1127.1 1.2450	0.1364 1224.6 1.3242	0.1583 1287.8 1.3734	0.1764 1338.2 1.4112	0.1922 1382.2 1.4430	0.2066 1422.2 1.4709	0.2200 1459.7 1.4962	0.2326 1495.5 1.5194	0.2447 1529.9 1.5412	0.2563 1563.6 1.5618	0.2784 1629.2 1.6002	0.2995 1693.6 1.6358	0.3198 1757.2 1.6691
3600	Sh v h s				0.0966 1108.6 1.2281	0.1296 1215.3 1.3148	0.1517 1281.2 1.3662	0.1697 1333.0 1.4050	0.1854 1377.9 1.4374	0.1996 1418.6 1.4658	0.2128 1456.5 1.4914	0.2252 1492.6 1.5149	0.2371 1527.4 1.5369	0.2485 1561.3 1.5576	0.2702 1627.3 1.5962	0.2908 1692.0 1.6320	0.3106 1755.9 1.6654
3800	Sh v h s				0.0799 1064.2 1.1888	0.1169 1195.5 1.2955	0.1395 1267.6 1.3517	0.1574 1322.4 1.3928	0.1729 1369.1 1.4265	0.1868 1411.2 1.4558	0.1996 1450.1 1.4821	0.2116 1487.0 1.5061	0.2231 1522.4 1.5284	0.2340 1556.8 1.5495	0.2549 1623.6 1.5886	0.2746 1688.9 1.6247	0.2936 1753.2 1.6584
4000	Sh v h s				0.0631 1007.4 1.1396	0.1052 1174.3 1.2754	0.1284 1253.4 1.3371	0.1463 1311.6 1.3807	0.1616 1360.2 1.4158	0.1752 1403.6 1.4461	0.1877 1443.6 1.4730	0.1994 1481.3 1.4976	0.2105 1517.3 1.5203	0.2210 1552.2 1.5417	0.2411 1619.8 1.5812	0.2601 1685.7 1.6177	0.2783 1750.6 1.6516
4200	Sh v h s				0.0498 950.1 1.0905	0.0945 1151.6 1.2544	0.1183 1238.6 1.3223	0.1362 1300.4 1.3686	0.1513 1351.2 1.4053	0.1647 1396.0 1.4366	0.1769 1437.1 1.4642	0.1883 1475.5 1.4893	0.1991 1512.2 1.5124	0.2093 1547.6 1.5341	0.2287 1616.1 1.5742	0.2470 1682.6 1.6109	0.2645 1748.0 1.6452
4400	Sh v h s				0.0421 909.5 1.0556	0.0846 1127.3 1.2325	0.1090 1223.3 1.3073	0.1270 1289.0 1.3566	0.1420 1342.0 1.3949	0.1552 1388.3 1.4272	0.1671 1430.4 1.4556	0.1782 1469.7 1.4812	0.1887 1507.1 1.5048	0.1986 1543.0 1.5268	0.2174 1612.3 1.5673	0.2351 1679.4 1.6044	0.2519 1745.3 1.6389

Sh = superheat, F
v = specific volume, cu ft per lb
h = enthalpy, Btu per lb
s = entropy, Btu per F per lb

TABLE A-2. Superheated Steam (*continued*)

Abs Press. Lb/Sq In. (Sat. Temp)		Sat. Water	Sat. Steam	Temperature — Degrees Fahrenheit												
				750	800	850	900	950	1000	1050	1100	1150	1200	1250	1300	1400	1500
4600	Sh v h s			0.0380 883.8 1.0331	0.0751 1100.0 1.2084	0.1005 1207.3 1.2922	0.1186 1277.2 1.3446	0.1335 1332.6 1.3847	0.1465 1380.5 1.4181	0.1582 1423.7 1.4472	0.1691 1463.9 1.4734	0.1792 1501.9 1.4974	0.1889 1538.4 1.5197	0.1982 1573.8 1.5407	0.2071 1608.5 1.5607	0.2242 1676.3 1.5982	0.2404 1742.7 1.6330
4800	Sh v h s			0.0355 866.9 1.0180	0.0665 1071.2 1.1835	0.0927 1190.7 1.2768	0.1109 1265.2 1.3327	0.1257 1323.1 1.3745	0.1385 1372.6 1.4090	0.1500 1417.0 1.4390	0.1606 1458.0 1.4657	0.1706 1496.7 1.4901	0.1800 1533.8 1.5128	0.1890 1569.7 1.5341	0.1977 1604.7 1.5543	0.2142 1673.1 1.5921	0.2299 1740.0 1.6272
5000	Sh v h s			0.0338 854.9 1.0070	0.0591 1042.9 1.1593	0.0855 1173.6 1.2612	0.1038 1252.9 1.3207	0.1185 1313.5 1.3645	0.1312 1364.6 1.4001	0.1425 1410.2 1.4309	0.1529 1452.1 1.4582	0.1626 1491.5 1.4831	0.1718 1529.1 1.5061	0.1806 1565.5 1.5277	0.1890 1600.9 1.5481	0.2050 1670.0 1.5863	0.2203 1737.4 1.6216
5200	Sh v h s			0.0326 845.8 0.9985	0.0531 1016.9 1.1370	0.0789 1156.0 1.2455	0.0973 1240.4 1.3088	0.1119 1303.7 1.3545	0.1244 1356.6 1.3914	0.1356 1403.4 1.4229	0.1458 1446.2 1.4509	0.1553 1486.3 1.4762	0.1642 1524.5 1.4995	0.1728 1561.3 1.5214	0.1810 1597.2 1.5420	0.1966 1666.8 1.5806	0.2114 1734.7 1.6161
5400	Sh v h s			0.0317 838.5 0.9915	0.0483 994.3 1.1175	0.0728 1138.1 1.2296	0.0912 1227.7 1.2969	0.1058 1293.7 1.3446	0.1182 1348.4 1.3827	0.1292 1396.5 1.4151	0.1392 1440.3 1.4437	0.1485 1481.1 1.4694	0.1572 1519.8 1.4931	0.1656 1557.1 1.5153	0.1736 1593.4 1.5362	0.1888 1663.7 1.5750	0.2031 1732.1 1.6109
5600	Sh v h s			0.0309 832.4 0.9855	0.0447 975.0 1.1008	0.0672 1119.9 1.2137	0.0856 1214.8 1.2850	0.1001 1283.7 1.3348	0.1124 1340.2 1.3742	0.1232 1389.6 1.4075	0.1331 1434.3 1.4366	0.1422 1475.9 1.4628	0.1508 1515.2 1.4869	0.1589 1552.9 1.5093	0.1667 1589.6 1.5304	0.1815 1660.5 1.5697	0.1954 1729.5 1.6058
5800	Sh v h s			0.0303 827.3 0.9803	0.0419 958.8 1.0867	0.0622 1101.8 1.1981	0.0805 1201.8 1.2732	0.0949 1273.6 1.3250	0.1070 1332.0 1.3658	0.1177 1382.6 1.3999	0.1274 1428.3 1.4297	0.1363 1470.6 1.4564	0.1447 1510.5 1.4808	0.1527 1548.1 1.5035	0.1603 1585.8 1.5248	0.1747 1657.4 1.5644	0.1883 1726.8 1.6008
6000	Sh v h s			0.0298 822.9 0.9758	0.0397 945.1 1.0746	0.0579 1084.6 1.1833	0.0757 1188.8 1.2615	0.0900 1263.4 1.3154	0.1020 1323.6 1.3574	0.1126 1375.7 1.3925	0.1221 1422.3 1.4229	0.1309 1465.4 1.4500	0.1391 1505.9 1.4748	0.1469 1544.6 1.4978	0.1544 1582.0 1.5194	0.1684 1654.2 1.5593	0.1817 1724.2 1.5960
6500	Sh v h s			0.0287 813.9 0.9661	0.0358 919.5 1.0515	0.0495 1046.7 1.1506	0.0655 1156.3 1.2328	0.0793 1237.8 1.2917	0.0909 1302.7 1.3370	0.1012 1358.1 1.3743	0.1104 1407.3 1.4064	0.1188 1452.2 1.4347	0.1266 1494.2 1.4604	0.1340 1534.1 1.4841	0.1411 1572.5 1.5062	0.1544 1646.4 1.5471	0.1669 1717.6 1.5844
7000	Sh v h s			0.0279 806.9 0.9582	0.0334 901.8 1.0350	0.0438 1016.5 1.1243	0.0573 1124.9 1.2055	0.0704 1212.6 1.2689	0.0816 1281.7 1.3171	0.0915 1340.5 1.3567	0.1004 1392.2 1.3904	0.1085 1439.1 1.4200	0.1160 1482.6 1.4466	0.1231 1523.7 1.4710	0.1298 1563.1 1.4938	0.1424 1638.6 1.5355	0.1542 1711.1 1.5735
7500	Sh v h s			0.0272 801.3 0.9514	0.0318 889.0 1.0224	0.0399 992.9 1.1033	0.0512 1097.7 1.1818	0.0631 1188.3 1.2473	0.0737 1261.0 1.2980	0.0833 1322.9 1.3397	0.0918 1377.2 1.3751	0.0996 1426.0 1.4059	0.1068 1471.0 1.4335	0.1136 1513.3 1.4586	0.1200 1553.7 1.4819	0.1321 1630.8 1.5245	0.1433 1704.6 1.5632
8000	Sh v h s			0.0267 796.6 0.9455	0.0306 879.1 1.0122	0.0371 974.4 1.0864	0.0465 1074.3 1.1613	0.0571 1165.4 1.2271	0.0671 1241.0 1.2798	0.0762 1305.5 1.3233	0.0845 1362.2 1.3603	0.0920 1413.0 1.3924	0.0989 1459.6 1.4208	0.1054 1503.1 1.4467	0.1115 1544.5 1.4705	0.1230 1623.1 1.5140	0.1338 1698.1 1.5533
8500	Sh v h s			0.0262 792.7 0.9402	0.0296 871.1 1.0037	0.0350 959.8 1.0727	0.0429 1054.5 1.1437	0.0522 1144.0 1.2084	0.0615 1221.9 1.2627	0.0701 1288.5 1.3076	0.0780 1347.5 1.3460	0.0853 1400.2 1.3793	0.0919 1448.2 1.4087	0.0982 1492.9 1.4352	0.1041 1535.3 1.4597	0.1151 1615.4 1.5040	0.1254 1691.7 1.5439
9000	Sh v h s			0.0258 789.3 0.9354	0.0288 864.7 0.9964	0.0335 948.0 1.0613	0.0402 1037.6 1.1285	0.0483 1125.4 1.1918	0.0568 1204.1 1.2468	0.0649 1272.1 1.2926	0.0724 1333.0 1.3323	0.0794 1387.5 1.3667	0.0858 1437.1 1.3970	0.0918 1482.9 1.4243	0.0975 1526.3 1.4492	0.1081 1607.9 1.4944	0.1179 1685.3 1.5349
9500	Sh v h s			0.0254 786.4 0.9310	0.0282 859.2 0.9900	0.0322 938.3 1.0516	0.0380 1023.4 1.1153	0.0451 1108.9 1.1771	0.0528 1187.7 1.2320	0.0603 1256.6 1.2785	0.0675 1318.9 1.3191	0.0742 1375.1 1.3546	0.0804 1426.1 1.3858	0.0862 1473.1 1.4137	0.0917 1517.3 1.4392	0.1019 1600.4 1.4851	0.1113 1679.0 1.5263
10000	Sh v h s			0.0251 783.8 0.9270	0.0276 854.5 0.9842	0.0312 930.2 1.0432	0.0362 1011.3 1.1039	0.0425 1094.2 1.1638	0.0495 1172.6 1.2185	0.0565 1242.0 1.2652	0.0633 1305.3 1.3065	0.0697 1362.9 1.3429	0.0757 1415.3 1.3749	0.0812 1463.4 1.4035	0.0865 1508.6 1.4295	0.0963 1593.1 1.4763	0.1054 1672.8 1.5180
10500	Sh v h s			0.0248 781.5 0.9232	0.0271 850.5 0.9790	0.0303 923.4 1.0358	0.0347 1001.0 1.0939	0.0404 1081.3 1.1519	0.0467 1158.9 1.2060	0.0532 1228.4 1.2529	0.0595 1292.4 1.2946	0.0656 1351.1 1.3371	0.0714 1404.7 1.3644	0.0768 1453.9 1.3937	0.0818 1500.0 1.4202	0.0913 1585.8 1.4677	0.1001 1666.7 1.5100

Sh = superheat, F
v = specific volume, cu ft per lb
h = enthalpy, Btu per lb
s = entropy, Btu per F per lb

TABLE A-2. Superheated Steam (*continued*)

Abs Press. Lb/Sq In. (Sat. Temp)		Sat. Water	Sat. Steam	Temperature — Degrees Fahrenheit													
				750	800	850	900	950	1000	1050	1100	1150	1200	1250	1300	1400	1500
11000	v h s			0.0245 779.5 0.9196	0.0267 846.9 0.9742	0.0296 917.5 1.0292	0.0335 992.1 1.0851	0.0386 1069.9 1.1412	0.0443 1146.3 1.1945	0.0503 1215.9 1.2414	0.0562 1280.2 1.2833	0.0620 1339.7 1.3209	0.0676 1394.4 1.3544	0.0727 1444.6 1.3842	0.0776 1491.5 1.4112	0.0868 1578.7 1.4595	0.0952 1660.6 1.5023
11500	v h s			0.0243 777.7 0.9163	0.0263 843.8 0.9698	0.0290 912.4 1.0232	0.0325 984.5 1.0772	0.0370 1059.8 1.1316	0.0423 1134.9 1.1840	0.0478 1204.3 1.2308	0.0534 1268.7 1.2727	0.0588 1328.8 1.3107	0.0641 1384.4 1.3446	0.0691 1435.5 1.3750	0.0739 1483.2 1.4025	0.0827 1571.8 1.4515	0.0909 1654.7 1.4949
12000	v h s			0.0241 776.1 0.9131	0.0260 841.0 0.9657	0.0284 907.9 1.0177	0.0317 977.8 1.0701	0.0357 1050.9 1.1229	0.0405 1124.5 1.1742	0.0456 1193.7 1.2209	0.0508 1258.0 1.2627	0.0560 1318.5 1.3010	0.0610 1374.7 1.3353	0.0659 1426.6 1.3662	0.0704 1475.1 1.3941	0.0790 1564.9 1.4438	0.0869 1648.8 1.4877
12500	v h s			0.0238 774.7 0.9101	0.0256 838.6 0.9618	0.0279 903.9 1.0127	0.0309 971.9 1.0637	0.0346 1043.1 1.1151	0.0390 1115.2 1.1653	0.0437 1184.1 1.2117	0.0486 1247.9 1.2534	0.0535 1308.8 1.2918	0.0583 1365.4 1.3264	0.0629 1418.0 1.3576	0.0673 1467.2 1.3860	0.0756 1558.2 1.4363	0.0832 1643.1 1.4808
13000	v h s			0.0236 773.5 0.9073	0.0253 836.3 0.9582	0.0275 900.4 1.0080	0.0302 966.8 1.0578	0.0336 1036.2 1.1079	0.0376 1106.7 1.1571	0.0420 1174.8 1.2030	0.0466 1238.5 1.2445	0.0512 1299.6 1.2831	0.0558 1356.5 1.3179	0.0602 1409.6 1.3494	0.0645 1459.4 1.3781	0.0725 1551.6 1.4291	0.0799 1637.4 1.4741
13500	v h s			0.0235 772.3 0.9045	0.0251 834.4 0.9548	0.0271 897.2 1.0037	0.0297 962.2 1.0524	0.0328 1030.0 1.1014	0.0364 1099.1 1.1495	0.0405 1166.3 1.1948	0.0448 1229.7 1.2361	0.0492 1291.0 1.2749	0.0535 1348.1 1.3098	0.0577 1401.5 1.3415	0.0619 1451.8 1.3705	0.0696 1545.2 1.4221	0.0768 1631.9 1.4675
14000	v h s			0.0233 771.1 0.9019	0.0248 832.6 0.9515	0.0267 894.3 0.9996	0.0291 958.0 1.0473	0.0320 1024.5 1.0953	0.0354 1092.3 1.1426	0.0392 1158.5 1.1872	0.0432 1221.4 1.2282	0.0474 1283.0 1.2671	0.0515 1340.2 1.3021	0.0555 1393.8 1.3339	0.0595 1444.4 1.3631	0.0670 1538.8 1.4153	0.0740 1626.5 1.4612
14500	v h s			0.0231 770.4 0.8994	0.0246 831.0 0.9484	0.0264 891.7 0.9957	0.0287 954.3 1.0426	0.0314 1019.6 1.0897	0.0345 1086.2 1.1362	0.0380 1151.4 1.1801	0.0418 1213.8 1.2208	0.0458 1275.4 1.2597	0.0496 1332.9 1.2949	0.0534 1386.4 1.3266	0.0573 1437.3 1.3560	0.0646 1532.6 1.4087	0.0714 1621.1 1.4551
15000	v h s			0.0230 769.6 0.8970	0.0244 829.5 0.9455	0.0261 889.3 0.9920	0.0282 950.9 1.0382	0.0308 1015.1 1.0846	0.0337 1080.6 1.1302	0.0369 1144.9 1.1735	0.0405 1206.8 1.2139	0.0443 1268.1 1.2525	0.0479 1326.0 1.2880	0.0516 1379.4 1.3197	0.0552 1430.3 1.3491	0.0624 1526.4 1.4022	0.0690 1615.9 1.4491
15500	v h s			0.0228 768.9 0.8946	0.0242 828.2 0.9427	0.0258 887.2 0.9886	0.0278 947.8 1.0340	0.0302 1011.1 1.0797	0.0329 1075.7 1.1247	0.0360 1139.0 1.1674	0.0393 1200.3 1.2073	0.0429 1261.1 1.2457	0.0464 1319.6 1.2815	0.0499 1372.8 1.3131	0.0534 1423.6 1.3424	0.0603 1520.4 1.3959	0.0668 1610.8 1.4433

Sh = superheat, F
v = specific volume, cu ft per lb
h = enthalpy, Btu per lb
s = entropy, Btu per F per lb

TABLE A-3. Specific Heat Capacities of Gases in the Ideal Gaseous State at Constant Pressure[†]

$c_p = a + bT + cT^2$ (cal/g-mole K)
298 to 1500 K

Compound	Formula	a	$b \times 10^3$	$c \times 10^6$
Normal paraffins:				
Methane	CH_4	3.381	18.044	−4.300
Ethane	C_2H_6	2.247	38.201	−11.049
Propane	C_3H_8	2.410	57.195	−17.533
n-Butane	C_4H_{10}	3.844	73.350	−22.655
n-Pentane	C_5H_{12}	4.895	90.113	−28.039
n-Hexane	C_6H_{14}	6.011	106.746	−33.363
n-Heptane	C_7H_{16}	7.094	123.447	−38.719
n-Octane	C_8H_{18}	8.163	140.217	−44.127
Increment per C atom above 8	......	1.097	16.667	−5.338
Normal monoolefins (1-alkenes):				
Ethylene	C_2H_4	2.830	28.601	−8.726
Propylene	C_3H_6	3.253	45.116	−13.740
1-Butene	C_4H_8	3.909	62.848	−19.617
1-Pentene	C_5H_{10}	5.347	78.990	−24.733
1-Hexene	C_6H_{12}	6.399	95.752	−30.116
1-Heptene	C_7H_{14}	7.488	112.440	−35.462
1-Octene	C_8H_{16}	8.592	129.076	−40.775
Increment per C atom above 8	......	1.097	16.667	−5.338
Miscellaneous materials:				
Acetaldehyde[‡]	C_2H_4O	3.364	35.722	−12.236
Acetylene	C_2H_2	7.331	12.622	−3.889
Ammonia	NH_3	6.086	8.812	−1.506
Benzene	C_6H_6	−0.409	77.621	−26.429
1,3-Butadiene	C_4H_6	5.432	53.224	−17.649
Carbon dioxide	CO_2	6.214	10.396	−3.545
Carbon monoxide	CO	6.420	1.665	−0.196
Chlorine	Cl_2	7.576	2.424	−0.965
Cyclohexane	C_6H_{12}	−7.701	125.675	−41.584
Ethyl alcohol	C_2H_6O	6.990	39.741	−11.926
Hydrogen	H_2	6.947	−0.200	0.481
Hydrogen chloride	HCl	6.732	0.433	0.370
Hydrogen sulfide	H_2S	6.662	5.134	−0.854
Methyl alcohol	CH_4O	4.394	24.274	−6.855
Nitric oxide	NO	7.020	−0.370	2.546
Nitrogen	N_2	6.524	1.250	−0.001
Oxygen	O_2	6.148	3.102	−0.923
Sulfur dioxide	SO_2	7.116	9.512	3.511
Sulfur trioxide	SO_3	6.077	23.537	−0.687
Toluene	C_7H_8	0.576	93.493	−31.227
Water	H_2O	7.256	2.298	0.283

[†]Selected mainly from values given by H. M. Spencer and coworkers, *J. Am. Chem. Soc.*, 56, 2311 (1934); 64, 2511 (1942); 67, 1859 (1945); and *Ind. Eng. Chem.*, 40, 2152 (1948). Also personal communication.
[‡] 298 to 1000 K

From *Introduction to Chemical Engineering Thermodynamics*, by J. M. Smith and H. C. Van Ness. Copyright 1959. Used with permission of McGraw-Hill Book Company.

TABLE A-3. Specific Heat Capacities of Solids at Constant Pressure[†] (*continued*)

$$c_p = a + bT + cT^2 \quad (cal/g\text{-}mole\ K)$$

Solid	a	$b \times 10^3$	$c \times 10^{-5}$	Range, K
CaO	11.67	1.08	−1.56	298–1800
CaCO$_3$	24.98	5.24	−6.20	298–1200
C (graphite)	4.10	1.02	−2.10	298–2300
Cu	5.41	1.50	. . . . .	298–1357
CuO	9.27	4.80	. . . . .	298–1250
Fe (α)	3.37	7.10	0.43	298–1033
FeO	12.38	1.62	−0.38	298–1200
Fe$_2$O$_3$	23.36	17.24	−3.08	298–1100
Fe$_3$O$_4$	39.92	18.86	−10.01	298–1100
FeS	15.20	. . . . .	. . . . .	298–412
NH$_4$Cl	11.80	32.00	. . . . .	298–458
Na	5.00	5.36	. . . . .	298–371
NaCl	10.98	3.90	. . . . .	298–1073
NaOH	19.2	. . . . .	. . . . .	298–595
S (rhombic)	3.58	6.24	. . . . .	298–369

[†] Selected from K. K. Kelley, *U. S. Bur. Mines, Bull.* 476, 1949.

From *Introduction to Chemical Engineering Thermodynamics*, by J. M. Smith and H. C. Van Ness. Copyright 1959. Used with permission of McGraw-Hill Book Company.

TABLE A-4. Critical Properties[15,16,17,18]

Compound	T_c, K	P_c, atm	$V_c \left(\dfrac{\text{liter}}{\text{g-mole}} \right)$	$z_c = \dfrac{P_c V_c}{RT_c}$	ω
Acetylene	308.3	61.7	.113	.276	.1984
*Ammonia	405.6	111.5	.073	.245	.2507
Argon	150.9	48.3	.075	.293	−.0046
Benzene	562.1	48.6	.258	.272	.2122
n-Butane	425.2	37.5	.255	.274	.1995
Carbon Dioxide	304.2	72.9	.094	.275	.2250
Carbon Monoxide	133.9	34.5	.093	.292	.0475
Cyclohexane	553.0	40.3	.311	.276	.2142
n-Decane	617.7	20.7	.602	.246	.4861
Ethane	305.4	48.2	.148	.285	.0980
*Ethanol	516.3	63.1	.167	.249	.6345
Ethylene	283.1	50.5	.131	.285	.0838
**Helium	5.25	2.26	.058	.304	
	(10.47)	(6.67)	(.0375)	(.291)	(0.0)
n-Heptane	540.3	27.0	.430	.262	0.3478
n-Hexane	507.5	29.7	.368	.262	.3006
**Hydrogen	33.3	12.8	.065	.304	
	(43.6)	(20.2)	(.0515)	(.291)	(0.0)
Hydrogen Sulfide	373.4	88.9	.098	.284	.1000
Methane	190.5	45.4	.0995	.289	.0109
*Methanol	512.6	79.9	.118	.224	.5659
*Methyl Choloride	416.4	65.8	.136	.262	.1638
**Neon	44.5	25.9	.042	.298	
	(45.5)	(26.9)	(.0403)	(.290)	(0.0)
Nitrogen	126.2	34.0	.090	.295	.0373
n-Nonane	594.6	22.6	.543	.252	0.4449
n-Octane	568.8	24.5	.486	.255	.3958
Oxygen	154.6	49.8	.075	.294	.0230
n-Pentane	469.7	33.3	.311	.269	.2519
Propane	369.8	41.9	.200	.276	.1533
Toluene	591.8	40.0	.316	.260	.2591
*Water	647.3	218.0	.057	.234	.3432

*Non-normal fluid
**Use effective critical constants in brackets with reduced correlations.

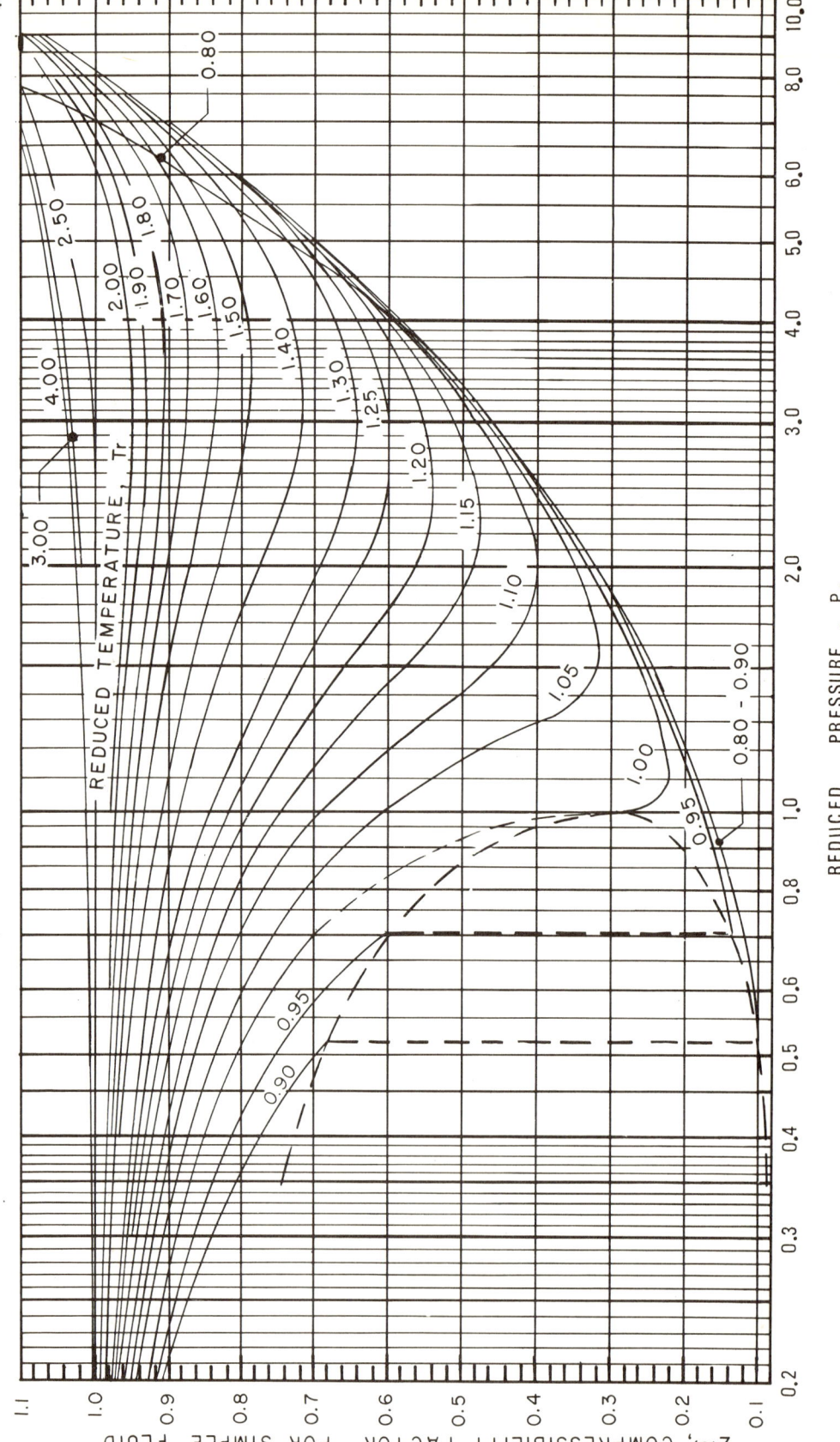

Fig. A-1A. *Simple fluid compressibility factor.*[15]

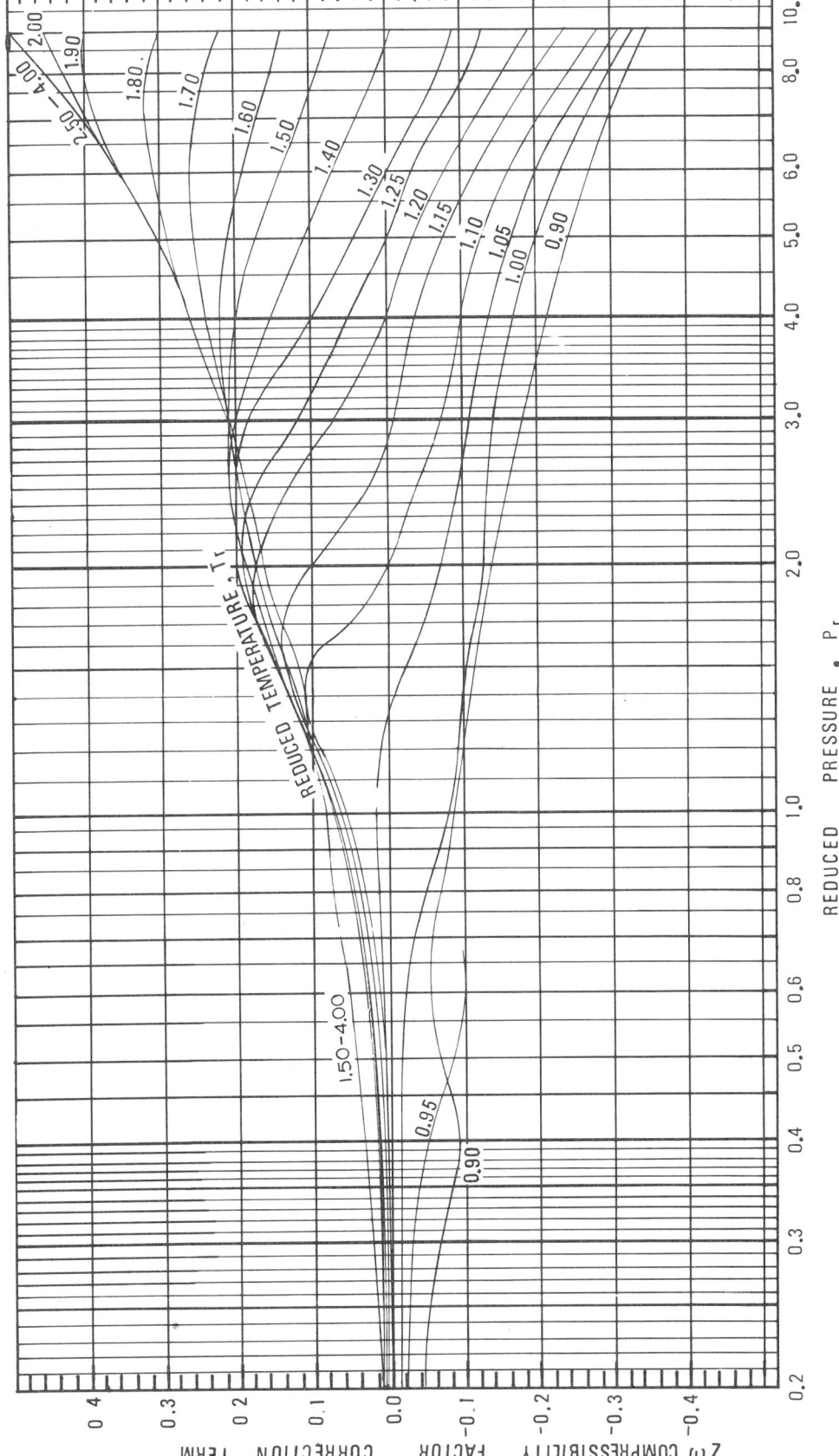

Fig. A-1B. *Compressibility factor correction term.*[15]

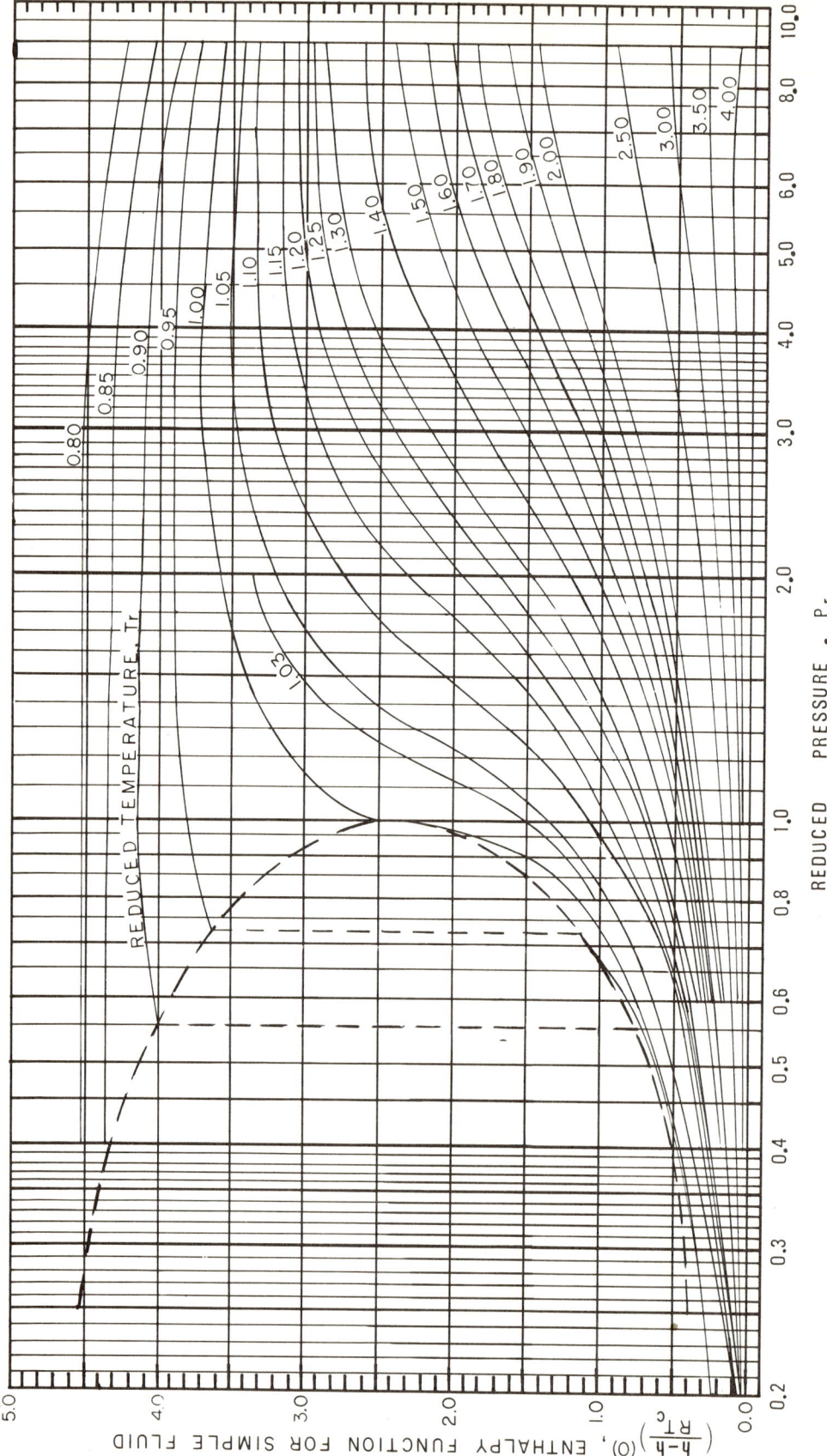

Fig. A-2A. *Simple fluid enthalpy function.*[27]

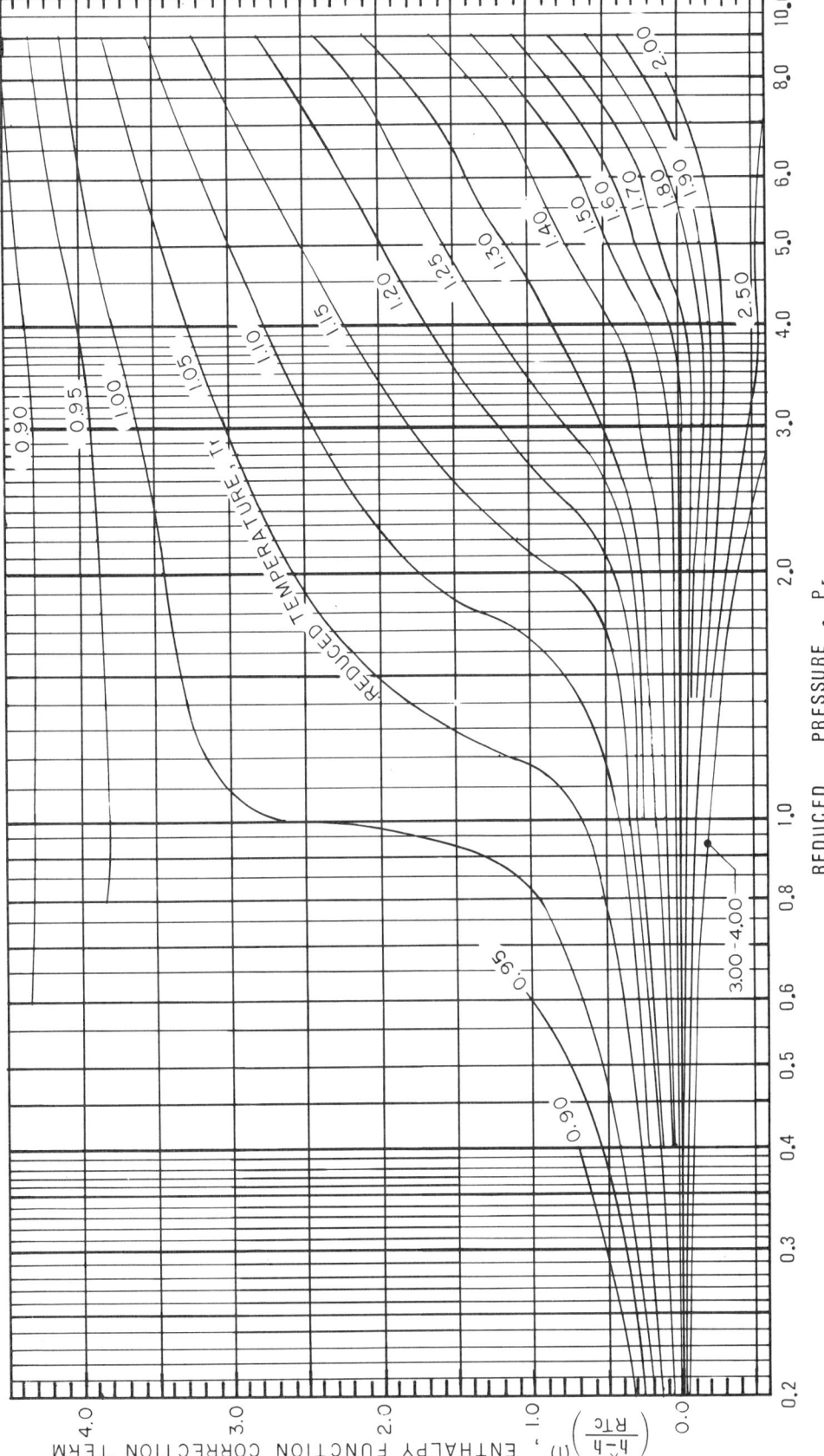

Fig. A-2B. *Enthalpy function correction term.*[27]

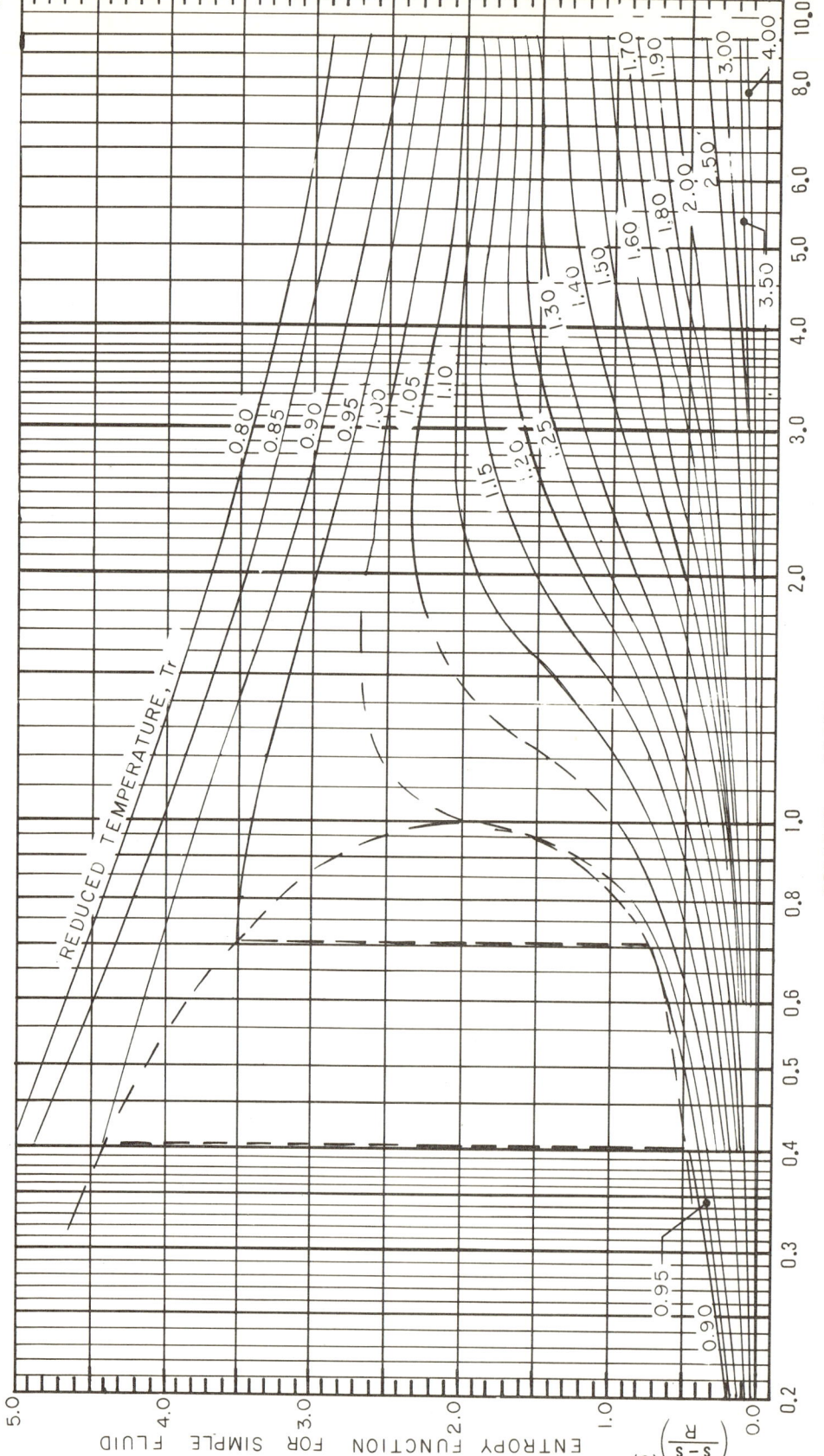

Fig. A-3A. *Simple fluid entropy function.*[27]

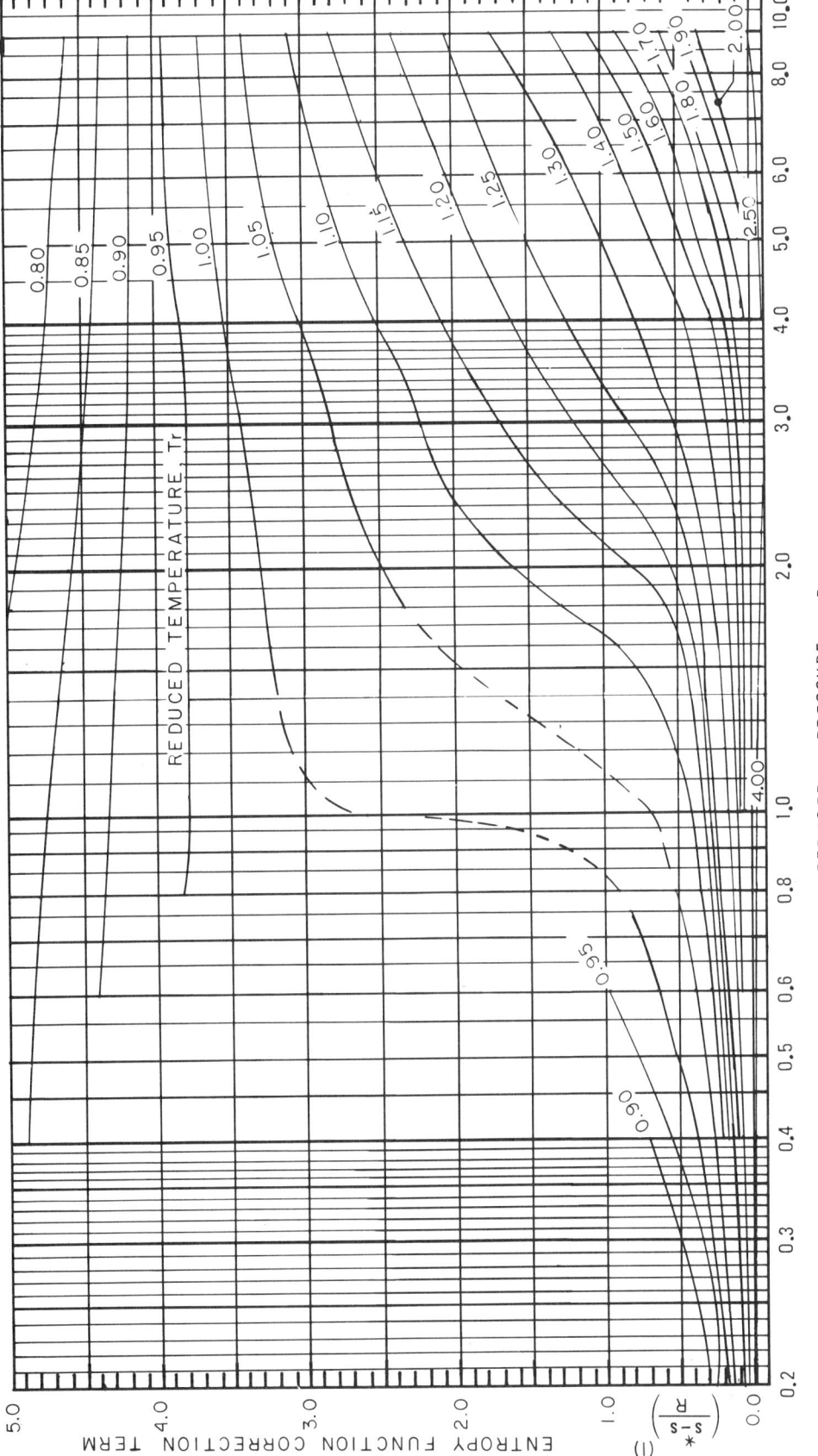

Fig. A-3B. *Entropy function correction term.*[27]

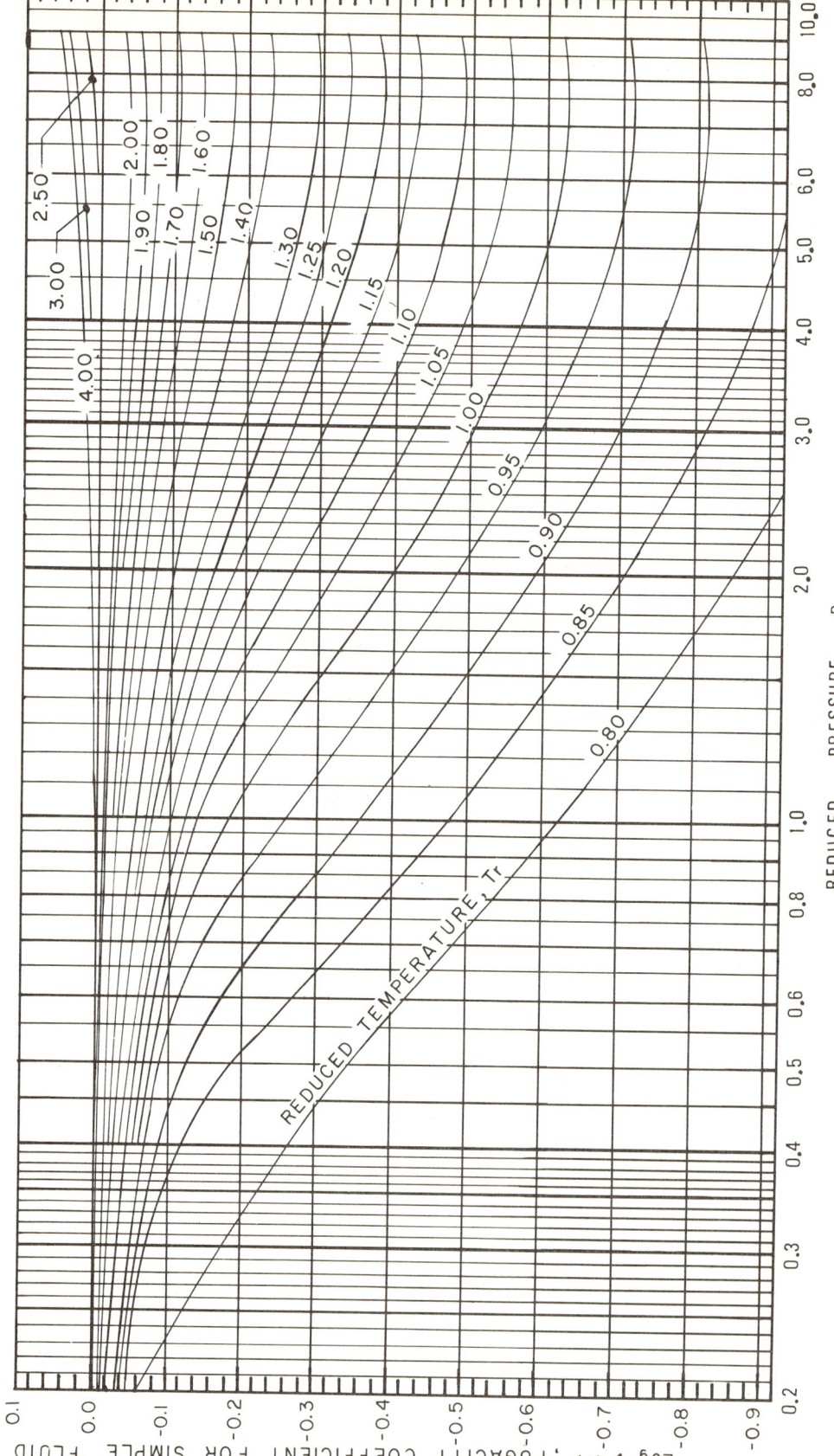

Fig. A-4A. *Simple fluid fugacity coefficient.*[27]

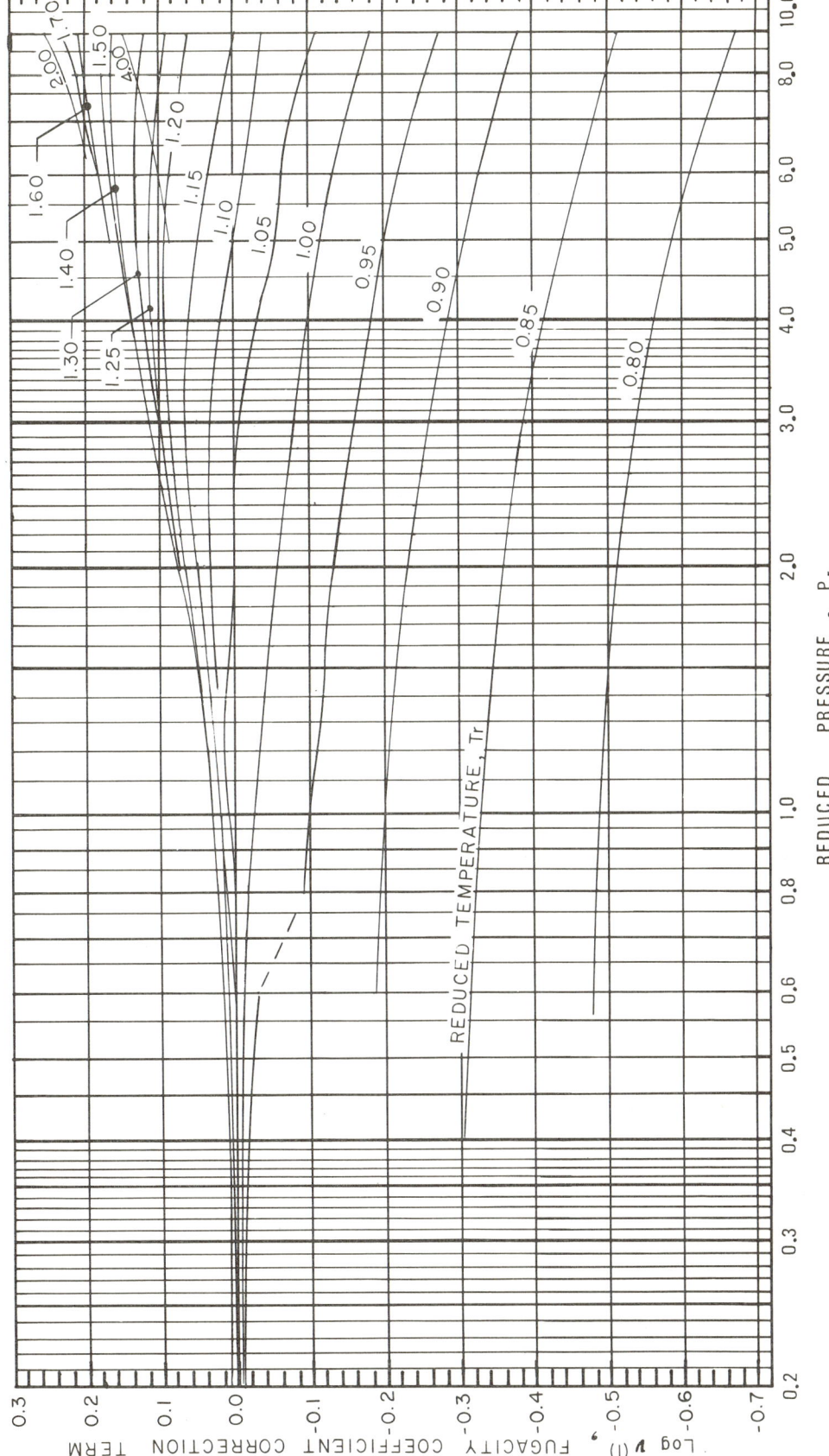

Fig. A-4B. *Fugacity coefficient correction term.*[27]

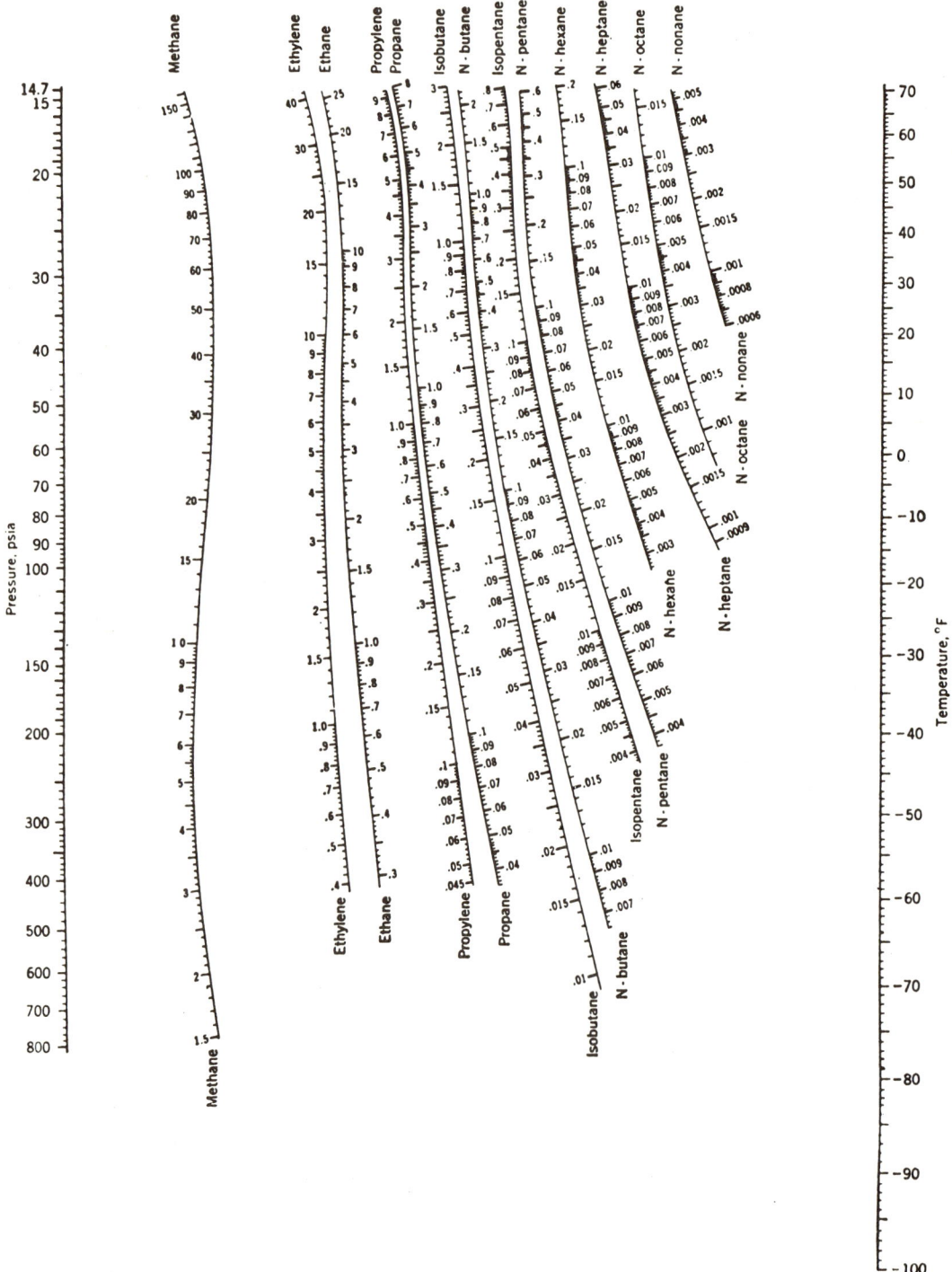

Fig. A-5. *Vapor-liquid equilibrium coefficients in ideal hydrocarbon solutions, low temperature range.*[31]

Reproduced by permission. *Chemical Engineering Progress Symposium Series*, copyright 1953.

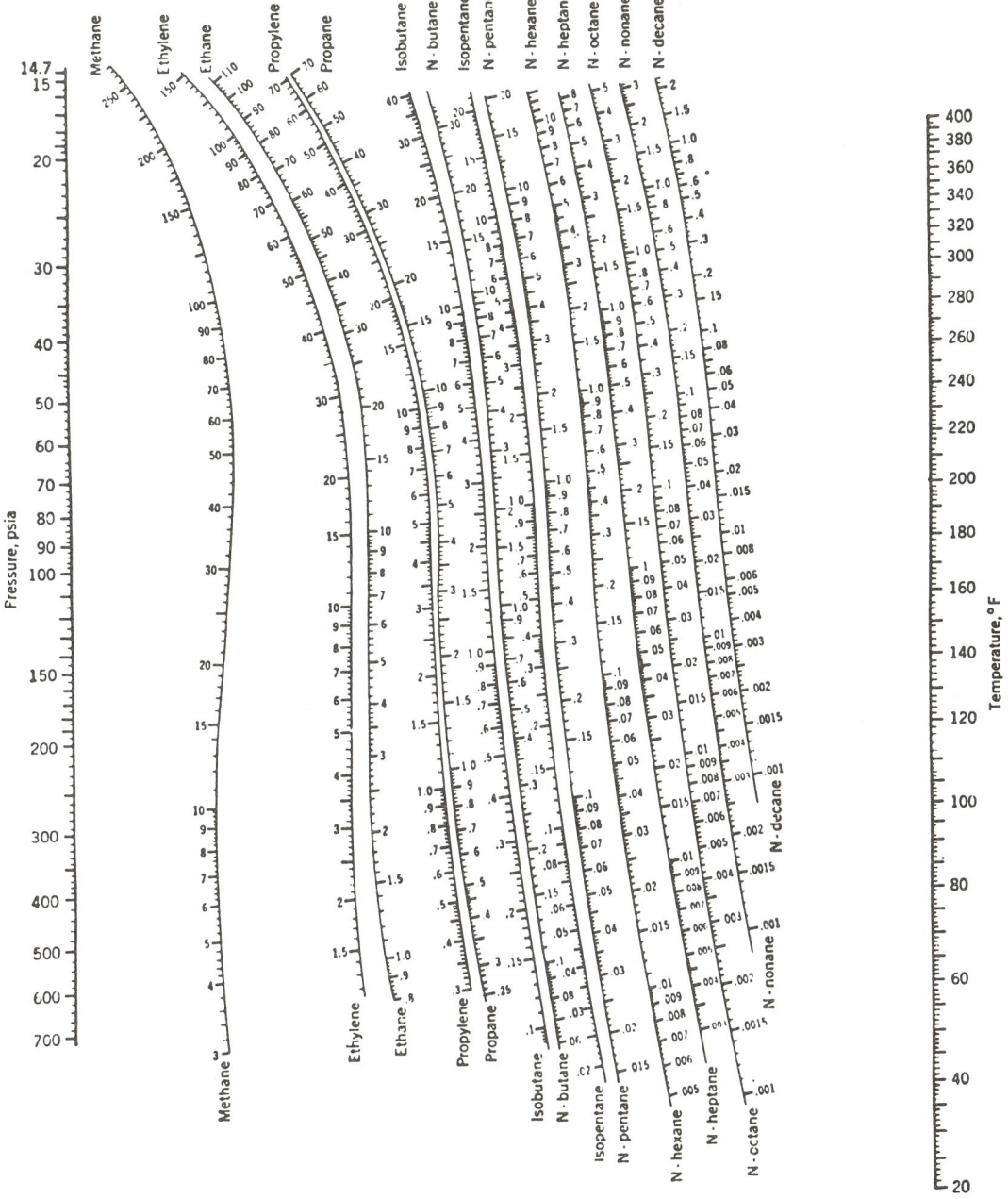

Fig. A-5. *Vapor-liquid equilibrium coefficients in ideal hydrocarbon solutions, high temperature range.*[31]

Reproduced by permission. *Chemical Engineering Progress Symposium Series*, copyright 1953.

Answers to Selected Problems

1-2. 6.9 hours
1-4. (a) 0.031 lb_f/poundal (b) 0.138 N/poundal
1-6. (a) 444,822 dynes/lb_f (b) 10^5 dynes/N (c) 13825.7 dynes/poundal
1-8. (a) $-40°$ (b) $0°$
1-10. (a) 0.7457 kW/hp (b) 42.408 BTU/min·hp (c) 56.869 BTU/min·kW
1-12. 760.0 mm of Hg
1-14. 62.43 lb_m/ft^3
1-16. (a) 67.005 lb_f/in^2 (b) 3465.1 mm of Hg (c) 461,982 N/m^2
1-18. (a) 0.0149 (b) 0.130 (c) 82.0 °Sh
1-20. 0.00749
1-22. 0.485

2-2. 199.4 GPM
2-4. 608 lb_m spent acid, 152 lb_m sulfuric acid, and 240 lb_m nitric acid
2-6. (a) 107 BTU (b) 113 × 10^3 joules
2-8. -4.28×10^{-9} BTU
2-10. +987.3 BTU/hr
2-14. (a) 1240.9 BTU/lb_m (b) 1269.0 BTU/lb_m
2-16. 1033.4 BTU
2-18. 61.53 °Sh, 2.5 BTU/lb_m
2-20. -2.34×10^4 joules
2-22. 375.4 K, 1.35×10^5 joules
2-24. 125.06 lb_f/in^2
2-26. 6.33×10^4 joules
2-28. 31.9 lb_m steam/hr   377.1 lb_m cold water/hr
2-30. 7.2 in.

Answers

2-32. (a) 895.4 °F (b) 600 °F
2-34. 360 lb_f/in^2
3-2. 0.255 joules/gm·K
3-4. Device feasible
3-6. $\eta_{therm} = 0.4906$ $W = 5.18 \times 10^5$ joules
3-8. 2.02 hp
3-10. $-\infty$ joules/hr
3-12. 135.8 BTU/lb_m
3-16. (c) $+5.0 \times 10^{-2}$ cal/K   (d) -15.0 cal
3-18. 706.2 BTU, 450 °F
3-20. (a) 13.97 lb_m (b) $+17,243$ BTU   (c) ΔS (universe) = 6.96 BTU/°R
3-22. -788.7 BTU/lb_m
3-24. 0.179 lb_m steam/lb_m feed

4-6. $\left(\dfrac{\partial v}{\partial T}\right)_S = -\dfrac{c_V}{T}\left(\dfrac{\partial T}{\partial P}\right)_V$

4-8. $dh = \left[c_V\left(\dfrac{\partial T}{\partial P}\right)_V + v\right]dP + c_P\left(\dfrac{\partial T}{\partial v}\right)_P dv$

4-10. (a) No (b) Yes
5-2. $\Delta h = c_P(T_2 - T_1)$
5-4. 659.5 °R, 239.8 ft/sec
5-6. -135.3 BTU/lb_m
5-8. -29.2 hp, 943 °R, 18,560 BTU/hr
5-10. 11.8 lb_m
6-2. 7 ft
6-6. 1459 m/sec
6-8. (a) 331 gram/sec   (b) 4.25 cm^2 (c) -16 °C   (d) 3.67 atm
6-10. (a) $T_0 = 638$ °C, $P_0 = 10.9$ bars   (b) 13.6 kg/sec
6-12. (a) 90.9 atm   (b) 57.5 atm
6-16. (a) $v = 124$ m/sec, $P_0 = 0.18$ bars
6-18. 3.23 cal/gm
6-20. $v = 421.5$ m/sec, $M = 1.17$
6-22. $P = 16.4$ psia, $\eta = 78$ percent
7-2. (a) 434 joule/gm   (b) 618 joule/gm
7-4. (a) 140 BTU/lb_m (b) 12%   (c) 38.8 psia

7-6. $mep = \dfrac{P_1 \, r(1 - T_1/T_4)\ln(v_1/v_2)}{r - 1}$

7-8. $\eta = 1 - \dfrac{T_4 - T_1}{T_3 - T_2 + \gamma(T_{3'} - T_3)}$

7-10. $\eta_{Diesel} = 0.61$, $\eta_{Otto} = 0.65$
7-12. (a) 200 psia   (b) 1400 psia
7-14. (b) $\eta_{Rankine} = 0.38$, $\eta_{Carnot} = 0.49$
7-16. (a) 600 °F   (b) 2.33×10^4 lb_m/hr   (c) 2.41×10^3 lb_m/hr
       (d) 3.32×10^3 lb_m/hr

7-18. (a) 0.17 (b) 0.34
7-20. 0.28 to 0.31 gal/min for atm temp between 0 and 120 °F
7-22. (a) 185 gram/sec (b) 5.60 kilowatts (c) No cooling would occur
8-4. (a) 16.86 atm (b) 16.96 atm
8-6. 1180 °R
8-10. 0.9967 BTU/lb_m °R
8-12. $W = -4.43$ joules, $Q = -166.2$ joules
9-2. 245.3 BTU/lb_m
9-6. $\Delta u_T = \dfrac{3a}{2T^{1/2}b} \ln \dfrac{v_2(v_1 + b)}{v_1(v_2 + b)}$
9-8. 390 °R, 65.0 ft/sec
9-10. 72.6 lb_m
9-12. 534 °R, 185.1 ft^3/lb mole, 384.3 ft/sec
10-2. $-102,600$ BTU/hr
10-4. $+282,950$ BTU/hr
10-6. $+282,600$ BTU/hr
10-8. (a) 43.9 atm (b) 44.1 atm (c) 43.2 atm
10-10. 4.253 ft^3/lb mole
10-12. 0.64 lb moles
10-14. $\overline{V}_E = 57.8$ cc/gm mole, $\overline{V}_W = 16.3$ cc/gm mole
11-2. 614 lb_f/in^2, 4.12 cc/gm, 119.2 cal/gm
11-6. (a) -0.5 °F (b) 20.4 lb_f/in^2 (c) 52.6 lb_f/in^2
11-8. (a) -0.5 °F (b) 20.4 lb_f/in^2 (c) 47.5 lb_f/in^2
11-10. (a) 16.26 atm (b) $Q = -151,830$ BTU, $W = -105,170$ BTU
11-16. $1.21/1000 gal water
11-18. (d) $b = 1.345, c = 0.686$ $A = 1.212, B = 681.8$
11-20. 0.51
12-2. $+19,823$ cal/gm mole benzene formed
12-4. $-64,100$ cal/gm mole H_2O formed
12-6. 1505 K
12-8. (a) 0.555 (b) 3200 cal/gm mole
12-10. 0
12-12. $\pi = 5$
12-14. (a) $-10,600$ cal/gm mole of CO (b) 650 K, 57.6 percent

Bibliography

1. *Steam Tables – Properties of Saturated and Superheated Steam*, Reprinted from 1967 ASME STEAM TABLES by Combustion Engineering, Inc., New York (1975).
2. J. H. Keenan, F. G. Keyes, P. G. Hill and J. G. Moore, *Steam Tables – Thermodynamic Properties of Water Including Vapor, Liquid and Solid Phases*, Wiley, New York (1969). (Available in either English Units or metric units.)
3. F. Din, *Thermodynamic Functions of Gases*, Vols. 1, 2 and 3, Butterworths, London (1961-62).
4. L. N. Canjar and F. S. Manning, *Thermodynamic Properties and Reduced Correlations for Gases*, Gulf Publishing Co., Houston, Texas (1967).
5. B. H. Sage and W. M. Lacey, *Thermodynamic Properties of the Lighter Paraffin Hydrocarbons and Nitrogen*, American Petroleum Institute, New York (1950).
6. K. E. Starling, "Thermo Data Refined for LPG – Part 2: Methane," *Hydrocarbon Processing*, 50, 4, 139 (1971).
7. J. T. Fanning, *A Practical Treatise on Hydraulic and Water Supply Engineering*, Van Nostrand, New York (1893).
8. L. F. Moody, "Friction Factors for Pipe Flow," *Trans. A.S.M.E.*, 66, 8, 671 (1944).
9. A. H. Shapiro, *The Dynamics and Thermodynamics of Compressible Fluid Flow*, Ronald, New York (1953).
10. J. H. Keenan and J. Kaye, *Gas Tables*, Wiley, New York (1948).

11. *Equations, Tables, and Charts for Compressible Flow,* NACA, Report 1135 (1953).
12. H. C. Hottel, G. C. Williams and C. N. Statterfield, *Thermodynamic Charts for Combustion Processes,* Wiley, New York (1949).
13. C. F. Taylor, *The Internal-Combustion Engine in Theory and Practice,* 2nd ed., M. I. T. Press, Cambridge, Mass. (1966).
14. *Tables of Thermodynamic Properties of Ammonia,* Circular of the Bureau of Standards, No. 142 (1923).
15. K. S. Pitzer, D. Z. Lippmann, R. F. Curl, Jr., C. M. Huggins and D. E. Petersen, "The Volumetric and Thermodynamic Properties of Fluids. II. Compressibility Factor, Vapor Pressure and Entropy of Vaporization," *I. E. C., 50,* 2, 265 (1958).
16. C. A. Passut and R. D. Danner, "Accentric Factor. A Valuable Correlating Parameter for the Properties of Hydrocarbons," *I. E. C., Process Des. Develop. 12,* 3, 365 (1973).
17. R. D. Gunn, P. L. Chueh and J. M. Prausnitz, "Prediction of Thermodynamic Properties of Dense Gas Mixtures Containing One or More of the Quantum Gases," *A. I. Ch. E. Jr., 12,* 5, 937 (1966). Also R. D. Gunn and S. P. Singh, personal communication.
18. R. C. Reid and T. K. Sherwood, *The Properties of Gases and Liquids,* 2nd ed., McGraw-Hill, New York (1966).
19. A. L. Lydersen, "Estimation of Critical Properties of Organic Compounds by the Method of Group Contributions," Univ. of Wisconsin Engineering Experiment Station Report N. 3, Madison, Wisconsin (1955).
20. J. J. Martin, "Equations of State – Applied Thermodynamics Symposium," *I. E. C., 59,* 12, 34 (1967); "Correction," *I. E. C., 60,* 6, 9 (1968).
21. J. J. Martin and Y. C. Hou, "Development of an Equation of State for Gases," *A. I. Ch. E. Jr., 1,* 2, 142 (June, 1955).
22. H. W. Cooper and J. C. Goldfrank, "BWR Constants and New Correlations," *Hydrocarbon Processing, 46,* 12, 141 (1967).
23. Hydrocarbon Research, Inc., Final Report to U.S.A.E.C., "Low Temperature Heavy Water Plant," Contract No. AT(30-1), 810, NYO-889 (March 15, 1951).
24. M. L. McGlashan and D. J. B. Porter, "An Apparatus for the Measurement of the Second Virial Coefficients of Some *n*-Alkanes and Some Mixtures of *n*-Alkanes," *Proc. Roy. Soc.* (London) A 267, 478 (1962).
25. K. S. Pitzer and R. F. Curl, "The Volumetric and Thermodynamic Properties of Fluids. III. Empirical Equation for the Second Virial Coefficient," *Jr. Am. Chem. Soc.., 79,* 2369 (1957).
26. J. M. Prausnitz, *Molecular Thermodynamics of Fluid-Phase Equilibrium,* Prentice-Hall, Englewood Cliffs, New Jersey (1969).

27. R. F. Curl, Jr., and K. S. Pitzer, "Volumetric and Thermodynamic Properties of Fluids — Enthalpy, Free Energy, and Entropy," *I. E. C., 50,* 2, 265 (1958).
28. M. Benedict, G. B. Webb and L. C. Rubin, "An Empirical Equation for Thermodynamic Properties of Light Hydrocarbons and Their Mixtures," *Chem. Eng. Prog., 47,* 8, 419 (1951).
29. B. F. Dodge, *Chemical Engineering Thermodynamics,* McGraw-Hill, New York (1944).
30. S. H. Fishtine, "Reliable Latent Heats of Vaporization," *I. E. C., 55,* 6, 47 (1963).
31. C. L. Depriester, "Light Hydrocarbon Vapor-Liquid Distribution Coefficients," *Chem. Eng. Prog., Sym. Ser., 49,* 7, 1 (1953).
32. A. K. Fischer and S. A. Johnson, "Liquid-Vapor Equilibrium in the Sodium-Lead System," *Jr. of Chemical and Engineering Data, 15,* 4, 492 (1970).
33. J. H. Hildebrand and R. L. Scott, *Solubility of Nonelectrolytes,* 3rd ed., Reinhold, New York (1950).
34. B. J. Zwolinski *et al.*, "Selected Values of Properties of Hydrocarbons and Related Compounds," American Petroleum Institute Research Project 44, Thermodynamics Research Center, Texas A & M University, College Station, Texas, 77843 (loose-leaf data sheets, extant 1973).
35. B. J. Zwolinski *et al.*, "Selected Values of Properties of Chemical Compounds," Thermodynamics Research Center Data Project, Thermodynamics Research Center, Texas A & M University, College Station, Texas, 77843 (loose-leaf data sheets, extant 1973).
36. D. D. Wagman, W. H. Evans, V. B. Parker, I. Halow, S. M. Bailey, and R. H. Schumm, NBS Technical Note 270-3 (January 1968) and 270-4 (May 1969); V. B. Parker, D. D. Wagman, and W. H. Evans, NBS Technical Note 270-6 (November 1971).
37. D. R. Stull and H. Prophet, JANAF Thermochemical Tables, 2nd ed., NSRDS-NBS 37 (June 1971).
38. International Council of Scientific Unions, Committee on Data for Science and Technology, CODATA Bulletin 5 (December 1971) and 10 (December 1973).
39. F. D. Rossini, D. D. Wagman, W. H. Evans, S. Levine, and I. Jaffe, "Selected Values of Chemical Thermodynamic Properties," NBS Circular 500 (1952).
40. H. P. Meissner, C. L. Kusik and W. H. Dalzell, "Equilibrium Composition with Multiple Reaction," *I. E. C. Fundamentals, 8,* 4, 659 (1969).
41. J. W. Andersen, G. H. Beyer and K. M. Watson, "Thermodynamic Properties of Organic Compounds — Estimation from Group Contributions," *National Petroleum News, 36,* 27, R-476 (July 5, 1944).
42. W. M. Latimer, *Oxidation Potentials,* 2nd ed., Prentice-Hall, New York (1952).

Index

A

Absolute temperature scale, 10, 98
Absolute zero of temperature, 10, 98
Accentric factor, 299, 533
Activity, 409, 412, 476, 509
  coefficient, 411, 424, 477, 509
  quotient, 475
Adiabatic process, 12
Air, composition of, 361
Air-standard cycle, 199
Amagat's law, 363
Anderson, Beyer and Watson method, 499
Anode, 505
Antoine equation, 391
Athermal solutions, 436
Atmosphere, 6, 10, 139
Available work, 82, 94, 229
Availability, 94
Avogadro, A., 130
Avogadro's number, 130, 142, 518
Azeotrope, 438

B

Balance equations
  energy, 29, 41, 457
  entropy, 62, 73, 83
  mass, 23, 455
  mechanical energy, 147
  momentum, 153
Benedict-Webb-Rubin equation of state, 311, 372, 423
Bernouilli, J., 27
Bernouilli's equation, 148
Binary system, 253
Black, J., 27
Boiler efficiency, 232, 240
Boltzmann, L., 143
Bow shock, 177
Boyle, R., 129, 154
Boyle's law, 129, 154, 294
Brake
  mean effective pressure, 203
  thermal efficiency, 203
Brayton cycle, 254
Bubble point, 404

C

Callendar equation of state, 295
Caloric, 27
Calorimetry, 36
Canjar, L., 16, 457

Carnot, S., 59, 196
Carnot cycle, 66, 211
   coefficient of performance, 69, 278, 279
   thermal efficiency, 68
Cathode, 506
Cell, electrochemical, 505
Celsius, A., 10
Celsius temperature scale, 10
Change in independent variable, 115
Charge, Faraday's equivalent, 507, 518
Charles' law, 130
Christin of Lyons, 10
Chemical potential, 394, 475
Chemical reaction, 24
   complex, 480
   enthalpy of, 457, 464, 510
   entropy of, 489, 510
   equilibrium of, 476
      and effect of temperature, 496
   extent of, 456, 474, 478, 482
   free energy (reaction potential), 473, 485
      estimation of, 485, 499
Choked flow, 166
Clapeyron Equation, 388
Clausius, R., 60, 61, 316
   equation, 391
   statement of second law, 60, 63
Clausius-Clapeyron Equation, 388
Closed system, 3
Coefficient of performance, 69, 278, 279
Combined cycles, 254, 272
Composition, measures of, 11
Compressible flow, 151
Compressibility
   factor, 297, 371
   isentropic, 184
   isothermal, 320
Compression ignition, 212
Compression ratio, 202
Compressor, 188, 240, 255
Condensor, 236
Conformal fluid, 299
Conservation
   of energy, 27
   of mass, 23
Control mass, 3
Control surface, 3
Control volume, 3
Conversion factors, 517
Corresponding states, 297

Corrosion, 505
Coulomb's inverse square law, 32
Critical point properties, 16, 533
   Estimation of, 301
   Compressibility factor, 299
Cycle
   air-standard, 199
   Brayton, 254
   Carnot, 66, 211
   combined, 254, 272
   Diesel, 214
   Ericsson, 289
   Lenoir, 196, 289
   Otto, 196
   Rankine, 236
   reciprocating compressor, 188
   Stirling, 220

D

Dalton's law, 362, 397
Degrees of freedom, 396, 484
Derivative
   ordinary, 108
   partial, 109
Derived dimension, 4
Deviation, 330
Dew point, 404, 420
Diesel, Rudolf, 212
Diesel Cycle, 214
Differential, total, 113
Differentiation
   analytical, 108
   graphical, 327, 427
Diffuser, 156, 180
Dimensions
   derived, 4
   fundamental, 4
Din, F., 16
Dulong's formula, 468

E

Economizer, 224
Edmister, W., 304
Effectiveness, 91
Efficiency
   mechanical, 79
   thermal, 68

Electric current, 4
  units of, 8
Electrochemical cells, 505
Endothermic reaction, 459
Energy, 2
  balance equation, 29, 41, 457, 459
  binding, 43
  conservation of, 27
  flow, 29, 40
  generation, 29
  Gibbs free, 114, 507
  Helmholtz free, 114
  internal—see Internal energy
  kinetic, 29, 37
  potential, 29, 38
  unavailable, 82
  units of, 7
Enthalpy, 42
  excess, 383, 430
  of combustion, 460, 466
  of formation, 460
  of ideal gas, 135, 365
  of ideal solution, 368, 381
  of mixing, 382
  of reaction, 457, 464
  of real gas, 338
  of vaporization, 391
  stagnation, 157
  static, 157
Entropy, 61
  balance equation, 62, 73, 83
  excess, 382
  generation, 62, 72
  increase, principle of, 61
  lost work, 72
  of formation, 489
  of ideal gas, 136, 365
  of ideal solution, 368
  of mixing, 365, 382
  of reaction, 489, 493
  of real gas, 340
  second law, 61
  statistical interpretation, 142
  unavailable energy, 84
Equations of state, 20
  corresponding state, 297, 371
  Benedict-Webb-Rubin, 311, 373, 423
  Callendar, 295
  ideal gas, 131, 293
  for liquids and solids, 319
  ideal gas, 131, 293
  Martin two constant, 310
  Redlich-Kwong, 310, 315, 372, 422
  van der Waals, 295, 307, 311, 371, 436
  virial, 316, 373, 423
Equilibrium, 2, 13
  chemical, 13, 476
  mechanical, 2, 13, 387
  phase, 388
  thermal, 6, 11, 13, 387
  thermodynamic, 13, 386
Equivalents, 507
Ericsson cycle, 289
Euler, L., 27
Exothermic reaction, 459
Expansion, coefficient of
  linear, 323
  thermal, 320
Extensive property, 11, 33
Extent of reaction, 456, 474, 478, 482
Excess functions, 380, 429, 433

F

Fahrenheit, D., 10
Fahrenheit temperature scale, 7, 10
Fanning, J. T., 147
Fanno line, 172, 177
Faraday, M., 507
Faraday's equivalent charge, 507, 518
Feedwater heater, 245
First law of thermodynamics, 27
Fishtine, S. H., 392
  equation, 392
Flame temperature, 468
Flow energy, 29, 40
Flow work, 41
Force, 2
  body, 32, 387
  contact, 32
  fields of, 32, 38
  generalized, 33
  units of, 5
Friction, 31, 70, 147, 177
Friction factor, 147
Frost-Kalkwarf equation, 391
Fuel cell, 505
Fugacity, 347

effect of P and T, 353
of component in mixture, 408, 411, 421
of pure component, 345
Fugacity coefficient
of component in mixture, 417, 421
of pure component, 349
Fusion, 16, 177

G

Galileo, G., 10, 27
Galvanic cell, 505
Gas constant, universal, 117, 131
Gas turbine, 254
Gay-Lussac, J., 98, 130
Gay-Lussac's law, 130
Generalized displacement, 33
Generalized property charts, 338, 534
Gibbs free energy, 114
excess, 382, 433
ideal gas mixture, 367
ideal solution, 368
of formation, 485
of mixing, 382
of real gas, 345
of reaction, 475, 507
Gibbs-Duhem Equation, 425
Gibbs phase rule, 392, 398, 405, 484
Guldberg, C. M., 301

H

Heat, 28, 35
capacity, specific, 36, 62, 116, 121, 122, 324, 531, 532
engines, 187
mechanical equivalent, 8, 28
of combustion, 205, 251, 460
of formation, 460
of fusion, 37
of reaction, 457
of sublimation, 37
of vaporization, 37
pump, 278
transfer, 50, 177
Heating value, 468
Helmholtz free energy, 114
excess, 382
ideal gas mixture, 368
ideal solution, 368

of mixing, 382
Henry's law coefficient, 418
Heterogeneous system, 3
High temperature gas reactor, 251, 269
Hildebrand, J., 436
Homogeneous system, 3
Hooke's law, 35
Humidity, 400
Huygens, C., 10

I

Ideal gas
equation of state, 131, 293
mixtures of, 361
properties of, 131
enthalpy, 135, 365
entropy, 136, 365
internal energy, 134, 363
specific heat capacity, 134, 364, 531
Ideal process, 49
Ideal solutions, 368, 379, 476, 478
Immiscible fluids, 424
Indicated
mean effective pressure, 203
thermal efficiency, 202
Indicator diagram, 188
Integration
graphical, 333
line, 134
Riemann, 132
Intensive property, 9, 33
Internal energy, 29, 41
ideal gas mixture, 134, 363
ideal solution, 368, 381
property relationship, 113
real gases, 340
Interstage cooling, 193, 268
Inverse lever arm principle, 19
Irreversibility, 82
Isentropic compression, 137, 193
Isochores, 309
Isolated system, 3
Isotherms, 16, 296
Isothermal compressibility, 321
Isothermal compression, 193

J

Joule, J., 28, 60
Joule-Thompson coefficient, 127

K

Kay, W., 371
Keenan, J., 16
Kelvin (William Thompson) Lord, 10, 60, 98
Kelvin-Planck statement of second law, 60, 64
Kelvin temperature scale, 6, 10, 98
Kinetic energy, 29, 37
Kistiakowsky, W., 392
 equation 392

L

Laplace, Marquis de, 155
Latimer, W. M., 509
Least square technique, 330, 438
Leduc's law, 363
Leibnitz, G., 27
Length, units of, 4
Lenoir cycle, 196, 289
Lewis, G. N., 347
Lewis-Randall approximation, 418, 476
Line integration, 134
Linear expansion, 127, 323
Lost work, 70, 73, 147
Luminous intensity, units of, 8
Lydersen, A. L., 302

M

McGlashan, M., 317
Mach number, 160
Manning, F., 457
Margules, M., 431
 equation, 432
Martin, J., 308
 -two constant equation, 310
 -Shinn-Kapoor equation, 391
Mass, 11
 balance equation, 24, 455
 conservation of, 23
 units of, 5
Maxwell relations, 117
Mayer, R., 28

Mean effective pressure, 203
Mechanical energy balance, 147
Meissner, H., 299, 463
Metric units, 4
Mixing rules, 371
Mixtures
 ideal gas, 361
 ideal solutions, 368
 non-ideal solutions, 376
 of phases, 17
Modulus of elasticity, 34, 127
Molality, 509
Molar property, 12
Mole, 11
Mole fraction, 12
Momentum balance, 153
Moody, L. F., 147

N

Nernst, W., 494
Nernst heat theorem, 495
Newcomen, T., 195
Newton, I., 10
 inverse square law, 38
 second law of motion, 2, 5, 37, 153
 Raphson method of solving equations, 444, 471
Newton, R., 298
Normal fluid, 300
Normal shock wave, 169
Nozzle flow, 156

O

Octane number, 202
Open system, 3
Otto cycle, 196
Oxidation reaction, 506

P

Parsons, C., 226, 236
Partial derivatives, 109
 relationships between, 110
Partial molal properties, 376
Partial pressure, 362, 396
Partial volume, 363
Pascal, B., 10

Path function, 30, 51, 106
Perpetual motion machine
  of first kind, 59
  of second kind, 60
Phase, 3, 387
Phase diagrams, 14, 296, 309
Phase distribution coefficient, 419
Phase rule, Gibbs, 392, 398, 405, 484
Pitzer, K., 299, 318
Planck, M., 495
  radiation law, 7
Point function, 106
  criteria for, 118
Polytropic process, 193
Pollution, 2
  air, 210, 219, 225, 251
  thermal, 251
Potential
  chemical, 394, 475, 485
  energy, 29, 38
  electrochemical, 508
  electrostatic, 29, 32
  gravitational, 29, 38
  magnetostatic, 29
  reaction, 474
Power, units of, 7
Poynting correction, 413, 419
Prausnitz, J. M., 436
Pressure, 9
  critical, 16, 533
  partial, 362, 396
  reduced, 298
  stagnation, 159
  vapor, 388
Prime movers, 188, 225
Processes, 12
  adiabatic, 12
  cyclic, 12, 66
  ideal, 49
  irreversible, 51, 70
  isentropic, 137, 193
  isobaric, 12
  isochoric, 12
  isothermal, 12, 193
  polytropic, 193
  quasi-static, 50
  reversible, 14, 31, 63
Properties, 3
  generalized charts, 338, 534
  of ideal gas, 131
  partial molal, 376
  relations between, 112
  tables of, 42, 519, 531, 533
Psychrometric chart, 402

Q

Quality, 19
Quasi-static process, 50

R

Rankine cycle, 236
Rankine temperature scale, 7
Raoult's law, 397
Ratio of specific heat capacities, 137
Rayleigh line, 173, 177
Raymond, C., 381
Reactions
  chemical—see Chemical reactions
  nuclear, 24, 29, 43
Redlich-Kwong equation of state, 310, 315, 372, 422
Reduced properties, 298
Reduction reaction, 506
Reference state, 41, 43, 61, 328, 353, 460
  temperature, 322, 460, 464, 466, 496
  pressure, 322, 340
Refrigeration cycles, 277
  absorption, 287
  Brayton, 284
  Carnot, 68, 277
  Rankine, 280
  Stirling, 283
  vapor-compression, 279
Regenerator, 224, 264
Regnault, H. V., 98
Regular solutions, 436
Reheater, 243, 246, 268
Reid, R. C., 301, 373, 505
Relative humidity, 401
Renaldine, C., 10
Residual volume, 294, 317, 421
Reynolds number, 148
Richards, T. W., 494
Riedel, L., 304
Riemann integration, 132
Rumford (Benjamin Thompson) Count, 27

S

Sage, B., 16
Saturated liquid, 18
Saturated vapor, 18, 399
Saturation pressure, 16
Scatchard, G., 381, 437
Second law of thermodynamics, 59
  and entropy, 61
  Carnot statement of, 59
  Clausius statement of, 60, 64
  Kelvin-Planck statement of, 60, 65
Seferian, R., 299
Shaft work, 32
Sherwood, T. K., 301, 373, 505
Shock wave, 169
Simple fluid, 299
Simpson's rule, 336
Solubility parameter, 437
Solutions—see Mixtures
Specific gas constant, 131
Specific heat capacity
  at constant pressure, 36, 62, 116, 121, 531, 532
  at constant volume, 36, 116, 121, 122
Specific property, 12
Speed of sound, 151
Stagnation properties, 156
Standard state, 61, 353, 408, 412, 460, 475, 485
State
  changes of, 12
  equations of—see Equations of state
  function, 51
  of system, 3
Static properties, 156
Steady state, 25
Steam engine, 195, 240
Steam turbine, 225
Stirling cycle, 220
Stoichiometric coefficients, 456, 458, 460, 466
Strain, 34
Stress, 34
Sublimation, 16
Supercharger, 208, 218, 259
Superheat, degrees, 19
Superheater, 241
Surface tension, 32
Surroundings, 3

System of units—see Units
Systems
  closed, 3
  heterogeneous, 3
  homogeneous, 3
  isolated, 3
  open, 3

T

Tait equation, 321
Temperature
  absolute zero, 10, 99
  Boyle, 294
  Celsius scale of, 7
  critical, 16, 533
  dry bulb, 401
  Fahrenheit scale of, 7, 10
  Kelvin scale of, 6, 11, 98
  Rankine scale of, 7
  reduced, 298
  stagnation, 156
  static, 156
  wet bulb, 401
Thermal efficiency, 68
Thermal equilibrium, 6, 11
Thermal expansion, 320
Third law of thermodynamics, 494
Thompson, B., 27
Thompson, W., 10, 60
Time, Units of, 5
Ton of refrigeration, 283
Torricelli, E., 10
Triple point, 6, 16, 99
Trouton, F., 391
Trouton's rule, 391
Tunnel, G., 348
Turbine, 225, 240, 254
Turbosupercharger, 210, 259

U

Unavailable energy, 82
Units
  energy, 7
  English, 4
  electric current, 8

force, 5
fundamental, 4
international, 4
length, 4
luminous intensity, 8
mass, 5
metric, 4
power, 7
temperature, 6
time, 5
work, 7
Universal gas constant, 117, 131

V

Valence, 507
van der Waals, J. D., 295
  equation of state, 295, 307, 311, 371, 436
van Laar equations, 436
Vapor-liquid equilibrium coefficient, 419
Vapor pressure, 16, 107, 388
Vaporization, 16, 391
Velocity of sound, 151
Virial equation of state, 316, 423
  coefficients, 317
  for mixture, 373
Voltaic cell, 505
Volume, 11
  critical, 16, 533
  excess, 382, 429
  partial, 363
  partial molal, 378
  specific, 11

W

Watson, K. M., 392
Watt, James, 189
Weight, 5
Weight fraction, 12
Work, 28, 30
  available, 82, 94, 229
  of compression, 30, 51, 71
  electrical, 32, 507
  flow, 41
  lost—see Lost work
  mechanical, 2
  reversible, 31, 34
  shaft, 32

Y

Young, S., 298
Young, T., 27

Z

Zero, absolute, 10, 99
Zeroth law of thermodynamics, 11